Nuclear Spin Relaxation in Liquids

Theory, Experiments, and Applications

Second Edition

Nuclear Spin Relaxation in Liquids

Theory, Experiments, and Applications

Second Edition

By

Jozef Kowalewski

Stockholm University, Sweden

&

Lena Mäler

Stockholm University, Sweden

CRC Press is an imprint of the
Taylor & Francis Group, an **informa** business

CRC Press
Taylor & Francis Group
6000 Broken Sound Parkway NW, Suite 300
Boca Raton, FL 33487-2742

First issued in paperback 2019

CRC Press is an imprint of Taylor & Francis Group, an Informa business

ISBN-13: 978-1-4987-8214-2 (hbk)
ISBN-13: 978-0-367-89006-3 (pbk)

Library of Congress Cataloging-in-Publication Data

Names: Kowalewski, Jozef, author. | Mäler, Lena, author.
Title: Nuclear spin relaxation in liquids : theory, experiments, and applications / Jozef Kowalewski & Lena Mäler.
Description: Second edition. | Boca Raton, FL : CRC Press, Taylor & Francis Group, [2018] | Includes bibliographical references and index.
Identifiers: LCCN 2017032734| ISBN 9781498782142 (hardback ; alk. paper) | ISBN 1498782140 (hardback ; alk. paper) | ISBN 9781498782159 (e-book) | ISBN 1498782159 (e-book)
Subjects: LCSH: Relaxation phenomena. | Nuclear spin.
Classification: LCC QC173.4.R44 K69 2018 | DDC 530.4/16--dc23
LC record available at https://lccn.loc.gov/2017032734

Visit the Taylor & Francis Web site at
http://www.taylorandfrancis.com

and the CRC Press Web site at
http://www.crcpress.com

Contents

8 Measuring T_1 and T_2 Relaxation Rates

9 Cross-Relaxation Measurements

10 Cross-Correlation and Multiple-Quantum Relaxation Measurements

11 Relaxation and Molecular Dynamics

Preface

Nuclear magnetic resonance (NMR) is a powerful branch of spectroscopy, with a wide range of applications within chemistry, physics, material science, biology and medicine. The importance of the technique has been recognised by the Nobel prizes in chemistry awarded to Professor Richard Ernst in 1991 for methodological development of high-resolution NMR spectroscopy and to Professor Kurt Wüthrich in 2002 for his development of NMR techniques for structure determination of biological macromolecules in solution. NMR employs a quantum mechanical property called *spin*, the origin of the characteristic magnetic properties of atomic nuclei, and its behaviour in an external magnetic field. NMR provides discriminating detail at the molecular level regarding the chemical environment of individual atomic nuclei. Through-bond or through-space connections involving pairs of nuclei, rates of certain chemical reactions, and water distribution and mobility in tissue typify the information commonly available from the NMR experiment. Perhaps the richest information is obtained from studies of nuclear spin relaxation (NMR relaxation) in solution, which investigate how the transfer of energy and order occurs among nuclear spins and between the spin system and its environment. Various aspects of NMR relaxation in liquids are the topic of this book.

The relaxation phenomena are described in all major NMR books, with a varying level of depth and detail. Nevertheless, we found it worthwhile to collect the relaxation theory, experimental techniques and illustrative applications in a single volume, and we go into more detail in these fields than most general NMR texts. The ambition is to explain relaxation in a way which would be possible to follow for readers familiar with the basics of NMR, such as advanced undergraduate and graduate students in chemistry, biochemistry, biophysics and related fields. Most of the book is not too demanding mathematically, with the objective to illustrate and explain the physical nature of the phenomena rather than their intricate details. Some of the theoretical chapters — Chapters 4–6 — go into more depth and contain more sophisticated mathematical tools. These chapters are more directed to specialists and may be omitted at the first reading of the book. Most of the key results obtained in these chapters are also given in the experimental or application sections.

Twelve years after the original publication of the book, we judge it appropriate to present an updated second edition. The outline of the second edition is similar to the first one and is as follows. An introduction in Chapters 1–3 presents a fairly simple version of relaxation theory, including the description of dipolar relaxation in a two-spin system at the level of Solomon equations, and also contains a sketchy presentation of intermolecular relaxation. Chapters 4 and 5 provide a more sophisticated and more general version of the theory as well as a discussion of other relaxation mechanisms. Various theoretical tools are presented and applied to appropriate problems and examples. Chapter 6 deals with spectral densities, the link connecting the relaxation theory with the liquid dynamics. Chapter 7 is an introduction to experimental tools of NMR, with emphasis on relaxation-related experimental techniques for solution state studies. The following chapters cover experimental methods for studying increasingly sophisticated phenomena of single-spin relaxation (Chapter 8), cross-relaxation (Chapter 9) and more general multispin phenomena (Chapter 10). In each chapter, a progression of techniques is described, starting with simple methods for simple molecules and moving on to advanced tools for studies of biological macromolecules. Chapters 11 and 12 are devoted to applications of relaxation studies as a source of information on molecular dynamics and molecular structure. The remaining five chapters deal with a variety of special topics: the effects of chemical (often conformational) exchange on relaxation phenomena and relaxation-related measurements (Chapter 13), relaxation processes in systems containing quadrupolar ($I \geq 1$) nuclei (Chapter 14), paramagnetic systems containing unpaired electron spin(s) (Chapter 15), singlet state NMR (Chapter 16, newly written for this edition) and, finally, a brief presentation of NMR relaxation in other aggregation states of matter (Chapter 17).

The book is not meant to be a comprehensive source of literature references on relaxation. We are rather selective in our choice of references, concentrating on reviews and original papers with pedagogic qualities that have been chosen at least partly according to our personal tastes. The references are collected after each chapter. Some of the NMR (and related) textbooks, useful in the context of more than one chapter, are collected in a special list called "Further Reading," provided immediately after this preface.

Many of the figures have been prepared especially for this book, while some are reproduced with permission of the copyright owners, who are acknowledged for their generosity. In the process of writing the second edition of this book, we have obtained assistance and support from several students and colleagues. Among these, we wish in particular to mention Hans Adolfsson, Jens Danielsson, Bertil Halle, Danuta Kruk, Malcolm Levitt, Arnold Maliniak, Arthur Palmer, Giacomo Parigi and Ernst Rössler. We are very grateful to Nina Kowalewska and Hans Adolfsson for their endurance and support during the 2 years it took to write this text. This edition was completed in June 2017.

Further Reading List

Abragam A 1961 *The Principles of Nuclear Magnetism* (Oxford: Oxford University Press)
Atkins P W and Friedman R S 1997 *Molecular Quantum Mechanics* (Oxford: Oxford University Press)
Bakhmutov V I 2004 *Practical NMR Relaxation for Chemists* (Chichester: Wiley)
Brink D M and Satchler G R 1993 *Angular Momentum* (Oxford: Clarendon Press)
Canet D 1996 *Nuclear Magnetic Resonance: Concepts and Methods* (Chichester: Wiley)
Cavanagh J, Fairbrother W J, Palmer A G, Rance M, and Skelton N J 2006 *Protein NMR Spectroscopy*, 2nd ed (San Diego: Academic Press)
Ernst R R, Bodenhausen G and Wokaun A 1987 *Principles of Nuclear Magnetic Resonance in One and Two Dimensions* (Oxford: Clarendon Press)
Goldman M 1988 *Quantum Description of High-Resolution NMR in Liquids* (Oxford: Clarendon Press)
Hennel J W and Klinowski J 1993 *Fundamentals of Nuclear Magnetic Resonance* (Harlow: Longman)
Keeler J 2010 *Understanding NMR Spectroscopy*, 2nd ed (Chichester: Wiley)
Kruk D 2016 *Understanding Spin Dynamics* (Singapore: Pan Stanford Publishing)
Levitt M H 2008 *Spin Dynamics*, 2nd ed (Chichester: Wiley)
McConnell J 1987 *The Theory of Nuclear Magnetic Relaxation in Liquids* (Cambridge: Cambridge University Press)
Neuhaus D and Williamson M P 1989 *The Nuclear Overhauser Effect in Structural and Conformational Analysis* (New York: VCH)
Slichter C P 1989 *Principles of Magnetic Resonance* (Berlin: Springer)
Van Kampen N G 1981 *Stochastic Processes in Physics and Chemistry* (Amsterdam: North-Holland)
Wüthrich K 1986 *NMR of Proteins and Nucleic Acids* (New York: Wiley)

1

Equilibrium and Non-Equilibrium States in NMR

The concepts of equilibrium and deviation from equilibrium have a central role in physical chemistry. Once a system is in a non-equilibrium state, it will tend to return to equilibrium in a process that does not occur instantaneously. This general phenomenon of development towards equilibrium is called *relaxation*. In the specific context of nuclear magnetic resonance, the equilibrium state is that of a macroscopic sample of nuclear spins in a magnetic field. In order to talk about a development towards equilibrium, we need to define and be able to create an initial state, which is different from the equilibrium state. Using experimental techniques of present-day nuclear magnetic resonance (NMR) spectroscopy, in the first place radiofrequency pulses, it is possible to create a large variety of non-equilibrium states. Starting from a given non-equilibrium situation, the spin systems evolve in a complicated way, in which the "return to equilibrium" processes compete with related phenomena, converting one type of non-equilibrium into another. In NMR, the processes by which the return to equilibrium is achieved are denoted *spin relaxation*. Relaxation experiments were among the earliest NMR applications of modern Fourier transform NMR, and the development of applications and experimental procedures has been enormous. Consequently, there is a vast literature on NMR relaxation, regarding both theoretical considerations and advances in methodology. Relaxation is important, since it has effects on how NMR experiments are carried out, but perhaps even more valuable is the information content derived from relaxation parameters. Information about the physical processes governing relaxation can be obtained from experimental NMR parameters.

In order to set the stage for theoretical descriptions of spin relaxation, we need to define the spin system and the means of manipulating the spin systems to obtain non-equilibrium states. In this chapter, we provide elementary tools for the description of spin systems and their equilibrium and non-equilibrium states.

1.1 Individual Spins: Elements of Quantum Mechanics

In order to arrive at a description of a large number of spins, we need to start with a single spin. The properties of spin are obtained using quantum mechanics. A brief account of the necessary quantum mechanical background is given in this section, and the reader is directed to *Further reading* (in particular the books by Atkins and Friedman, Keeler, Levitt and Slichter) for a more complete presentation. The basis of quantum mechanics can be formulated as a series of postulates, fundamental statements assumed to be true without proof. The proof of these postulates lies in the fact that they successfully predict the outcome of experiments. One field in which this works extremely well is nuclear magnetic resonance. The first postulate of quantum mechanics states that the state of a physical system is described, as fully as it is possible, by a *wave function* for the system. The wave function is a function of relevant coordinates and of time. In the case of spin, the wave functions are defined in terms of a spin coordinate. The important property of the spin wave functions is the fact that they can be formulated exactly in terms of linear combinations of a finite number of functions with known properties. The second postulate of quantum mechanics states that for every measurable quantity (called *observable*) there exists an associated *operator*. An operator, $\hat{Q}$ (we are going to denote operators, with certain exceptions specified below, with a "hat"), is a mathematical object that operates on a function and creates another function. In certain cases, the action of an operator on a function yields the same function, multiplied by a constant. The functions having this property are called the *eigenfunctions* of the operator and the corresponding constants the *eigenvalues*.

In the context of spins, the most important operators are the *spin angular momentum* operators. Every nuclear species is characterised by a nuclear spin quantum number, I, which can attain integer or half-integer values. The nuclear spin quantum number is related to the eigenvalue of the total spin angular momentum operator, $\hat{I}^2$, through:

$$\hat{I}^2\psi = I(I+1)\psi \tag{1.1}$$

where ψ is an eigenfunction of the operator $\hat{I}^2$. The third postulate of quantum mechanics states that if a system is described by a wave function, which is an eigenfunction of a quantum mechanical operator, then the result of every measurement of the corresponding observable will be equal to that eigenvalue. Eq. (1.1) assumes that the spin operators are dimensionless, a convention followed in this book. A natural unit for angular momentum in quantum mechanics is otherwise $\hbar$, the Planck constant, divided by $2\pi(\hbar = 1.05457 \cdot 10^{-34}$ Js). The magnitude of the angular momentum – which is obtained in every measurement – in these units becomes $\hbar\sqrt{I(I+1)}$. If $I = 0$, then the nucleus has no spin angular momentum and is not active in NMR.

The z-component of the spin angular momentum vector is another important operator. This operator is related to the second of two angular momentum quantum numbers. Besides I, there is also m, which specifies the z-component of the spin angular momentum vector:

$$\hat{I}_z\psi = m\psi \tag{1.2}$$

The quantum number m can attain values ranging between $-I$ and I, in steps of one. We can label the eigenfunctions ψ to the operators $\hat{I}^2$ and $\hat{I}_z$ with the corresponding quantum numbers, $\psi_{I,m}$. The eigenfunctions are normalised in the sense that the integral over all space of the product of the eigenfunction $\psi_{I,m}$ with its complex conjugate $\psi^*_{I,m}$ is equal to unity: $\int \psi^*_{I,m}\psi_{I,m}d\sigma = 1$. σ is a variable of integration in the spin space (the spin coordinate). Functions used in quantum mechanics are often complex and the symbol ψ^* denotes the complex conjugate of the function ψ, *i.e.* a corresponding function where all the imaginary symbols i are replaced by $-i$ (we note that $i = \sqrt{-1}$). The normalisation requirement is a property of quantum mechanical wave functions. The square of the absolute value of a wave function, $\psi^*\psi$, is related to the probability density of finding the system at a certain value of a relevant coordinate. Consequently, the probability density has to integrate to unity over all space. If we integrate, on the other hand, a product of two eigenfunctions corresponding to different values of the quantum number m, the integral vanishes, $\int \psi^*_{I,m}\psi_{I,m'}d\sigma = 0$ if $m \neq m'$, and we say that the functions are *orthogonal*. The set of functions fulfilling the requirements of normalisation and orthogonality is called *orthonormal*. An alternative way of describing these requirements is through the relation $\int \psi^*_{I,m}\psi_{I,m'}d\sigma = \delta_{mm'}$, where the symbol $\delta_{mm'}$ is called the *Kronecker delta* and is equal to unity if $m = m'$ and to zero otherwise.

It is often convenient to use the bra-ket (bracket) notation, whereby the eigenfunctions are treated as unit vectors, denoted by a "ket" $|I, m\rangle$. In the bra-ket notation, the normalisation condition becomes $\langle I, m|I, m\rangle = 1$, in which the "bra," $\langle I, m|$, is associated with $\psi^*_{I,m}$. The star on the first symbol in $\langle I, m|I, m\rangle$ is excluded, because the bra symbol, $\langle\ |$, in itself implies complex conjugation. The integration is replaced by the scalar product of a bra, $\langle I, m|$, and a ket, $|I, m\rangle$.

For simplicity, we concentrate at this stage on nuclei with the nuclear spin quantum number $I = ½$ (which, in fact, we shall work with throughout most of this book). In this case, there are two possible eigenvalues m of the z-component of the spin, −1/2 and +1/2. The spin eigenfunctions corresponding to these eigenvalues are denoted β and α, respectively. Using the bra-ket formalism, we shall use the notation $|1/2,-1/2\rangle = |\beta\rangle$ and $|1/2,-1/2\rangle = |\alpha\rangle$. The orthogonality of the eigenfunctions can be formulated very compactly in the bracket notation: $\langle\alpha|\beta\rangle = 0$. If we represent functions as vectors, an operator transforms one vector into another. A set of orthogonal vectors defines a vector space, denoted as *Hilbert space*, and an operator performs a transformation of one vector in this space into another. Such an operation can be represented by a matrix in the Hilbert space. The elements of such a matrix, called *matrix elements of the operator* $\hat{Q}$, are defined as

$$Q_{ij} = \langle i|\hat{Q}|j\rangle \tag{1.3}$$

The symbols i and j can refer to different eigenstates of the quantum system or to other vectors in the Hilbert space, $\hat{Q}|j\rangle$ is the ket obtained as a result of $\hat{Q}$ operating on $|j\rangle$ and Q_{ij} can be considered as a scalar product of the bra $\langle i|$ and that ket. For example, the matrix representation of the z-component of the spin angular momentum in the space defined by the eigenvectors $|\alpha\rangle$ and $|\beta\rangle$ is:

$$\hat{I}_z = \begin{pmatrix} \frac{1}{2} & 0 \\ 0 & -\frac{1}{2} \end{pmatrix} \tag{1.4a}$$

The matrix is diagonal as a result of our choice of the eigenvectors of the operator $\hat{I}_z$ as the *basis set*. We can also obtain matrix representations for the x- and y-components of spin in the basis of $|\alpha\rangle$ and $|\beta\rangle$:

$$\hat{I}_x = \begin{pmatrix} 0 & \frac{1}{2} \\ \frac{1}{2} & 0 \end{pmatrix} \tag{1.4b}$$

$$\hat{I}_y = \begin{pmatrix} 0 & -\frac{i}{2} \\ \frac{i}{2} & 0 \end{pmatrix} \tag{1.4c}$$

The quantum mechanical operators corresponding to physically measurable quantities are *Hermitian*, *i.e.* their matrix elements conform to the relation $\langle i|\hat{Q}|j\rangle = \langle j|\hat{Q}|i\rangle^*$ (note how this functions for the $\hat{I}_y$!). The matrix representations of $\hat{I}_x$, $\hat{I}_y$ and $\hat{I}_z$ in Eq. (1.4) are called the *Pauli spin matrices*.

In quantum mechanics, one often needs to operate on a function (vector) with two operators after each other. We can in this case, formally, speak about a product of two operators. The order in which the operators act may be important: we say that the multiplication of operators is in general non-commutative. From this, we can introduce the concept of a *commutator* of two operators, $\hat{Q}$ and $\hat{P}$:

$$[\hat{Q}, \hat{P}] = \hat{Q}\hat{P} - \hat{P}\hat{Q} \tag{1.5}$$

For certain pairs of operators, the commutator vanishes and we say that the operators $\hat{Q}$ and $\hat{P}$ commute. The components of spin, $\hat{I}_x$, $\hat{I}_y$ and $\hat{I}_z$, do not commute with each other, but each of them commutes with $\hat{I}^2$. Important theorems about commuting operators state that they have a common set of eigenfunctions and that the corresponding observables can simultaneously have precisely defined values. The proofs of these theorems can, for example, be found in the book by Atkins and Friedman (*Further reading*).

TABLE 1.1 Properties of Some Common Nuclear Spin Species

Nucleus	Spin Quantum Number	Natural Abundance, %	Magnetogyric Ratio, $\gamma_I/(10^7 \text{ rad T}^{-1} \text{ s}^{-1})$	Larmor Frequency at 9.4T, $\lvert\omega_I\rvert/(10^8 \text{ rads}^{-1})$	Quadrupole Moment, $Q/(10^{-31} \text{ m}^2)$
^1H	1/2	99.98	26.752	25.13	–
^2H	1	0.02	4.107	3.858	2.9
^{13}C	1/2	1.11	6.728	6.321	–
^{14}N	1	99.63	1.934	1.817	15.6
^{15}N	1/2	0.37	−2.713	2.548	–
^{17}O	5/2	0.04	−3.628	3.408	−25.8
^{19}F	1/2	100	25.167	23.64	–
^{31}P	1/2	100	10.839	10.18	–
^{23}Na	3/2	100	7.080	6.652	103
^{35}Cl	3/2	75.53	2.624	2.465	−82.5
^{79}Br	3/2	50.54	6.726	6.318	293

The spin angular momentum is simply related to the nuclear magnetic moment. In terms of vector components, we can write, for example,

$$\mu_z = \gamma_I I_z \tag{1.6a}$$

$$\hat{\mu}_z = \gamma_I \hat{I}_z \tag{1.6b}$$

where μ_z and I_z denote measurable quantities and the quantities with "hats" represent the corresponding quantum mechanical operators. The quantity γ_I is called the *magnetogyric ratio* (the notation *gyromagnetic ratio* is also used). Magnetogyric ratios for some common nuclear species are summarised in Table 1.1, together with the corresponding quantum numbers and natural abundances.

When a nuclear spin is placed in a magnetic field, **B,** with the magnitude B_0, the field interacts with the magnetic moment (the *Zeeman interaction*). The symbols **B** and B_0 are, strictly speaking, the magnetic induction vector and its magnitude, but we are going to refer to these terms as *magnetic field*, in agreement with a common practice. The magnetic field in NMR is always assumed to define the laboratory-frame z-axis. In classical physics, the energy of the interaction is written as minus the scalar product of the magnetic moment vector and the magnetic field vector: $E = -\boldsymbol{\mu} \cdot \mathbf{B} = -\mu_z B_0$. In quantum mechanics, the Zeeman interaction is described by the Zeeman Hamiltonian:

$$\hat{H}_z = -\gamma_I B_0 \hat{I}_z \tag{1.7}$$

The Hamiltonian – or the total energy operator – is a very important operator in quantum mechanics. The eigenfunctions (or eigenvectors) of a quantum system with a general Hamiltonian, $\hat{H}$, fulfil the *time-independent Schrödinger equation* $\hat{H}\psi_j = E_j\psi_j$ or $\hat{H}\,|\,j\rangle = E_j\,|\,j\rangle$ in the bra-ket notation. For $I = 1/2$, the Zeeman Hamiltonian has two eigenvalues, $E_{1/2} = -\frac{1}{2}\gamma_I B_0$ and $E_{-1/2} = \frac{1}{2}\gamma_I B_0$, corresponding to the eigenvectors $|\alpha\rangle$ and $|\beta\rangle$, respectively. Both the Hamiltonian and its eigenvalues are expressed in angular frequency units, a convention followed throughout this book. The eigenvalues can easily be converted into "real" energy units by multiplying with $\hbar$. If the magnetogyric ratio is positive, which for example is the case for protons, ^1H, and ^{13}C nuclei, the α state corresponds to the lowest energy. The difference between the two eigenvalues, $\omega_0 = -\gamma_I B_0$, is called the *Larmor frequency* (while we are going to use the symbol E_j for energy eigenvalues in angular frequency units, we shall use the symbol ω with appropriate index or indices for the energy differences in the same units).

Quantum mechanics does not require the system to be in a specific eigenstate of the Hamiltonian. A spin-1/2 system can also exist in a *superposition state*, *i.e.* a superposition of the two states $|\alpha\rangle$ and $|\beta\rangle$. The superposition state is described by a wave function

$$|\psi\rangle = c_\alpha|\alpha\rangle + c_\beta|\beta\rangle \tag{1.8}$$

which is subject to the normalisation condition, $|c_\alpha|^2 + |c_\beta|^2 = 1$. The coefficients c_α and c_β are complex numbers and the squares of the absolute values of the coefficients provide the weights (probabilities) of the two eigenstates in the superposition state.

In NMR, the time evolution of quantum systems is of primary interest and the *time-dependent* form of the *Schrödinger equation* is very important. Another postulate of quantum mechanics states that the time evolution of the wave function for a quantum system is given by the time-dependent Schrödinger equation:

$$\frac{\partial\psi(t)}{\partial t} = -i\hat{H}\psi(t) \tag{1.9}$$

which explains the central role of the Hamilton operator or the Hamiltonian. Note that, following the convention that the Hamiltonian is in angular frequency units, the symbol $\hbar$, usually present on the left-hand side of the time-dependent Schrödinger equation, is dropped. If the system is initially in one of the eigenstates, and the Hamiltonian is independent of time, as is the case for the Zeeman Hamiltonian of Eq. (1.7), then the solutions to Eq. (1.9) are simply related to the eigenfunctions and eigenvalues of the time-independent Schrödinger equation

$$\psi(t) = \psi_j \exp(-iE_j t) \tag{1.10}$$

Thus, the system described originally by an eigenstate remains in that eigenstate. The eigenstates are therefore also denoted as stationary solutions to the time-dependent Schrödinger equation or *stationary states*. The complex exponential factor is called the *phase factor* (a complex exponential can be expressed in terms of cosine and sine functions: $\exp(i\alpha) = \cos\alpha + i\sin\alpha$). A more general time-dependent solution can be expressed as a linear combination of functions given in Eq. (1.10):

$$\psi(t) = \sum_{m=-I}^{I} c_m \psi_{I,m} \exp(-iE_m t) \tag{1.11a}$$

or, using the bracket notation:

$$|\psi(t)\rangle = \sum_{m=-I}^{I} c_m \exp(-iE_m t)|I,m\rangle \tag{1.11b}$$

Another important quantity required for the discussion of quantum systems is the *expectation value*. Through still another of the postulates of quantum mechanics, the expectation value of an operator corresponding to an observable is equal to the average value of a large number of measurements of the observable under consideration for a system that is not in the eigenstate of that operator. Consider, for example, the x-component of the nuclear magnetic moment; the expectation value of this quantity at time t is defined by

$$\langle\hat{\mu}_x\rangle(t) = \int \psi^*(t)\hat{\mu}_x\psi(t)d\tau \tag{1.12a}$$

or

$$\langle \hat{\mu} \rangle(t) = \langle \psi(t) | \hat{\mu}_x | \psi(t) \rangle \tag{1.12b}$$

We note that the expectation value of the x-component of the magnetic moment, $\langle \hat{\mu}_x \rangle(t)$, is explicitly time dependent, while the operator $\hat{\mu}_x$ is not. Thus, the time dependence of the expectation value originates from the time dependence of the wave functions. This way of expressing the time dependence is denoted the *Schrödinger representation*.

Using the definitions of Eq. (1.11), as well as the relation $\hat{\mu}_x = \gamma_I \hat{I}_x$, we can evaluate Eq. (1.12b):

$$\langle \hat{\mu}_x \rangle(t) = \gamma_I \sum_{m}^{I} \sum_{m'}^{I} c_{m'}^{*} c_m \langle I,m' | \hat{I}_x | I,m \rangle \exp\left(i(E_{m'} - E_m)t\right) \tag{1.13}$$

From elementary properties of angular momentum (see, for example, Atkins and Friedman, *Further reading*), we know that expressions for time-independent matrix elements $\langle I,m' | I_x | I,m \rangle = \int \psi_{I,m'}^{*} I_x \psi_{I,m} d\sigma$ vanish unless $m' = m \pm 1$. For the case of $I = 1/2$, we thus have non-vanishing elements for $m' = 1/2$ if $m = -1/2$ or vice versa, corresponding to the matrix representation of $\hat{I}_x$ in Eq. (1.4b). When this condition is fulfilled, we can see that the argument of the exponential function becomes $\pm i\omega_0 t$, where $\omega_0 = -\gamma_I B_0$ is the Larmor frequency. Noting both these relations, one can demonstrate (the reader can look it up in Slichter's book (*Further reading*) or do it as an exercise) that the expectation value of $\hat{\mu}_x$ oscillates with time at the Larmor frequency. In the same way, one can show that the expectation value of $\hat{\mu}_y = \gamma_I \hat{I}_y$ also oscillates at the Larmor frequency, while the expectation value of the z-component is constant with time. A detailed analysis (see Slichter's book) shows that the quantum mechanical expectation value of the nuclear magnetic moment vector operator $\hat{\boldsymbol{\mu}}$ behaves in analogy with a vector of length $\gamma_I/2$ moving on a conical surface around the z-axis, the direction of the magnetic field, in analogy to the classical *Larmor precession*. Another important result is that the angle between the conical surface and the magnetic field can have an arbitrary value and that the time-independent z-component of the magnetic moment of a nuclear spin in a superposition state can hold any arbitrary value between $-\gamma_I/2$ and $+\gamma_I/2$. We emphasise that this result is consistent with the notion that the spins can exist in superposition states and are not confined to pointing either parallel or antiparallel to the external field. In the same vein, we cannot associate an individual spin in a superposition state with either of the two energy eigenstates (energy levels) of the Hamiltonian, even though the motion is clearly related to the energy difference between the Zeeman states.

1.2 Ensembles of Spins: The Density Operator

So far, we have only considered the properties of one single spin. The concepts of relaxation and equilibrium are closely connected to the behaviour of macroscopic samples of spins. A theoretical tool we need to use is that of an *ensemble* of spins, a large collection of identical and independent systems. For simplicity, we deal here with an ensemble of spin 1/2 particles interacting with the magnetic field through the Zeeman interaction but not interacting with each other.

The *density operator* method is an elegant way to deal with a very large number (on the order of 10^{21}) of quantum systems, corresponding to a macroscopic sample. We present here a very brief summary of the technique and recommend the reader to turn to *Further reading* (especially the books by Abragam, Ernst *et al.*, Keeler, Levitt and Slichter) for more comprehensive presentations. The density operator approach allows us to calculate expectation values of operators for ensembles of quantum systems rather than for individual systems. Let us assume that a certain individual spin is described by a superposition state wave function, according to Eq. (1.8). We disregard for the moment the time dependence. The expectation value of an operator $\hat{Q}$ is for that spin given by:

$$\begin{aligned}\langle\hat{Q}\rangle &= \sum_{m=-I}^{I}\sum_{m'=-I}^{I} c_{m'}^{*}c_m\langle I,m'|\hat{Q}|I,m\rangle \\ &= c_\alpha^{*}c_\alpha\langle\alpha|\hat{Q}|\alpha\rangle + c_\beta^{*}c_\alpha\langle\beta|\hat{Q}|\alpha\rangle + c_\alpha^{*}c_\beta\langle\alpha|\hat{Q}|\beta\rangle + c_\beta^{*}c_\beta\langle\beta|\hat{Q}|\beta\rangle\end{aligned} \tag{1.14}$$

in analogy with the case of the x-component of the magnetic moment (Eq. (1.12b)). If we wish to make a similar calculation for another spin, the coefficients c_α and c_β will be different, but the matrix elements of the operator $\hat{Q}$ will be the same. Averaging over all spins in the ensemble, we obtain

$$\begin{aligned}\langle\hat{Q}\rangle &= \overline{c_\alpha^{*}c_\alpha}\langle\alpha|\hat{Q}|\alpha\rangle + \overline{c_\beta^{*}c_\alpha}\langle\beta|\hat{Q}|\alpha\rangle + \overline{c_\alpha^{*}c_\beta}\langle\alpha|\hat{Q}|\beta\rangle + \overline{c_\beta^{*}c_\beta}\langle\beta|\hat{Q}|\beta\rangle \\ &= \sum_m\sum_{m'}\overline{c_{m'}^{*}c_m}\langle m'|\hat{Q}|m\rangle = \sum_m\sum_{m'}\rho_{mm'}\langle m'|\hat{Q}|m\rangle \\ &= \sum_m\sum_{m'}\langle m|\hat{\rho}|m'\rangle\langle m'|\hat{Q}|m\rangle = \mathrm{Tr}(\hat{\rho}\hat{Q})\end{aligned} \tag{1.15}$$

The bar over the products of coefficients denotes ensemble average. Here, we have introduced the symbol $\hat{\rho}$, which denotes the density operator, with the matrix representation $\langle m|\hat{\rho}|m'\rangle = \rho_{m,m'} = \overline{c_{m'}^{*}c_m}$. $\mathrm{Tr}(\hat{\rho}\hat{Q})$ represents taking the trace (summing the diagonal elements in a matrix representation) of the product of the two operators or the two matrices. The definition above shows that the density operator is Hermitian. If the density operator for an ensemble is known, then the expectation value of any operator corresponding to observable quantities can be computed. We may wish to calculate the expectation value as a function of time. This can be done by expressing the superposition states in terms of time-dependent coefficients (and absorbing the phase factors $\exp(-iEt)$ into them). More practically, we can calculate the time dependence of the density operator directly, using the Liouville–von Neumann equation:

$$\frac{d}{dt}\hat{\rho}(t) = -i\left[\hat{H}(t),\hat{\rho}(t)\right] = i\left[\hat{\rho}(t),\hat{H}(t)\right] \tag{1.16}$$

where the concept of a commutator of the two operators, introduced in Eq. (1.5), is used. The Liouville–von Neumann equation can be derived in a straightforward way from the time-dependent Schrödinger equation. Using the time-dependent density operator, $\hat{\rho}(t)$, the time-dependent form of the expectation value of Eq. (1.15) can be written as:

$$\begin{aligned}\langle\hat{Q}\rangle(t) &= \sum_m\sum_{m'}\overline{c_{m'}^{*}(t)c_m(t)}\langle m'|\hat{Q}|m\rangle = \sum_m\sum_{m'}\rho_{mm'}(t)\langle m'|\hat{Q}|m\rangle \\ &= \mathrm{Tr}(\hat{\rho}(t)\hat{Q})\end{aligned} \tag{1.17}$$

The matrix representation of the density operator is called the *density matrix*. If we assume that the time dependence resides in the density matrix rather than in the operator, the formulation of the time-dependent expectation values in Eq. (1.17) is denoted as the *Schrödinger representation*, in analogy with the single-spin case of Eq. (1.12). The elements of the density matrix have a straightforward physical interpretation. The diagonal elements, ρ_{mm}, represent the probabilities that a spin is in the eigenstate specified by the quantum number m, or the relative *population* of that state. At the thermal equilibrium, these populations are given by the Boltzmann distribution:

$$\rho_{mm} = \frac{\exp(-\hbar E_m/k_B T)}{\sum_j \exp(-\hbar E_j/k_B T)} \tag{1.18}$$

T is the absolute temperature and k_B is the Boltzmann constant, $k_B = 1.38066 \cdot 10^{-23}$ JK^{-1}. Nuclear spins are quantum objects and their distribution among various quantum states should in principle be obtained using the Fermi–Dirac statistics (for half-integer spins) or by Bose–Einstein statistics (for integer spins) rather than from Eq. (1.18). However, for anything but extremely low temperatures, the Boltzmann statistics is an excellent approximation.

The energy differences involved in NMR are tiny, which results in very small population differences. For an ensemble of N spin-1/2 particles with a positive magnetogyric ratio, we can write

$$n_\alpha^{eq}/N = \rho_{\alpha\alpha}^{eq} = \tfrac{1}{2}\left(\exp\left(\tfrac{1}{2}\gamma_I \hbar B_0/k_B T\right)\right) \approx \tfrac{1}{2}\left(1+\tfrac{1}{2}\left(\gamma_I \hbar B_0/k_B T\right)\right) = \tfrac{1}{2}\left(1+\tfrac{1}{2}b_I\right) \tag{1.19a}$$

and

$$n_\beta^{eq}/N = \rho_{\beta\beta}^{eq} = \tfrac{1}{2}\left(\exp\left(-\tfrac{1}{2}\gamma_I \hbar B_0/k_B T\right)\right) \approx \tfrac{1}{2}\left(1-\tfrac{1}{2}\left(\gamma_I \hbar B_0/k_B T\right)\right) = \tfrac{1}{2}\left(1-\tfrac{1}{2}b_I\right) \tag{1.19b}$$

where we expand the exponential in a power series, retain only the linear term and introduce the quantity $b_I = \gamma_I \hbar B_0 / k_B T$, called the *Boltzmann factor*. N is the total number of spins. Retaining only the linear term is valid as long as $\gamma_I \hbar B_0 \ll k_B T$, a condition easily fulfilled for anything but extremely low temperatures. The approximation of retaining only the linear term is called the *high temperature approximation*. For protons, which have the magnetogyric ratio $\gamma_H = 26.7522 \cdot 10^7$ T^{-1}s^{-1}, at 300 K and in the magnetic field of 9.4 T (corresponding to the ^{1}H frequency of 400 MHz), we obtain $n_\alpha^{eq}/N = 0.500016$ and $n_\beta^{eq}/N = 0.499984$. At 4 K, the liquid helium temperature, the corresponding numbers are $n_\alpha^{eq}/N = 0.5024$ and $n_\beta^{eq}/N = 0.4976$. Clearly, the natural Boltzmann polarisation of the two nuclear spin levels is very low, which is the origin of the poor sensitivity of NMR compared with other spectroscopic techniques. The population difference between the spin energy levels determines the expectation value of the z-component of the ensemble-averaged nuclear magnetic moment – which we call the *magnetisation* vector – of an ensemble of spins. This can be seen (the reader is recommended to prove it as an exercise) by inspecting Eq. (1.15) and recognising the fact that the matrix representation of $\hat{\mu}_z$ or $\hat{I}_z$ (see Eq. (1.4a)) has only diagonal elements.

The off-diagonal elements of the density matrix are called *coherences*. The coefficients c_m can be written as products of an amplitude $|c_m|$ and a phase factor $\exp(ia_m)$. The coherence between the eigenstates m and m' is given by:

$$\overline{c_{m'}^{*} c_m} = \overline{|c_m^{*}||c_{m'}|\exp\left(i\left(a_m - a_{m'}\right)\right)} \tag{1.20}$$

The coherences are closely related to the magnetisation components perpendicular to the magnetic field (the transverse magnetisation). The phase factor $\exp(i(a_m - a_{m'}))$ for an individual spin specifies the direction of the transverse magnetic moment. For a large number of spins at thermal equilibrium, there is no physical reason to assume any of the directions perpendicular to the field to be more probable than any other, which amounts to a random distribution of the phase factors and vanishing coherences.

1.3 Simple NMR: The Magnetisation Vector

According to the previous section, an ensemble of non-interacting nuclear spins at the thermal equilibrium can be represented by a magnetisation vector, **M**, oriented along the direction of the external magnetic field. As we shall see later, the spins do interact with each other, but the interactions are usually very weak compared with the Zeeman energies. In addition, the interactions tend to average to zero because of molecular motions in isotropic fluids. For all practical purposes, the spins in isotropic liquids can be considered as non-interacting if their NMR spectra do not show spin-spin splittings (J-couplings). The magnitude of the magnetisation vector at thermal equilibrium, M_0, is proportional to the Boltzmann factor and depends, in addition, on the magnetic moment of an individual spin and on the number of spins:

$$M_0 = \frac{N\gamma_I^2\hbar^2 I(I+1)B_0}{3k_B T} \tag{1.21}$$

The magnetisation vector is a macroscopic quantity and its motion can be described using classical physics. Classically, if a magnetic moment is not aligned along the magnetic field, it will precess around the field direction, the same motion that we found above quantum mechanically for the magnetic moment of an individual spin and which we refer to as *Larmor precession.*

The concept of a magnetisation vector is very useful for describing NMR experiments in systems of non-interacting spins. We can use it, for example, to describe the effect of radiofrequency pulses. To describe an NMR experiment, we need to consider the presence of a static magnetic field in the z-direction with the magnitude B_0 as well as the time-dependent magnetic field corresponding to the magnetic component of electromagnetic radiation (the radiofrequency field), $\mathbf{B}_1$. The role of the radiofrequency field is simplest to consider in a frame rotating around the B_0 field with the angular velocity corresponding to the radiofrequency. The reader can find the derivation and discussion of the *rotating frame* in several books, *e.g.* Slichter or Levitt (*Further reading*). For our purposes, it is sufficient to say that if a radiofrequency field is applied near resonance (meaning that the radiofrequency is close to the Larmor frequency), the nuclear spins move as if it were the only field present, *i.e.* they precess around it with the angular velocity $\gamma_I B_1$. In this rotating frame, the applied radiofrequency field appears to be static. If we keep the radiofrequency field switched on for a time τ, the magnetisation will rotate by the angle $\gamma_I B_1 \tau$. By adjusting the magnitude of the radiofrequency field, and/or the time during which we apply it, in such a way that the angle becomes $\pi/2$, we obtain a 90° *pulse* or a $\pi/2$ pulse, which has the effect of turning the magnetisation, originally in the z-direction, to the transverse plane according to Figure 1.1. Setting the condition $\gamma_I B_1 \tau = \pi$ results in a 180° pulse or π pulse, which would invert the magnetisation.

Based on classical equations of motion for the dynamics of the magnetisation vector, the transverse magnetisation in the rotating frame will, after a 90° pulse on resonance, retain its direction and magnitude. If the radiofrequency pulse is not exactly on resonance, *i.e.* if there is a frequency offset, ω_{off}, between the applied radiofrequency and the Larmor frequency, the transverse magnetisation after a 90° pulse is expected to precess in the rotating frame with the frequency corresponding to the offset. Thus, the rotating frame serves to make the frequency offset the only precession frequency. Due to chemical shifts, the offset will vary for chemically inequivalent nuclear spins of the same species, *e.g.* for inequivalent protons. The motion of the magnetisation vector generates an oscillating signal in the detector of the NMR spectrometer, the *free induction decay* (FID). The FID in this simple picture is expected to last forever, which we know is not in agreement with experimental facts, and the reason for the decay of the FID is *nuclear spin relaxation*.

The description of even the very simplest NMR experiment, the detection of an FID after a 90° pulse, requires that we take into consideration two types of motion: the *coherent* motion or precession around the effective fields and the *incoherent* motion or relaxation. A very simple model containing these two elements is described by the *Bloch equations*, named after one of the inventors of NMR, who proposed them in a seminal paper from 1946.[1] The Bloch equations can be written in the following, slightly simplified, form:

$$\frac{dM_z}{dt} = \frac{M_0 - M_z}{T_1} \tag{1.22a}$$

$$\frac{dM_x}{dt} = M_y \omega_{off} - \frac{M_x}{T_2} \tag{1.22b}$$

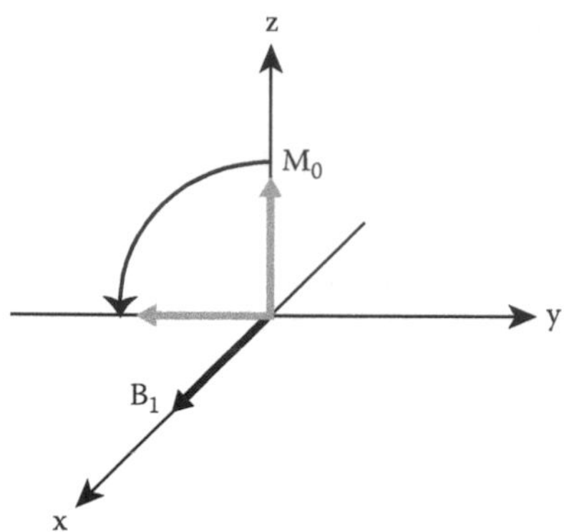

FIGURE 1.1 Illustration of the effect of radiofrequency pulses, here a 90° pulse.

$$\frac{dM_y}{dt} = -M_x\omega_{off} - \frac{M_y}{T_2} \tag{1.22c}$$

The Bloch equations are phenomenological, *i.e.* they aim at providing a simple description of the observed NMR phenomenon without the requirement for a strict derivation. We shall show, later on, that the relaxation behaviour in the form of the Bloch equations can be derived in some situations, while more complicated relaxation phenomena are predicted in other cases.

Eq. (1.22a) describes the time variation of the *longitudinal* (along the external field) component of the magnetisation vector. The equation predicts that the magnetisation component along $\mathbf{B_0}$ will relax exponentially to its equilibrium value, M_0. The time constant for that process is called the *spin-lattice* or *longitudinal* relaxation time, and is denoted T_1. The rate of exponential recovery of M_z to equilibrium is given by the inverse of T_1, sometimes denoted R_1, and called *spin-lattice relaxation rate* (the notation *rate constant* rather than rate would be more logical but is rarely used). We will use both T_1 and R_1 to describe longitudinal relaxation throughout this book. The solution of the Bloch equation for M_z, after the initial inversion of the magnetisation (corresponding to the application of a 180° pulse at the time $t = 0$) can be written as:

$$M_z(t) = M_0\left(1 - 2\exp\left(-t/T_1\right)\right) \tag{1.23}$$

The process of recovery of M_z is illustrated in Figure 1.2. The reader is encouraged to prove that Eq. (1.23) is a solution to Eq. (1.22a).

Eqs. (1.22b and c) describe the motion of the transverse components of the magnetisation vector. The first part of the expressions corresponds to the coherent motion of **M** in the rotating frame. The second part introduces the concept of the *transverse*, or *spin-spin relaxation time*, T_2, describing the exponential decay of the xy-magnetisation to its equilibrium value of zero. One interesting observation we can make in Eqs. (1.22b and c) is related to the units. Clearly, the factors multiplied by the magnetisation components in the two terms should have the same dimensions. Since the natural unit for the angular frequency is radians per second, the relaxation rate, or the inverse relaxation time, $R_2 = 1/T_2$, should indeed also be expressed in these units. Usually, relaxation times are given in seconds (the rates are given in s^{-1}), which tacitly implies that radians can be omitted; we note in parenthesis that the radian is in physics considered a dimensionless unit. Assuming that the pulse is on resonance, the evolution of the transverse components of the magnetisation vector after a 90° pulse is expressed as:

$$M_{x,y} = M_0\exp\left(-t/T_2\right) \tag{1.24}$$

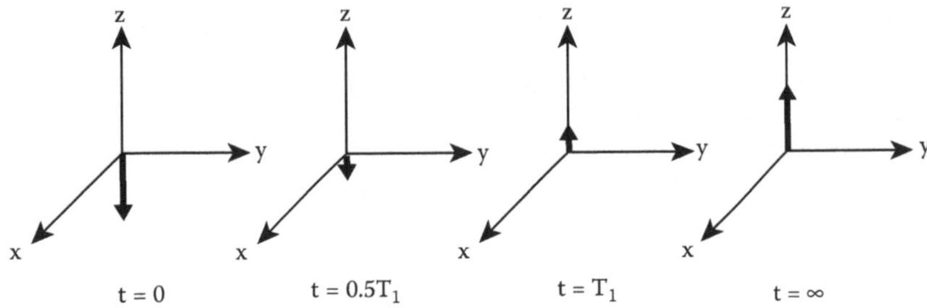

FIGURE 1.2 Illustration of the recovery of the M_z magnetisation after the 180° pulse in the inversion-recovery experiment.

The decay of the transverse magnetisation for the on-resonance situation is illustrated in Figure 1.3. We recall that the NMR spectrum for a system of non-interacting spins is the Fourier transform of the transverse magnetisation (see *Further reading*). The Fourier transform of an exponential decay is a Lorentzian, centred at zero frequency, with the full width at half-height (in Hz) equal to $\Delta\nu = 1/\pi T_2$; *cf.* Figure 1.4.

The reason for introducing two different relaxation times is that the return to the equilibrium is a physically different process for the longitudinal and transverse magnetisation components. The longitudinal relaxation changes the energy of the spin system and thus involves the energy exchange between the spins and the other degrees of freedom in the surrounding matter (the notion of spin-lattice relaxation originates from early NMR work on solids, where these other degrees of freedom were those of the crystal lattice). The transverse relaxation involves, on the other hand, the loss of phase coherence in the motion of individual spins.

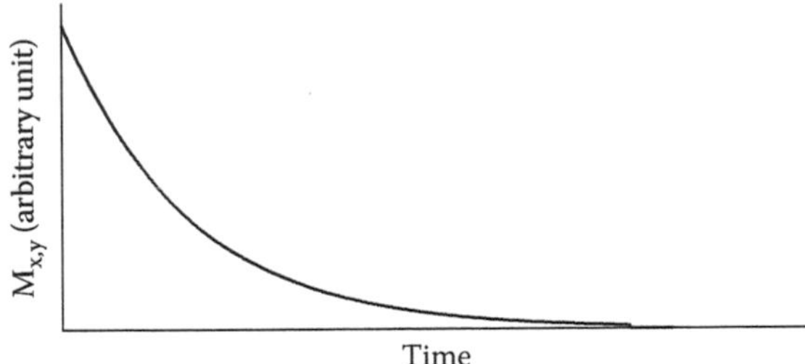

FIGURE 1.3 The decay of on-resonance transverse magnetisation as a function of time.

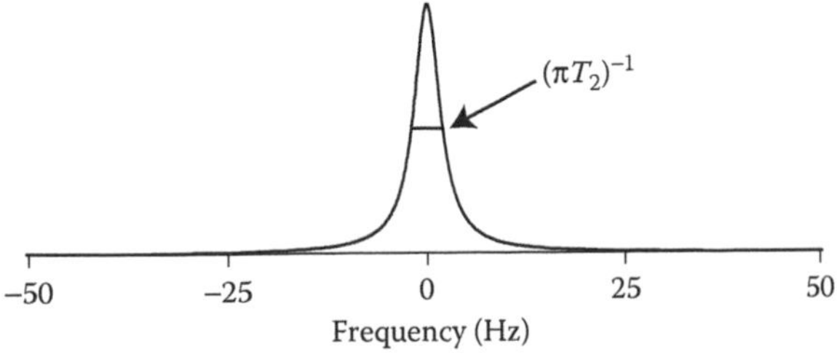

FIGURE 1.4 Relationship between the Lorentzian lineshape and the transverse relaxation time constant, T_2.

1.4 Coupled (Interacting) Spins: The Product Operator Formalism

The physical description of a spin system using magnetisation vectors is useful for an ensemble of non-interacting spins. For the case of coupled spins, more refined tools are necessary to describe NMR experiments and NMR relaxation processes. These tools have their basis in the concept of the density operator. In Section 1.2, we introduced the density operator by means of its matrix representation, the density matrix. For ensembles of interacting spins, working with the matrix representation of the density operator quickly becomes quite difficult to handle, and therefore it is often useful to work directly with the expansion of the density operator in another type of vector space, called the *Liouville space*. The concept of the Liouville space is discussed, among others, in the book by Ernst, Bodenhausen and Wokaun (*Further reading*) and in a review by Jeener[2]. A basis set in the Liouville space can be formulated in terms of the basis set in the Hilbert space in the following way. Consider the case of our isolated spin 1/2 nucleus, with its basis set of $|\alpha\rangle$ and $|\beta\rangle$. As we discussed earlier, the scalar products of a bra and a ket, such as $\langle\alpha|\beta\rangle$, are numbers. We can also define *outer products* of these vectors, in which we have a ket to the left and a bra to the right. There are four such possible constructs:

$$|\alpha\rangle\langle\alpha|, |\alpha\rangle\langle\beta|, |\beta\rangle\langle\alpha|, |\beta\rangle\langle\beta| \tag{1.25}$$

These objects are operators, which can be illustrated by the following example:

$$|\alpha\rangle\langle\beta|\beta\rangle = |\alpha\rangle 1 = \langle\alpha| \tag{1.26}$$

The operator $|\alpha\rangle\langle\beta|$ acts on the ket $|\beta\rangle$. $\langle\beta|\beta\rangle$ is a number (unity) and we thus obtain that the result of the operation is another ket, $|\alpha\rangle$. The operators created as outer products of the basis vectors in the Hilbert space can be used to form an operator basis set in the Liouville space. If the dimensionality of the Hilbert space is n ($n = 2$ in our example of an isolated spin 1/2, with $|\alpha\rangle$ and $|\beta\rangle$ as the basis vectors), then the dimensionality of the corresponding Liouville space is n^2 (which in our case is four, corresponding to the four basis vectors $|\alpha\rangle\langle\alpha|, |\alpha\rangle\langle\beta|, |\beta\rangle\langle\alpha|, |\beta\rangle\langle\beta|$). The operators expressed as outer products will be an exception from the rule of decorating the operators with "hats."

It is easy to obtain the matrix representations of these four operators in the Hilbert space. For example, the matrix representation of the $|\alpha\rangle\langle\alpha|$ operator is $\begin{pmatrix} 1 & 0 \\ 0 & 0 \end{pmatrix}$. The reader is advised to confirm this and to derive the matrices corresponding to the other three operators. Comparing the matrix representations of the operators of Eq. (1.25) with the Pauli spin matrices introduced in Eq. (1.4), we can note that:

$$\tfrac{1}{2}\left(|\alpha\rangle\langle\alpha| + |\beta\rangle\langle\beta|\right) = \tfrac{1}{2}\hat{1}_{op} \tag{1.27a}$$

$$\tfrac{1}{2}\left(|\alpha\rangle\langle\alpha| - |\beta\rangle\langle\beta|\right) = \hat{I}_z \tag{1.27b}$$

$$\tfrac{1}{2}\left(|\alpha\rangle\langle\beta| + |\beta\rangle\langle\alpha|\right) = \hat{I}_x \tag{1.27c}$$

$$\tfrac{i}{2}\left(|\beta\rangle\langle\alpha| - |\alpha\rangle\langle\beta|\right) = \hat{I}_y \tag{1.27d}$$

where we have introduced the unit operator, $\hat{1}_{op}$, with the property that it leaves whatever function or vector comes after it unchanged. The unit operator is represented by a unit matrix, a matrix with ones on the diagonal and zeroes elsewhere. Eq. (1.27) implies an orthogonal transformation of one set of vectors in the Liouville space into another.

The operators in the outer product form, containing a ket and a bra corresponding to the same eigenstate, can be thought of as representing the population of that eigenstate. Thus, $|\alpha\rangle\langle\alpha|$ and $|\beta\rangle\langle\beta|$ are operators describing the populations of the α and β states. We shall make use of this property in Chapter 16.

Once we have defined an appropriate Liouville space basis, we can expand any other operator in that space. For example, the density operator at thermal equilibrium for an ensemble of isolated spins can be expressed as

$$\hat{\rho}^{eq} = \tfrac{1}{2}\hat{1}_{op} + \tfrac{1}{2}b_I\hat{I}_z \tag{1.28}$$

where the Boltzmann factor, b_I, was defined together with Eq. (1.19b). The idea of expanding the density operator into an operator basis set related to the spin operators is easily generalised to more complicated spin systems. For a system of two spins, I and S, with the components $\hat{I}_x,\hat{I}_y,\hat{I}_z$, and $\hat{S}_x,\hat{S}_y,\hat{S}_z$, we can form an appropriate basis set by including the unit operator for each spin. The *product operator* basis for a two-spin system will thus consist of 4×4 operators, given by the product of one operator for the spin I and one operator for spin S, with suitable normalisation. This product operator basis will consist of $\tfrac{1}{2}\hat{1}_{op}$, $\hat{I}_x,\hat{I}_y,\hat{I}_z$, $\hat{S}_x,\hat{S}_y,\hat{S}_z$, $2\hat{I}_x\hat{S}_x$, $2\hat{I}_x\hat{S}_y$, $2\hat{I}_x\hat{S}_z$, $2\hat{I}_y\hat{S}_x$, $2\hat{I}_y\hat{S}_y$, $2\hat{I}_y\hat{S}_z$, $2\hat{I}_z\hat{S}_x$, $2\hat{I}_z\hat{S}_y$, $2\hat{I}_z\hat{S}_z$, a total of 16 operators. A two-spin system is characterised by the occurrence of four eigenstates of the Zeeman Hamiltonian, $|\alpha\alpha\rangle, |\alpha\beta\rangle, |\beta\alpha\rangle, |\beta\beta\rangle$, implying the dimensionality of 4 for the Hilbert space and the dimensionality of $4^2 = 16$ for the Liouville space. Clearly, we retain the dimensionality of the Liouville space moving between the basis set consisting of ket-bra products, $|\alpha\alpha\rangle\langle\alpha\alpha|$ *etc.*, and the product operator basis. Contrary to the density matrix calculations, the product operators provide a method to easily describe NMR experiments, and are invaluable for evaluating NMR pulse sequences on interacting spins. The product operator formalism has in particular been used for improving and designing new experiments. We shall explore this further in the chapters describing experimental techniques and applications.

In the same way as one can define operators in the Hilbert space, it is possible to construct their analogues in the Liouville space. These are called *superoperators*. The superoperator analogue of the Hamiltonian is called the *Liouville superoperator* or *Liouvillian*, $\hat{\hat{L}}$ (we shall use the "double hat" symbol for superoperators). It is defined as commutator with the Hamiltonian, $\hat{\hat{L}} = [\hat{H},]$. Operating with the Liouvillian on another operator thus means taking the commutator of the Hamiltonian with that operator. This superoperator formalism can be used to re-write the Liouville–von Neumann equation (Eq. (1.16)) in the form

$$\frac{d}{dt}\hat{\rho} = -i\left[\hat{H},\hat{\rho}\right] = -i\hat{\hat{L}}\hat{\rho} \tag{1.29}$$

Eq. (1.29) illustrates one of the reasons for using the superoperator formalism: it allows a very compact notation. Another important superoperator we shall meet is the *relaxation superoperator*, which operates on the density operator and describes its evolution towards equilibrium. In the same way as operators are represented in the Hilbert space by matrices, the superoperators have matrix representations in the Liouville space, denoted as *supermatrices*. The concept of a relaxation supermatrix is very important in relaxation theory and we shall come back to it in Chapter 4. Yet another type of superoperators are various *rotation superoperators*, introduced in Chapter 7. In some situations, we need to consider more complicated evolution of the density operator and the Liouville space formalism is then very useful; we shall see examples of that application of the formalism in Chapters 10, 15 and 16.

1.5 Exercises for Chapter 1

1. Show that the quantum mechanical expectation value of the nuclear magnetic moment vector undergoes motion on a conical surface around the direction of the magnetic field.

2. Show that the z-component of the magnetisation vector for a system of isolated nuclear spins is determined by the population difference between the spin energy levels.
3. Prove that Eq. (1.23) is a solution to Eq. (1.22a).
4. Show that the matrix $\begin{pmatrix} 1 & 0 \\ 0 & 0 \end{pmatrix}$ is a representation of the operator $|\alpha\rangle\langle\alpha|$. Derive the matrix representations of the operators $|\alpha\rangle\langle\beta|$, $|\beta\rangle\langle\alpha|$, $|\beta\rangle\langle\beta|$.

References

1. Bloch, F., Nuclear induction. *Phys. Rev.* 1946, 70, 460–474.
2. Jeener, J., Superoperators in magnetic resonance. *Adv. Magn. Reson.* 1982, 10, 1–51.

2

Simple Relaxation Theory

In order to establish a foundation for relaxation theory, we shall in this chapter provide a background by introducing the concepts of spin-lattice and spin-spin relaxation. We begin this chapter by looking at a simple example, which captures the basic principles of the nuclear spin relaxation without being directly applicable to any real physical situation. We use this simple example to introduce the important basic concepts within the theory of random processes, and then to calculate transition probabilities, driven by random processes. We look briefly at the prediction of the simple model and introduce the important finite temperature corrections to the theory.

2.1 An Introductory Example: Spin-Lattice Relaxation

Among the two types of relaxation processes, the spin-lattice relaxation is easier to explain in simple conceptual terms, and therefore, we shall start with describing this relaxation process. Let us consider a system of $I=1/2$ spins with a positive magnetogyric ratio in a magnetic field B_0. We deal thus with a two-level system with populations n_α and n_β, with the energy spacing $\Delta E = -\hbar\omega_0 = \hbar\gamma B_0$, and with the Larmor frequency ω_0. At thermal equilibrium, the relative populations of the two levels are given by Eq. (1.19).

Let us assume that the spin system is not in equilibrium. The non-equilibrium situation can be produced in different ways. One way is to quickly change the magnetic field. The equilibrium population distribution at the original field does not correspond to the equilibrium at the new field. An everyday example of such an experiment is putting a nuclear magnetic resonance (NMR) sample into the NMR magnet. A better-controlled variety of such an experiment is the field-cycling experiment, in which the magnet current is switched rapidly and to which we shall return in Chapter 8. An important feature of an experiment of this kind is the fact that the Hamiltonian – and thus the Zeeman splitting – changes immediately, while the density operator – for example, populations – requires some time to adjust. This is the simplest example of spin-lattice relaxation. Another way of creating non-equilibrium states in NMR, mentioned in Section 1.3, is to use radiofrequency pulses. For example, a 180° pulse (or a π-pulse) inverts the populations in a two-level system.

We assume that the changes of the populations of the two levels follow the simple kinetic scheme:

$$\frac{dn_\alpha}{dt} = \left(n_\beta - n_\beta^{eq}\right)W_{\beta\alpha} - \left(n_\alpha - n_\alpha^{eq}\right)W_{\alpha\beta} = W_I\left(n_\beta - n_\beta^{eq} - n_\alpha + n_\alpha^{eq}\right) \tag{2.1a}$$

$$\frac{dn_\beta}{dt} = W_I\left(n_\alpha - n_\alpha^{eq} - n_\beta + n_\beta^{eq}\right) \tag{2.1b}$$

Eq. (2.1) implies that the α- and β-levels are populated (and depopulated) by first-order kinetic processes, with the rates (in the sense of chemical kinetics) proportional to the deviations of the populations from the equilibrium values. The proportionality constants are *transition probabilities*, assumed for the time being to be equal in both directions: $W_{\beta\alpha} = W_{\alpha\beta} = W_I$. We shall return to this point in Section 2.5.

Instead of discussing the changes of the populations n_α and n_β, we can introduce new variables: the difference in populations, $n=n_\alpha-n_\beta$, and the sum of the populations, $N=n_\alpha+n_\beta$. In terms of these variables, Eq. (2.1) can be rewritten as:

$$\frac{dN}{dt}=0 \tag{2.2a}$$

$$\frac{dn}{dt}=-2W_I\left(n-n^{eq}\right) \tag{2.2b}$$

Eq. (2.2a) tells us that the total number of spins is constant, Eq. (2.2b) that the population difference returns to equilibrium in an exponential process. We note that the population difference is proportional to the longitudinal component of the magnetisation vector, $n(t)\sim\gamma_I\langle\hat{I}_z\rangle(t)=M_z(t)$, and by inspection of the first of the Bloch equations, Eq. (1.22a), we can relate W_I to the T_1 according to:

$$T_1^{-1}=2W_I \tag{2.3}$$

The relaxation rate is thus proportional to a transition probability. The transitions giving rise to NMR relaxation are non-radiative, *i.e.* they do not arise through emission or absorption of radiation from radiofrequency fields. Instead, they occur as a result of weak magnetic interactions, with the origin in the sample itself, if those oscillate in time with frequency components at the Larmor frequency. The weakness of the relevant interactions for spin 1/2 nuclei results in small transition probabilities and the spin-lattice relaxation processes being slow, typically on the millisecond to second time scale. More specifically, the time dependence of the interactions has its origin in random molecular motions. The effect of molecular motions can be explained as follows. Many interactions in NMR are anisotropic, *i.e.* they depend on the orientation of the spin-carrying molecule in the magnetic field. A prominent example, to be discussed in detail in the next chapter, is the dipole-dipole interaction. Thus, the interaction changes when the molecule reorients. We concentrate at this stage on the case of ordinary, isotropic liquids. As opposed to the situation in the gas phase, molecules in a liquid are surrounded by neighbours and are not able to rotate freely. Rather, the rotational motion of a molecule-fixed axis in a molecule immersed in a liquid can be pictured as a sequence of small angular steps (a random walk on the surface of a sphere). This is illustrated in Figure 2.1. The combination of the

FIGURE 2.1 Illustration of a random walk on the surface of a sphere.

random walk and the anisotropic interactions gives rise to Hamiltonians for nuclear spins varying randomly with time. The effect of these random – or *stochastic* – interactions is to cause transitions, which are intimately connected with the nuclear spin-relaxation processes. Therefore, we need to expand on this issue and we explain how this happens in two steps. Since the random nature of the interactions is essential, we introduce first some basic ideas and concepts from the theory of random processes. Second, we present a quantum mechanical treatment of transition probabilities caused by random motions.

2.2 Elements of Statistics and Theory of Random Processes

The results of many physical experiments on molecular systems are best described by using statistical methods. This is because many processes are random in nature and we need a way to characterise expectation values and averages. In this section we provide an introduction to the theory of random processes. A more complete description of the subject can, for example, be found in the book of Van Kampen (see *Further reading*).

2.2.1 Stochastic Variables

Consider a quantity that can be measured and assigned a numerical value x. Assume that the numerical values vary within a certain interval between different realisations of the measurement in an unpredictable way. We then call X a *stochastic variable.* The values, x, that X adopts are called *numerical realisations.* An example from daily life is the length of an individual person, in which X is a symbolic notation for the length of a person and x is the value of the length. This, of course, varies stochastically within a certain group of people. An example relevant for NMR is the orientation of a molecule-fixed vector with respect to the laboratory z-axis. The orientation can be specified in terms of an angle θ, indicated in Figure 2.2, which can (at least in principle) be measured and thus assigned a numerical realisation.

A stochastic variable can be described by a *probability density, p(x)*:

$$p(x)dx = P(x \leq X \leq x + dx) \tag{2.4}$$

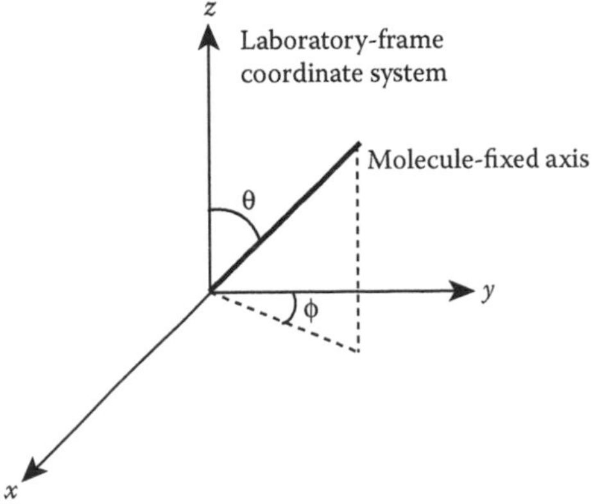

FIGURE 2.2 Illustration of the orientation of a molecule-fixed vector with respect to the laboratory coordinate frame. The orientation can be specified in terms of the angles θ and ϕ.

where $P(x \le X \le x + dx)$ denotes the probability of the numerical realisation taking on a value within the indicated, infinitesimally small interval between x and $x + dx$. Rather than working in terms of infinitesimal intervals, we can use an integrated form:

$$\int_{x_1}^{x_2} p(x)dx = P(x_1 \le X \le x_2) \tag{2.5}$$

which describes the distribution of values x that the variable X adopts. The probability density can be established by a long series of measurements. It is easy to imagine this being done for the length of an individual in a certain population, but perhaps not so for the second example. For that example, the orientation of a molecule-fixed vector in an isotropic liquid with respect to an external frame, one would, rather, make an assumption that the distribution of the angles is isotropic, *i.e.* that all angles are equally probable. The value of the probability density is then determined by the normalisation condition, *i.e.* by the requirement that the integral of the probability density over all angles has to yield unity. We shall return to this point in Chapter 6.

Stochastic variables often occur in pairs. For example, we can measure both the length and the weight of an individual within a certain population. We call two such variables X and Y and denote the corresponding numerical realisations x and y. We introduce a probability density in two dimensions:

$$p(x;y)dxdy = P(x \le X \le x + dx; y \le Y \le y + dy) \tag{2.6}$$

where the symbol ";" means "and." Alternatively, the corresponding integrated form can be used:

$$\int_{x_1}^{x_2}\int_{y_1}^{y_2} p(x;y)dxdy = P(x_1 \le X \le x_2; y_1 \le Y \le y_2) \tag{2.7}$$

If the two stochastic variables are statistically independent (uncorrelated), then:

$$p(x;y) = p(x)p(y) \tag{2.8}$$

An important property of a stochastic variable is its *average value* (mean) and this is given by:

$$\langle X \rangle = \int_{-\infty}^{\infty} xp(x)dx \tag{2.9}$$

We prefer to use the symbol $\langle X \rangle$ rather than $\bar{X}$ for average; the bar is used when the other notation can be confusing. Analogously, we can define an average value, *expectation value*, of a function of X. For example, the average value of X^n (the nth power of X), called the *nth moment*, is defined as:

$$m_n = \langle X^n \rangle = \int_{-\infty}^{\infty} x^n p(x)dx \tag{2.10}$$

Another important average is the average of $(X - c)^n$, where c is a constant. This is called the nth moment around c. Moments around mean values are often used in statistics. They are denoted μ_n, and the second moment, μ_2, is particularly important:

$$\mu_2 = \langle (X - m_1)^2 \rangle = \langle X^2 \rangle - 2m_1 \langle X \rangle + m_1^2 = \langle X^2 \rangle - m_1^2 = \sigma^2 \tag{2.11}$$

and is called *variance*. The square root of the variance is called *standard deviation*, σ.

Average values can also be defined for cases involving more than one stochastic variable, *e.g.* for a product:

$$m_{11} = \langle XY \rangle = \int_{-\infty}^{\infty}\int_{-\infty}^{\infty} xyp(x;y)dxdy \tag{2.12}$$

It can be useful to express $p(x;y)$ as a product:

$$p(x;y) = p(x)p(x|y) \tag{2.13}$$

where $p(x|y)$ means the probability density for Y acquiring the value y, provided that X assumes the value x. It is called *conditional probability density.* If X and Y are statistically independent, then $p(x|y) = p(y)$ and:

$$m_{11} = \langle XY \rangle = \langle X \rangle \langle Y \rangle \tag{2.14}$$

We can also define a mixed second moment:

$$\mu_{11} = \langle (X - \langle X \rangle)(Y - \langle Y \rangle) \rangle = m_{11} - m_{10}m_{01} \tag{2.15}$$

which vanishes if X and Y are statistically independent. A convenient way to express the correlation between two variables is to define a *correlation coefficient*, ρ, between X and Y. This is defined as:

$$\rho = \frac{\mu_{11}}{\sigma_X \sigma_Y} \tag{2.16}$$

where σ_X and σ_Y are the standard deviations for the two stochastic variables. For statistically independent X and Y, $\rho = 0$. For the opposite limiting case, $X = Y$, it is easily seen that $\mu_{11} = \mu_2 = \sigma^2$, and thus we have $\rho = 1$.

2.2.2 Stochastic Functions of Time

In NMR relaxation theory, *stochastic processes* (stochastic functions of time) are important. As was mentioned in the preceding section, the combination of random walk and anisotropic interactions leads to stochastic interactions. Stochastic processes, $Y(t)$, are those that give rise to a time-dependent stochastic variable, which means a quantity that at every point, t, in time behaves as a stochastic variable. The stochastic process is characterised by a probability density, $p(y,t)$, in general also dependent on time. An example of such a process might be the depth of water at a certain point along an ocean beach measured on a windy day. Because of the waves, the water level changes with time, within certain limits and in a random way. The NMR-relevant example mentioned earlier, the angle between a molecule-fixed vector and the external magnetic field, is a typical stochastic function of time because of random motions (*cf.* Figure 2.1). The average value of $Y(t)$ at time t is defined:

$$\langle Y(t) \rangle = \int_{-\infty}^{\infty} yp(y,t)dy \tag{2.17}$$

The properties of a stochastic function, $Y(t)$, at different times t are in general not independent. We are going to investigate the correlation between $Y(t)$ at times t_1 and t_2. It is rather easy to imagine that such a correlation exists if the time points t_1 and t_2 are close to each other on the time scale defined by the

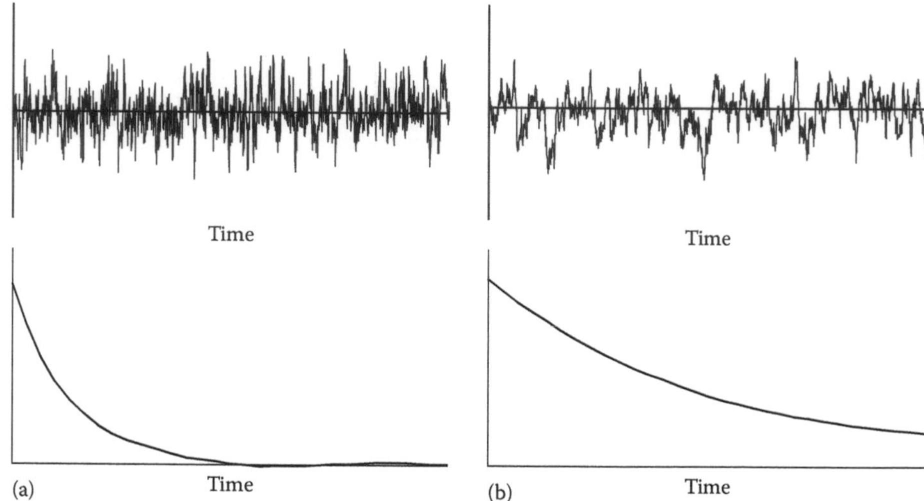

FIGURE 2.3 Illustration of a stochastic function of time and the corresponding correlation function for a rapidly vanishing correlation, or short correlation time (a) and a more persistent correlation, long correlation time (b).

random oscillation of the process $Y(t)$. This is shown in Figure 2.3a and b, illustrating a rapidly decaying and a more persistent correlation. We introduce $p(y_1,t_1;\, y_2,t_2)$, the probability density for $Y(t)$ acquiring value y_1 at t_1 and y_2 at t_2. Also, in this case, we can use the concept of conditional probability:

$$p(y_1,t_1;\, y_2,t_2) = p(y_1,t_1)\,p(y_1,t_1 \mid y_2,t_2) \tag{2.18}$$

A stochastic process is called *stationary* if the probability density $p(y,t)$ does not vary with time. If this is the case, the conditional probability density simplifies to:

$$p(y_1,t_1 \mid y_2,t_2) = p(y_1,0 \mid y_2,t_2 - t_1) = p(y_1 \mid y_2,\tau) \tag{2.19}$$

where we have introduced the length $\tau = t_2 - t_1$ of the time interval between t_1 and t_2. For the average value of a product of a stationary process $Y(t)$ at different times, t_1 and t_2, we have, from Eq. (2.12):

$$\begin{aligned} \langle Y(t_1)Y(t_2)\rangle &= \iint y_1(t_1)y_2(t_2)\,p(y_1,t_1)\,p(y_1,t_1 \mid y_2,t_2)\,dy_1dy_2 \\ &= \iint y_1y_2\,p(y_1,0)\,p(y_1,0 \mid y_2,t_2 - t_1)\,dy_1dy_2 \end{aligned} \tag{2.20}$$

The expression for the average value is dependent only on the time difference $\tau = t_2 - t_1$ and we can define:

$$\langle Y(t_1)Y(t_2)\rangle = G(t_2 - t_1) = G(\tau) \tag{2.21}$$

The quantity $G(\tau)$ is called the *time-correlation function* (tcf). Since it correlates a stochastic process with itself at different points in time, it is called the *autocorrelation function*. The autocorrelation functions for the random processes in Figure 2.3a and b are also shown there. *Cross-correlation* functions can also be defined and we shall return to such functions later. In relaxation theory, we often deal with complex functions $Y(t)$. For such cases, the definition of the autocorrelation function has to be modified slightly:

$$G(\tau) = \langle Y(t)Y^*(t+\tau)\rangle \tag{2.22}$$

For $G(\tau)$ defined in this way, we have:

$$G(\tau) = G^*(\tau) = G(-\tau) \tag{2.23}$$

which means that the time-correlation function is real and an even function of time. For $\tau = 0$, we obtain then:

$$G(0) = \langle Y(t)Y^*(t)\rangle = \langle |Y(t)|^2 \rangle = \sigma^2 \tag{2.24}$$

i.e. the autocorrelation function of Y at zero time, is equal to the variance of Y. Let us further assume that $\langle Y(t)\rangle = 0$. This assumption does not really cause any loss of generality for a stationary process, because we can always subtract the time-independent average from our stochastic function of time. For the limit of very long time, $\tau \to \infty$, it is reasonable to assume that $Y(t)$ and $Y(t+\tau)$ become uncorrelated and we can write:

$$\lim_{\tau\to\infty} G(\tau) = \langle Y(t)\rangle^2 = 0 \tag{2.25}$$

provided that $\langle Y(t)\rangle = 0$. Thus, we expect a general time-correlation function to be a decaying function of time, with an initial value given by the variance of Y. The function:

$$G(\tau) = G(0)\exp(-|\tau|/\tau_c) \tag{2.26}$$

might be a reasonable choice. The symbol τ_c is called *correlation time*. In Chapter 6, we are going to use a simple model for the random walk, depicted in Figure 2.1, to demonstrate that our choice of time-correlation function of the form given in Eq. (2.26) can indeed be obtained for the spherical harmonics of the angles specifying molecular orientation in a liquid. The correlation time has many possible simple interpretations: it is a measure of the time scale of oscillations of the random process or a measure of the persistence of the correlation between values of $Y(t)$ at different points in time. The two correlation functions in Figure 2.3 can thus be identified as being characterised by different correlation times. For the case of molecular reorientation in a liquid, we can treat τ_c as an average time for a molecular axis to change its direction by one radian.

The quantities of prime interest in relaxation theory are *spectral density functions*, which are Fourier transforms of the tcfs:

$$J(\omega) = \int_{-\infty}^{\infty} G(\tau)\exp(-i\omega\tau)d\tau \tag{2.27}$$

Since the concept of negative time is awkward, it may be more convenient to define the spectral densities as twice the one-sided Fourier transform of the time-correlation function:

$$J(\omega) = 2\int_{0}^{\infty} G(\tau)\exp(-i\omega\tau)d\tau \tag{2.28}$$

According to the *Wiener–Khinchin* theorem in the theory of stochastic processes, the spectral density has a straightforward physical interpretation: it is a measure of the distribution of the fluctuations of $Y(t)$ among different frequencies. The spectral density associated with the exponentially decaying tcf is Lorentzian:

$$J(\omega) = G(0)\frac{2\tau_c}{1+\omega^2\tau_c^2} \tag{2.29}$$

The shape of the Lorentzian spectral density is indicated in Figure 2.4.

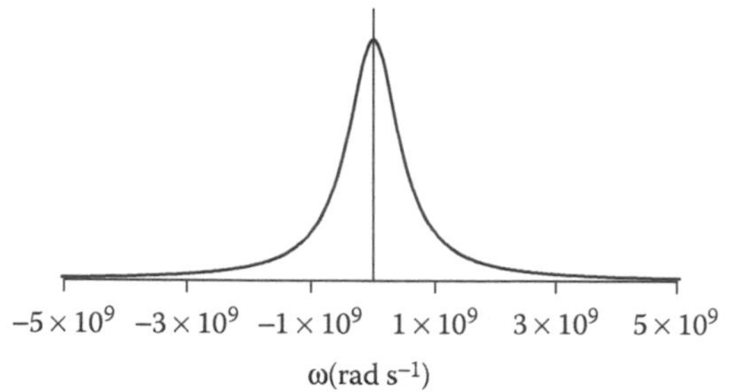

FIGURE 2.4 Spectral density function calculated with a correlation time $\tau_c = 2$ ns.

2.3 Time-Dependent Perturbation Theory and Transition Probabilities in NMR Relaxation Theory

Very few problems in quantum mechanics can be solved exactly, so therefore techniques of approximation are needed. If we can somehow relate the problem at hand to a system with a known solution, we can use this to get further. If the most important part of the system can be characterised by analogy with a known system, described by the Hamiltonian $\hat{H}_0$, and with known solutions, we can focus on the small addition to the main part. In general, perturbation theory is based on the assumption that the Hamiltonian for the system under consideration can be expressed as a sum of the main, *unperturbed* part, $\hat{H}_0$ (usually independent of time), and a smaller term (or terms), called *perturbation*. The eigenstates of $\hat{H}_0$ are assumed to be known and the theory is applied to find how these are modified by the presence of the perturbation. The perturbation may be constant in time or time dependent. The approximation technique is referred to as *perturbation theory*. Quantum mechanical calculations of transition probabilities are based on the *time-dependent perturbation theory*, in which the perturbation is assumed to be time dependent. The somewhat sketchy derivation here is based on the book of Carrington and McLachlan.[1] For a more complete and formal presentation, the reader is referred to *e.g.* the book of Atkins and Friedman (*Further reading*).

Consider a simple case of a two-level quantum system (energy levels E_a and E_b, eigenfunctions ψ_a and ψ_b), which is acted on by a Hamiltonian containing a sum of the main part, $\hat{H}_0$, and time-dependent perturbation $\hat{V}(t)$. We look for approximate solutions to the time-dependent Schrödinger equation, formulated in Eq. (1.9), in the form:

$$\psi(t) = c_a(t)\psi_a e^{-iE_at} + c_b(t)\psi_b e^{-iE_bt} \tag{2.30}$$

Eq. (2.30) is formally very similar to Eq. (1.11). The physical meaning of Eq. (2.30) is that the perturbation does not change the nature of the problem in any fundamental way, *i.e.* that the eigenstates of $\hat{H}_0$ still provide a useful set of functions in which the approximate solutions to $\hat{H} = \hat{H}_0 + \hat{V}(t)$ can be expanded. In NMR, this is never really a problem: the eigenstates to the Zeeman Hamiltonian form a complete set of functions in the spin space, *i.e.* any arbitrary spin function for $I = 1/2$ can be expressed exactly in the form of Eq. (2.30). Assume vanishing diagonal elements of $\hat{V}(t)$, *i.e.* the Hamiltonian represented by the matrix:

$$\begin{bmatrix} E_a & V_{ab}(t) \\ V_{ba}(t) & E_b \end{bmatrix} \tag{2.31}$$

with the matrix elements $V_{ab}(t) = \langle a|\hat{V}(t)|b\rangle = V_{ba}^*(t)$ (the perturbation is a Hermitian operator). It is easy to show that the coefficients fulfil the equations:

$$i\frac{dc_a(t)}{dt} = V_{ab}(t)e^{i(E_a - E_b)t}c_b(t) = V_{ab}(t)e^{i\omega_{ab}t}c_b(t) \tag{2.32a}$$

$$i\frac{dc_b(t)}{dt} = V_{ba}(t)e^{i(E_b - E_a)t}c_a(t) = V_{ba}(t)e^{-i\omega_{ab}t}c_a(t) \tag{2.32b}$$

where $\omega_{ab} = (E_a - E_b)$ and all energy-related quantities are in angular frequency units. Note that the equations for the coefficients are coupled. Suppose that the system is in the state a at $t=0$ ($c_a(0) = 1$, $c_b(0) = 0$) and that the perturbation is weak, *i.e.* the coefficients change slowly. We can then obtain an approximate expression for the coefficient $c_b(t)$ as:

$$c_b(t) = -i\int_0^t V_{ba}(t')e^{-i\omega_{ab}t'}dt' \tag{2.33a}$$

$$c_b^*(t) = i\int_0^t V_{ba}^*(t')e^{i\omega_{ab}t'}dt' \tag{2.33b}$$

The probability that the system is in state b is expressed by $|c_b(t)|^2 = c_b(t)\, c_b^*(t)$. The transition probability per unit time is the rate of change of this quantity:

$$W_{ab} = \frac{d}{dt}|c_b(t)|^2 = \frac{dc_b(t)}{dt}c_b^*(t) + c_b(t)\frac{dc_b^*(t)}{dt} \tag{2.34}$$

The expression for $dc_b(t)/dt$ in Eq. (2.32b) can be rearranged slightly to:

$$\frac{dc_b(t)}{dt} = -ie^{-i\omega_{ab}t}V_{ba}(t) \tag{2.35}$$

and we can use this result together with Eq. (2.34) to obtain:

$$\begin{aligned} W_{ab} &= e^{-i\omega_{ab}t}V_{ba}(t)\int_0^t V_{ba}^*(t')e^{i\omega_{ab}t'}dt' + c.c. \\ &= \int_0^t V_{ba}(t)V_{ba}^*(t')e^{i\omega_{ab}(t'-t)}dt' + c.c. \end{aligned} \tag{2.36}$$

where *c.c.* denotes complex conjugate.

When discussing relaxation-related transitions in a liquid, we need to know an average behaviour of a large number of systems with stochastic $\hat{V}(t)$, *i.e.* with the perturbation varying from one member in the spin ensemble to another in a random way. To describe the average behaviour in an ensemble of spins, we make the variable substitution $\tau = t' - t$ and take at the same time the ensemble average of Eq. (2.36):

$$W_{ab} = \int_0^t \langle V_{ba}(t)V_{ba}^*(t+\tau)\rangle e^{i\omega_{ab}\tau}d\tau + c.c. \tag{2.37}$$

We can note, in passing, that according to the *ergodic hypothesis* of statistical mechanics, the ensemble average is equivalent to the time average taken for a single system over a long time. We note that the integral in Eq. (2.37) contains the tcf of $V_{ba}(t)$, $\langle V_{ba}(t)V_{ba}^*(t+\tau)\rangle$. Making use of the properties of tcfs, we can, following Eq. (2.23), write $G_{ba}^*(\tau)=\langle V_{ba}(t)V_{ba}^*(t+\tau)\rangle$ and we obtain:

$$W_{ab}=\int_{-t}^{t} G_{ba}(\tau)e^{i\omega_{ab}\tau}d\tau \tag{2.38}$$

Let us now assume that we wish to study the transition probabilities on a time scale much larger than the correlation time characterising the decay of the time-correlation function, $t \gg \tau_c$. Under this condition, $G(\tau)$ vanishes at $\pm t$ and the integration limits can be extended to $\pm\infty$. Using Eq. (2.23) once more, we obtain for the transition probability:

$$\begin{aligned} W_{ab} &= \int_{-\infty}^{\infty} G_{ba}(\tau)e^{i\omega_{ab}\tau}d\tau = \int_{-\infty}^{\infty} G_{ba}(\tau)e^{-i\omega_{ab}\tau}d\tau \\ &= 2\int_{0}^{\infty} G_{ba}(\tau)e^{-i\omega_{ab}\tau}d\tau = \mathcal{J}_{ba}(\omega_{ab}) \end{aligned} \tag{2.39}$$

Thus, we obtain the very important result that the transition probability, induced by a randomly fluctuating interaction, is equal to the spectral density of the random perturbation, evaluated at the frequency corresponding to the relevant energy level spacing. We introduce here a notation convention to be followed throughout this book. We use the symbol $J(\omega)$ for a spectral density of a purely *classical* random function, while the script symbol with appropriate indices, here $\mathcal{J}_{ba}(\omega)$, denotes a spectral density for matrix elements of a *stochastic operator*. Through the Wiener–Khinchin theorem, the transition probability is therefore connected to the power available at the transition frequency. This tells us that the relaxation processes are in a sense related to the Einstein formulation of transition probabilities for stimulated absorption and emission of radiation, both quantities being proportional to the power available at the transition frequency. In passing, we can note that the spontaneous emission processes, which in principle could contribute to depopulating the upper spin state, can be completely neglected in NMR, because of the low frequencies (energy differences) involved. For a more complete discussion of Einstein transition probabilities, the reader is referred to the book by Atkins and Friedman (*Further reading*).

2.4 Predictions of the Simple Model

We are now in a position to examine what the results mean for the relaxation rates. Combining Eq. (2.3) for the spin-lattice relaxation rate with Eq. (2.39) for the transition probability and Eq. (2.29) for the spectral density, we arrive at a simple expression for T_1:

$$T_1^{-1}=2G(0)\frac{2\tau_c}{1+\omega_0^2\tau_c^2} \tag{2.40}$$

Here, the frequency corresponding to the difference in energy levels according to the previous section is denoted by ω_0 and is known as the *Larmor frequency*, or *resonance frequency*, introduced in Chapter 1. According to Eq. (2.24), $G(0)$ is the variance or the mean-square amplitude of the interaction leading to relaxation. The average value of the interaction itself is assumed to be zero. Let us assume a simple (and physically not quite realistic) relaxation mechanism, which we can call the *randomly reorienting field*. We assume thus that the spins are subject to a local magnetic field (a vector) with a constant magnitude b, whose direction with respect to the large external field $\mathbf{B}_0$ (or to the laboratory z-direction) fluctuates randomly in small angular steps in agreement with Figure 2.1. This model is a simplified version of the effects of dipolar local fields, discussed in detail in Chapter 3. The model is also related to the *fluctuating random field* description as used in books by Slichter, Canet and Levitt (*Further reading*). The mean-square amplitude of the fluctuating Zeeman interaction is then $G(0) = \gamma_I^2 b^2$ and we obtain:

$$T_1^{-1} = 4\gamma_I^2 b^2 \frac{\tau_c}{1+\omega_0^2\tau_c^2} \tag{2.41}$$

Magnetic field has the units of tesla (T) and γ_I has the units of rad s^{-1} T^{-1}. Thus, the square of the interaction strength expression in front of the Lorentzian has the units rad^2 s^{-2}. As mentioned in Section 1.3, the radians are tacitly omitted, which leads to both the correlation time and the relaxation time being expressed in seconds. In full analogy with this case, in all other expressions in relaxation theory the interaction strength is expressed in radians per second, the correlation time in seconds and the relaxation rate in s^{-1}.

The meaning of the correlation time in Eq. (2.41) requires some reflection. The Zeeman interaction between the z-component of the nuclear magnetic moment and the local field can be written:

$$\hat{V} = -\gamma_I \hat{I}_z b_z = -\gamma_I \hat{I}_z b\cos\theta \tag{2.42}$$

where we have used the fact that if the local field vector is oriented at angle θ with respect to the laboratory z-axis, then its z-component can be expressed as $b\cos\theta$. The angle θ, and thus its cosine, is according to the model a random function. The correlation time is related to the rate of decay of the correlation between $\cos\theta(t)$ and $\cos\theta(t+\tau)$. We note, for future reference, that $\cos\theta$ is, except for a normalisation constant, identical to the rank-one *spherical harmonics* function, $Y_{1,0}$. More about spherical harmonics can be found in the books by Atkins and Friedman and Brink and Satchler (*Further reading*). We shall come to correlation functions for spherical harmonics in Chapter 6. It will be demonstrated there that the correlation function of $Y_{1,0}$ is indeed an exponential decay as in Eq. (2.26). We shall call the correlation time in Eq. (2.41) *rotational correlation time* (to be strict, we should really add "for a rank-one spherical harmonics"), related to the reorientation of the molecule carrying the spins. Indeed, using the results to be obtained in Section 6.1, we should insert a factor $1/4\pi$ into the right-hand side of Eq. (2.41). This does not matter too much because we are here only interested in following the variation of the T_1^{-1} with certain physical quantities and not in its absolute magnitude.

We can use the simple form of Eq. (2.41) to predict the dependence of the spin-lattice relaxation rate on the strength of the external magnetic field, B_0 (through the Larmor frequency), and on the molecular size and temperature, through the correlation time. Let us begin by looking at the dependence of T_1^{-1} on the magnetic field at a constant value of the correlation time. A plot of the relaxation rate versus ω_0 is shown in Figure 2.5 for two physically reasonable values of τ_c: 200 ps and 2 ns. According to the plot, the relaxation rate is expected to be constant at low field and decrease dramatically in the

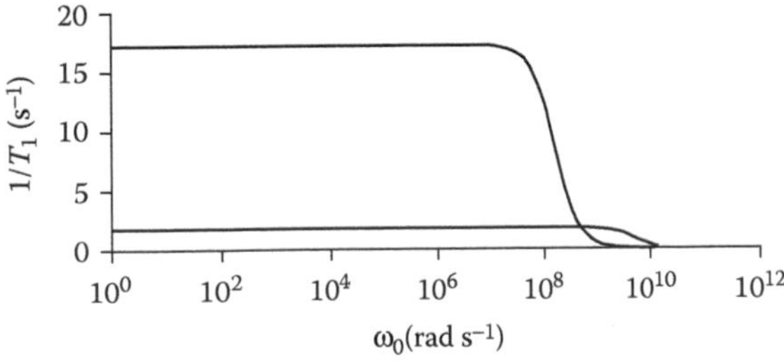

FIGURE 2.5 $1/T_1$ calculated as a function of Larmor frequency for two correlation times, $\tau_c = 0.2$ ns (lower curve) and $\tau_c = 2$ ns, using Eq. (2.41). The value $\gamma_I{}^2 b^2 = 2.15\times10^9$ s^{-2} was used.

vicinity of the condition $\omega_0\tau_c = 1$. The rapid reduction of the relaxation rate is sometimes called *dispersion*. The curves in Figure 2.5 map exactly the plot of the spectral density versus frequency, as shown in Figure 2.4 (note the logarithmic scale on the horizontal axis in Figure 2.5), which of course reflects the fact that the two quantities are proportional to each other for the simple case at hand. The flat region of the plot of T_1^{-1} against Larmor frequency, *i.e.* the region where the relaxation rate is independent of the magnetic field, is called the *extreme narrowing* region. In quantitative terms, the extreme narrowing range corresponds to the condition $\omega_0^2\tau_c^2 \ll 1$. The extreme narrowing regime extends up to a certain value of the Larmor frequency (magnetic field), the range being smaller for a longer correlation time. One more observation we can make from Figure 2.5 is that, for a given correlation time, the nuclei with lower magnetogyric ratio will come out of the extreme narrowing regime at a higher field. Thus, protons, with their high magnetogyric ratio, come out of the extreme narrowing regime at a lower field than ^{13}C or ^{15}N (*cf.* Table 1.1).

The second interesting variable – besides the Larmor frequency or the magnetic field – in Eq. (2.41) is the rotational correlation time. We shall arrive at a stringent definition of this quantity later, but we wish already here to state its dependence on molecular size, solution viscosity and temperature. Using hydrodynamic arguments for a spherical particle with the hydrodynamic radius a and volume $V = 4\pi a^3/3$, reorienting in a viscous medium, one can derive the Stokes–Einstein–Debye (SED) relation, introduced in the NMR context in the classical paper by Bloembergen, Purcell and Pound (BPP)[2]:

$$\tau_c(l=1) = \frac{4\pi\eta a^3}{k_B T} = \frac{3V\eta}{k_B T} \tag{2.43a}$$

Here, η is the viscosity of the solution (in the units kg s^{-1} m^{-1}) and V is the volume of the molecule. Clearly, the volume or the a^3 factor in the numerator indicates that the rotational correlation time is expected to increase with the molecular size. Indeed, there are "rule of thumb" relations for aqueous protein solutions, relating the correlation time to the molecular weight. We should notice that Eq. (2.43a) applies for the correlation time for $l=1$ spherical harmonics. Several of the physically more realistic relaxation mechanisms (*e.g.* the dipole–dipole relaxation) depend on the rotational correlation time for $l=2$ spherical harmonics; in that case, Eq. (2.43a) is modified to:

$$\tau_c(l=2) = \frac{4\pi\eta a^3}{3k_B T} = \frac{V\eta}{k_B T} \tag{2.43b}$$

The correlation time depends on temperature in two ways. First, the viscosity is strongly temperature dependent, which is commonly described by an Arrhenius-type expression (proposed for the first time more than 100 years ago by de Guzman[3]):

$$\eta = \eta_0 \exp\left(E_a^\eta / k_B T\right) \tag{2.44}$$

where E_a^η is the activation energy for viscous flow and η_0 is a constant without too deep physical significance. Second, there is the $1/T$ dependence originating from the presence of temperature in the denominator of Eq. (2.43). The exponential factor overruns the $1/T$ dependence, and it is common to express the temperature dependence of the correlation time by an analogous Arrhenius-type expression:

$$\tau_c = \tau_0 \exp\left(E_a^\tau / k_B T\right) \tag{2.45}$$

where the activation energy E_a^τ is related to the barrier hindering the reorientation process and not necessarily the same as E_a^η. The symbol τ_0 is a constant, again without too much physical significance.

A plot of T_1 versus correlation time, for a given magnetic field B_0 or Larmor frequency ω_0, is shown in Figure 2.6. When molecular motions are rapid, so that $\tau_c^2\omega_0^2 \ll 1$, the extreme narrowing region prevails,

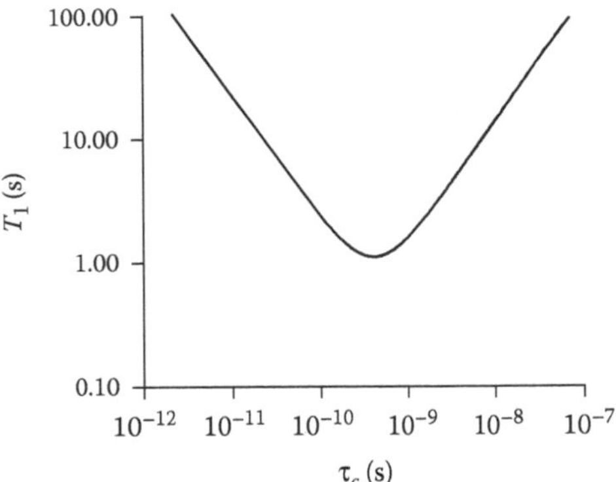

FIGURE 2.6 Plot of T_1 versus correlation time, τ_c. The plot was calculated using Eq. (2.41), assuming $\gamma_I^2 b^2 = 1.13 \times 10^9$ for a proton at a magnetic field strength of 9.4 T (corresponding to a ^{1}H resonance frequency of 400 MHz).

the frequency dependence in the denominators of Eqs. (2.41) and (2.29) vanishes and we get the result that T_1^{-1} is proportional to the correlation time (the left-hand side of the diagram in Figure 2.6). Turning to longer correlation times, we can see in the figure that the relaxation is most efficient (the T_1^{-1} is largest or the T_1 shortest) when $\tau_c = 1/\omega_0$. At a typical NMR field of 9.4 T (400 MHz proton resonance frequency or $|\omega_0| = 2\pi \cdot 400\ 10^6$ rad s^{-1}), this happens for protons at $\tau_c = 400$ ps, which is a typical rotational correlation time for a medium-sized organic molecule in an aqueous solution at room temperature. When we study even larger molecules or solutions of high viscosity, where the motions are more sluggish, the opposite condition, $\tau_c^2\omega_0^2 \gg 1$, applies and the relaxation rate becomes inversely proportional to the correlation time (the right-hand side of the diagram in Figure 2.6). According to what we stated concerning the dependence of the correlation time on molecular size and temperature, the horizontal axis in Figure 2.6 can be thought of as corresponding to the molecular weight. Similarly, we can think of temperature increasing to the left in the diagram.

The case of the spin-spin relaxation time (transverse relaxation time) is a little more complicated. It turns out that the corresponding rate for the simple example at hand is proportional to the sum of spectral densities at frequencies zero and ω_0:

$$T_2^{-1} = 2\gamma_I^2 b^2 \left(\frac{\tau_c}{1 + \omega_0^2 \tau_c^2} + \tau_c \right) \tag{2.46}$$

The relation between $T_1^{-1} = R_1$ and $T_2^{-1} = R_2$ as a function of the correlation time is summarised in Figure 2.7. We can see that the two relaxation rates are equal when the extreme narrowing conditions hold and that R_2, as opposed to R_1, continues to increase (the NMR signal becomes broader; compare Figure 1.4) with increasing correlation time.

Already at this stage, we can see that very important results are obtained from the relaxation rate parameters. To summarise, we have seen that relaxation can provide detailed information about molecular motion and size through the field dependence of the relaxation providing the rotational correlation time. This will be discussed in detail in Chapters 11 and 12, where applications of relaxation measurements are discussed.

2.5 Finite Temperature Corrections to the Simple Model

Before leaving our simple model, we wish to discuss one principally important complication that occurs also in physically more realistic situations. The issue is the assumed equality of the transition

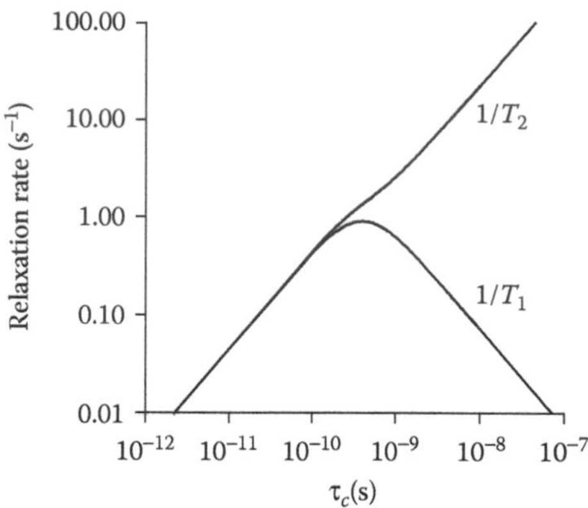

FIGURE 2.7 Plot of $1/T_1$ and $1/T_2$ relaxation rates versus correlation time, τ_c. The plot was calculated using Eqs. (2.41) and (2.46), assuming $\gamma_I^2 b^2 = 1.13 \times 10^9$ for a proton at a magnetic field strength of 9.4 T (corresponding to a ^{1}H resonance frequency of 400 MHz).

probabilities, $W_{\alpha\beta} = W_{\beta\alpha} = W_I$, which was introduced at the beginning of this chapter. If this indeed were true, *i.e.* if the transition probabilities up and down in the two-level system were equal, then the spin system would evolve towards a situation with equal populations of the two levels. According to the Boltzmann distribution, this would correspond to infinite temperature. If we want our spin system to evolve towards thermal equilibrium at a finite temperature, we need to introduce small correction terms to our transition probabilities. For a spin with positive magnetogyric ratio, implying the α state being lower in energy than the β state, we need to write:

$$W_{\alpha\beta} = W_I\left(1 - b_I\right) \tag{2.47a}$$

$$W_{\beta\alpha} = W_I\left(1 + b_I\right) \tag{2.47b}$$

The symbol W_I is an average transition probability, given in our two-level model by Eq. (2.39), and b_I is the Boltzmann factor introduced in Eq. (1.19). We remind the reader that the Boltzmann factors in NMR are so tiny that the practical implications of the correction terms are small. However, when we use the assumption of Eq. (2.47), the flow of populations in both directions becomes equal:

$$n_\alpha^{eq} W_{\alpha\beta} = n_\beta^{eq} W_{\beta\alpha} \tag{2.48}$$

as indeed it should be at the thermal equilibrium. In the way Eq. (2.47) is formulated, the thermal correction terms can be considered as inserted *ad hoc* without any more profound motivation than that they lead to the correct results of Eq. (2.48). It is possible to derive Eq. (2.47) starting from the fundamental physical principles, if one assumes that the environment of spins (the lattice) is also of quantum mechanical nature and recognises that one should consider simultaneous transitions in the spin system and the lattice. The details of that problem are, however, beyond the scope of this book and we recommend the interested reader to consult the book of Abragam (*Further reading*).

References

1. Carrington, A.; and McLachlan, A. D., *Introduction to Magnetic Resonance*. Harper and Row: New York, 1967.
2. Bloembergen, N.; Purcell, E. M.; and Pound, R. V., Relaxation effects in nuclear magnetic resonance absorption. *Phys. Rev.* 1948, 73, 679–712.
3. de Guzman, J., Relation between fluidity and heat of fusion. *An. Soc. Espan. Fis. Quim.* 1913, 11, 353–362.

3

Relaxation through Dipolar Interactions

Until now, we have only considered relaxation through a simple approach using a model relaxation mechanism. In this chapter, we will deal with the dipole-dipole (DD) relaxation, which has proven to be one of the most important sources for obtaining molecular dynamics information. The discussion on this relaxation mechanism is based on the Solomon equations, which, in turn, are based on rates of transitions between spin states. For many spin 1/2 nuclei, the dipole-dipole interaction between nuclear magnetic moments of spins in spatial vicinity of each other is the most important relaxation mechanism. This was recognised already in the seminal work by Bloembergen, Purcell and Pound (BPP) from 1948 (mentioned in the previous chapter), which created the ground for all subsequent development of relaxation theory. What was not correct in the BPP paper was the assumption that one of the spins acts as a source of a kind of random field for the other spin, on which interest was concentrated. As shown by Solomon in 1955,[1] the dipolar relaxation must be dealt with in terms of a four-level system with two mutually interacting spins. In this chapter, we go through the Solomon theory for the dipolar spin-lattice relaxation and its consequences. Good descriptions of dipolar interaction and dipolar relaxation can be found in the books by Abragam and by Levitt (see *Further reading*).

3.1 The Nature of the Dipolar Interaction

Consider a situation where we have two nuclear magnetic moments or magnetic dipoles, $\boldsymbol{\mu}_1$ and $\boldsymbol{\mu}_2$, close to each other in space. Each of the magnetic dipoles generates around itself a local magnetic field. The local magnetic field is a vector field, *i.e.* it is at every point characterised by a magnitude and a direction *cf.* Figure 3.1. The form of the vector field generated by a magnetic dipole is rather complicated. Let us consider the field created by the dipole $\boldsymbol{\mu}_2$; as discussed by Atkins and Friedman (*Further reading*), the magnetic field at the point $\mathbf{r}$ (with respect to $\boldsymbol{\mu}_2$ at the origin) can be expressed by

$$\mathbf{B}_{loc}(\boldsymbol{\mu}_2) = -\frac{\mu_0}{4\pi r^3}\left(\boldsymbol{\mu}_2 - 3\frac{\mathbf{rr}}{r^2}\cdot\boldsymbol{\mu}_2\right) \tag{3.1}$$

The symbol μ_0 is the permeability of vacuum, $\mu_0/4\pi$ is in the SI units equal to 10^{-7} $\mathrm{Js^2\,C^{-2}\,m^{-1}}$ or $\mathrm{T^2\,J^{-1}\,m^3}$, r is distance from the origin (a scalar quantity), while $\mathbf{r}$ is a vector with Cartesian components x, y, z; $\mathbf{rr}$ is a tensor, an outer product of two vectors.

A tensor has the property that when it acts on a vector, the result is another vector with, in general, different magnitude and orientation. A Cartesian tensor can be represented by a 3×3 matrix. An outer

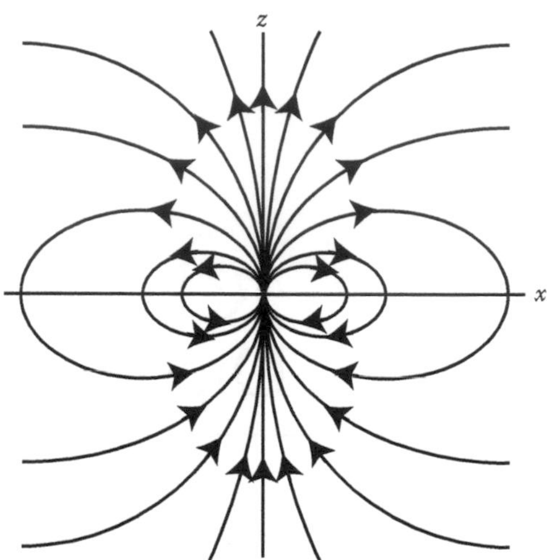

FIGURE 3.1 The local magnetic field from a magnetic point dipole at origin.

product of two vectors, **a** and **b**, with Cartesian components a_x, a_y, a_z and b_x, b_y, b_z, respectively, can be represented by a matrix with elements

$$\mathbf{ab} = \begin{pmatrix} a_x \\ a_y \\ a_z \end{pmatrix} \begin{pmatrix} b_x & b_y & b_z \end{pmatrix} = \begin{pmatrix} a_x b_x & a_x b_y & a_x b_z \\ a_y b_x & a_y b_y & a_y b_z \\ a_z b_x & a_z b_y & a_z b_z \end{pmatrix} \tag{3.2a}$$

Thus, the **rr** tensor can be expressed as

$$\mathbf{rr} = \begin{pmatrix} xx & xy & xz \\ yx & yy & yz \\ zx & zy & zz \end{pmatrix} = \begin{pmatrix} x^2 & xy & xz \\ yx & y^2 & yz \\ zx & zy & z^2 \end{pmatrix} \tag{3.2b}$$

The operation of a tensor on a vector can then be represented as a multiplication of that matrix with a vector. The second term in the expression in Eq. (3.1) is thus a vector whose magnitude and direction depend on the point that we are interested in (as an exercise, the reader is advised to derive the z component of the local field at **r**). The local magnetic field created by the dipole $\boldsymbol{\mu}_2$ interacts with $\boldsymbol{\mu}_1$, and, in the same way, the local field of $\boldsymbol{\mu}_1$ interacts with $\boldsymbol{\mu}_2$. The classical dipole-dipole interaction energy, E_{DD}, is

$$E_{DD} = \frac{\mu_0}{4\pi r^3}\left(\boldsymbol{\mu}_1 \cdot \boldsymbol{\mu}_2 - 3\boldsymbol{\mu}_1 \cdot \frac{\mathbf{rr}}{r^2} \cdot \boldsymbol{\mu}_2\right) \tag{3.3}$$

where $\mathbf{r} = x\mathbf{i} + y\mathbf{j} + z\mathbf{k}$ (**i**, **j**, and **k** are the unit vector along the three Cartesian axes) now denotes the vector connecting the two dipoles.

The quantum mechanical counterpart of the classical dipole-dipole energy expression is the dipole-dipole Hamiltonian, which we obtain by replacing the magnetic dipoles by $\gamma_I \hbar \hat{\mathbf{I}}$ and $\gamma_S \hbar \hat{\mathbf{S}}$, corresponding to the two spins denoted as I and S.

$$\hat{H}_{DD} = -\frac{\mu_0 \gamma_I \gamma_S \hbar}{4\pi r^3}\left(3\hat{\mathbf{I}} \cdot \frac{\mathbf{rr}}{r^2} \cdot \hat{\mathbf{S}} - \hat{\mathbf{I}} \cdot \hat{\mathbf{S}}\right) = b_{IS} \hat{\mathbf{I}} \cdot \mathbf{D} \cdot \hat{\mathbf{S}} \tag{3.4}$$

If the distance between spins (the length of the **r** vector) is constant and is equal to r_{IS}, then the quantity b_{IS}, given by

$$b_{IS} = -\frac{\mu_0 \gamma_I \gamma_S \hbar}{4\pi r_{IS}^3} \tag{3.5}$$

is also a constant, denoted as the *dipole-dipole coupling constant.* Note that the Hamiltonian in Eq. (3.4) is in the same units as the dipole-dipole coupling constant, *i.e.* in the angular frequency units (rad s^{-1}). The dipolar tensor **D** can, in analogy with Eq. (3.2), be represented by a 3×3 matrix:

$$\mathbf{D} = \begin{pmatrix} \frac{3x^2}{r_{IS}^2}-1 & \frac{3xy}{r_{IS}^2} & \frac{3xz}{r_{IS}^2} \\ \frac{3xy}{r_{IS}^2} & \frac{3y^2}{r_{IS}^2}-1 & \frac{3yz}{r_{IS}^2} \\ \frac{3xz}{r_{IS}^2} & \frac{3yz}{r_{IS}^2} & \frac{3z^2}{r_{IS}^2}-1 \end{pmatrix}$$
$$= \begin{pmatrix} 3\sin^2\theta\cos^2\phi-1 & 3\sin^2\theta\cos\phi\sin\phi & 3\sin\theta\cos\theta\cos\phi \\ 3\sin^2\theta\cos\phi\sin\phi & 3\sin^2\theta\sin^2\phi-1 & 3\sin\theta\cos\theta\sin\phi \\ 3\sin\theta\cos\theta\cos\phi & 3\sin\theta\cos\theta\sin\phi & 3\cos^2\theta-1 \end{pmatrix} \tag{3.6}$$

In the second line of Eq. (3.6), we express the elements of the Cartesian tensor **D** in spherical polar coordinates (r, θ, ϕ) using standard relations: $x = r\sin\theta\cos\phi$, $y = r\sin\theta\sin\phi$, $z = r\cos\theta$. When working with the Hamiltonian of Eq. (3.4) or with the dipolar tensor of Eq. (3.6), we have to be careful with the choice of the coordinate system. The tensor takes on a particularly simple form in the coordinate frame where the z-axis coincides with the vector **r**. We then have $x = y = 0$ ($\theta = \phi = 0$) and $z = r_{IS}$. All the off-diagonal elements of **D** vanish, and the diagonal elements become –1, –1, and 2. In the presence of a strong magnetic field, the main interaction the spins sense is the Zeeman interaction. Therefore, the natural choice of the coordinate frame for the spin operators is the laboratory frame, where the z-direction is defined by the magnetic field. The orientation of the vector **r** should thus also be specified in the laboratory frame *cf.* Figure 3.2.

Before turning to the role that the dipole-dipole interaction has in relaxation, let us have a quick look at how it demonstrates itself in solid crystals and powders. Let us begin by assuming that the spins I and S are different species (*e.g.* a carbon-13 and a proton) and that we deal with isolated spin pairs,

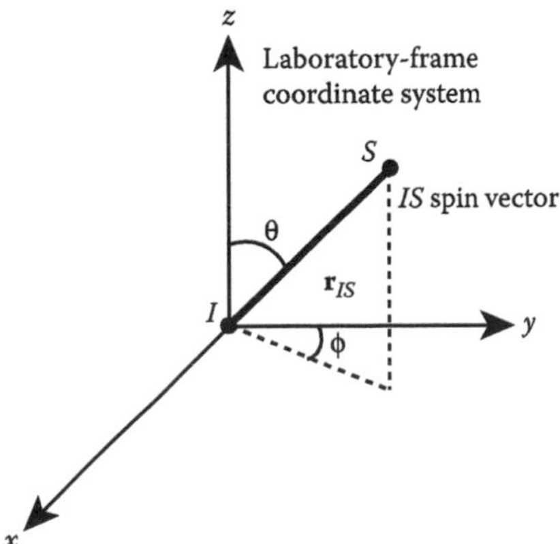

FIGURE 3.2 The orientation of IS spin vector, **r**, in the laboratory coordinate system, defined by the angles θ and ϕ.

i.e. that each spin interacts with one and only one other spin. In a high magnetic field, where the Zeeman interactions dominate, we can make use of the *secular approximation*. This approximation means that we only retain the terms in H_{DD} that commute with the Zeeman Hamiltonian, which for the case of two spins can be written, in analogy with Eq. (1.7), as

$$\hat{H}_z = -(\gamma_I \hat{I}_z + \gamma_S \hat{S}_z)B_0 = \omega_I \hat{I}_z + \omega_S \hat{S}_z \tag{3.7}$$

The dipole-dipole Hamiltonian then attains a simplified form:

$$\hat{H}_{DD}^{\text{sec}} = b_{IS}(3\cos^2\theta - 1)\hat{I}_z \hat{S}_z \tag{3.8}$$

Here, θ is the angle between the axis joining the two dipolar-coupled spins (fixed in a crystal) and the magnetic field direction, assumed to be the laboratory-frame z-direction. The angle θ between the laboratory z-direction and the internuclear axis, the **r** vector, is defined in Figure 3.2. The result of the presence of the dipolar Hamiltonian is that the I- and S-spin resonance lines split into doublets in the same way as they do in the presence of the scalar spin-spin coupling (indirect spin-spin coupling or J-coupling). The splitting, which for the indirectly coupled spins is equal to $2\pi J$ (note the angular frequency units!), is here $b_{IS}(3\cos^2\theta - 1)$. The splitting is thus dependent on the orientation of the crystal in the magnet. Rotating the crystal in the magnet in such a way that the θ angle varies between 0 and π results in the splitting changing as in Figure 3.3. When comparing the dipole-dipole coupling with the J-coupling, we should be aware of the interaction strengths involved: the J-coupling for a directly bonded carbon-13–proton pair is typically about 800 rad s^{-1} (130 Hz), while the corresponding dipolar coupling constant, b_{IS}, is about 140 000 rad s^{-1} (22 kHz).

Let us then turn to a solid powder, *i.e.* a system of a large number of microcrystallites with random orientation with respect to each other and to the magnetic field. Every orientation with respect to the laboratory-frame z-axis contributes to the spectrum with a particular splitting according to Figure 3.3, and the total spectrum is a superposition of all the doublets. The intensities of lines with various splittings depend on the probabilities of the corresponding θ angles. The assumption that all orientations of the microcrystallites are equally probable does not imply that all the θ angles are equally probable. Using a "hand-waving" argument, we can imagine that the perpendicular orientations, $\theta = \pi/2$, are much more numerous, corresponding to all possible vector orientations in the xy-plane, and can be realised much more often than the parallel orientations with only two possibilities, $\theta = 0$ or π. In mathematical terms, we need to think about the angular volume element being $\sin\theta\, d\theta\, d\phi$, where ϕ is the angle specifying the orientation of the projection of the IS axis on the xy-plane. The powder spectrum of the I-spin, dipole-dipole coupled to the S-spin, is shown in Figure 3.4. The pattern is called *powder*

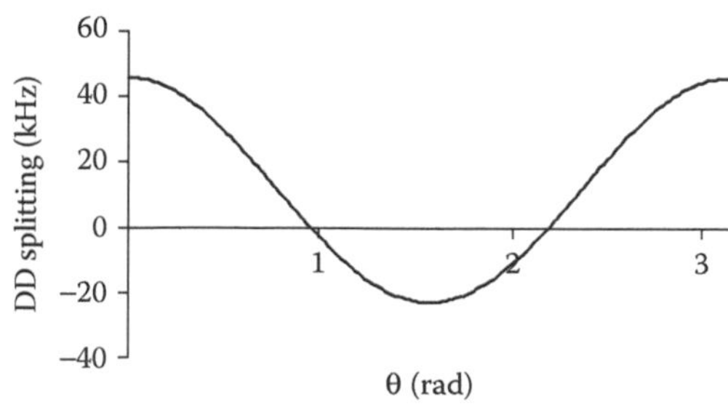

FIGURE 3.3 The variation of the dipolar splitting for a ^{1}H-^{13}C spin pair, with $b_{CH} = 144\ 10^3$ rad s^{-1} ($b_{CH}/2\pi = 22.9$ kHz). The value for b_{CH} was obtained assuming an internuclear distance of $r_{CH} = 1.09$ Å.

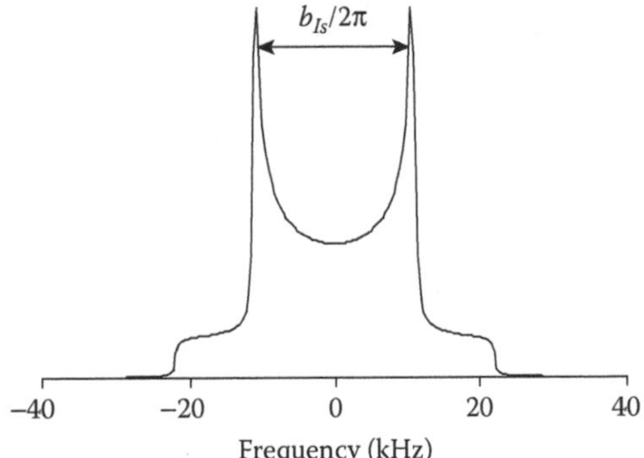

FIGURE 3.4 Powder spectrum, or Pake pattern, of the *I*-spin, dipole-dipole coupled with the *S*-spin with b_{IS} = 144 10^3 rad s^{-1} ($b_{IS}/2\pi$ = 22.9 kHz).

pattern or *Pake-doublet*, after the scientist who reported it for the first time.[2] The splitting between the maxima in the powder pattern is equal to $|b_{IS}|$, which corresponds to $\theta = \pi/2$.

We can note, before leaving the solids, that the dipolar splitting vanishes if $3\cos^2\theta - 1 = 0$. This happens when $\theta = \arccos(1/\sqrt{3}) = 54.74°$. This angle is called the *magic angle*. An important technique in solid-state NMR is to spin the powder sample around an axis oriented at the magic angle with respect to the magnetic field. If the spinning motion imposed by the experimenter is sufficiently rapid, the dipole-dipole interaction (as well as other interactions with the same θ dependence) is "projected" on the spinning axis, and the dipole-dipole interaction vanishes from the spectrum.

The dipole-dipole splittings vanish also in isotropic liquids. This is the effect of motion also here, but now it is imposed by the nature of the sample rather than the experimenter. Molecular tumbling in ordinary liquids changes the orientation of the *IS*-spin axis on a time scale that is very fast compared with the dipole-dipole couplings. The rotational correlation times are on the order of pico- or nanoseconds, corresponding to reorientation rate constants of 10^7–10^{11} rad^{-1}s^{-1}, which should be compared with the dipolar couplings on the order of 10^5 rad s^{-1} (compare the example above). Thus, the dipole-dipole coupling in a liquid represents an average of $b_{IS}(3\cos^2\theta - 1)$ over the isotropic distribution of the orientations, which amounts to zero:

$$\begin{aligned} b_{IS}(ave) &= b_{IS}\int_0^{2\pi}\int_0^{\pi}(3\cos^2\theta - 1)p(\theta,\phi)\sin\theta d\theta d\phi \\ &= \frac{b_{IS}}{4\pi}\int_0^{2\pi}\int_0^{\pi}(3\cos^2\theta - 1)\sin\theta d\theta d\phi = 0 \end{aligned} \tag{3.9}$$

The symbol $p(\theta,\phi)$ is the probability density for the angles θ and ϕ. The probability density is uniform and equal to $1/4\pi$ because of normalisation condition. The reader is encouraged to perform the integration in the second line of Eq. (3.9). Thus, the dipole-dipole interaction does not give splittings in isotropic liquids, and the fine structure of NMR spectra in such systems is determined only by the *J*-couplings. This does not mean, however, that the dipole-dipole interaction cannot be effective as a relaxation mechanism. The ensemble-averaged dipole-dipole Hamiltonian (Eq. (3.4)) vanishes at every point in time, $\langle \hat{H}_{DD}(t)\rangle = 0$ (we have explicitly shown that this is true for the secular part, Eq. (3.8), but the same applies to the other terms), but this does not imply that the ensemble averages of the type $\langle \hat{H}_{DD}(t)\hat{H}_{DD}(t+\tau)\rangle$, relevant in the relaxation theory, need to vanish.

3.2 The Solomon Relaxation Theory for a Two-Spin System

Let us consider a liquid consisting of molecules containing two nuclei with spins I and S, both characterised by the spin quantum number 1/2. We assume that they are distinguishable, *i.e.* either belong to different nuclear species or have different chemical shifts. We assume that there are no J-couplings but allow for the dipole-dipole interaction between the spins. The system of such two spins is characterised by four Zeeman energy levels and by a set of transition probabilities between the levels (*cf.* Figure 3.5).

3.2.1 The Dipolar Alphabet

The Zeeman Hamiltonian of Eq. (3.7) can be treated as the main Hamiltonian, with the eigenstates αα, αβ, βα, ββ (the first symbol indicates the state of the I-spin, the second that of the S-spin), and the dipole-dipole Hamiltonian of Eq. (3.4) as a perturbation. We shall find it useful to express the dipole-dipole Hamiltonian as a sum of six terms:

$$\hat{H}_{IS}^{DD} = b_{IS}\left[\hat{A}+\hat{B}+\hat{C}+\hat{D}+\hat{E}+\hat{F}\right] \tag{3.10}$$

The dipole-dipole coupling constant is for the time being assumed to be constant in time. This amounts to treating the molecules carrying the two spins as rigid objects. We shall return to the *intermolecular* dipolar relaxation in the final section of this chapter and to the *intramolecular* dipolar relaxation in the case of flexible molecules in Chapter 6.

The form of the dipole-dipole Hamiltonian as given in Eq. (3.10) is often referred to as the "dipolar alphabet." All the terms in Eq. (3.10) consist of products of spin operators and angular terms, proportional to the rank-two spherical harmonics:

$$\hat{A} = \hat{I}_z\hat{S}_z(3\cos^2\theta - 1) \tag{3.11a}$$

$$\hat{B} = -\frac{1}{4}\left[\hat{I}_+\hat{S}_- + \hat{I}_-\hat{S}_+\right](3\cos^2\theta - 1) \tag{3.11b}$$

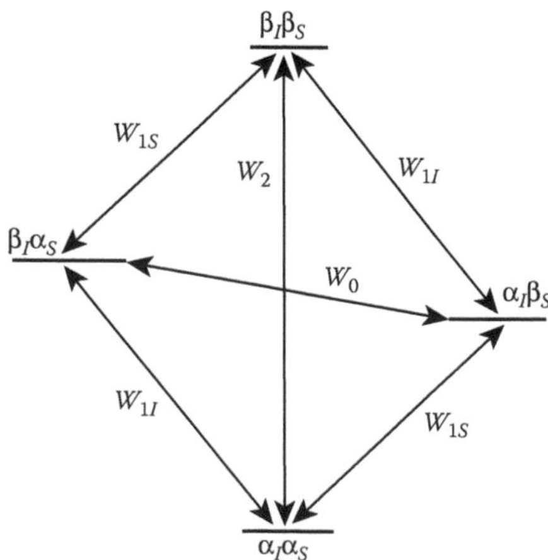

FIGURE 3.5 Energy level diagram associated with the IS spin system.

$$\hat{C} = \frac{3}{2}\left[\hat{I}_+\hat{S}_z + \hat{I}_z\hat{S}_+\right]\sin\theta\cos\theta\exp(-i\phi) \tag{3.11c}$$

$$\hat{D} = \frac{3}{2}\left[\hat{I}_-\hat{S}_z + \hat{I}_z\hat{S}_-\right]\sin\theta\cos\theta\exp(i\phi) \tag{3.11d}$$

$$\hat{E} = \frac{3}{4}\hat{I}_+\hat{S}_+\sin^2\theta\exp(-2i\phi) \tag{3.11e}$$

$$\hat{F} = \frac{3}{4}\hat{I}_-\hat{S}_-\sin^2\theta\exp(2i\phi) \tag{3.11f}$$

One can easily derive Eqs. (3.10) and (3.11) from Eqs. (3.4) and (3.6), using the relation $\exp(i\alpha) = \cos\alpha + i\sin\alpha$ and the step-up and step-down operators, $\hat{I}_+$, $\hat{S}_+$ and $\hat{I}_-$, $\hat{S}_-$, respectively, defined as

$$\hat{I}_\pm = \hat{I}_x \pm i\hat{I}_y \tag{3.12}$$

We recognise that the term Â is identical to the secular part of the dipolar Hamiltonian, as given in Eq. (3.8), while the remaining terms do not commute with the Zeeman Hamiltonian.

When the step operators operate on the $|I,m\rangle$ kets, they produce other kets, with the m quantum number increased ($\hat{I}_+$) respectively decreased ($\hat{I}_-$) by one. Thus, the terms $\hat{B}-\hat{F}$ correspond to the non-secular part of $\hat{H}_{DD}$ in the sense that the spin operators do not commute with the Zeeman Hamiltonian. We can also note that the terms $\hat{C}$ and $\hat{D}$ are similar to each other: the spin operators contain products of a step operator and a *z*-component operator, and the angular parts are complex conjugates of each other. In the same sense, the $\hat{E}$ and $\hat{F}$ terms are related to each other. One can make use of these relations and formulate the DD-Hamiltonian in terms of an interaction strength constant, certain combinations of spin operators, called *rank-two irreducible spherical tensor operators,* $\hat{T}_{2,m}$, and suitably normalised spherical harmonics $Y_{2,m}$:

$$\hat{H}_{IS}^{DD} = \xi_{IS}^{DD}\sum_{m=-2}^{2}(-1)^m Y_{2,m}\hat{T}_{2,-m} \tag{3.13}$$

The interaction strength constant, $\xi_{IS}^{DD} = b_{IS}\sqrt{24\pi/5}$, plays the same role as the dipole-dipole coupling constant; it is introduced in Eq. (3.13) instead of b_{IS} so that the other factors can have appropriate normalisation (see Section 4.2). Here, $\hat{T}_{2,m}$ represents the time-independent tensor operator acting on spin variables only, corresponding to the products of spin operators in Eq. (3.11). The formulations of Eqs. (3.10) and (3.13) are fully equivalent; the choice of one or another is a matter of convenience. In the context of the Solomon theory, we choose to work in terms of the dipolar alphabet, while we shall use the formulation of Eq. (3.13) later on. The reader interested in general aspects of various formulations of Hamiltonians for NMR may find it useful to consult the early paper by Spiess[3] and the extensive series of reviews by Smith and co-workers.[4–6]

Because of the angular (θ, ϕ) dependency, the dipole-dipole interaction is influenced (and averaged to zero on a sufficiently long time scale) by molecular tumbling. Taking this motional variation into consideration, it is not particularly difficult to use time-dependent perturbation theory to derive transition probabilities between the pairs of levels in Figure 3.5. The transition probabilities are obtained in a way similar to the case discussed in Chapter 2. They are related to time-correlation functions $\left\langle\left(\hat{H}_{DD}(t)\right)_{ij}\left(\hat{H}_{DD}(t+\tau)\right)_{ij}\right\rangle$, where $\left(\hat{H}_{DD}(t)\right)_{ij}$ denotes the *ij* matrix element of the dipolar Hamiltonian, and the corresponding spectral density functions, which reflect the intensity of the local magnetic fields fluctuating at the frequencies corresponding to the energy level differences in Figure 3.5.

Because the $\hat{H}_{DD}(t)$ is expressed as a sum of six terms $\left(\hat{A}-\hat{F}\right)$, a total of 36 terms will in principle be present in the product $\left(\hat{H}_{DD}(t)\right)_{ij}\left(\hat{H}_{DD}(t+\tau)\right)_{ij}$. Since all the terms in $\hat{H}_{DD}(t)$ are products of time-independent spin operators and time-dependent angular functions, the calculation of transition probabilities involves two steps.

First, we need to calculate the matrix elements of the time-independent spin operators in the basis of the eigenkets to the Zeeman Hamiltonian. Doing that, we find that different pairs of levels are connected by matrix elements of different operators. In other words, different transitions are induced by different terms of the dipolar alphabet. The relations between pairs of levels and the operators connecting them are summarised in Table 3.1.

Only the $\hat{A}$-symbol (secular terms) occurs on the diagonal. The occurrence of the other terms can be illustrated by examining one of the matrix elements, for instance $\langle\alpha\alpha|\hat{H}_{DD}|\beta\beta\rangle$. Operation of the term $\hat{A}$ in the dipolar alphabet of Eq. (3.11) on the $|\beta\beta\rangle$ ket leaves it unchanged, and the matrix element vanishes because of the orthogonality of the spin functions. The terms containing a step-down operator $\left(\hat{B},\ \hat{D},\ \text{and}\ \hat{F}\right)$ operating on $|\beta\beta\rangle$ yield zero, since the ket with $m=-1/2$ and $I=1/2$ cannot have its m-quantum number reduced further. The term $\hat{C}$ operating on $|\beta\beta\rangle$ yields a linear combination of $|\alpha\beta\rangle$ and $|\beta\alpha\rangle$, both of which are orthogonal to the $|\alpha\alpha\rangle$, and the matrix element vanishes for the same reason as that of the $\hat{A}$-operator. So, the only term left is $\hat{E}$; the spin operator $\hat{I}_+\hat{S}_+$ acting on $|\beta\beta\rangle$ raises the m-quantum number of both spins and thus yields a ket proportional to $|\alpha\alpha\rangle$. Taking the scalar product with $\langle\alpha\alpha|$ results in a non-zero matrix element corresponding to the $\hat{E}$-symbol in the lower-left corner of Table 3.1.

In the second step of the calculations, we deal with the time-correlation functions of the angular functions in Eq. (3.11). These are dependent on the nature of random reorientations and thereby on the model we choose to describe these motions. We shall derive an expression for the time-correlation function for some simple models, including isotropic reorientation of a rigid body in small, diffusional steps, in Chapter 6. For the time being, we assume that the relevant time-correlation function for a normalised $Y_{2,0}$ spherical harmonics

$$Y_{2,0}=\sqrt{\frac{5}{16\pi}}(3\cos^2\theta-1) \tag{3.14}$$

is

$$\left\langle Y_{2,0}(t)Y_{2,0}(t+\tau)\right\rangle=\frac{1}{4\pi}\exp(-\tau/\tau_c) \tag{3.15}$$

TABLE 3.1 The Mixing of *IS* Spin States by Various Terms of the Dipolar Hamiltonian

Spin State	ββ	αβ	βα	αα
ββ	$\hat{A}$	$\hat{D}$	$\hat{D}$	$\hat{F}$
αβ	$\hat{C}$	$\hat{A}$	$\hat{B}$	$\hat{D}$
βα	$\hat{C}$	$\hat{B}$	$\hat{A}$	$\hat{D}$
αα	$\hat{E}$	$\hat{C}$	$\hat{C}$	$\hat{A}$

Source: Adapted from Hennel, J.W. and Klinowski, J. *Fundamentals of Nuclear Magnetic Resonance*, Longman, Harlow, 1993.

i.e. it has the same form as that of Eq. (2.26). The corresponding spectral densities are again Lorentzian, similar to those of Eq. (2.29):

$$J(\omega)=\frac{1}{4\pi}\frac{2\tau_c}{(1+\omega^2\tau_c^2)}=\frac{1}{2\pi}\frac{\tau_c}{(1+\omega^2\tau_c^2)} \tag{3.16}$$

Note that the factor 1/4π, corresponding to G(0), is carried over from Eq. (3.15) to Eq. (3.16). The correlation time in Eqs. (3.15) and (3.16) is similar, but not identical, to that introduced in Chapter 2. In the small-step rotational diffusion model, the correlation time for spherical harmonics with a given l value is inversely proportional to $l(l+1)$. This relation will be derived in Chapter 6. Here, we only note that the correlation time relevant for the dipole-dipole relaxation ($l=2$) is expected to be three times smaller than the correlation time for the randomly reorienting local field, $l=1$. This relation was reflected in Chapter 2 (Eq. (2.43)).

3.2.2 The Solomon Equations

In analogy with the simple model for the spin-lattice relaxation, we can set up a set of equations describing the populations' kinetics in the four-level system of Figure 3.5. For the αα level, we can write:

$$\begin{aligned}\frac{dn_{\alpha\alpha}}{dt}&=-(W_{1I}+W_{1S}+W_2)(n_{\alpha\alpha}-n_{\alpha\alpha}^{eq})\\&\quad+W_2(n_{\beta\beta}-n_{\beta\beta}^{eq})+W_{1I}(n_{\beta\alpha}-n_{\beta\alpha}^{eq})+W_{1S}(n_{\alpha\beta}-n_{\alpha\beta}^{eq})\end{aligned} \tag{3.17}$$

For the other levels, analogous equations can be obtained. The relations between the population differences and expectation values of z-magnetisation for the two spins are given by

$$\langle\hat{I}_z\rangle\propto n_I=(n_{\alpha\alpha}-n_{\beta\alpha})+(n_{\alpha\beta}-n_{\beta\beta}) \tag{3.18a}$$

$$\langle\hat{S}_z\rangle\propto n_S=(n_{\alpha\alpha}-n_{\alpha\beta})+(n_{\beta\alpha}-n_{\beta\beta}) \tag{3.18b}$$

Using Eq. (3.18), we can move on from the population kinetics to the magnetisation kinetics. Solomon has shown that the relaxation of the longitudinal magnetisation components, proportional to the expectation values of $\hat{I}_z$ and $\hat{S}_z$ operators, is related to the populations of the four levels and can be described by a set of two coupled equations:

$$\frac{d\langle\hat{I}_z\rangle}{dt}=-(W_0+2W_{1I}+W_2)\left(\langle\hat{I}_z\rangle-I_z^{eq}\right)-(W_2-W_0)\left(\langle\hat{S}_z\rangle-S_z^{eq}\right) \tag{3.19a}$$

$$\frac{d\langle\hat{S}_z\rangle}{dt}=-(W_2-W_0)\left(\langle\hat{I}_z\rangle-I_z^{eq}\right)-(W_0+2W_{1S}+W_2)\left(\langle\hat{S}_z\rangle-S_z^{eq}\right) \tag{3.19b}$$

or

$$\frac{d\langle\hat{I}_z\rangle}{dt}=-\rho_I\left(\langle\hat{I}_z\rangle-I_z^{eq}\right)-\sigma_{IS}\left(\langle\hat{S}_z\rangle-S_z^{eq}\right) \tag{3.20a}$$

$$\frac{d\langle\hat{S}_z\rangle}{dt}=-\sigma_{IS}\left(\langle\hat{I}_z\rangle-I_z^{eq}\right)-\rho_S\left(\langle\hat{S}_z\rangle-S_z^{eq}\right) \tag{3.20b}$$

or more compactly as

$$\frac{d}{dt}\begin{pmatrix}\langle \hat{I}_z\rangle \\ \langle \hat{S}_z\rangle\end{pmatrix} = -\begin{pmatrix}\rho_I & \sigma_{IS} \\ \sigma_{IS} & \rho_S\end{pmatrix}\begin{pmatrix}\langle \hat{I}_z\rangle - I_z^{eq} \\ \langle \hat{S}_z\rangle - S_z^{eq}\end{pmatrix} \tag{3.21}$$

I_z^{eq} and S_z^{eq} are the equilibrium longitudinal magnetisations for the two spins (originating from the equilibrium populations in Eq. (3.17)), and ρ_I and ρ_S are the corresponding decay rates (spin-lattice auto-relaxation rates or, briefly, spin-lattice relaxation rates). The symbol σ_{IS} denotes the cross-relaxation rate. By comparing Eqs (3.19) and (3.20), we can make the following identifications:

$$\rho_I = W_0 + 2W_{1I} + W_2 \tag{3.22a}$$

$$\rho_S = W_0 + 2W_{1S} + W_2 \tag{3.22b}$$

$$\sigma_{IS} = W_2 - W_0 \tag{3.22c}$$

We can re-write Eq. (3.20) in an even more compact vector-matrix expression:

$$\frac{d}{dt}\mathbf{V} = -\mathbf{R}(\mathbf{V} - \mathbf{V}^{eq}) \tag{3.23}$$

V is a column vector with the expectation values of $\hat{I}_z$ and $\hat{S}_z$ operators as the elements, while $\mathbf{V}^{eq}$ is its thermal equilibrium counterpart

The matrix **R** in the right-hand side of Eqs. (3.21) and (3.23) is called *relaxation matrix*. The general solutions of the Solomon equations for $\langle \hat{I}_z\rangle$ or $\langle \hat{S}_z\rangle$ are sums of two exponentials:

$$\langle \hat{I}_z\rangle = v_{11}\exp(-\lambda_1 t) + v_{12}\exp(-\lambda_2 t) \tag{3.24a}$$

$$\langle \hat{S}_z\rangle = v_{21}\exp(-\lambda_1 t) + v_{22}\exp(-\lambda_2 t) \tag{3.24b}$$

where λ_1 and λ_2 are eigenvalues of the relaxation matrix, *i.e.* the values obtained when the relaxation matrix is transformed into a form with non-zero elements only on the diagonal by replacing the original vector basis set $\langle \hat{I}_z\rangle$, $\langle \hat{S}_z\rangle$ by suitable linear combinations. The parameters v_{ij} depend on the linear combination coefficients and the initial conditions. The eigenvalues λ_1 and λ_2 would be equal to the spin-lattice auto-relaxation rates, ρ_I and ρ_S, if the cross-relaxation rate, σ_{IS}, were equal to zero. If this were the case, the relaxation matrix would then be diagonal when expressed in terms of the original basis set. The cross-relaxation phenomenon is, however, in principle always present in dipolar-relaxed systems (except for certain combinations of Larmor frequencies and correlation times, which result in cancellation of terms in the expression for σ_{IS}), and the longitudinal magnetisations in a two-spin system do not follow the Bloch equations. Solomon also demonstrated that, besides the odd situation when $\sigma_{IS} = 0$, the simple exponential relaxation behaviour of the longitudinal magnetisation is recovered under certain limiting conditions:

1. The two spins are identical (*e.g.* the two proton spins in a water molecule). In this case, only the sum of the two magnetisations is an observable quantity. Moreover, we have $W_{II} = W_{IS} = W_I$, and the spin-lattice relaxation rate is given by

$$T_1^{-1} = 2(W_1 + W_2) \tag{3.25}$$

The reader is encouraged to demonstrate that this indeed is the case.

2. One of the spins, say S, is characterised by another, and faster, relaxation mechanism. We can then say that the S-spin remains in thermal equilibrium on the time scale of the I-spin relaxation. This situation occurs, for example, in paramagnetic systems, where S is an electron spin. The spin-lattice relaxation rate for the I-spin is then given by

$$\rho_I = T_{1I}^{-1} = W_0 + 2W_{1I} + W_2 \tag{3.26a}$$

3. One of the spins, say I, is saturated by an intense radiofrequency (r.f.) field at its resonance frequency. These are the conditions applying for, for example, carbon-13 (treated as the S-spin) under broadband decoupling of protons (I-spins). The relaxation rate is also in this case given by an S-spin analogue of Eq. (3.26a):

$$\rho_S = T_{1S}^{-1} = W_0 + 2W_{1S} + W_2 \tag{3.26b}$$

and is called the *spin-lattice relaxation rate.*

When the I-spin is saturated, $\langle \hat{I}_z \rangle = 0$, and the equation for the evolution of the $\langle \hat{S}_z \rangle$ becomes

$$\frac{d\langle \hat{S}_z \rangle}{dt} = -\rho_S\left(\langle \hat{S}_z \rangle - S_z^{eq}\right) - \sigma_{IS}\left(-I_z^{eq}\right) \tag{3.27}$$

The steady-state solution $\left(d\langle \hat{S}_z \rangle / dt = 0\right)$ is given by

$$\langle \hat{S}_z \rangle_{steady\text{-}state} = S_z^{eq} + \frac{\sigma_{IS}}{\rho_S} I_z^{eq} = S_z^{eq}\left(1 + \frac{\gamma_I}{\gamma_S}\frac{\sigma_{IS}}{\rho_S}\right) = S_z^{eq}(1+\eta) \tag{3.28}$$

where $\eta = \gamma_I \sigma_{IS} / \gamma_S \rho_S$ has been introduced. Because of the simultaneous action of double-resonance irradiation and dipole-dipole interaction, the steady-state solution for $\langle \hat{S}_z \rangle$ is modified. The steady state $\langle \hat{S}_z \rangle$ can be turned into a detectable signal by a 90° pulse, and the intensity of this steady state is multiplied by the factor $1+\eta$. The phenomenon is referred to as the *nuclear Overhauser enhancement* (NOE). Before further discussion of the NOE (Section 3.3) and dipolar relaxation (Section 3.4), we wish to point out the formal similarity between the cross-relaxation (the *magnetisation* exchange) and the *chemical* exchange. The Solomon equations expressed in the form of Eq. (3.20) are formally very similar to McConnell's modification of the Bloch equation in the presence of two-site chemical exchange. We shall come back to the Bloch–McConnell equations in Chapter 13. Moreover, we are going to discuss in more detail, in Chapters 9 and 13, how the cross-relaxation and the chemical exchange can be observed in experiments of the same type.

3.3 The Nuclear Overhauser Enhancement

The nuclear Overhauser effect is one of the most interesting phenomena in NMR, as it can be related to both structural and dynamical properties of a molecule, and studies of NOE are very important and common. We can discern two different situations:

1. The spins I and S are different species, *e.g.* ^1H and ^{13}C. In this case, we talk about *heteronuclear* NOE. This relaxation parameter carries information on relaxation mechanisms and on molecular dynamics. We shall return to this point later.

2. The spins I and S are both protons (*homonuclear* NOE). While measurements of proton relaxation rates, ρ_I or ρ_S, are not very common, due to the fact that their interpretation is difficult, the proton NOE is today a very important parameter in NMR. The reason is that measurements of the homonuclear NOEs allow determination of molecular structure through the relation $\sigma_{IS} \sim b_{IS}^2 \sim r_{IS}^{-6}$. From the identification provided in Eq. (3.22c), we see that under certain conditions, leading to $W_2 = W_0$, the cross-relaxation rate vanishes. Depending on the dynamics of the reorientation, the cross-relaxation rate and the homonuclear NOE can therefore be either positive or negative.

The homonuclear NOE effect can be explained in the following way. We can redraw the energy level diagram of Figure 3.5 to yield Figure 3.6, where the effect of saturation and relaxation has been indicated (adapted after Hore[7]). The saturating r.f. field at the resonance frequency of the I-spin tends to equalise population across the I-spin transitions, *i.e.* from the $\alpha\alpha$ to the $\beta\alpha$ state and from the $\alpha\beta$ to the $\beta\beta$ state (Figure 3.6b). Looking at the population differences across the S-spin transitions, the saturation of the I-spin does not yield any change by itself. Variation of the steady-state population difference and, consequently, of the signal intensity arises if the transition probabilities W_2 and W_0 are allowed to act in an unbalanced way. When the W_2 dominates, and the population transfer from $\beta\beta$ to $\alpha\alpha$ is efficient, then the net result of the saturation and relaxation is to increase the population differences across the S-spin transitions, a situation corresponding to Figure 3.6c. If, on the other hand, the W_0 processes dominate, then the population difference across the S-spin transitions is reduced, and the signal intensity follows Figure 3.6d.

The DD transition probabilities can be evaluated in terms of dipolar coupling constants and spectral densities corresponding to the angular variables in Eq. (3.11):

$$W_0 = \frac{\pi}{5} b_{IS}^2 J(\omega_I - \omega_S) \tag{3.29a}$$

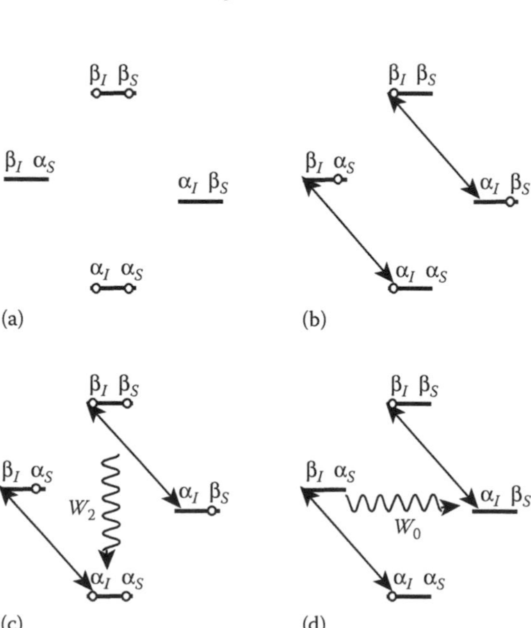

FIGURE 3.6 Schematic diagram of the NOE effect. Filled circles indicate a population excess, and open circles indicate a population deficit. (a) shows the situation at thermal equilibrium, (b) after r.f irradiation on the I-spins, leading to saturation, (c) shows the effect of W_2 relaxation, and (d) the effect of W_0 relaxation. Differences in population levels are proportional to signal intensities. (Adapted from Hore, P.J., *Nuclear Magnetic Resonance*, Oxford University Press, Oxford, 1995.)

$$W_2 = \frac{6\pi}{5} b_{IS}^2 J(\omega_I + \omega_S) \tag{3.29b}$$

The different numerical prefactors in the two expressions have their origin in the appropriate matrix elements of the spin operators occurring in the dipolar alphabet and in the normalisation of spherical harmonics. If we assume that the spectral densities in Eq. (3.29) are given by Eq. (3.16), then the relation between W_2 and W_0 at a given magnetic field depends only on the rotational correlation time, τ_c.

From inspection of Eqs. (3.16), (3.22) and (3.29) we can thus see that the NOE depends on the correlation time of the molecules. This is shown in the diagrams of Figure 3.7. For small molecules with short correlation times, $W_0 < W_2$ (because of the larger numerical prefactor in W_2), and the NOE is positive. For large molecules, with long correlation times, the spectral density is frequency dependent and has a much higher value at the close-to-zero frequency ($\omega_S - \omega_I$) than at the high frequency ($\omega_S + \omega_I$); consequently, $W_0 > W_2$, and a negative NOE is obtained. The diagrams are based on the specific and very simple motional model discussed so far: isotropic reorientation of the dipole-dipole axis in small, diffusional steps. We shall discuss more sophisticated dynamic models in Chapter 6. The gross feature of Figure 3.7 can, however, be believed to survive also under more complicated motional conditions.

For the sake of completeness, we also provide here the other two transition probabilities, obtained in a similar way:

$$W_{1I} = \frac{3\pi}{10} b_{IS}^2 J(\omega_I) \tag{3.30a}$$

$$W_{1S} = \frac{3\pi}{10} b_{IS}^2 J(\omega_S) \tag{3.30b}$$

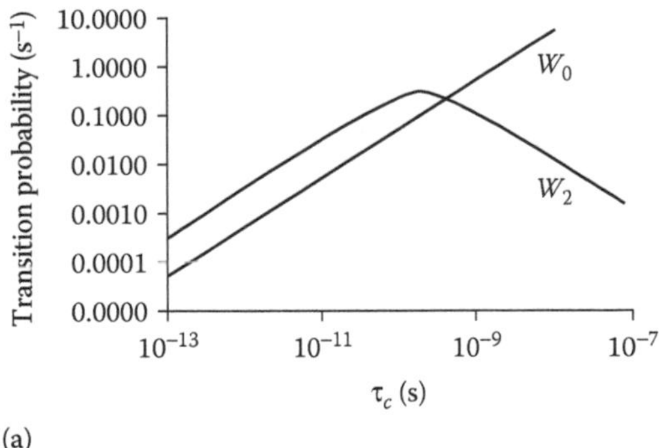

(a)

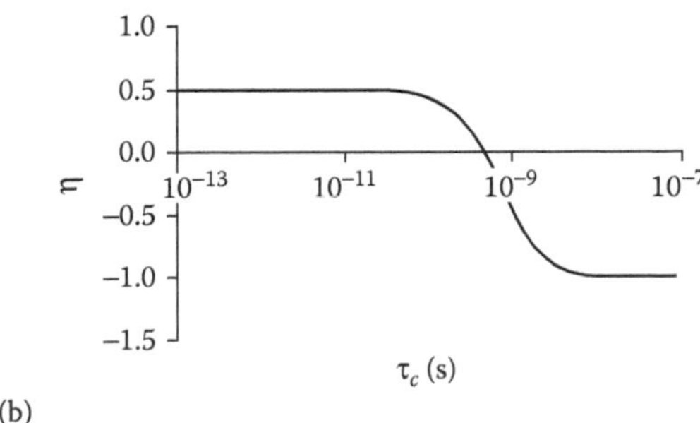

(b)

FIGURE 3.7 The dependence of the cross-relaxation rate on the correlation time. In (a), the transition probabilities W_2 and W_0 are shown. In (b), the η parameter as a function of the correlation time is displayed. A dipolar interaction strength $b_{IS}/2\pi = 15$ kHz was used, based on an interproton distance of 2 Å.

3.4 Carbon-13 Relaxation

As mentioned in Section 3.3, the information content of proton T_1 data is rather limited. The main reason for this fact is that many neighbouring protons contribute to the relaxation process, the relaxation is not single exponential (unless under special conditions), and the resulting relaxation rates are not easily related to specific dynamic or structural parameters. The situation is different for carbon-13 and nitrogen-15. The natural abundance of these nuclides is low (1.1% for ^{13}C, 0.37% for ^{15}N), and, under the conditions of proton broadband irradiation, every carbon-13 or nitrogen-15 gives rise to a single line. For carbon-13 nuclei directly bonded to one or more protons, the dipole-dipole interaction with these protons is normally the dominant relaxation mechanism, while other mechanisms can be significant for nitrogen-15. Application of proton broadband irradiation (spin-decoupling) removes not only the J-couplings but also the coupling between relaxation equations, Eqs. (3.19) or (3.20). Thus, the longitudinal carbon-13 magnetisation is characterised by simple exponential relaxation (well-defined T_1) and by an NOE factor, which is always positive. From Eqs. (3.26b) and (3.28), we obtain

$$T_{1C}^{-1} = \rho_C = \frac{\pi}{5} b_{CH}^2 [J(\omega_H - \omega_C) + 3J(\omega_C) + 6J(\omega_H + \omega_C)] \tag{3.31a}$$

$$\eta = \left(\frac{\gamma_H}{\gamma_C}\right) \frac{6J(\omega_H + \omega_C) - J(\omega_H - \omega_C)}{J(\omega_H - \omega_C) + 3J(\omega_C) + 6J(\omega_H + \omega_C)} \tag{3.31b}$$

We use here the same normalisation of spectral densities as in Eq. (3.16), and the spectral densities as given in that equation can be inserted into Eq. (3.31). More sophisticated models, based on more complicated molecular motion, discussed in Chapter 6, can also be applied. Measurements of carbon-13 relaxation parameters outside of the extreme narrowing range, *i.e.* where they depend on the applied magnetic field, allow the testing of dynamic models and are in general an excellent source of information on molecular dynamics in solution. We shall return to the application of the experiments of this type in Chapter 11.

Under extreme narrowing conditions, which here means that the product of the correlation time and the angular frequency has to be much smaller than unity for all relevant frequencies (note that the highest frequency occurring in Eq. (3.31) is $\omega_H + \omega_C$), for a carbon-13 spin with a single directly bonded proton, and assuming pure dipole-dipole relaxation, the situation becomes very simple indeed, and the relaxation parameters are given by

$$T_{1C}^{-1} = b_{CH}^2 \tau_c \tag{3.32a}$$

$$\eta = \frac{\gamma_H}{2\gamma_C} = 1.99 \tag{3.32b}$$

For nitrogen-15, the situation is a little more complicated, because it often has another relaxation mechanism (relaxation by chemical shift anisotropy; see Chapter 5), competing with the dipole-dipole relaxation. In addition, the magnetogyric ratio for nitrogen-15 is negative, which under certain conditions can lead to a signal intensity reduction by NOE. Due to the low sensitivity of nitrogen-15, natural-abundance measurements are difficult and rarely performed. Nitrogen-15 relaxation measurements are, however, extremely important and useful for investigations of peptides and proteins, but measurements are in this case almost always carried out on isotopically enriched samples.

Even though this section deals with heteronuclear dipolar relaxation, we should note that Eq. (3.31) is also valid for two unlike homonuclear spins. If we perform a selective I-decoupling experiment on a system of two protons, I and S, the longitudinal relaxation of the S-spin will follow an expression analogous to Eq. (3.31), with the carbon and proton frequencies replaced by the S and I proton Larmor frequencies, respectively. It is instructive to compare the spin-lattice relaxation rate in Eq. (3.31a) and the corresponding homonuclear relaxation rate for the case of identical spins, which is obtained by setting the transition probability expressions of Eqs. (3.29) and (3.30) into Eq. (3.25). This yields

$$T_{1H}^{-1}(\text{ident}) = 2\frac{3\pi}{10}b_{HH}^2[J(\omega_H) + 4J(2\omega_H)] \tag{3.33}$$

which in the extreme narrowing limit takes the form

$$T_{1II}^{-1}(\text{ident}) = \frac{3}{2}b_{HH}^2\tau_c \tag{3.34}$$

Comparing the results for identical spins, Eq. (3.34), with the previous situation with unlike spins, Eq. (3.32a), we notice that the numerical pre-factor is 3/2 times larger in the case of identical spins. This is the so-called "3/2 effect", which implies that, everything else being equal, the relaxation efficiency of identical spins is larger than for non-identical spins. The effect occurs because of the cross-relaxation between the identical spins, no longer directly observable, but still physically present.

So far, we have only discussed the spin-lattice relaxation. Solomon also derived expressions for the spin-spin, or transverse, relaxation in the *IS* spin system with dipole-dipole coupling. The derivation is not fully as elegant as for the longitudinal relaxation, because one has to work with the eigenstates of $\hat{I}_x$ instead of $\hat{I}_z$. We are going to limit our interest here to the heteronuclear case and skip the details of the derivation (the derivation can be found in the original work[1] or in the book by Hennel and Klinowski (*Further reading*)). Also in this case, the relaxation of the *I*- and S-spins is coupled (biexponential). Under appropriate experimental conditions, discussed in Chapter 8, the transverse relaxation for the case of a carbon-13 relaxed by a directly bonded proton can become single exponential with the rate constant:

$$T_{2C}^{-1} = \frac{\pi}{10}b_{CH}^2[4J(0) + J(\omega_H - \omega_C) + 3J(\omega_C) + 6J(\omega_H) + 6J(\omega_H + \omega_C)] \tag{3.35}$$

We can see that the transverse relaxation rate contains the spectral density function taken at other frequencies than those present in the expressions for T_1 and NOE, Eq. (3.31). It is therefore common that carbon-13 or nitrogen-15 relaxation studies also include the T_2 measurements. Under extreme narrowing conditions (we note that also here the highest relevant frequency, for which the condition $\omega^2\tau_c^2 << 1$ should be fulfilled, is $\omega_H + \omega_C$), we obtain $T_1 = T_2$. The correlation-time dependence of carbon-13 relaxation rates at two magnetic fields, based on the simple isotropic and rigid model, is shown in Figure 3.8.

Eq. (3.35) describes the correlation-time dependence of the natural linewidth for the ^{13}C NMR signal, assuming an ideal, homogeneous magnetic field and neglecting spin couplings. As will be discussed later, the experimental determination of the spin-spin relaxation rate is usually not done by linewidth measurement; rather, one uses multipulse techniques. The scalar carbon-proton couplings cannot in these experiments be suppressed in the same simple way as in the measurement of longitudinal relaxation, and therefore, special techniques are required. The relaxation rate corresponding to Eq. (3.35) is then obtained in certain limiting situations and provided that certain precautions have been taken. We shall return to experimental techniques for measuring both longitudinal and transverse relaxation in Chapter 8.

3.5 Intermolecular Dipolar Relaxation

The dipolar interaction discussed so far in this chapter has been of intramolecular nature. Furthermore, the relaxation (for carbon-13) was assumed to be dominated by the dipole-dipole interaction with a directly bonded proton spin. This led, in a natural way, to the assumption of a constant internuclear distance and

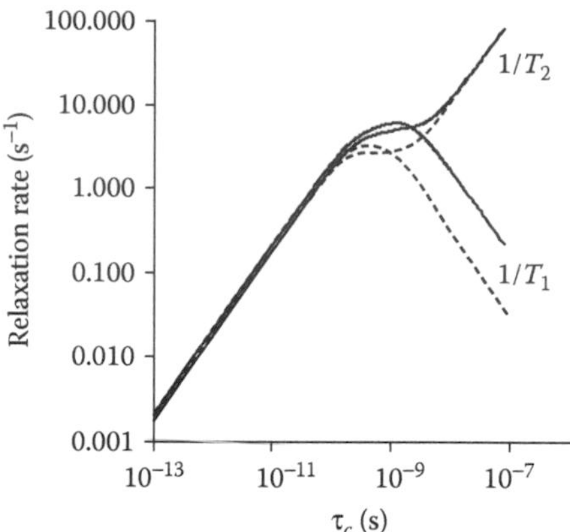

FIGURE 3.8 $1/T_1$ and $1/T_2$ relaxation rates calculated for a ^{1}H-^{13}C spin-pair using a rigid isotropic reorientation model for the spectral density function. Relaxation rates at 9.4 T are shown as solid lines, and relaxation rates at 18.8 T are shown as dashed lines. A value for the dipole-dipole coupling constant of 144×10^3 rad s^{-1} was used, which was obtained assuming an internuclear distance of 1.09 Å.

to the concept of the dipole-dipole coupling constant in the derivation of the time-correlation functions for the elements of the dipole-dipole Hamiltonian. The consequences of the fact that molecules are not rigid objects will be discussed in Chapter 6. We wish, however, to introduce already here the concept of intermolecular dipolar interactions and intermolecular dipolar relaxation, which implies that the relevant internuclear distances are not constant.

Consider a situation when a spin 1/2 nucleus has no close dipolar neighbour in the same molecule. This would be the situation, for example, for protons in chloroform ($CHCl_3$) with the low natural abundance of carbon-13. In liquid chloroform, there is obviously a fairly high density of protons, which can interact with each other through the intermolecular dipolar interaction. The axis connecting protons in nearby molecules changes its orientation with respect to the magnetic field with time, in a similar way as in the intramolecular case, but in addition, the relevant internuclear proton-proton distance, or the strength of the dipole-dipole interaction, changes randomly in time through translational diffusion. Another and more important example of intermolecular relaxation is encountered in solutions containing paramagnetic substances characterised by the presence of electron spin S (*e.g.* nitroxide radical or transition metal complexes), if they interact only weakly with the molecular species (solvent or solute) carrying the nuclear spins. The relevant translational motion in that case is the *mutual diffusion* of the molecular species containing the electron spins S and the nuclear spins I.

When dealing with a dipolar interaction modulated by translational diffusion, the time-correlation functions that we need to evaluate become more complicated than the one encountered in Eq. (3.15). Instead, we have to work with

$$G_{inter}(\tau) = \left\langle \frac{Y_{2,0}\left[\Omega_{IS}^{L}(0)\right]}{r_{IS}^{3}(0)} \frac{Y_{2,0}\left[\Omega_{IS}^{L}(\tau)\right]}{r_{IS}^{3}(\tau)} \right\rangle \tag{3.36}$$

The symbol $\Omega_{IS}^{L}(\tau)$ stands for the angles θ and ϕ specifying the orientation of the IS axis, at time τ, in the laboratory (L) coordinate frame.

The symbol $r_{IS}(\tau)$ denotes the distance between the spins, the length of the interspin vector $\mathbf{r}_{IS}(\tau)$. The general approach to calculating the $G_{inter}(\tau)$ is based on Eq. (2.20) and requires formulating the conditional probability density for the vector $\mathbf{r}_{IS}(\tau)$ attaining a certain length and direction at a certain point in time, provided it took on given values at some other time. We shall go through an example of such calculation in Chapter 6 for the rotational motion. For the translational case, we limit ourselves here to a sketch of a derivation. The reader is referred to the books by Abragam, Kruk or McConnell (*Further reading*) for more detailed presentations. Briefly, the stationary conditional probability density, $p(\mathbf{r}_0, 0|\mathbf{r}, \tau)$, of the interspin vector being $\mathbf{r}$ at time τ, provided that it was $\mathbf{r}_0$ at time zero, is obtained starting from the Fick's law for translational diffusion in three dimensions. The relevant diffusion process is actually a mutual diffusion of the two spin-carrying molecules. The coefficient of mutual diffusion, D_{12}, is equal to the sum of the diffusion coefficients for the two species, which for the case of identical molecules is twice the "ordinary" translational diffusion coefficient. Abragam gives the conditional probability density as proportional to a Gaussian function: $\exp\left(-(\mathbf{r}-\mathbf{r}_0)^2/4D_{12}\tau\right)$. As discussed independently by Hwang and Freed[8] and by Ayant and coworkers[9] in 1975, this expression is only valid for large $\mathbf{r}-\mathbf{r}_0$. The correction for the proper behaviour at small $\mathbf{r}-\mathbf{r}_0$ (short diffusion time) is in both these works carried over to the next step, the calculation of the time-correlation function of Eq. (3.36). After some rather tedious manipulations for the case of identical spins (*i.e.* a system similar to chloroform in our example above), the following expression for the time-correlation function can be obtained for the simple case of molecules modelled as non-interacting, rigid, spherical particles with radii a_I and with the spins in the centres of the spheres:

$$G_{inter}(\tau) = \frac{N_I}{d^3}\frac{18}{\pi}\int_0^{\infty}\frac{u^2}{81+9u^2-2u^4+u^6}\exp\left(-\frac{D_{12}}{d^2}u^2\tau\right)du \tag{3.37}$$

where N_I is the number density of spins I and d is the distance of closest approach between the spins, $d=2a_I$. It is useful to introduce a *translational diffusion correlation time*, τ_D, defined by

$$\tau_D = \frac{d^2}{D_{12}} \tag{3.38}$$

A convenient expression for the spectral density can be derived,[10] corresponding to the time-correlation function of Eq. (3.37), here in a dimensionless form:

$$J_{inter}(\omega) = \frac{1+5z/8+z^2/8}{1+z+z^2/2+z^3/6+4z^4/81+z^5/81+z^6/648} \tag{3.39}$$

where $z=(2\omega\tau_D)^{1/2}$. One should notice that the shape of the spectral density in Eq. (3.39) is distinctly non-Lorentzian. The shape of the intermolecular spectral density is compared with the Lorentzian-like shape with the same correlation time, $1/(1+\omega^2\tau_D^2)$ in Figure 3.9. We can see in the figure that the shape of $J_{inter}(\omega)$ is characterised by a slower decay at high frequency.

We proceed by investigating the behaviour of the intermolecular dipolar spectral density at low frequencies, so that $\omega\tau_D \ll 1$. The issue has been dealt with by numerous authors over the years; here, we choose to refer the interested reader to the work of Sholl[11] and Fries.[12] The important result is that the frequency dependence of the spectral density is given by the simple relation

$$J_{inter}(\omega) = J_{inter}(0) - B\omega^{1/2} \tag{3.40a}$$

Here, the coefficient

$$B = \frac{1}{18}(2/D_{12})^{3/2} \tag{3.40b}$$

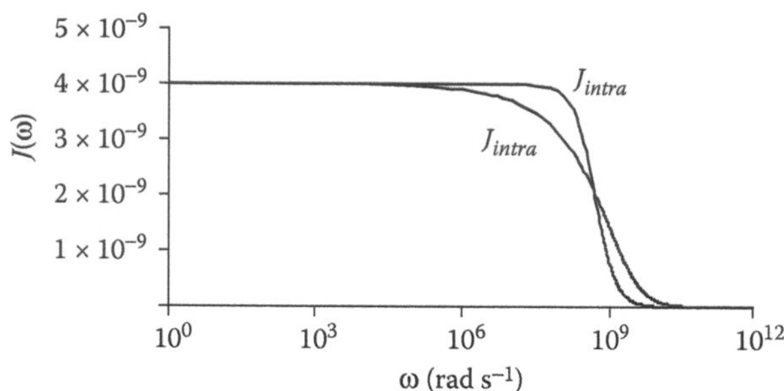

FIGURE 3.9 Comparison of the intermolecular spectral density function (Eq. (3.39)) and the Lorentzian spectral density given by Eq. (3.16). A correlation time of 2 ns was used in both cases.

(normalised in the same way as the rotational spectral density in Eq. (3.16)) is independent of the detailed characteristics within a broad family of models corresponding to the conditional probability $p(\mathbf{r}_0, 0|\mathbf{r}, \tau)$ satisfying the diffusion equation for large times and distances. Contrary to that property of the coefficient of the $\omega^{1/2}$ term, the frequency-independent term $J_{inter}(0)$ is model dependent and thus, at least in a certain sense, less useful. At low magnetic fields (low resonance frequencies), the extreme narrowing conditions are often fulfilled for the rotational motion, which means that the rotational spectral densities are frequency independent, and all the frequency dependence in the total dipolar spectral density, $J(\omega) = J_{intra}(\omega) + J_{inter}(\omega)$, comes from the intermolecular term. The frequency dependence of spectral densities in Eq. (3.40) translates into Larmor frequency dependence of spin-lattice relaxation rates through relations similar to Eqs. (3.25), (3.29b) and (3.30), but referring to the total dipolar spectral density. Thus, measurements of field dependence of the spin-lattice relaxation rate through the fast field-cycling methodology (to be described in Chapter 8) allow the determination of the translational diffusion coefficient. This approach has been tested against alternative NMR methods to determine the diffusion coefficient (based on a combination of spin-echo and magnetic field gradients; see Chapter 8) and extensively exploited by Kruk and co-workers.[13–15] An example of linear plots of relaxation rate versus square root of Larmor frequency, obtained by Kruk et al.[15] for xylitol at different temperatures, is shown in Figure 3.10.

The intermolecular spectral density given by Eq. (3.39) can also be used for the case of nuclear spins, I, being relaxed by electron spins, S. The electron spin relaxation is assumed to be much faster than the

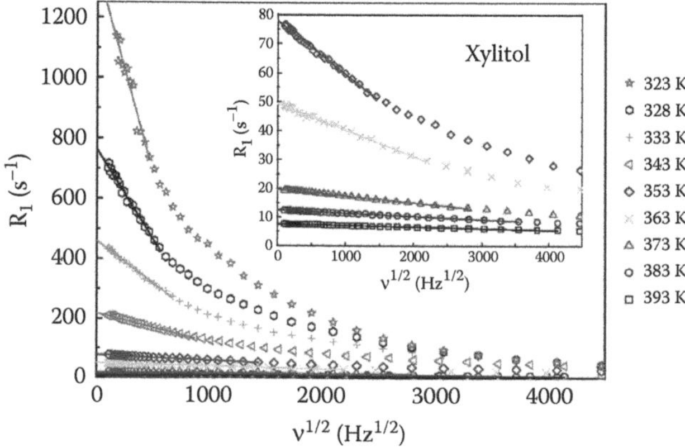

FIGURE 3.10 Proton relaxation dispersion of the liquid xylitol at different temperatures plotted versus $\nu^{1/2}$. The slope of the linear part at low frequency provides the diffusion coefficient. The inset shows enlarged high-temperature data. (Adapted with permission from Kruk, D., Meier, R. and Rössler, E.A., *Physical Review E*, 85, 020201, 2012. Copyright (2012) by the American Physical Society.)

nuclear spin relaxation (which allows us to neglect the cross-relaxation effects) and yet much slower than the diffusion. The spin-carrying molecules are characterised by the radii a_I and a_S and the diffusion coefficients D_I and D_S. The distance of closest approach becomes $d=a_I+a_S$ and the mutual diffusion coefficient $D_{12}=D_I+D_S$. The intermolecular relaxation of I-spins can be shown to follow Bloch equations with the relaxation rates

$$\frac{1}{T_{1I}^{inter}}=\frac{32\pi}{405}\left(\frac{\mu_0}{4\pi}\right)^2 1000N_A[M_S]\frac{\gamma_I^2\gamma_S^2\hbar^2 S(S+1)}{dD_{12}}\{7J_{inter}(\omega_S)+3J_{inter}(\omega_I)\} \tag{3.41a}$$

$$\frac{1}{T_{2I}^{inter}}=\frac{16\pi}{405}\left(\frac{\mu_0}{4\pi}\right)^2 1000N_A[M_S]\frac{\gamma_I^2\gamma_S^2\hbar^2 S(S+1)}{dD_{12}}\{4J_{inter}(0)+13J_{inter}(\omega_S)+3J_{inter}(\omega_I)\} \tag{3.41b}$$

N_A is the Avogadro number and $[M_S]$ is the molar concentration of the species carrying unpaired electrons. The factor $1000N_A[M_S]$ replaces here the number density of spins (in m^{-3}) in Eq. (3.37).

There are several possible factors complicating the simple description of intermolecular dipolar relaxation by Eq. (3.41). First, the spins may reside off-centre in their molecules, and the coupling between molecular rotational and translational motion then needs to be considered. This issue, sometimes referred to as *spin-eccentricity*, has been studied by Ayant and co-workers[16]. Second, the assumption of diffusing non-interacting hard spheres (called *force-free diffusion*) may not be a good approximation of dynamics in a real liquid. A more realistic description of a liquid is based on the statistical-mechanical concepts of distribution functions: in the simplest case of spherical symmetry, a *radial distribution function* $g(r)$, related to the potential of mean force, $w(r)$, through

$$w(r)=-k_BT\ln g(r) \tag{3.42}$$

The origin of this relation can be found in any book on statistical mechanics of liquids. Eqs. (3.37) and (3.39) are based on the $g(r)$ having the shape of a step function or the potential of mean force changing between zero and infinity at the distance of closest approach, d. The concept of the radial distribution function was first introduced to the field of intermolecular relaxation by Harmon and Muller[17] and explored further by Hwang and Freed.[8] Hwang and Freed replaced Fick's equation as a tool to obtain the conditional probability density by the Smoluchowski equation, which describes diffusion in a potential, such as that given in Eq (3.42). They proposed also an algorithm for implementing the effects of non-uniform distribution function on the intermolecular relaxation. It should be mentioned that the corrections due to spin eccentricity and radial distribution functions do not affect the validity of the frequency dependence of the spectral density at low frequencies, as given by Eq. (3.40).

The last complication to the simple description provided by Eq. (3.41) that we wish to mention at this point is relevant for the case of paramagnetic solutions. As mentioned earlier, the electron spin relaxation is often very rapid. In some cases, it may be fast enough to compete with translational diffusion as the source of modulation of the intermolecular dipolar interaction. We shall return to the difficult question of electron spin relaxation and its effects on nuclear spin relaxation in Chapter 15.

3.6 Exercises for Chapter 3

1. Derive the expression for the z component of the local field at position **r**.
2. Perform the integration in the second line of Eq. (3.9).
3. Demonstrate that the spin-lattice relaxation time for a pair of identical spins is given by Eq. (3.25).

References

1. Solomon, I., Relaxation processes in a system of two spins. *Phys. Rev.* 1955, *99*, 559–565.
2. Pake, G. E., Nuclear resonance absorption in hydrated crystals: Fine structure of the proton line. *J. Chem. Phys.* 1948, *16*, 327–336.
3. Spiess, H. W., Rotation of molecules and nuclear spin relaxation. *NMR Basic Principles Progr.* 1978, *15*, 59–214.
4. Smith, S. A.; Palke, W. E.; and Gerig, J. T., The Hamiltonians of NMR. Part I. *Concepts Magn. Reson.* 1992, *4*, 107–144.
5. Smith, S. A.; Palke, W. E.; and Gerig, J. T., The Hamiltonians of NMR. Part II. *Concepts Magn. Reson.* 1992, *4*, 181–204.
6. Smith, S. A.; Palke, W. E.; and Gerig, J. T., The Hamiltonians of NMR. Part III. *Concepts Magn. Reson.* 1993, *5*, 151–177.
7. Hore, P. J., *Nuclear Magnetic Resonance*. Oxford University Press: Oxford, 1995.
8. Hwang, L.-P.; and Freed, J. H., Dynamic effects of pair correlation functions on spin relaxation by translational diffusion in liquids. *J. Chem. Phys.* 1975, *63*, 4017–4025.
9. Ayant, Y.; Belorizky, E.; Alizon, J.; and Gallice, J., Calculation of the spectral densities for relaxation resulting from random translational motion by magnetic dipolar coupling in liquids. *J. Phys. (Paris)* 1975, *36*, 991–1004.
10. Polnaszek, C. F.; and Bryant, R. G., Nitroxide radical induced solvent proton relaxation: Measurement of localized translational diffusion. *J. Chem. Phys.* 1984, *81*, 4038–4045.
11. Sholl, C. A., Nuclear spin relaxation by translational diffusion in liquids and solids: High- and low-frequency limits. *J. Phys. C* 1981, *14*, 447–464.
12. Fries, P. H., Dipolar nuclear spin relaxation in liquids and plane fluids undergoing chemical reactions. *Mol. Phys.* 1983, *48*, 503–526.
13. Kruk, D.; Meier, R.; and Rössler, E. A., Translational and rotational diffusion of glycerol by means of field cycling H-1 NMR relaxometry. *J. Phys. Chem. B* 2011, *115*, 951–957.
14. Meier, R.; Kruk, D.; Gmeiner, J.; and Rössler, E. A., Intermolecular relaxation in glycerol as revealed by field cycling H-1 NMR relaxometry dilution experiments. *J. Chem. Phys.* 2012, *136*, 034508.
15. Kruk, D.; Meier, R.; and Rössler, E. A., Nuclear magnetic resonance relaxometry as a method of measuring translational diffusion coefficients in liquids. *Phys. Rev. E* 2012, *85*, 020201.
16. Ayant, Y.; Belorizky, E.; Fries, P. H.; and Rosset, J., Effect of intermolecular dipolar magnetic interactions on the nuclear relaxation of polyatomic molecules in liquids. *J. Phys. (Paris)* 1977, *38*, 325–337.
17. Harmon, J. F.; and Muller, B. H., Nuclear spin relaxation by translational diffusion in liquid ethane. *Phys. Rev.* 1969, *182*, 400–410.

4

The Redfield Relaxation Theory

So far, the description of relaxation has been focused on simple T_1 and T_2 relaxation processes and the somewhat more complicated phenomena of cross-relaxation and the nuclear Overhauser effect. A common feature of all these effects is that they are not related to the multiplet structure of nuclear magnetic resonance (NMR) spectra. A more general formulation of relaxation theory, suitable also for systems with scalar spin-spin couplings (*J*-couplings), is known as the Wangsness, Bloch and Redfield (WBR) theory or the Redfield theory[1–4]. The Redfield formulation of relaxation theory has the advantage of being applicable to any type of non-equilibrium system and any relaxation mechanism. In analogy with the Solomon theory, the Redfield theory is also based on the second-order perturbation theory, which in certain situations (unusual for nuclear spin systems in liquids) can be a limitation. Rather than dealing with the concepts of magnetisations or energy level populations as in the Solomon formulation, the Redfield theory is given in terms of density operator. This allows a more general description of the non-equilibrium states and their evolution under random perturbation.

The Redfield relaxation theory is often denoted as *semiclassical* theory. This means that the spin system is treated quantum mechanically, but the surroundings of the spins – sometimes called the *lattice* – with its stochastic processes are dealt with classically. Three formulations of the Redfield theory are commonly used and we shall go through them in the three following sections. We finish this chapter by investigating the effects of radiofrequency (r.f.) fields on the nuclear spin relaxation.

4.1 The Eigenstate Formulation

In this section, we begin by formulating the equation of motion for the density operator for a system where the Hamiltonian can be written as a sum of a large time-independent part and a smaller time-dependent perturbation.

4.1.1 Equation of Motion for the Density Operator

In the semiclassical Redfield theory, the Hamiltonian is given as a sum of the main, time-independent part, $\hat{H}_0$, and a time-dependent perturbation, $\hat{H}_1(t)$. The equation of motion of the density operator (the Liouville–von Neumann equation) (*cf.* Eq. (1.16)) is expressed by:

$$\frac{d\hat{\rho}(t)}{dt} = -i\left[\hat{H}_0 + \hat{H}_1(t), \hat{\rho}(t)\right] = i\left[\hat{\rho}(t), \hat{H}_0 + \hat{H}_1(t)\right] \tag{4.1}$$

The density operator is explicitly written as time dependent. If $\hat{H}_1(t) = 0$, the solution of Eq. (4.1) can be written:

$$\hat{\rho}(t) = \exp\left(-i\hat{H}_0 t\right)\hat{\rho}(0)\exp\left(+i\hat{H}_0 t\right) \tag{4.2}$$

The *exponential operators* occurring in Eq. (4.2) are defined by the series expansion:

$$\exp\left(i\hat{F}\right)=1+\left(i\hat{F}\right)+\frac{\left(i\hat{F}\right)^2}{2!}+\frac{\left(i\hat{F}\right)^3}{3!}+\cdots \tag{4.3}$$

in full analogy to the series expansion of the exponential function. We shall return to a physical interpretation of the expression in Eq. (4.2) a little later.

The influence of the time-independent part can be removed by transforming the Liouville–von Neumann equation to a new reference frame, the *interaction frame*. We shall arrive at an expression for the time-dependent Hamiltonian in this frame. Let us define an operator $\tilde{\hat{\rho}}(t)$ such that:

$$\tilde{\hat{\rho}}(t)=\exp\left(+i\hat{H}_0t\right)\hat{\rho}(t)\exp\left(-i\hat{H}_0t\right) \tag{4.4a}$$

or

$$\hat{\rho}(t)=\exp\left(-i\hat{H}_0t\right)\tilde{\hat{\rho}}(t)\exp\left(+i\hat{H}_0t\right) \tag{4.4b}$$

By comparing Eqs. (4.4b) and (4.2), we note that $\tilde{\hat{\rho}}(t)$ is constant and equal to $\hat{\rho}(0)$ if $\hat{H}_1(t)=0$. Moreover, $\tilde{\hat{\rho}}_1(0)=\hat{\rho}(0)$, independently of $\hat{H}_1(t)$. For small $\hat{H}_1(t)$, it is thus reasonable to expect that $\tilde{\hat{\rho}}(t)$ will change with time rather slowly; the transformation to the interaction representation removes the rapid variations of the density operator caused by $\hat{H}_0$. Substituting Eq. (4.4b) into the left-hand side of Eq. (4.1) and comparing with the right-hand side, we obtain:

$$\begin{aligned}\frac{d\hat{\rho}(t)}{dt}&=\frac{d}{dt}\left[\exp\left(-i\hat{H}_0t\right)\tilde{\hat{\rho}}(t)\exp\left(+i\hat{H}_0t\right)\right]=\frac{d}{dt}\left(\exp\left(-i\hat{H}_0t\right)\right)\tilde{\hat{\rho}}(t)\exp\left(+i\hat{H}_0t\right)\\&+\exp\left(-i\hat{H}_0t\right)\frac{d}{dt}\tilde{\hat{\rho}}(t)\exp\left(+i\hat{H}_0t\right)+\exp\left(-i\hat{H}_0t\right)\tilde{\hat{\rho}}(t)\frac{d}{dt}\left(\exp\left(+i\hat{H}_0t\right)\right)\\&=-i\hat{H}_0\hat{\rho}(t)+\exp\left(-i\hat{H}_0t\right)\frac{d\tilde{\hat{\rho}}(t)}{dt}\exp\left(+i\hat{H}_0t\right)+i\hat{\rho}(t)\hat{H}_0\\&=-i\left[\hat{H}_0,\,\hat{\rho}(t)\right]+\exp\left(-i\hat{H}_0t\right)\frac{d\tilde{\hat{\rho}}(t)}{dt}\exp\left(+i\hat{H}_0t\right)=-i\left[\hat{H}_0+\hat{H}_1(t),\,\hat{\rho}(t)\right]\end{aligned} \tag{4.5}$$

The commutator $\left[\hat{H}_0,\,\hat{\rho}(t)\right]$ occurs now on both sides of the equation and can therefore be omitted. In this way, we get

$$\exp\left(-i\hat{H}_0t\right)\frac{d\tilde{\hat{\rho}}(t)}{dt}\exp\left(+i\hat{H}_0t\right)=-i\left[\hat{H}_1(t),\,\hat{\rho}(t)\right] \tag{4.6}$$

If we multiply Eq. (4.6) by $\exp\left(+i\hat{H}_0t\right)$ from the left and by $\exp\left(-i\hat{H}_0t\right)$ from the right and use the fact that we can insert the product $\exp\left(-i\hat{H}_0t\right)\exp\left(+i\hat{H}_0t\right)\equiv 1$ between any pair of operators, we get a differential equation for $d\tilde{\hat{\rho}}(t)/dt$:

$$\begin{aligned}\frac{d\tilde{\hat{\rho}}(t)}{dt}&=-i\left[\exp\left(+i\hat{H}_0t\right)\hat{H}_1(t)\exp\left(-i\hat{H}_0t\right),\ \exp\left(+i\hat{H}_0t\right)\hat{\rho}(t)\exp\left(-i\hat{H}_0t\right)\right]\\&=-i\left[\tilde{\hat{H}}_1(t),\,\tilde{\hat{\rho}}(t)\right]\end{aligned} \tag{4.7}$$

where the following definition was introduced:

$$\hat{\tilde{H}}_1(t) = \exp\left(+i\hat{H}_0 t\right)\hat{H}_1(t)\exp\left(-i\hat{H}_0 t\right) \tag{4.8}$$

Eq. (4.8) describes the transformation of $\hat{H}_1(t)$ to the interaction representation. In the same way, $\hat{\tilde{\rho}}(t)$ is the interaction representation counterpart of the density operator $\hat{\rho}(t)$. The tilde above an operator (or its matrix element) denotes the interaction frame (or the interaction representation). It can be shown that the matrix elements of the operators follow the relations

$$\tilde{\rho}_{\alpha\alpha'}(t) = \left\langle \alpha \middle| \hat{\tilde{\rho}}(t) \middle| \alpha' \right\rangle = \exp\left[i(E_\alpha - E_{\alpha'})t\right]\left\langle \alpha \middle| \hat{\rho}(t) \middle| \alpha' \right\rangle \tag{4.9a}$$

and

$$\left(\tilde{H}_1(t)\right)_{\alpha\alpha'} = \left\langle \alpha \middle| \hat{\tilde{H}}_1(t) \middle| \alpha' \right\rangle = \exp\left[i(E_\alpha - E_{\alpha'})t\right]\left\langle \alpha \middle| \hat{H}_1(t) \middle| \alpha' \right\rangle \tag{4.9b}$$

where $|\alpha\rangle$ and E_α are an eigenfunction and an eigenvalue, respectively, of the main Hamiltonian, $\hat{H}_0$. By transforming the operators $\hat{\rho}(t)$ and $\hat{H}_1(t)$ to the interaction representation, we remove the rapid oscillations in the off-diagonal matrix elements caused by $\hat{H}_0$, the main Hamiltonian. The diagonal elements in the two representations are identical.

The transformation to the interaction representation has the same function as the transformation to the rotating frame, mentioned in the context of Bloch equations in Section 1.3. It separates the relatively un-interesting motion under $\hat{H}_0$ from the effects of the perturbation. If we assume that we deal with a system of isolated spins and that the unperturbed Hamiltonian is given by $\hat{H}_0 = -\gamma_I B_0 \hat{I}_z$ (Eq. (1.7)), the transformation in Eq. (4.8) becomes:

$$\exp\left(-i\gamma_I B_0 t\hat{I}_z\right)\hat{H}_1(t)\exp\left(i\gamma_I B_0 t\hat{I}_z\right) = \exp\left(+i\omega_I t\hat{I}_z\right)\hat{H}_1(t)\exp\left(-i\omega_I t\hat{I}_z\right) \tag{4.10}$$

We note that the expression in Eq. (4.10), generally written as:

$$\hat{A}' = \exp\left(-i\theta\hat{I}_\alpha\right)\hat{A}\exp\left(i\theta\hat{I}_\alpha\right) \tag{4.11}$$

(where $\hat{A}$ is an arbitrary operator) can be interpreted as a counterpart of the operator $\hat{A}$, rotated by the angle θ around the α-axis (see Levitt or Ernst, Bodenhausen and Wokaun, *Further reading*). Thus, Eq. (4.10) describes a rotation of $\hat{H}_1(t)$ around the z-axis by the angle $\gamma_I B_0 t = -\omega_I t$, *i.e.* a rotation around the field direction with the angular velocity $|\omega_I| = \gamma_I B_0$.

The differential equation for $d\hat{\tilde{\rho}}(t)/dt$, Eq. (4.7), can formally be integrated from zero to time t:

$$\hat{\tilde{\rho}}(t) = \hat{\tilde{\rho}}(0) - i\int_0^t \left[\hat{\tilde{H}}_1(t'), \hat{\tilde{\rho}}(t')\right]dt' \tag{4.12}$$

This is only a solution in the formal sense, because the $\hat{\tilde{\rho}}(t')$ under the integral sign is not known. We can, however, try to obtain an approximate solution by setting $\hat{\tilde{\rho}}(t') = \hat{\tilde{\rho}}(0)$. The validity of this may again be rationalised by noting that, for small $\hat{H}_1(t)$, the variation of $\hat{\tilde{\rho}}(t)$ with time is slow. This yields:

$$\hat{\tilde{\rho}}(t) = \hat{\tilde{\rho}}(0) - i\int_0^t \left[\hat{\tilde{H}}_1(t'), \hat{\tilde{\rho}}(0)\right]dt' \tag{4.13}$$

In fact, it turns out (see below) that the term under the integral sign in Eq. (4.13) vanishes and Eq. (4.13) is not a useful result as an approximate solution. A better approximation can, however, be obtained if we set $\tilde{\hat{\rho}}(t)$ given by Eq. (4.13) into the formal solution, Eq. (4.12):

$$\begin{aligned}\tilde{\hat{\rho}}(t) &= \tilde{\hat{\rho}}(0) - i\int_0^t \left[\tilde{\hat{H}}_1(t'), \left(\tilde{\hat{\rho}}(0) - i\int_0^{t'}\left[\tilde{\hat{H}}_1(t''), \tilde{\hat{\rho}}(0)\right]dt''\right)\right]dt' \\ &= \tilde{\hat{\rho}}(0) - i\int_0^t\left[\tilde{\hat{H}}_1(t'), \tilde{\hat{\rho}}(0)\right]dt' + (i)^2\int_0^t\int_0^{t'}\left[\tilde{\hat{H}}_1(t'), \left[\tilde{\hat{H}}_1(t''), \tilde{\hat{\rho}}(0)\right]\right]dt''dt'\end{aligned} \tag{4.14}$$

At this point, it is useful to find, once again, an expression for $d\tilde{\hat{\rho}}(t)/dt$:

$$\frac{d\tilde{\hat{\rho}}(t)}{dt} = -i\left[\tilde{\hat{H}}_1(t), \tilde{\hat{\rho}}(0)\right] + (i)^2\int_0^t\left[\tilde{\hat{H}}_1(t), \left[\tilde{\hat{H}}_1(t'), \tilde{\hat{\rho}}(0)\right]\right]dt' \tag{4.15}$$

After making the variable substitution $\tau = t' - t$, Eq. (4.15) becomes:

$$\frac{d\tilde{\hat{\rho}}(t)}{dt} = -i\left[\tilde{\hat{H}}_1(t), \tilde{\hat{\rho}}(0)\right] - \int_0^t\left[\tilde{\hat{H}}_1(t), \left[\tilde{\hat{H}}_1(t+\tau), \tilde{\hat{\rho}}(0)\right]\right]d\tau \tag{4.16}$$

Eq. (4.16) describes in principle the time evolution of the density matrix, which can be converted into an expression for derivatives of its matrix elements, $d\langle\alpha|\tilde{\hat{\rho}}(t)|\alpha'\rangle/dt = d\tilde{\rho}_{\alpha\alpha'}(t)/dt$ (compare Eq. (4.9)).

4.1.2 Introducing the Ensemble of Ensembles

We begin by inspecting the contribution from the first commutator:

$$\begin{aligned}\left\langle\alpha\left|\left[\tilde{\hat{H}}_1(t), \tilde{\hat{\rho}}(0)\right]\right|\alpha'\right\rangle &= \sum_\beta\left\langle\alpha\left|\tilde{\hat{H}}_1(t)\right|\beta\right\rangle\left\langle\beta\left|\tilde{\hat{\rho}}(0)\right|\alpha'\right\rangle \\ &\quad - \sum_\beta\left\langle\alpha\left|\tilde{\hat{\rho}}(0)\right|\beta\right\rangle\left\langle\beta\left|\tilde{\hat{H}}_1(t)\right|\alpha'\right\rangle\end{aligned} \tag{4.17}$$

Here, we employed the fact that, for a complete set of functions β, $\Sigma_\beta|\beta\rangle\langle\beta| = 1$ (the closure relation). Furthermore, the sum over the set of ket-bra products can be inserted between any pair of operators.

Let us now consider an ensemble of systems (really an ensemble of ensembles; since we work with the density operator, we implicitly assume that our system is indeed an ensemble), which are identical at $t=0$ (we assume that the perturbation is absent prior to $t=0$), but which are characterised by different $\hat{H}_1(t)$. We will assume that the average of $\hat{H}_1(t)$ over the ensemble of ensembles is zero at every point in time. If this requirement is not fulfilled for a particular choice of the perturbation Hamiltonian, we can always modify the $\hat{H}_1(t)$ by moving the average to $\hat{H}_0$. As an example, let us consider the dipole-dipole interaction acting as a perturbation. In an isotropic liquid, the average vanishes. In a weakly anisotropic liquid, however, there is a non-zero average of the $\hat{A}$-term in the dipolar alphabet, known as *residual dipolar couplings*. This averaged part of the interaction is not important from the point of view of relaxation and can be moved to $\hat{H}_0$ (where it modifies the splitting term, in which the residual dipolar couplings are added to *J*-couplings). Consequently, we can write:

$$\overline{\langle \alpha | \tilde{\hat{H}}(t) | \beta \rangle} = 0 \tag{4.18}$$

In this particular case, we find it more convenient to use the horizontal bar to denote the ensemble average, because the symbol $\langle \cdots \rangle$ denotes a matrix element. Because of the result of Eq. (4.18), the first commutator in Eq. (4.16) does not contribute to $d\hat{\rho}_{\alpha\alpha'}(t)/dt$. We can consider Eq. (4.16) as the first two terms in a series expansion in $\hat{H}(t)$. In fact, by a similar argument, one can prove that all the odd terms in the expansion vanish and only even terms remain.

We proceed by expanding the double commutator in Eq. (4.16):

$$\begin{aligned}\left[\tilde{\hat{H}}_1(t), \left[\tilde{\hat{H}}_1(t+\tau), \tilde{\hat{\rho}}(0)\right]\right] &= \left[\tilde{\hat{H}}_1(t), \left(\tilde{\hat{H}}_1(t+\tau)\tilde{\hat{\rho}}(0) - \tilde{\hat{\rho}}(0)\tilde{\hat{H}}_1(t+\tau)\right)\right] \\ &= \tilde{\hat{H}}_1(t)\tilde{\hat{H}}_1(t+\tau)\tilde{\hat{\rho}}(0) - \tilde{\hat{H}}_1(t)\tilde{\hat{\rho}}(0)\tilde{\hat{H}}_1(t+\tau) \\ &\quad - \tilde{\hat{H}}_1(t+\tau)\tilde{\hat{\rho}}(0)\tilde{\hat{H}}_1(t) + \tilde{\hat{\rho}}(0)\tilde{\hat{H}}_1(t+\tau)\tilde{\hat{H}}_1(t)\end{aligned} \tag{4.19}$$

The corresponding expression for the derivative of the matrix elements $d\hat{\rho}_{\alpha\alpha'}(t)/dt$ can be obtained by using Eqs. (4.18) and (4.19) and the closure relation. For the ensemble of ensembles, we obtain:

$$\begin{aligned}\frac{d\tilde{\rho}_{\alpha\alpha'}(t)}{dt} = -\sum_{\beta,\beta'} \int_0^t \Big\{ & \overline{\langle \alpha | \tilde{\hat{H}}_1(t) | \beta \rangle \langle \beta | \tilde{\hat{H}}_1(t+\tau) | \beta' \rangle \langle \beta' | \tilde{\hat{\rho}}(0) | \alpha' \rangle} \\ & - \overline{\langle \alpha | \tilde{\hat{H}}_1(t) | \beta \rangle \langle \beta | \tilde{\hat{\rho}}(0) | \beta' \rangle \langle \beta' | \tilde{\hat{H}}_1(t+\tau) | \alpha' \rangle} \\ & - \overline{\langle \alpha | \tilde{\hat{H}}_1(t+\tau) | \beta \rangle \langle \beta | \tilde{\hat{\rho}}(0) | \beta' \rangle \langle \beta' | \tilde{\hat{H}}_1(t) | \alpha' \rangle} \\ & + \overline{\langle \alpha | \tilde{\hat{\rho}}(0) | \beta \rangle \langle \beta | \tilde{\hat{H}}_1(t+\tau) | \beta' \rangle \langle \beta' | \tilde{\hat{H}}_1(t) | \alpha' \rangle} \Big\} dt'\end{aligned} \tag{4.20}$$

Now, let us see how we can use this result to compute transition probabilities. Before proceeding with the general case, let us first assume that we deal with a system with two possible states (k and m), which is initially in the state $|k\rangle$, *i.e.* that we have an ensemble of pure states $|k\rangle$. This implies that the initial density matrix elements are:

$$\langle k | \tilde{\hat{\rho}}(0) | k \rangle = \langle k | \hat{\rho}(0) | k \rangle = 1 \tag{4.21a}$$

and

$$\langle m | \tilde{\hat{\rho}}(0) | n \rangle = 0 \text{ unless } m = n = k \tag{4.21b}$$

Now, we want to compute the transition probability $W_{k \to m}$. The transition probability is given by the rate of change of the matrix element $\langle m | \tilde{\hat{\rho}}(t) | m \rangle$ (compare Eq. (4.9a)):

$$W_{k \to m} = \frac{d}{dt} \langle m | \tilde{\hat{\rho}}(t) | m \rangle = \frac{d\tilde{\rho}_{mm}(t)}{dt} \tag{4.22}$$

Eq. (4.20) can be used to evaluate Eq. (4.22), given the conditions stated in Eq. (4.21). These conditions have the effect of making the first and last terms of Eq. (4.20) vanish. The remainder is:

$$\begin{aligned} W_{k\to m} &= \int_0^t \left\{ \overline{\langle m|\tilde{\hat{H}}_1(t)|k\rangle\langle k|\tilde{\hat{H}}_1(t+\tau)|m\rangle} + \overline{\langle m|\tilde{\hat{H}}_1(t+\tau)|k\rangle\langle k|\tilde{\hat{H}}_1(t)|m\rangle} \right\} d\tau \\ &= \int_0^t \overline{\langle m|\hat{H}_1(t)|k\rangle\langle k|\hat{H}_1(t+\tau)|m\rangle} \exp(i\omega_{mk}\tau)\,d\tau \\ &+ \int_0^t \overline{\langle m|\hat{H}_1(t+\tau)|k\rangle\langle k|\hat{H}_1(t)|m\rangle} \exp(-i\omega_{mk}\tau)\,d\tau \\ &= \int_0^t \overline{V_{mk}(t)V_{km}(t+\tau)} \exp(i\omega_{km}\tau)\,d\tau + c.c. \end{aligned} \tag{4.23}$$

where the identification $E_m - E_k = \omega_{mk}$, and $V_{mk}(t) = \langle m|\hat{H}_1(t)|k\rangle$ has been made. The averaging implied by the overbar is performed over different perturbations, $\hat{H}_1(t)$. The expression for the transition probability is identical to the result of the second-order time-dependent perturbation theory (Eq. (2.37)). The operator $\hat{V} = \hat{H}_1(t)$ is Hermitian; *i.e.* $V_{mk} = V^*_{km}$, and *c.c.* in Eq. (4.23) means complex conjugate also here.

4.1.3 Approximations Leading to the Redfield Equation

We now return to Eq. (4.16) for a description of a more general case. In order to proceed, we need to make an important assumption. We assume that motions in the environment, or surrounding, of the spins (the "lattice" or the "bath") are characterised by a short correlation time τ_c. The correlation time should be sufficiently short to make its product with the root-mean square fluctuation of the perturbation, here somewhat loosely defined as $\omega_{SB} = |\overline{H_1^2}|^{1/2}$, much less than unity:

$$\omega_{SB}\tau_c \ll 1 \tag{4.24}$$

The inequality in Eq. (4.24) defines the conditions of motional narrowing (not to be confused with extreme narrowing). We focus on the behaviour of the density operator on the time scale t, such that:

$$\tau_c \ll t \ll \omega_{SB}^{-1} \tag{4.25}$$

Under these conditions, it is valid to stop the expansion of $d\tilde{\hat{\rho}}(t)/dt$ after the second commutator, as in Eq. (4.16). Each further term would be related to the foregoing one roughly as $\omega_{SB}\tau_c$, which, according to (4.24), is much less than 1.

In order to arrive at useful expressions, we need some further simplifying assumptions.

1. We replace $\tilde{\hat{\rho}}(0)$ in the right-hand side of Eq. (4.16) by $\tilde{\hat{\rho}}(t)$. We can do this because of the assumed slow variation in $\tilde{\hat{\rho}}(t)$ and the shortness of the period t.
2. We neglect the possible correlations between $\tilde{\hat{\rho}}(t)$ and $\tilde{\hat{H}}(t)$. This makes sense because of the separation of time scales, again resulting from Eq. (4.25).
3. We extend the upper limit of the integration to infinity. We can do this since the short correlation time has the effect of making the integrand vanish for long times.

Upon these assumptions, we obtain the following "master equation".

$$\frac{d\tilde{\hat{\rho}}(t)}{dt} = -\int_0^{\infty} \overline{\left[\tilde{\hat{H}}_1(t), \left[\tilde{\hat{H}}_1(t+\tau), \tilde{\hat{\rho}}(t)\right]\right]} d\tau \tag{4.26}$$

Strictly speaking, we should also use an overbar over the density matrix element in the left-hand side of Eqs. (4.20) and (4.26). This would complicate the notation in the rest of this chapter and we choose not to do it, assuming instead implicitly that the density operator and its matrix elements refer to the average density operator, defined in the ensemble of ensembles.

Eq. (4.26) can be re-written in terms of matrix elements of $\tilde{\hat{\rho}}(t)$ using the basis set of the eigenstates to $\hat{H}_0$. After somewhat tedious calculations, one obtains:

$$\frac{d\tilde{\rho}_{\alpha\alpha'}(t)}{dt} = \sum_{\beta,\beta'} \exp\left[i(\alpha - \alpha' - \beta + \beta')t\right] R_{\alpha\alpha'\beta\beta'} \tilde{\rho}_{\beta\beta'}(t) \tag{4.27}$$

where $\tilde{\rho}_{\beta\beta'}(t) = \left\langle \beta \middle| \tilde{\hat{\rho}}(t) \middle| \beta' \right\rangle$, $\alpha \equiv E_\alpha$ (in angular frequency units) and the exponential factor has its origin in Eq. (4.9b).

The symbols $R_{\alpha\alpha'\beta\beta'}$ are called *relaxation supermatrix elements* and are given by:

$$\begin{aligned} R_{\alpha\alpha'\beta\beta'} = \frac{1}{2}\Big[& \mathcal{J}_{\alpha\beta\alpha'\beta'}(\alpha' - \beta') + \mathcal{J}_{\alpha\beta\alpha'\beta'}(\alpha - \beta) \\ & - \delta_{\alpha'\beta'} \sum_{\gamma} \mathcal{J}_{\gamma\beta\gamma\alpha}(\gamma - \beta) - \delta_{\alpha\beta} \sum_{\gamma} \mathcal{J}_{\gamma\alpha'\gamma\beta'}(\gamma - \beta') \Big] \end{aligned} \tag{4.28}$$

where we again explicitly average over the ensemble of different perturbation Hamiltonians:

$$\mathcal{J}_{\alpha\alpha'\beta\beta'}(\omega) = 2\int_0^{\infty} \overline{\left\langle \alpha \middle| H_1(t) \middle| \alpha' \right\rangle \left\langle \beta' \middle| H_1(t+\tau) \middle| \beta \right\rangle} \exp(-i\omega\tau) d\tau \tag{4.29}$$

In fact, the functions $\mathcal{J}_{\alpha\alpha'\beta\beta'}(\omega)$ as defined in Eq. (4.29) may be complex. The real part is normally much larger than the imaginary part. The imaginary part gives rise to small shifts in line frequencies, called *dynamic frequency shifts*. We shall return to this issue in the next section.

As a further approximation, we reduce the number of relaxation supermatrix elements by retaining only those for which $\alpha - \alpha' - \beta + \beta' = 0$, *i.e.* only those for which $\exp[i(\alpha - \alpha' - \beta + \beta')t] = 1$. These terms are called *secular* and the approximation is known as the *secular approximation* (note that the term *secular* is used here in a different meaning than in Section 3.1). The non-secular terms are less efficient in causing relaxation, because they oscillate rapidly, which effectively averages them to (almost) zero. In particular, the secular approximation prevents coupling of density matrix elements corresponding to different *orders of coherence*, $p_{\alpha\alpha'} = M_\alpha - M_{\alpha'}$ (where M_α and $M_{\alpha'}$ are the total magnetic quantum numbers for the states α and α'). The issue of the order of coherence is important in modern NMR and we shall return to it in Chapter 7. Within a block of a given coherence order, the secular approximation should not be used if one has nearly degenerate transitions, $\alpha - \alpha' \approx \beta - \beta'$, *i.e.* overlapping lines.

Making use of the secular approximation, Eq. (4.27) takes the form

$$\frac{d\tilde{\rho}_{\alpha\alpha'}}{dt} = \sum_{\beta,\beta'} R_{\alpha\alpha'\beta\beta'} \tilde{\rho}_{\beta\beta'} \tag{4.30a}$$

where the summation is restricted to the secular terms. Eq. (4.30a) may be re-written in a form making explicit use of the relaxation superoperator:

$$\frac{d\hat{\tilde{\rho}}(t)}{dt} = \hat{\hat{R}}\hat{\tilde{\rho}}(t) \tag{4.30b}$$

If required, one can formulate an equation, similar to Eq. (4.30a), for the density matrix elements in the laboratory frame rather than in the interaction representation. This can be done without invoking the secular approximation. We use the facts that:

$$\tilde{\rho}_{\alpha\alpha'} = \exp\big(i(\alpha - \alpha')t\big)\rho_{\alpha\alpha'} \tag{4.31}$$

and

$$\begin{aligned}\frac{d\tilde{\rho}_{\alpha\alpha'}}{dt} &= i(\alpha - \alpha')\tilde{\rho}_{\alpha\alpha'} + \exp\big(i(\alpha - \alpha')t\big)\frac{d\rho_{\alpha\alpha'}}{dt} \\ &= \sum_{\beta,\beta'} R_{\alpha\alpha'\beta\beta'} \exp\big[i(\alpha - \alpha')t\big]\rho_{\beta\beta'}\end{aligned} \tag{4.32}$$

In the second equality, we make use of Eq. (4.27). The laboratory-frame counterpart of Eq. (4.27) thus becomes:

$$\frac{d\rho_{\alpha\alpha'}}{dt} = i\big(\alpha' - \alpha\big)\rho_{\alpha\alpha'} + \sum_{\beta,\beta'} R_{\alpha\alpha'\beta\beta'}\rho_{\beta\beta'} \tag{4.33}$$

We note that the expressions for the time evolution of diagonal elements of the density matrix (populations), $\alpha = \alpha'$, in the interaction representation and in the laboratory frame are identical. Differences arise in the expressions for the time evolution of coherences. The laboratory-frame formulation has the advantage of not invoking the secular approximation. This feature becomes particularly useful in the context of simulations of NMR lineshapes in systems where the frequency splittings in the spectra are not very much larger than the relaxation supermatrix elements. Example of such a treatment can be found in a paper by Werbelow and Kowalewski[5]. We shall come back to this point in Chapter 14.

4.1.4 The Physics of the Redfield Equation

Consider now the physical meaning of the various relaxation supermatrix elements. $R_{\alpha\alpha\beta\beta}$ are simply the transition probabilities between the levels α and β. The spin-lattice relaxation (T_1 processes) is related to these elements. The symbols $R_{\alpha\alpha\beta\beta}$ describe the relaxation of the off-diagonal elements of the density operator, *i.e.* coherences between states α and β or T_2-type processes. As a consequence of the secular approximation, we obtain that the relaxation of non-degenerate (and non-overlapping) lines is a simple exponential process and the lines are Lorentzian. Summarising the properties of various relaxation supermatrix elements, we come to the conclusion that the matrix representation of the relaxation superoperator can be given a pictorial description shown in Figure 4.1 (based on Ernst, Bodenhausen and Wokaun's book [*Further reading*]), known as the *Redfield kite*.

Before leaving this section, we wish to discuss two rather subtle but important questions. The first one is related to the condition $t \ll \omega_{SB}^{-1}$ that we assumed in order to obtain the master equation. The relaxation times in NMR ($R_{\alpha\alpha'\beta\beta'}^{-1}$) are typically much longer than ω_{SB}^{-1} and we are normally interested in following the behaviour of the density operator on this rather long time scale. This dilemma can be resolved if we consider our long time period as a sequence of short intervals τ_B, such that $\tau_B \ll t \ll \omega_{SB}^{-1}$.

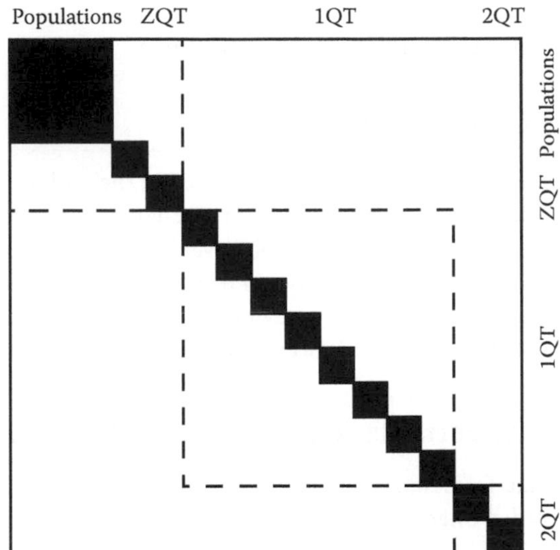

FIGURE 4.1 The Redfield kite. The Redfield relaxation matrix has a block structure due to the neglect of non-secular terms. Solid squares indicate non-zero relaxation rate constants between different magnetisation orders. Only populations have non-zero cross-relaxation rate constants. The dashed boxes indicate additional non-zero cross-relaxation rates if transitions are degenerate.

Since $\hat{H}_1(t)$ in the consecutive intervals can safely be assumed to be uncorrelated, we can generalise the validity of the master equation to longer t. The remaining validity condition for the Redfield theory is then Expression (4.24). If this condition is fulfilled, then we always find:

$$R^{-1}_{\alpha\alpha'\beta\beta'} \gg \tau_c \tag{4.34}$$

i.e. that the relaxation times are always longer than the correlation time. Some authors consider this relation to be the validity condition of the Redfield theory, which is not fully correct.

The second point is related to the effect of non-equal equilibrium populations of different levels on transition probabilities, an issue mentioned already in Section 2.5. Expression (4.28) leads to $R_{\alpha\alpha\beta\beta} = R_{\beta\beta\alpha\alpha}$ or to the transition probabilities $\alpha \longrightarrow \beta$ and $\beta \longrightarrow \alpha$ being equal. This in turn leads to an equal distribution of populations between different levels, which corresponds to the equilibrium distribution at infinite temperature. As stated earlier, to get a correct description at finite temperatures, one should in principle treat the lattice quantum mechanically. In practical terms, this leads to a modification of Eq. (4.30a):

$$\frac{d\tilde{\rho}_{\alpha\alpha'}}{dt} = \sum_{\beta,\beta'}^{\text{sec.}} R_{\alpha\alpha'\beta\beta'}\left(\tilde{\rho}_{\beta\beta'} - \tilde{\rho}^{eq}_{\beta\beta'}(T)\right) \tag{4.35}$$

where the summation is restricted to secular terms and $\tilde{\rho}^{eq}_{\beta\beta'}(T)$ denotes the equilibrium density matrix element at temperature T, given by Eq. (1.18) for the diagonal elements (populations) and equal to zero if $\beta \neq \beta'$. Since the diagonal $\beta = \beta'$ matrix elements are identical in the interaction representation and the laboratory frame, we can write:

$$\rho^{eq}_{\beta\beta'}(T) = \delta_{\beta\beta'}\frac{\exp(-\hbar\beta/k_B T)}{\sum_{\beta''}\exp(-\hbar\beta''/k_B T)} \tag{4.36}$$

It is worth noting the similarity of Eq. (4.35) to Eq. (3.21), as it provides a link between the simple, Solomon-like approach and the more complete Redfield treatment.

4.2 The Operator Formulation

In practical calculations, we are usually not so much interested in the evolution of the density matrix as in that of observable quantities, represented by quantum mechanical operators. Therefore, a formulation of the Redfield theory in terms of evolution of observables has certain advantages. An extensive presentation of the operator formulation of the Redfield relaxation theory has been given in Abragam's and Goldman's books (*Further reading*) and in a review by Goldman.[6] A shorter presentation and a comparison with the eigenstate formulation has been given by Murali and Krishnan[7].

4.2.1 Irreducible Tensor Formulation of the Perturbation Hamiltonian

In analogy with the eigenstate formulation, the master equation, Eq. (4.26), is the starting point for further discussion. To proceed, we here introduce a general expression for $\hat{H}_1(t)$, similar to Eq. (3.13):

$$\hat{H}_1(t)=\xi\sum_{q=-1}^{l}(-1)^q V_{l,q}(t)\hat{T}_{l,-q} \tag{4.37a}$$

The perturbation is expressed as a sum of $2l+1$ terms, where l is called the tensor rank (for the DD interaction, described in Chapter 3, we have $l=2$). Every term is a product of a classical random function of time, $\xi V_{l,q}(t)$, and an expression $\hat{T}_{l,-q}$ dependent only on the spin operators. The strength of the interaction, ξ, has here been factored out from the function of time. For a DD interaction between the spins I and S, we shall use the notation ξ_{IS}^{DD}. Thus, Eq. (4.37a) formulates clearly the semiclassical nature of the theory. The functions $V_{l,q}(t)$ are in general complex, fulfil $V_{l,q}^*(t)=(-1)^q V_{l,-q}(t)$, and depend on the space variables, physical constants and time. The operator $\hat{T}_{l,q}$ is called the qth component of the rank l irreducible tensor operator and fulfils $\hat{T}_{l,q}^\dagger=(-1)^q\hat{T}_{l,-q}$, where $\hat{T}_{l,q}^\dagger$ is called the *Hermitian conjugate* of $\hat{T}_{l,q}$. We find it useful to express $\hat{H}_1(t)$ also in terms of the complex and Hermitian conjugates of the quantities in Eq. (4.37a):

$$\hat{H}_1(t)=\xi\sum_{q=-1}^{l}(-1)^q V_{l,q}^*(t)\hat{T}_{l,-q}^\dagger \tag{4.37b}$$

The reason for commonly choosing to work in terms of the irreducible tensor operators is that they have convenient behaviour under rotation of the axes of the coordinate system. In fact, the irreducible tensor operators are defined according to their properties under such rotations:

$$\hat{T}_{l,q}'(\text{new axes})=\sum_{p=-l}^{l}\hat{T}_{l,p}(\text{old axes})D_{p,q}^l(\alpha,\beta,\gamma) \tag{4.38}$$

The prime refers to the new set of axes. The symbols α, β, γ are the three Euler angles necessary to define an arbitrary rotation of the coordinate frame. The Euler angles are defined in Figure 4.2. For further discussion of the subject, the reader should, for example, consult the book by Brink and Satchler (*Further reading*) or a review by Mueller[8]. The symbol $D_{p,q}^l(\alpha,\beta,\gamma)$ denotes the elements of the Wigner rotation matrix. The matrix elements have a simple dependence on the angles α and γ:

$$D_{p,q}^l(\alpha,\beta,\gamma)=\exp\left(-i(\alpha p+\gamma q)\right)d_{p,q}^l(\beta) \tag{4.39}$$

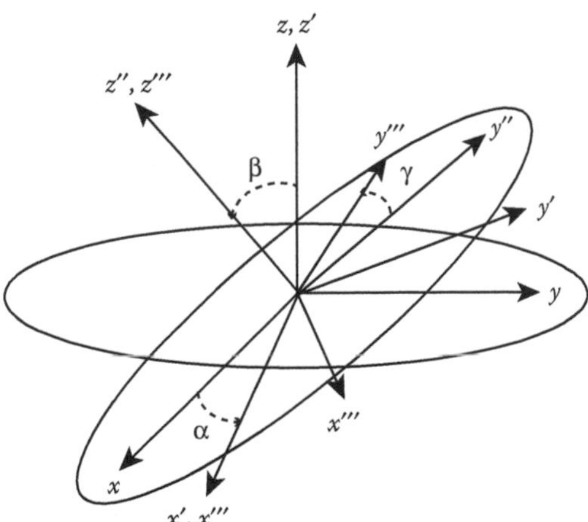

FIGURE 4.2 Illustration of the rotations in three dimensions by the three Euler angles, α, β, γ. The original axes are denoted *x*, *y*, *z*. The rotations are defined according to the convention that uses first a rotation about the *z*-axis by an angle α. This leads to the new positions/axes (x', y', z'). Next, one performs a rotation by β about the x'-axis (leading to the x'', y'', z'' positions) and, finally, a rotation by γ around the new z''-axis yields the final positions x''', y''', z'''.

while the β-dependence is more complicated. The $d^{l}_{p,q}(\beta)$ functions for $l=2$ are listed in Table 4.1. For further details, the reader is referred also in this case to the book by Brink and Satchler.

A second important property of irreducible tensor operators is a consequence of the relation between rotations of the coordinate systems and angular momentum operators, illustrated in Eq. (4.11). As a consequence of the transformation properties under rotations, the irreducible spherical tensor operators follow the relation:

$$\left[\hat{I}_z, \hat{T}_{l,q}\right] = q\hat{T}_{l,q} \tag{4.40a}$$

It is easy to convince oneself (readers are advised to do so) that Eq. (4.40a) is fulfilled by the operators

$$\hat{I}_{10} = \hat{I}_z \tag{4.41a}$$

$$\hat{I}_{1\pm1} = \mp\frac{1}{\sqrt{2}}\hat{I}_{\pm} \tag{4.41b}$$

where $\hat{I}_{\pm}$ are defined in Eq. (3.12) and we recall the relations:

$$\left[\hat{I}_z, \hat{I}_{\pm}\right] = \pm\hat{I}_{\pm} \tag{4.40b}$$

which are derived in Atkins and Friedman's book (*Further reading*). We note that $\hat{I}_+$ and $\hat{I}_-$ are Hermitian conjugates of each other. One should add that the choice of the operators in Eq. (4.41) is not unique; here (and in the following) we follow the convention of Brink and Satchler (*Further reading*).

The third useful property of the irreducible tensor operators is the fact that operators with ranks l_1 and l_2 can be coupled together to form operators of rank $l=l_1+l_2, l_1+l_2-1, \ldots, |l_1-l_2|$. Consider two

TABLE 4.1 The $d^l_{p,q}(\beta)$ Functions for $l=2$

p	q	$d^l_{p,q}(\beta)$
−2	−2	$\cos^4(\beta/2)$
−2	−1	$\frac{1}{2}\sin\beta(1+\cos\beta)$
−2	0	$\sqrt{3/8}\sin^2\beta$
−2	1	$-\frac{1}{2}\sin\beta(\cos\beta-1)$
−2	2	$\sin^4(\beta/2)$
−1	−2	$-\frac{1}{2}\sin\beta(1+\cos\beta)$
−1	−1	$\frac{1}{2}(2\cos\beta-1)(\cos\beta+1)$
−1	0	$\sqrt{3/2}\sin\beta\cos\beta$
−1	1	$\frac{1}{2}(2\cos\beta+1)(1-\cos\beta)$
−1	2	$-\frac{1}{2}\sin\beta(\cos\beta-1)$
0	−2	$\sqrt{3/8}\sin^2\beta$
0	−1	$-\sqrt{3/2}\sin\beta\cos\beta$
0	0	$\frac{1}{2}\left(3\cos^2\beta-1\right)$
0	1	$\sqrt{3/2}\sin\beta\cos\beta$
0	2	$\sqrt{3/8}\sin^2\beta$
1	−2	$\frac{1}{2}\sin\beta(\cos\beta-1)$
1	−1	$\frac{1}{2}(2\cos\beta+1)(1-\cos\beta)$
1	0	$-\sqrt{3/2}\sin\beta\cos\beta$
1	1	$\frac{1}{2}(2\cos\beta-1)(\cos\beta+1)$
1	2	$\frac{1}{2}\sin\beta(1+\cos\beta)$
2	−2	$\sin^4(\beta/2)$
2	−1	$\frac{1}{2}\sin\beta(\cos\beta-1)$
2	0	$\sqrt{3/8}\sin^2\beta$
2	1	$-\frac{1}{2}\sin\beta(1+\cos\beta)$
2	2	$\cos^4(\beta/2)$

irreducible tensor operators with elements $\hat{R}_{l_1,q_1}$ and $\hat{P}_{l_2,q_2}$. The elements $\hat{T}_{l,q}$ of the irreducible tensor operator of rank l (corresponding to one of the values given above) are given by:

$$\hat{T}_{l,q}=\sum_{q_1=-l_1}^{l_1}\sum_{q_2=-l_2}^{l_2}\hat{R}_{l_1,q_1}\hat{P}_{l_2,q_2}C\left(l_1l_2q_1q_2|lq\right) \tag{4.42}$$

where $C(l_1l_2q_1q_2|lq)$ is a vector coupling coefficient or a Clebsch–Gordan coefficient. Exactly the same coefficients occur in the theory of coupling of angular momenta when the subsystem eigenfunctions are combined to form eigenfunctions for the total angular momentum. The reader is again referred to the book by Brink and Satchler (*Further reading*) for further information and explicit expressions for the Clebsch–Gordan coefficients.

Let us identify $\hat{R}_{l_1,q_1}$ with the operators given in Eq. (4.41) and $\hat{P}_{l_2,q_2}$ with the analogous operators for the S-spin. The rank two irreducible tensor operators obtained using Eq. (4.42) and the appropriate Clebsch–Gordan coefficients are as follows:

$$\hat{T}_{2,0}=\frac{1}{\sqrt{6}}\left(3\hat{I}_z\hat{S}_z-\hat{\mathbf{I}}\cdot\hat{\mathbf{S}}\right)=\frac{1}{\sqrt{6}}\left[2\hat{I}_z\hat{S}_z-\frac{1}{2}\left(\hat{I}_+\hat{S}_-+\hat{I}_-\hat{S}_+\right)\right] \tag{4.43a}$$

$$\hat{T}_{2,\pm1} = \mp\frac{1}{2}\left(\hat{I}_{\pm}\hat{S}_z + \hat{I}_z\hat{S}_{\pm}\right) \tag{4.43b}$$

$$\hat{T}_{2,\pm2} = \frac{1}{2}\hat{I}_{\pm}\hat{S}_{\pm} \tag{4.43c}$$

To construct the Hamiltonian, $\hat{H}_1(t)$, with which we started this section, and which was defined in Eqs. (3.13) and (4.37), we use this form of the irreducible tensor operators together with the functions $V_{l,q}$ equal to normalised spherical harmonics, fulfilling the definitions of irreducible tensors:

$$Y_{2,0} = \left(\frac{5}{16\pi}\right)^{1/2}\left(3\cos^2\theta - 1\right) \tag{4.44a}$$

$$Y_{2,\pm1} = \mp\left(\frac{15}{8\pi}\right)^{1/2}\cos\theta\sin\theta\exp\left(\pm i\phi\right) \tag{4.44b}$$

$$Y_{2,\pm2} = \left(\frac{15}{32\pi}\right)^{1/2}\sin^2\theta\exp\left(\pm 2i\phi\right) \tag{4.44c}$$

In this way, we can obtain the relation between the dipolar interaction strength constant ξ_{IS}^{DD} and b_{IS}, defined in Chapter 3. By multiplying together the normalisation constants in Eqs. (4.43) and (4.44) and comparing the pre-factors of Eqs. (3.10) and (3.11) with the DD version of Eq. (4.37a), we obtain the relationship:

$$b_{IS} = \xi_{IS}^{DD}\left(\frac{1}{6}\right)^{1/2} 2\left(\frac{5}{16\pi}\right)^{1/2} = \xi_{IS}^{DD}\left(\frac{5}{24\pi}\right)^{1/2} \tag{4.45}$$

Let us now transform $\hat{H}_1(t)$, given in Eq. (4.37a), to the interaction representation, according to Eq. (4.8):

$$\begin{aligned}\tilde{\hat{H}}_1(t) &= \exp\left(i\hat{H}_0 t\right)\hat{H}_1(t)\exp\left(-i\hat{H}_0 t\right)\\ &= \xi\sum_q (-1)^q V_{l,q}(t)\exp\left(i\hat{H}_0 t\right)\hat{T}_{l,-q}\exp\left(-i\hat{H}_0 t\right)\end{aligned} \tag{4.46}$$

The expression $\exp\left(i\hat{H}_0 t\right)\hat{T}_{l,-q}\exp\left(-i\hat{H}_0 t\right)$ requires careful consideration. Let us consider $\hat{H}_1(t)$ being the dipolar Hamiltonian of Eq. (3.13), with the spin operators in the form of Eq. (4.43), and $\hat{H}_0$ consisting of the Zeeman interaction for the spins I and S, as given in Eq. (3.7). We note that the operators $\hat{I}_z$ and $\hat{S}_z$ commute with each other, which has the consequence that:

$$\exp\left(i\hat{H}_0 t\right) = \exp\left(i\omega_I t\hat{I}_z\right)\exp\left(i\omega_S t\hat{S}_z\right) \tag{4.47}$$

The expression $\exp\left(i\hat{H}_0 t\right)\hat{T}_{l,-q}\exp\left(-i\hat{H}_0 t\right)$ becomes thus:

$$\begin{aligned}&\exp\left(i\hat{H}_0 t\right)\hat{T}_{l,-q}\exp\left(-i\hat{H}_0 t\right)\\ &= \exp\left(i\omega_I t\hat{I}_z\right)\exp\left(i\omega_S t\hat{S}_z\right)\hat{T}_{l,-q}\exp\left(-i\omega_I t\hat{I}_z\right)\exp\left(-i\omega_S t\hat{S}_z\right)\end{aligned} \tag{4.48}$$

At this point, it is again useful to recall the expression given in Eq. (4.11). According to that expression, Eq. (4.48) implies that the operators $\hat{T}_{l,-q}$ are rotated around the *z*-axis. More specifically, the *I*-spin operators in $\hat{T}_{l,-q}$ are rotated according to $\hat{A}' = \exp(i\omega_I t\hat{I}_z)\hat{A}\exp(-i\omega_I t\hat{I}_z)$, while the *S*-spin operators in $\hat{T}_{l,-q}$ are rotated by the Zeeman Hamiltonian of the *S*-spin. The exponential operators containing $\hat{I}_z$ do not influence the *S*-spin operators (and vice versa), because *S*-spin operators and the *I*-spin operators always commute. As a result, the exponential operators containing $\hat{I}_z$ introduce, according to Eq. (4.48), terms rotating at the positive and negative *I*-spin Larmor frequency as well as zero-frequency terms. In the same way, the exponential operators containing $\hat{S}_z$ lead to the occurrence of positive and negative *S*-spin Larmor frequencies. For example, explicit calculations for $q=0$ yield:

$$\exp\left(i\hat{H}_0 t\right)\hat{T}_{2,0}\exp\left(-i\hat{H}_0 t\right)$$
$$=\sqrt{\frac{1}{6}}\left[2\hat{I}_z\hat{S}_z-\frac{1}{2}\hat{I}_+\hat{S}_-\exp\left(i(\omega_I-\omega_S)t\right)-\frac{1}{2}\hat{I}_-\hat{S}_+\exp\left(-i(\omega_I-\omega_S)t\right)\right] \tag{4.49}$$

The expression $\exp\left(i\hat{H}_0 t\right)\hat{T}_{l,-q}\exp\left(-i\hat{H}_0 t\right)$ for any q becomes:

$$\exp\left(i\hat{H}_0 t\right)\hat{T}_{l,-q}\exp\left(-i\hat{H}_0 t\right)=\sum_p \hat{A}_p^{(-q)}\exp\left(i\omega_p^{(-q)}t\right) \tag{4.50}$$

Here, $\hat{A}_p^{(-q)}$ are a set of operators and the frequencies $\omega_p^{(-q)}$ correspond to either zero or linear combinations of the *I*- and *S*-spin Larmor frequencies. The symbols on the right-hand side of Eq. (4.50) are summarised in Table 4.2.

The expressions given in Eqs. (4.46) and (4.50) can be inserted in the "master equation" (4.26). We make use of Eq. (4.37a) for $\hat{H}_1(t)$ and of Eq. (4.37b) for $\hat{H}_1(t+\tau)$ and note that in an ensemble of systems, all the operators $\hat{T}_{l,-q}$ or $\hat{A}_p^{(q)}$ are identical, while the functions $V_{l,q}(t)$ are random. After ensemble-averaging, this leads to:

$$\frac{d\tilde{\hat{\rho}}(t)}{dt}=-\sum_{q,q',p,p'}(-1)^{q+q'}\exp\left(i\left(\omega_p^{(-q)}+\omega_{p'}^{(-q')}\right)t\right)\left[\hat{A}_p^{(-q)},\left[\hat{A}_{p'}^{\dagger(-q')},\tilde{\hat{\rho}}(t)\right]\right]$$
$$\times\xi^2\int_0^\infty\left\langle V_{l,q}(t)V_{l,q'}^*(t+\tau)\right\rangle\exp\left(i\omega_{p'}^{(-q')}\tau\right)d\tau \tag{4.51}$$

We recognise the term under the integral sign (where the symbol $\langle\cdots\rangle$ has the same meaning as the overbar in Eqs. (4.18), (4.20), (4.23), (4.26) and (4.29)) as a time-correlation function. From the fact that

TABLE 4.2 Results of Calculations of Symbols in the Right-Hand Side of Eq. (4.50)

	$A_p^{(q)}$			$\omega_p^{(q)}$		
q	$p=1$	$p=2$	$p=3$	$p=1$	$p=2$	$p=3$
−2	$\frac{1}{2}\hat{I}_-\hat{S}_-$	–	–	$-\omega_I-\omega_S$	–	–
−1	$\frac{1}{2}\hat{I}_-\hat{S}_z$	$\frac{1}{2}\hat{I}_z\hat{S}_-$	–	$-\omega_I$	$-\omega_S$	–
0	$\sqrt{\frac{2}{3}}\hat{I}_z\hat{S}_z$	$-\sqrt{\frac{1}{24}}\hat{I}_+\hat{S}_-$	$-\sqrt{\frac{1}{24}}\hat{I}_-\hat{S}_+$	0	$\omega_I-\omega_S$	$-\omega_I+\omega_S$
1	$-\frac{1}{2}\hat{I}_+\hat{S}_z$	$-\frac{1}{2}\hat{I}_z\hat{S}_+$	–	ω_I	ω_S	–
2	$\frac{1}{2}\hat{I}_+\hat{S}_+$	–	–	$\omega_I+\omega_S$	–	–

$V_{1,q}(t)$ are defined as irreducible spherical tensors, the corresponding time-correlation functions follow the symmetry relation[9]:

$$\left\langle V_{l,q}(t)V^{*}_{1,q'}(t+\tau)\right\rangle = G_{l,q,q'}(\tau) = \delta_{q,q'}G_{l}(\tau) \tag{4.52}$$

We shall return to this relation in Chapter 6.

Using the relation (4.52) and neglecting non-secular terms, Eq. (4.51) simplifies to:

$$\frac{d\tilde{\hat{\rho}}(t)}{dt} = -\frac{1}{2}\xi^2\sum_{q,p} j_l\left(-\omega_p^{(-q)}\right)\left[\hat{A}_p^{(-q)},\left[\hat{A}_p^{\dagger(-q)},\tilde{\hat{\rho}}(t)\right]\right] \tag{4.53}$$

We call this formulation *master equation in operator form*. Note that we deal here with the spectral densities, $j_l(\omega_p^{(-q)})$, related to the classical random functions, $V_{l,q}(t)$. These are not the same as the spectral densities occurring in Eqs. (4.28) and (4.29), which contain the matrix elements of the full perturbation Hamiltonian. The matrix elements of the spin operators, present in the spectral densities in those equations, occur in Eq. (4.53) in the double commutator. In analogy with the spectral densities $\mathcal{J}(\omega)$, also the $j_l(\omega)$ are indeed complex quantities: $j_l(\omega) = J_l(\omega) + iL_l(\omega)$. The real part is important for relaxation and usually much larger than the imaginary part, which leads to dynamic frequency shifts, small shifts in line positions. The real part of the spectral density is an even function of frequency, $J_l(\omega) = J_l(-\omega)$, while the imaginary part is an odd function of frequency, $L_l(\omega) = -L_l(-\omega)$. In most of the discussions to follow, we will neglect the dynamic frequency shifts and use the symbol $J_l(\omega)$, which we simply call *spectral density*. The reader interested in the dynamic frequency shifts is advised to consult the reviews by Werbelow and London[10,11].

4.2.2 Equation of Motion of Expectation Values

The operator approach becomes really useful when we move from looking at the time derivative of the density operator to studying the time dependence of an expectation value of an operator corresponding to a measurable quantity. The relation between the time-dependent expectation value of an operator $\hat{Q}$, denoted $q(t)$, and the density operator was formulated already in Chapter 1 (Eq. (1.17)), stating that $q(t) = \langle\hat{Q}\rangle(t) = \mathrm{Tr}(\hat{\rho}(t)\hat{Q})$ (in the Schrödinger representation). The interaction representation counterpart reads:

$$\tilde{q}(t) = \mathrm{Tr}\left(\tilde{\hat{\rho}}(t)\hat{Q}\right) = \mathrm{Tr}\left(\hat{Q}\tilde{\hat{\rho}}(t)\right) \tag{4.54}$$

where we have applied one of many useful relations between traces of products of operators: $\mathrm{Tr}(\hat{A}\hat{B}) = \mathrm{Tr}(\hat{B}\hat{A})$. Note that Eq. (1.17) is equally valid in the laboratory frame and in the interaction frame. Thus, in order to obtain the expectation value of $\hat{Q}$ in the interaction frame, we express the density operator in that frame. Most of the time, the expectation value $\tilde{q}(t)$ can be interpreted as an expectation value in the rotating frame. Taking the time derivative of both sides of Eq. (4.54), we obtain

$$\frac{d}{dt}\tilde{q}(t) = Tr\left\{\hat{Q}\frac{d}{dt}\tilde{\hat{\rho}}(t)\right\} \tag{4.55}$$

where we have used the fact that the operator $\hat{Q}$ itself is not dependent on time. By substituting the master equation (4.53) into the right-hand side of Eq. (4.55), while neglecting the imaginary part of the spectral density, the following expression is obtained:

$$\frac{d}{dt}\tilde{q}(t) = -\frac{1}{2}\xi^2\mathrm{Tr}\left\{\sum_{q,p} J_l\left(-\omega_p^{(-q)}\right)\hat{Q}\left[\hat{A}_p^{(-q)},\left[\hat{A}_p^{\dagger(-q)},\tilde{\hat{\rho}}(t)\right]\right]\right\} \tag{4.56}$$

We proceed by making use of another relation between traces:

$$\mathrm{Tr}\left\{\hat{A}\left[\hat{B},\hat{C}\right]\right\}=\mathrm{Tr}\left\{\left[\hat{A},\hat{B}\right]\hat{C}\right\} \tag{4.57}$$

Applying this twice to Eq. (4.56), we obtain:

$$\begin{aligned}\frac{d}{dt}\tilde{q}(t)&=-\frac{1}{2}\xi^2\mathrm{Tr}\left\{\sum_{q,p}J_I\left(-\omega_p^{(-q)}\right)\left[\hat{Q},\hat{A}_p^{(-q)}\right]\left[\hat{A}_p^{\dagger(-q)},\tilde{\hat{\rho}}(t)\right]\right\}\\&=-\frac{1}{2}\xi^2\mathrm{Tr}\left\{\sum_{q,p}J_I\left(-\omega_p^{(-q)}\right)\left[\left[\hat{Q},\hat{A}_p^{(-q)}\right],\hat{A}_p^{\dagger(-q)}\right]\tilde{\hat{\rho}}(t)\right\}\\&=-\mathrm{Tr}\left\{\hat{P}\tilde{\hat{\rho}}(t)\right\}=-\tilde{p}(t)\end{aligned} \tag{4.58}$$

where the operator $\hat{P}$ (with a time-dependent expectation value in the rotating frame, $\tilde{p}(t)$) has been defined as:

$$\hat{P}=\frac{1}{2}\xi^2\sum_{q,p}J\left(-\omega_p^{(-q)}\right)\left[\left[\hat{Q},\hat{A}_p^{(-q)}\right],\hat{A}_p^{\dagger(-q)}\right] \tag{4.59}$$

Eq. (4.58) states that the time derivative of an expectation value of an operator corresponding to an observable is equal to an expectation value of another operator. As stated in Eq. (4.59), this other operator is a linear combination of double commutators $\left[\left[\hat{Q},\hat{A}_p^{(-q)}\right],\hat{A}_p^{\dagger(-q)}\right]$ with the spectral densities, $J_I(-\omega_p^{(-q)})$, as the coefficients.

Eq. (4.58) has the same deficiency as Eqs. (4.27) and (4.28): it requires the correction for finite temperature (compare Section 2.5). We define the expectation value of the operator $\hat{P}$ at thermal equilibrium at finite temperature as:

$$p^{eq}=\langle\hat{P}\rangle^{eq}=\mathrm{Tr}\left(\hat{P}\hat{\rho}^{eq}\right)=\mathrm{Tr}\left\{\frac{1}{2}\xi^2\sum_{q,p}J_I\left(-\omega_p^{(-q)}\right)\left[\left[\hat{Q},\hat{A}_p^{(-q)}\right],\hat{A}_p^{\dagger(-q)}\right]\hat{\rho}^{eq}\right\} \tag{4.60}$$

and modify Eq. (4.58) to read:

$$\frac{d}{dt}\tilde{q}(t)=-\left(\tilde{p}(t)-p^{eq}\right) \tag{4.61}$$

As an illustration of the formalism, let us go back to the specific case of the dipole-dipole interaction between unlike spins I and S (both with the spin quantum number ½). Let us consider $\hat{Q}=\hat{I}_z$. We need to compute double commutators $[[\hat{I}_z,\hat{A}_p^{(-q)}],\hat{A}_p^{\dagger(-q)}]$ for all possible combinations of p and q (these indices should not be confused with the time-dependent expectation values $p(t)$ and $q(t)$!) (*cf.* Table 4.2). We use the operator relations in Eq. (4.40b) and perform the calculation for the case of $q=2, p=1$, corresponding to the term in Eq. (4.59) with the coefficient $J(\omega_I+\omega_S)$:

$$\begin{aligned}\left[\left[\hat{I}_z,\frac{1}{2}\hat{I}_-\hat{S}_-\right],\frac{1}{2}\hat{I}_+\hat{S}_+\right]&=-\frac{1}{4}\left[\hat{I}_-\hat{S}_-,\hat{I}_+\hat{S}_+\right]=\frac{1}{4}\left[\hat{I}_+\hat{S}_+,\hat{I}_-\hat{S}_-\right]\\&=\frac{1}{2}\left(\hat{I}^2-\hat{I}_z^2\right)\hat{S}_z+\frac{1}{2}\left(\hat{S}^2-\hat{S}_z^2\right)\hat{I}_z\end{aligned} \tag{4.62}$$

Evaluating the expectation value of this expression, we note that $\langle \hat{I}^2 - \hat{I}_z^2 \rangle = \langle \hat{S}^2 - \hat{S}_z^2 \rangle = \frac{1}{2}$ and obtain

$$\left\langle \left[\left[\hat{I}_z, \frac{1}{2} \hat{I}_- \hat{S}_- \right], \frac{1}{2} \hat{I}_+ \hat{S}_+ \right] \right\rangle = \frac{1}{4} \left(\langle \hat{S}_z \rangle + \langle \hat{I}_z \rangle \right) \tag{4.63}$$

Eq. (4.63) tells us that the $q=2, p=1$ term in Eq. (4.59), or the spectral density taken at the sum of the two Larmor frequencies, $J(\omega_I + \omega_S)$, couples the time evolution of $\langle \hat{S}_z \rangle$ and $\langle \hat{I}_z \rangle$ (note that we have dropped the subscript $l=2$ on the symbol of spectral density). This is nothing new – the spectral density $J(\omega_I + \omega_S)$ occurs in the expression for W_2 (Eq. (3.29b)), which indeed appears in both the decay terms and the cross-relaxation terms in the Solomon equations, Eqs. (3.19)–(3.21). The reader is encouraged to use Table 4.2 and obtain full expressions for all the rate constants in the Solomon equations.

4.2.3 Transverse Relaxation

Let us now turn to the case of transverse relaxation. The relevant operators, which give rise to detectable NMR signals and whose time evolution we wish to study, are $\hat{I}_+$ and $\hat{S}_+$. If a large difference between the Larmor frequencies is assumed, explicit calculations of the double commutators in this case yield simple exponential relaxation for both expectation values, $\langle \hat{S}_+ \rangle$ and $\langle \hat{I}_+ \rangle$, and no (transverse) cross-relaxation. The transverse relaxation rate for the I-spin is given by Eq. (3.35) and the corresponding rate for the S-spin can be obtained by exchanging the roles of the two Larmor frequencies.

When the difference between the Larmor frequencies is small, for example, for two protons with only slightly different chemical shifts, transverse relaxation should be treated with care, as the secular approximation may not apply. This case is discussed in some detail in Goldman's book (*Further reading*). Here, we just present a very brief summary. Using the interaction representation with the average Larmor frequency, ω_0, in the unperturbed Hamiltonian, one obtains that the time evolution of the density operator consists of both an oscillation term and a damping. The evolution of the two relevant expectation values is described by:

$$\frac{d}{dt} \langle \hat{I}_+ \rangle = i\delta \langle \hat{I}_+ \rangle - \lambda \langle \hat{I}_+ \rangle - \mu \langle \hat{S}_+ \rangle \tag{4.64a}$$

$$\frac{d}{dt} \langle \hat{S}_+ \rangle = -i\delta \langle \hat{S}_+ \rangle - \lambda \langle \hat{S}_+ \rangle - \mu \langle \hat{I}_+ \rangle \tag{4.64b}$$

Here, $\pm\delta$ correspond to the deviations of the individual Larmor frequencies from the average frequency. The decay rate λ and the cross-relaxation rate μ are given by:

$$\lambda = \frac{\pi}{10} b_{IS}^2 \left[5J(0) + 9J(\omega_0) + 6J(2\omega_0) \right] \tag{4.65a}$$

which, in fact, is a limiting form of Eq. (3.35) when $\omega_I - \omega_S \to 0$, and:

$$\mu = \frac{\pi}{5} b_{IS}^2 \left[2J(0) + 3J(\omega_0) \right] \tag{4.65b}$$

The FID corresponding to the solution of Eq. (4.64) can be computed. When $\mu > \delta$, one obtains a FID corresponding to a coalesced signal, consisting of two Lorentzians with different widths and slightly different frequencies. When $\mu < \delta$, the spectrum (the Fourier transform of the FID) is a sum of two resolved resonances with a frequency separation somewhat smaller than 2δ. When $\mu \ll \delta$, the effect of the transverse cross-relaxation vanishes, the secular approximation holds and we recover the simple situation in the form of two well-separated and Lorentzian lines. Simulated lineshapes for the three cases are shown in Figure 4.3.

Before leaving this section, we wish to deal with a complication (with respect to the simple case considered by Solomon) caused by the presence of scalar coupling (*J*-coupling) between *I* and *S*. The unperturbed Hamiltonian needs to be modified to

$$\hat{H}_0 = \omega_I \hat{I}_z + \omega_S \hat{S}_z + 2\pi J_{IS} \hat{I}_z \hat{S}_z \tag{4.66}$$

where we have assumed that the spin-spin coupling constant ($2\pi J_{IS}$ in angular frequency units) is much smaller than the difference between the Larmor frequencies. This allows us to neglect the $\hat{I}_x\hat{S}_x$ and $\hat{I}_y\hat{S}_y$ terms in the coupling Hamiltonian. The corresponding spectrum now consists of four lines, in the form of two doublets with the splitting equal to $2\pi J_{IS}$, centred at the two Larmor frequencies. The relaxation-generating perturbation is still the dipole-dipole interaction. Under these conditions, the discussion of the longitudinal relaxation is essentially identical to the Solomon case. The relaxation of $\langle \hat{S}_z \rangle$ and $\langle \hat{I}_z \rangle$ is still described by a 2 × 2 relaxation matrix, with the same decay and cross-relaxation rate constants.

The four lines in the spectrum correspond to four detectable (-1)-quantum coherences (see the books of Ernst *et al.* and Levitt, *Further reading*). The four coherences can be thought of as four off-diagonal elements of the density operator in the eigenstate representation, but other representations are also possible. In order to apply the Redfield approach in the operator formulation, a convenient choice is to investigate the time evolution of four expectation values: $\langle \hat{S}_+ \rangle$, $\langle 2\hat{S}_+\hat{I}_z \rangle$, $\langle \hat{I}_+ \rangle$ and $\langle 2\hat{I}_+\hat{S}_z \rangle$. $\langle \hat{S}_+ \rangle$ and $\langle \hat{I}_+ \rangle$ correspond to in-phase doublets at the Larmor frequencies of the *S*- and *I*-spins, respectively, while the other two expectation values correspond to the anti-phase doublets (*cf.* Figure 4.4). The individual lines in the spectrum can be expressed as $\langle \hat{S}_+ \rangle \pm \langle 2\hat{S}_+\hat{I}_z \rangle$, and $\langle \hat{I}_+ \rangle \pm \langle 2\hat{I}_+\hat{S}_z \rangle$. Also, this case has been described in Goldman's book (*Further reading*). The result is that the in-phase coherences are not coupled to each other by relaxation, as in the case of vanishing scalar interaction. On the other hand, the individual lines

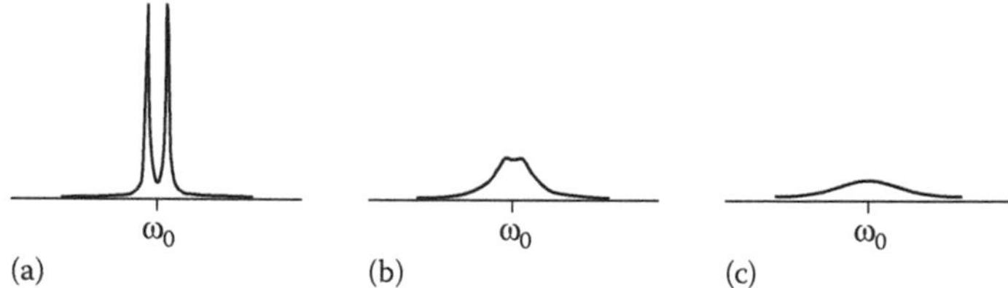

FIGURE 4.3 Simulated lineshapes, obtained from Eq. (4.64), corresponding to $\delta < \mu$ (a), $\mu < \delta$ (b), and $\mu \ll \delta$ (c).

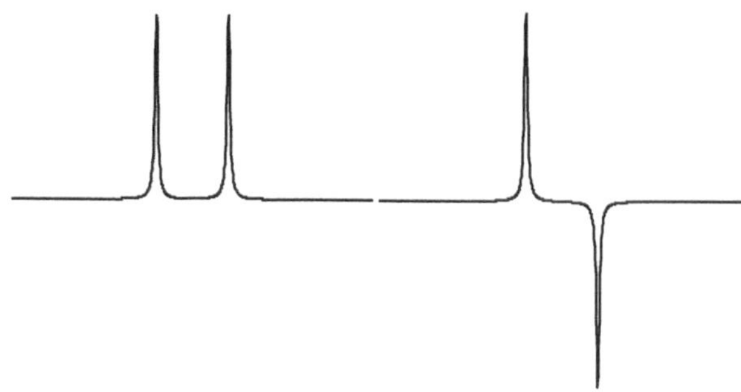

FIGURE 4.4 In-phase (left panel) and anti-phase (right panel) doublets.

within the doublets behave in a similar way as in the case, discussed above, of small chemical shift difference. The relaxation properties in a coupled two-spin system are also sensitive to the presence of anisotropic chemical shielding, another interaction leading to relaxation. We return to this issue in Section 5.2.

4.3 The Liouvillian Formulation of the WBR Theory

Some authors choose to formulate the WBR theory using the Liouville space formulation, as described, for example, by Jeener[12] (mentioned already in Chapter 1) or Mayne[13]. One of the advantages of this formulation is that it facilitates full use of symmetry properties of the spin systems[14–16]. We choose here to present a simplified description of the theory, similar to that in the book by Cavanagh *et al.* (*Further reading*). The readers interested in full detail are referred to the paper by Szymanski *et al.*[15] or the review by Mayne and Smith[17].

The Liouville space formulation of relaxation theory makes use of the Liouvillian, a superoperator counterpart of the Hamiltonian. As mentioned in Chapter 1, Eq. (1.29), the Liouvillian is the commutator with the Hamiltonian: $\hat{\hat{L}} = [\hat{H},]$. Commutators with Hamiltonians are abundant in the BWR theory and use of superoperators simplifies the notation. Making use of the superoperator formulation, Eq. (4.1) can be re-written as:

$$\frac{d\hat{\rho}(t)}{dt} = -i\left[\hat{H}_0 + \hat{H}_1(t),\hat{\rho}(t)\right] = -i\left(\hat{\hat{L}}_0 + \hat{\hat{L}}_1(t)\right)\hat{\rho}(t) \tag{4.67}$$

The evolution of the density operator under $\hat{H}_0$, given in Eq. (4.2) in the form of a "sandwich" of exponential operators, now takes the form:

$$\hat{\rho}(t) = \exp\left(i\hat{\hat{L}}_0 t\right)\hat{\rho}(0) \tag{4.68}$$

In an analogous way, the density operator in the interaction frame (given earlier by Eq. (4.4a)) is defined as:

$$\tilde{\hat{\rho}}(t) = \exp\left(i\hat{\hat{L}}_0 t\right)\hat{\rho}(t) \tag{4.69}$$

The differential equation for the density operator in the interaction frame, Eq. (4.7), becomes:

$$\frac{d\tilde{\hat{\rho}}(t)}{dt} = -i\tilde{\hat{\hat{L}}}_1(t)\tilde{\hat{\rho}}(t) \tag{4.70}$$

where the perturbation Liouvillian in the interaction frame is defined as:

$$\tilde{\hat{\hat{L}}}(t) = \exp\left(i\hat{\hat{L}}_0(t)\right)\hat{\hat{L}}_1(t) \tag{4.71}$$

Leaving out a few derivation steps, we can reformulate Eq. (4.14) as:

$$\tilde{\hat{\rho}}(t) = \tilde{\hat{\rho}}(0) - i\int_0^t \tilde{\hat{\hat{L}}}_1(t')\tilde{\hat{\rho}}(0)dt' - \int_0^t\int_0^{t'} \tilde{\hat{\hat{L}}}_1(t')\tilde{\hat{\hat{L}}}_1(t'')\tilde{\hat{\rho}}(0)dt''dt' \tag{4.72}$$

After introducing the simplifying assumptions specified immediately before Eq. (4.26), we obtain the Liouvillian analogue of this master equation:

$$\frac{d\tilde{\hat{\rho}}(t)}{dt} = -\int_0^\infty \overline{\tilde{\hat{\hat{L}}}(t)\tilde{\hat{\hat{L}}}_1(t+\tau)\tilde{\hat{\rho}}(t)d\tau} \tag{4.73}$$

From here, one can continue to formulate expressions for time derivatives of the elements of the density operators in terms of the relaxation superoperator/supermatrix or proceed according to Section 4.2. An important fact to remember is that the Liouville space formulation is indeed natural for the relaxation superoperator. Rather than working with the awkward four-indexed objects, the representation of the relaxation superoperator is here a normal matrix, with elements $R_{\mu\nu}$, where μ and ν stand for pairs of basis operators. As an example, we might set $\mu = \nu = |\alpha\rangle\langle\beta|$ and the $R_{\mu\nu}$ would then correspond to $R_{\alpha\beta\alpha\beta}$ or the transverse relaxation rate for the coherence connecting the pair of eigenstates (to the unperturbed Hamiltonian) α and β.

4.4 Relaxation in the Presence of Radiofrequency Fields

In many cases, it is useful to consider relaxation in the presence of radiofrequency fields. Several experiments are designed for such measurements and these are described in Chapters 8 through 10. We have already encountered the effects of radiofrequency fields on nuclear spin relaxation in this book in the context of the simplification of the Solomon equations through broadband decoupling of I-spins (often protons) when observing the S-spins (such as carbon-13 or nitrogen-15). The effect of the irradiation was in that case to replace the equilibrium state of a spin system by a steady state with the populations of the I-spins equalised over its energy levels. This led to single exponential S-spin relaxation and to the manifestation of cross-relaxation through the nuclear Overhauser enhancement. Here, we deal with the case where the irradiation more directly affects the observed spins themselves and their dynamics.

Qualitatively, the key features introduced by radiofrequency fields can be explained as follows. First, introducing an r.f. field can change the quantisation axis of nuclear spins. One can say that spins can be locked along the new quantisation axis and the effect is sometimes called *spin-locking*. The second effect is related to cross-relaxation processes. As mentioned on p. 67, cross-relaxation between two coherences characterised by different frequencies (with the frequency difference much larger than the linewidths) is inefficient, because the corresponding Redfield matrix elements are non-secular. On the contrary, cross-relaxation between the $\hat{I}_z$- and $\hat{S}_z$-operators in a two-spin system is secular (all populations and their combinations correspond to zero frequency) and efficient (see Figure 4.1). When a spin-locking radiofrequency field is introduced, the various magnetisation components become aligned along that field and do not precess in the rotating frame defined by the first r.f. field. From the point of view of the laboratory frame, all these components precess at the same frequency and the cross-relaxation between them becomes secular.

A quantitative discussion of relaxation in the presence of radiofrequency fields can be found in several reviews[6,18–23], and the following presentation is based on that work, in particular the paper by Goldman[6]. Consider a heteronuclear system of two spins, I and S, without scalar coupling in a static magnetic field B_0. The spins are subjected to a linearly polarised radiofrequency field, perpendicular to the static field, with the frequency ω in the vicinity of the Larmor frequency of the S-spins, and with the amplitude B_1 ($\omega_1 = -\gamma_S B_1$ in the angular velocity units). We assume that the radiofrequency field is weak compared with the static field, $B_1 \ll B_0$, but strong enough to fulfil the relation $R_{relax} \ll \omega_1$, where R_{relax} is an arbitrary relaxation matrix element. The linearly polarised r.f. field can be expressed as a sum of circularly polarised r.f. fields rotating in opposite directions with frequencies $+\omega$ and $-\omega$ (see, for example, Slichter, *Further reading*). One of the rotating components will be very far from $\omega_S = -\gamma_S B_0$ and it is sufficient to retain the one that is close to the Larmor frequency, let us say $+\omega$. In the same way, the rotating field is assumed not to interact with the I-spin.

The unperturbed Hamiltonian now contains the Zeeman interactions between both spins and the static field, as well as between the spin S and the rotating field:

$$\begin{aligned}\hat{H}_0 &= \omega_I \hat{I}_z + \omega_S \hat{S}_z + \omega_1\left(\hat{S}_x \cos\omega t + \hat{S}_y \sin\omega t\right)\\ &= \omega_I \hat{I}_z + \omega_S \hat{S}_z + \omega_1 \exp\left(-i\omega \hat{S}_z t\right)\hat{S}_x \exp\left(i\omega \hat{S}_z t\right)\\ &= \exp\left(-i\omega \hat{S}_z t\right)\left(\omega_I \hat{I}_z + \omega_S \hat{S}_z + \omega_1 \hat{S}_x\right)\exp\left(i\omega \hat{S}_z t\right)\end{aligned} \tag{4.74}$$

In the second line, we have expressed the rotation around the z-axis using exponential operators as previously (Section 4.1.1), and in the third line, we have exploited the fact that rotation around the z-axis leaves the operators $\hat{I}_z$ and $\hat{S}_z$ unchanged. In addition to the $\hat{H}_0$ term, we have the perturbation $\hat{H}_1(t)$, causing relaxation.

Following the procedure of Section 4.1.1, we want to transform all operators to the interaction representation. We choose to do it in two steps. In the first step, we transform to the frame rotating at the frequency ω around the z-direction of the static field, introduced in Section 1.3. This amounts to the transformation of operators according to

$$\hat{Q} \rightarrow \tilde{\hat{Q}} = \exp\left(i\omega \hat{S}_z t\right)\hat{Q}\exp\left(-i\omega \hat{S}_z t\right) \tag{4.75}$$

By a calculation similar to Eq. (4.5), we obtain the Liouville–von Neumann equation of motion for the density operator in this interaction representation:

$$\frac{d}{dt}\tilde{\hat{\rho}}(t) = i\left[\tilde{\hat{\rho}}(t),\, \hat{H}_{eff} + \omega_I \hat{I}_z\right] \tag{4.76}$$

with the *effective* S-spin Hamiltonian:

$$\hat{H}_{eff} = \left(\omega_S - \omega\right)\hat{S}_z + \omega_1 \hat{S}_x = \Delta \hat{S}_z + \omega_1 \hat{S}_x = \Omega \hat{S}_Z \tag{4.77}$$

In the second equality, we have introduced the offset frequency, $\Delta = \omega_S - \omega$, characterising the off-resonance condition of the applied radiofrequency. As indicated by the last part of Eq. (4.77), the effective Hamiltonian is a Zeeman Hamiltonian with a static field along the $0Z$ direction, which is tilted with respect to the laboratory z-axis by an angle Θ such that:

$$\tan\Theta = \frac{\omega_1}{\Delta} \tag{4.78a}$$

The relation between the laboratory frame and the tilted rotating frame is illustrated in Figure 4.5. The Larmor frequency in the new frame is:

$$\Omega = \left(\omega_1^2 + \Delta^2\right)^{1/2} \tag{4.78b}$$

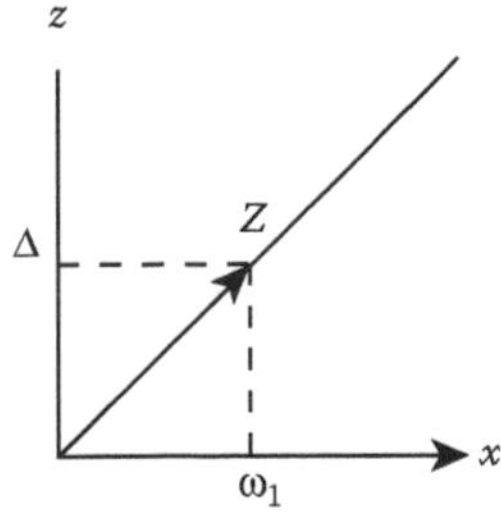

FIGURE 4.5 Illustration of the relation between the laboratory frame and the tilted rotating frame.

The perturbation Hamiltonian, in the form of the dipolar *IS* interaction, needs also to be transformed to the tilted rotating frame in the same way as we did with the density operator:

$$\tilde{\hat{H}}(t) = \exp(i\omega\hat{S}_z t)\hat{H}_1(t)\exp(-i\omega\hat{S}_z t) \tag{4.79}$$

In the second step, we go over to a *doubly rotating* frame by:

$$\tilde{\hat{Q}} \rightarrow \tilde{\tilde{Q}} = \exp\left(i\left(\Omega\hat{S}_Z + \omega_I\hat{I}_z\right)t\right)\tilde{\hat{Q}}\exp\left(-i\left(\Omega\hat{S}_Z + \omega_I\hat{I}_z\right)t\right) \tag{4.80}$$

The symbol $\tilde{\tilde{Q}}$ should contain, for the consistency of notation, also an operator "hat." This would, however, be awkward and we hope that the reader will not find the notation confusing. In this interaction representation, we can derive a master equation, analogous to Eq. (4.26):

$$\frac{d\tilde{\tilde{\rho}}}{dt} = -\int_0^\infty \left[\tilde{\tilde{H}}_1(t), \left[\tilde{\tilde{H}}_1(t+\tau), \tilde{\tilde{\rho}}(t)\right]\right]d\tau \tag{4.81}$$

Eq. (4.81), or its operator counterpart (*cf.* Goldman[6]), needs a correction for the finite temperature of the lattice, as discussed earlier. Using the perturbation Hamiltonian in the form of Eq. (4.37), the relaxation properties of the system can be evaluated in a reasonably straightforward way and the results can be found in the review by Desvaux and Berthault[22]. Here, we only discuss the results for some limiting cases.

The first case of interest is an experiment in which we perform on-resonance irradiation of the S-spin. In this case, $\Delta = 0$, $\Theta = \pi/2$ and $\Omega = \omega_1$. This condition leads to a situation where the expectation value of the Z-component of the S magnetisation, $\langle \hat{S}_Z \rangle$, decays exponentially with a time constant denoted $T_{1\rho}$ (which stands for T_1 in the rotating frame). The expression for the corresponding rate is:

$$T_{1\rho}^{-1} = \frac{\pi}{10} b_{IS}^2 \left[4J(\omega_1) + J(\omega_I - \omega_S) + 3J(\omega_S) + 6J(\omega_I) + 6J(\omega_I + \omega_S)\right] \tag{4.82}$$

which is almost identical to the expression for T_2^{-1}, given for the specific case of the carbon-13–proton system, in Eq. (3.35). The difference lies in the fact that the zero-frequency spectral density is replaced by the spectral density taken at the frequency ω_1. Since we assumed $B_1 \ll B_0$, the condition $|\omega_1| \ll |\omega_S|$ will be fulfilled, and the difference between $T_{1\rho}^{-1}$ and T_2^{-1} can be neglected in the absence of very slow motions. The difference can in fact be exploited to probe very slow motions and we shall return to this point in the discussion of experimental methods in Chapter 13. The close similarity of the two relaxation rates is a consequence of the fact that in the case of on-resonance irradiation, the quantisation axis for the I-spin is changed to the transverse plane. In other words, $\langle \hat{S}_Z \rangle$ becomes identical to the transverse magnetisation, $\langle \hat{S}_x \rangle$ or $\langle \hat{S}_y \rangle$.

A second interesting situation arises if we apply the radiofrequency off-resonance with $\Delta \neq 0$, $0 < \Theta < \pi/2$ and $\Omega > \omega_1$. Physically, this case corresponds to a measurement of the relaxation of a magnetisation component in between the longitudinal (with the decay rate T_1^{-1}, assuming that the I-spin is either rapidly relaxing or saturated) and the transverse magnetisation. The relaxation rate, $(T_{1\rho}^{off-res})^{-1}$, is in this case given by

$$\begin{aligned}\left(T_{1\rho}^{off-res}\right)^{-1} = \frac{\pi}{10} b_{IS}^2 \Big[&4\sin^2\Theta J(\omega_1) + \left(2 - \sin^2\Theta\right)J(\omega_I - \omega_S) \\ &+ \left(6 - 3\sin^2\Theta\right)J(\omega_S) + 6\sin^2\Theta J(\omega_I) + 6\left(2 - \sin^2\Theta\right)J(\omega_I + \omega_S)\Big]\end{aligned} \tag{4.83}$$

It is easy to see that Eq. (4.83) simplifies to Eq. (4.82) if $\Theta = \pi/2$ and to Eq. (3.31a) when $\Theta = 0$ (the reader is encouraged to check this statement).

The final limiting case is that of a homonuclear *IS* spin system, with the two spins (*e.g.* two protons) having different chemical shifts but the radiofrequency field strong enough to fulfil $\Delta \ll \omega_1$ and $\Theta \approx \pi/2$ for both spins, *i.e.* to change the quantisation axis of both spins to the transverse plane of the laboratory frame. The terms $\omega_I \hat{I}_z$ in Eq. (4.76) is then included in the effective Hamiltonian, which becomes:

$$\hat{H}_{eff} = (\omega_I - \omega)\hat{I}_z + (\omega_S - \omega)\hat{S}_z + \omega_I(\hat{I}_x + \hat{S}_x) = \Omega_I \hat{I}_Z + \Omega_S \hat{S}_Z \tag{4.84}$$

where the Larmor frequencies for the two spins are allowed to be different, both in the laboratory frame and in the common tilted rotating frame, with its unique *Z*-axis. The consequence of both spins being locked along the common axis in the rotating frame is that the cross-relaxation between the two *Z*-components becomes secular. In other words, the relaxation of the two *Z*-components (which indeed are transverse components in the laboratory frame) becomes a coupled process, following an equation system similar to the Solomon case. Bothner-By[23] gave the following equations for an isolated spin pair:

$$-\frac{d}{dt}\begin{pmatrix} \langle \hat{I}_Z \rangle \\ \langle \hat{S}_Z \rangle \end{pmatrix} = \begin{pmatrix} R_{2I} & \sigma_{IS}^{ROE} \\ \sigma_{IS}^{ROE} & R_{2S} \end{pmatrix} \begin{pmatrix} \langle \hat{I}_Z \rangle \\ \langle \hat{S}_Z \rangle \end{pmatrix} \tag{4.85}$$

with

$$R_{2I} = R_{2S} = \frac{\pi}{10} b_{IS}^2 \left[5J(0) + 9J(\omega_I) + 6J(2\omega_I) \right] \tag{4.86a}$$

and

$$\sigma_{IS}^{ROE} = \frac{\pi}{10} b_{IS}^2 \left[4J(0) + 6J(\omega_I) \right] \tag{4.86b}$$

The transverse relaxation rates, R_{2I} and R_{2S}, are identical to Eq. (4.82) at the limit of vanishing frequency difference between the *I*- and *S*-spins and ω_1 approximated by zero. They are also identical to the relaxation rate, λ, introduced in the context of transverse relaxation in systems with small chemical shift differences. The transverse cross-relaxation rate, σ_{IS}^{ROE}, is the same as μ, introduced in Eq. (4.65b) and it is an important quantity. As opposed to the longitudinal cross-relaxation rate, *cf.* Eqs. (3.22c) and (3.29), which can be both positive and negative and crosses zero at a certain value of the correlation time, the transverse rate is always positive. This has important consequences for determining which nuclei are spatially close to each other, and thus determination of structure, in medium-size molecules, which often fall in the regime of having close-to-zero longitudinal cross-relaxation rates. We shall return to the experimental procedures for determining the transverse cross-relaxation rates in Chapter 9.

Exercises for Chapter 4

1. Show that the relation in Eq. (4.40a) is fulfilled by the operators given in Eqs. (4.41a) and (4.41b).
2. Use Table 4.2 and derive all the rate constants in the Solomon equation (Eq. (3.22)).
3. Prove that Eq. (4.83) simplifies to Eq. (4.82) if $\Theta = \pi/2$ and to Eq. (3.31a) when $\Theta = 0$.

References

1. Wangsness, R. K.; and Bloch, F., The dynamical theory of nuclear induction. *Phys. Rev.* 1953, 89, 728–739.
2. Redfield, A. G., The theory of relaxation processes. *Adv. Magn. Reson.* 1965, 1, 1–32.
3. Redfield, A. G., *Relaxation Theory: Density Matrix Formulation.* Wiley: eMagRes, 2007.

4. Hubbard, P. S., Quantum-mechanical and semiclassical forms of the density operator theory of relaxation. *Rev. Mod. Phys.* 1961, 33, 249–264.
5. Werbelow, L. G.; and Kowalewski, J., Nuclear spin relaxation of spin one-half nuclei in the presence of neighboring higher-spin nuclei. *J. Chem. Phys.* 1997, 107, 2775–2781.
6. Goldman, M., Formal theory of spin-lattice relaxation. *J. Magn. Reson.* 2001, 149, 160–187.
7. Murali, N.; and Krishnan, V., A primer for nuclear magnetic relaxation in liquids. *Concepts Magn. Reson. Part A.* 2003, 17A, 86–116.
8. Mueller, L. J., Tensors and rotations in NMR. *Concepts Magn. Reson. Part A.* 2011, 38A, 221–235.
9. Hubbard, P. S., Some properties of correlation functions of irreducible tensor operators. *Phys. Rev.* 1969, 180, 319–326.
10. Werbelow, L.; and London, R. E., Dynamic frequency shift. *Concepts Magn. Reson.* 1996, 8, 325–338.
11. Werbelow, L. G., *Dynamic Frequency Shift*. Wiley: eMagRes, 2011.
12. Jeener, J., Superoperators in magnetic resonance. *Adv. Magn. Reson.* 1982, 10, 1–51.
13. Mayne, C. L., *Liouville Equation of Motion*. Wiley: eMagRes, 2007.
14. Pyper, N. C., Theory of symmetry in nuclear magnetic relaxation including applications to high resolution NMR line shapes. *Mol. Phys.* 1971, 21, 1–33.
15. Szymanski, S.; Gryff-Keller, A. M.; and Binsch, G., A Liouville space formulation of Wangsness-Bloch-Redfield theory of nuclear spin relaxation suitable for machine computation. I. Fundamental aspects. *J. Magn. Reson.* 1986, 68, 399–432.
16. Chang, Z. W.; and Halle, B., Longitudinal relaxation in dipole-coupled homonuclear three-spin systems: Distinct correlations and odd spectral densities. *J. Chem. Phys.* 2015, 143, 234201.
17. Mayne, C. L.; and Smith, S. A., *Relaxation Processes in Coupled-Spin Systems*. Wiley: eMagRes, 2007.
18. Bull, T. E., Relaxation in the rotating frame in liquids. *Prog. NMR Spectr.* 1992, 24, 377–410.
19. Bax, A.; and Grzesiek, S., *ROESY*. Wiley: eMagRes, 2007.
20. Schleich, T., *Rotating Frame Spin-Lattice Relaxation Off-Resonance*. Wiley: eMagRes, 2007.
21. Palke, W. E.; and Gerig, J. T., Relaxation in the presence of an rf field. *Concepts Magn. Reson.* 1997, 9, 347–353.
22. Desvaux, H.; and Berthault, P., Study of dynamic processes in liquids using off-resonance rf irradiation. *Progr. NMR Spectrosc.* 1999, 35, 295–340.
23. Bothner-By, A. A.; Stephens, R. L.; and Lee, J. M., Structure determination of a tetrasaccharide: Transient nuclear Overhauser effects in the rotating frame. *J. Am. Chem. Soc.* 1984, 106, 811–813.

5

Applications of Redfield Theory to Systems of Spin 1/2 Nuclei

In this chapter, we present applications and results of the Redfield theory. We restrict ourselves to systems of spin 1/2 nuclei; nuclei with spin 1 or higher (quadrupolar nuclei) will be the subject of Chapter 14. So far, the discussion has only been concerned with dipolar relaxation involving two coupled spins. There are, however, also other interactions that vary with time through stochastic processes related to molecular motion. In particular, relaxation through the anisotropic chemical shift is important in many spin 1/2 systems. We begin this chapter by looking at some features of dipolar relaxation in systems consisting of more than two spins and proceed to deal with relaxation effects originating from other interactions involving spin 1/2 nuclei.

5.1 Multispin Dipolar Relaxation: Magnetisation Modes

An interesting complication of the Solomon-like description arises in a spin system consisting of more than two spins (all with the spin quantum number of 1/2) with mutual dipolar interactions. The unperturbed Hamiltonian consists of the Zeeman interactions between all the spins and the magnetic field and, possibly, scalar spin-spin couplings. The perturbation is a sum of dipole-dipole (DD) interactions. The summation over η, running over all pairs of dipolar nuclei, leads to a generalisation of Eq. (4.37a):

$$\hat{H}_1(t) = \sum_{\eta} \hat{H}_{\eta}(t) = \sum_{\eta} \xi_{\eta} \sum_{q=-l}^{l} (-1)^q V_{l,q}^{\eta}(t) \hat{T}_{l,-q}^{\eta} \tag{5.1}$$

We can use this expression for the dipole-dipole Hamiltonian to compute the spectral density $\mathcal{J}_{\alpha\alpha'\beta\beta'}(\omega)$, using the formulation given in Section 4.1 by substituting this form of the perturbation Hamiltonian into Eq. (4.29). The spectral density thus becomes:

$$\begin{aligned}
\mathcal{J}_{\alpha\alpha'\beta\beta'}(\omega) &= 2\int_0^{\infty} \overline{\langle \alpha | \hat{H}_1(t) | \alpha' \rangle \langle \beta' | \hat{H}_1(t+\tau) | \beta \rangle} \exp(-i\omega t) d\tau \\
&= 2\sum_{\eta}\sum_{\eta'} \xi_{\eta}\xi_{\eta'} \sum_{q}\sum_{q'} (-1)^{q+q'} \langle \alpha | \hat{T}_{l,-q}^{\eta} | \alpha' \rangle \langle \beta' | \hat{T}_{l,-q'}^{\dagger\eta'} | \beta \rangle \\
&\quad \times \int_0^{\infty} \langle V_{l,q}^{\eta}(t) V_{l,q'}^{*\eta'}(t+\tau) \rangle \exp(-i\omega\tau) d\tau \\
&= \sum_{\eta}\sum_{\eta'} \xi_{\eta}\xi_{\eta'} \sum_{q} \langle \alpha | \hat{T}_{l,-q}^{\eta} | \alpha' \rangle \langle \beta' | \hat{T}_{l,-q'}^{\dagger\eta'} | \beta \rangle J_l^{\eta\eta'}(\omega)
\end{aligned} \tag{5.2}$$

Since this Hamiltonian now contains several terms, and the summation runs over all possible η, η' indices, we obtain "cross-terms" between different components of the Hamiltonian. The quantity $G_{l,q,q'}^{\eta\eta'}(\tau)=\langle V_{l,q}^{\eta}(t)V_{l,q'}^{*\eta'}(t+\tau)\rangle$ for $\eta=\eta'$ is our familiar time-autocorrelation function, while its counterpart for $\eta\neq\eta'$ is called the *cross-correlation function.* It describes the correlation between a classical random function contained in the dipolar Hamiltonian $\hat{H}_{\eta}(t)$ and its counterpart related to $\hat{H}_{\eta'}(t+\tau)$. The Fourier transform of $G_{l,q,q'}^{\eta\eta'}(\tau)$, denoted $J_l^{\eta\eta'}(\omega)$, is called the *cross-correlation spectral density.* In the last equality in Eq. (5.2), we have used the fact that the symmetry relation of Eq. (4.52) also applies to the cross-correlation functions.

5.1.1 Magnetisation Modes

The cross-correlation spectral densities do influence relaxation properties in multispin systems, at least under certain conditions. These effects have been subject of reviews by Kumar and co-workers[1] and by other authors[2,3]. Eq. (4.58) or (4.61) can be generalised to accommodate the more complicated $\hat{H}_1(t)$ of Eq. (5.1), but this becomes rather messy. Therefore, we are going to take a different approach. We shall examine only cross-correlation effects in longitudinal relaxation, since this is a bit easier than transverse relaxation. We are going to introduce the multispin effects in longitudinal relaxation (evolution of diagonal elements of the density matrix) using the concept of *magnetisation modes.* This approach to describe multispin relaxation phenomena has been discussed extensively by Werbelow and Grant[4] and Canet[5]. We use the expressions for the time evolution of the density matrix element, Eq. (4.30a) or (4.33), as a starting point. The elements of the relaxation supermatrix were obtained in the previous chapter as linear combinations of the spectral density functions at different frequencies (Eq. (4.28)). For a multispin system, we need to use this formulation in combination with Eq. (5.2). We then introduce a new operator basis, in the form of linear combinations of the $|\alpha\rangle\langle\alpha|$ operators, according to:

$$\hat{v}_i=\sum_{\alpha}Q_{i,\alpha}|\alpha\rangle\langle\alpha| \tag{5.3}$$

The matrix Q with elements $Q_{i,\alpha}$ describes the transformation between the two basis sets. The density operator can be expanded in the new basis, according to:

$$\hat{\rho}=\sum_{i}c_i\hat{v}_i=\sum_{i}\sum_{\alpha}Q_{i,\alpha}\rho_{\alpha\alpha}\hat{v}_i \tag{5.4a}$$

i.e. the expansion coefficients, c_i, for the density operator in the new basis can be obtained through

$$c_i=\sum_{\alpha}Q_{i,\alpha}\rho_{\alpha\alpha} \tag{5.4b}$$

Using this basis set and applying the correction for finite temperature, Eq. (4.30a) or (4.33) can be written as:

$$\frac{d}{dt}\langle\hat{v}_i\rangle(t)=-\sum_{j}\Gamma_{ij}\left(\langle\hat{v}_j\rangle(t)-\langle\hat{v}_j\rangle^{eq}\right) \tag{5.5}$$

where Γ_{ij} are the relaxation supermatrix elements in the new basis:

$$-\Gamma_{ij}=\sum_{\alpha,\beta}Q_{i,\alpha}R_{\alpha\alpha\beta\beta}Q_{j,\beta} \tag{5.6}$$

We shall examine one example of this kind of analysis, following the presentation by Werbelow and Grant[4]. In the example, the system consists of three different spin 1/2 nuclei, denoted AMX in the common notation. We assume that the three spins are coupled through mutual scalar couplings, denoted by J_{AM}, J_{AX} and J_{MX}, and that there are no degeneracies, in the sense that each line in the spectrum corresponds to a single transition between eigenstates of $\hat{H}_0$. In addition, we assume that all the coupling constants are much smaller than the differences in chemical shifts, the so-called *weak coupling* limit. The nuclear magnetic resonance (NMR) spectrum of such a spin system consists of three groups of signals, each group being a symmetric doublet of doublets (*cf.* Figure 5.1). We begin the analysis by setting up a new, useful form of the vector basis and then discuss its properties and its relation to the spectrum in Figure 5.1.

The unperturbed Hamiltonian can be expressed in analogy with Eq. (4.66) by adding one more Zeeman term and two more scalar coupling terms. The eigenvectors are characterised by the three quantum numbers $|m_A m_M m_X\rangle$:

$$\begin{aligned} &|1\rangle \equiv |+++\rangle,\ |2\rangle \equiv |++-\rangle,\ |3\rangle \equiv |+-+\rangle,\ |4\rangle \equiv |-++\rangle, \\ &|5\rangle \equiv |+--\rangle,\ |6\rangle \equiv |-+-\rangle,\ |7\rangle \equiv |--+\rangle,\ |8\rangle \equiv |---\rangle \end{aligned} \tag{5.7}$$

where the shorthand notation $|+++\rangle$ implies $m_A = +1/2$, $m_M = +1/2$, $m_X = +1/2$. We perform the transformation of Eq. (5.3) or (5.4) in two steps. It is useful to combine the ket-bra products in such a way that the resulting operators have definite symmetry properties with respect to spin-inversion:

$$\begin{bmatrix} {}^{a}\hat{v}'_1 \\ {}^{a}\hat{v}'_2 \\ {}^{a}\hat{v}'_3 \\ {}^{a}\hat{v}'_4 \\ {}^{s}\hat{v}'_5 \\ {}^{s}\hat{v}'_6 \\ {}^{s}\hat{v}'_7 \\ {}^{s}\hat{v}'_8 \end{bmatrix} = \begin{bmatrix} |1\rangle\langle 1| - |8\rangle\langle 8| \\ |2\rangle\langle 2| - |7\rangle\langle 7| \\ |3\rangle\langle 3| - |6\rangle\langle 6| \\ |4\rangle\langle 4| - |5\rangle\langle 5| \\ |1\rangle\langle 1| + |8\rangle\langle 8| \\ |2\rangle\langle 2| + |7\rangle\langle 7| \\ |3\rangle\langle 3| + |6\rangle\langle 6| \\ |4\rangle\langle 4| + |5\rangle\langle 5| \end{bmatrix} \tag{5.8}$$

Spin-inversion refers to changing the sign of all the magnetic quantum numbers. Clearly, the first four of the new vectors are antisymmetric with respect to spin-inversion, *i.e.* they change sign as a result of this operation, while the last four vectors are symmetric. We indicate these symmetry properties by the superscripts *a* and *s*. In the second step, we perform a transformation of the vectors v' into v in such a way that the spin-inversion symmetry is retained:

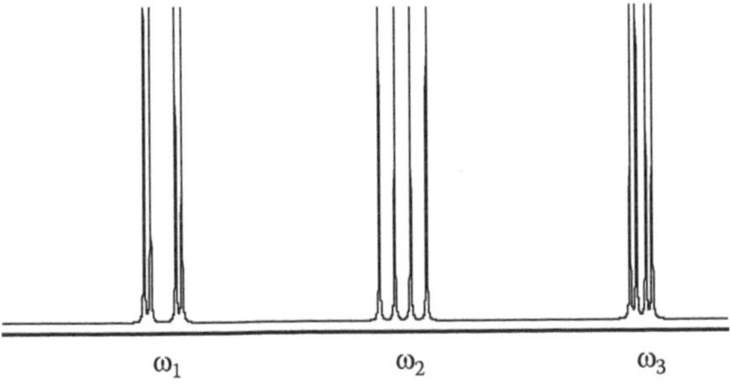

FIGURE 5.1 The AMX spectrum. The spectrum corresponds to the case $J_{12} > J_{23} > J_{13}$.

$$\begin{bmatrix} {}^a\hat{v}_1 \\ {}^a\hat{v}_2 \\ {}^a\hat{v}_3 \\ {}^a\hat{v}_4 \\ {}^s\hat{v}_1 \\ {}^s\hat{v}_2 \\ {}^s\hat{v}_3 \\ {}^s\hat{v}_4 \end{bmatrix} = \frac{1}{2}\begin{bmatrix} 1 & 1 & 1 & -1 & 0 & 0 & 0 & 0 \\ 1 & 1 & -1 & 1 & 0 & 0 & 0 & 0 \\ 1 & -1 & 1 & 1 & 0 & 0 & 0 & 0 \\ 1 & -1 & -1 & -1 & 0 & 0 & 0 & 0 \\ 0 & 0 & 0 & 0 & 1 & 1 & -1 & -1 \\ 0 & 0 & 0 & 0 & 1 & -1 & 1 & -1 \\ 0 & 0 & 0 & 0 & 1 & -1 & -1 & 1 \\ 0 & 0 & 0 & 0 & 1 & 1 & 1 & 1 \end{bmatrix}\begin{bmatrix} {}^a\hat{v}'_1 \\ {}^a\hat{v}'_2 \\ {}^a\hat{v}'_3 \\ {}^a\hat{v}'_4 \\ {}^s\hat{v}'_5 \\ {}^s\hat{v}'_6 \\ {}^s\hat{v}'_7 \\ {}^s\hat{v}'_8 \end{bmatrix} \tag{5.9}$$

This transformation has the advantage that all the elements of v have a straightforward physical interpretation. The magnetisation modes are the expectation values of the new operator set. The first three magnetisation modes simply correspond to the expectation values of the $\hat{I}_z$ operators for the three spins:

$$^a v_1(t) = \mathrm{Tr}\left[\hat{\rho}(t)\hat{I}_z^A\right] \tag{5.10a}$$

$$^a v_2(t) = \mathrm{Tr}\left[\hat{\rho}(t)\hat{I}_z^M\right] \tag{5.10b}$$

$$^a v_3(t) = \mathrm{Tr}\left[\hat{\rho}(t)\hat{I}_z^X\right] \tag{5.10c}$$

Note that the expectation values are explicitly time dependent, while the operators $\hat{v}_i$ are not. The reader is advised to check that the relations (5.10) indeed are fulfilled. Physically, the magnetisation mode $^a v_1(t)$ represents (after subtraction of the corresponding equilibrium value) the summed deviations from thermal equilibrium of the four lines in the multiplet centred at the resonance frequency of the A-spins. The modes $^a v_2(t)$ and $^a v_3(t)$ have the same meaning for the multiplets at the resonance frequency of M and X, respectively. The fourth element, $^a v_4(t)$, can also be related to a trace of an operator expression:

$$^a v_4(t) = \mathrm{Tr}\left[\hat{\rho}(t)4\hat{I}_z^A\hat{I}_z^M\hat{I}_z^X\right] \tag{5.11}$$

The relation between the product operator $4\hat{I}_z^A\hat{I}_z^M\hat{I}_z^X$ and the spectrum is less apparent. Still, its expectation value is again a measurable quantity: the deviation from thermal equilibrium for the sum of the outer lines minus the sum of the inner lines (or *vice versa*, depending on the signs of the scalar coupling constants) for each of the multiplets. We can say that $^a v_4(t)$ describes one kind of *multiplet asymmetry*, the difference in the evolution of the individual lines in a multiplet split by the scalar coupling.

The first three symmetric magnetisation modes, $^s v_1(t)$, $^s v_2(t)$, $^s v_3(t)$, can also be related to the expectation values of product operators:

$$^s v_1(t) = \mathrm{Tr}\left[\hat{\rho}(t)2\hat{I}_z^A\hat{I}_z^M\right] \tag{5.12a}$$

$$^s v_2(t) = \mathrm{Tr}\left[\hat{\rho}(t)2\hat{I}_z^A\hat{I}_z^X\right] \tag{5.12b}$$

$$^s v_3(t) = \mathrm{Tr}\left[\hat{\rho}(t)2\hat{I}_z^M\hat{I}_z^X\right] \tag{5.12c}$$

and correspond to other multiplet asymmetries in the AMX spectrum. The last symmetric element, ${}^{s}v_4(t) = \mathrm{Tr}[\hat{\rho}(t)\frac{1}{2}\hat{E}]$, corresponds to the trace of the density operator multiplied with the identity operator, or to the sum of the populations of all spin states, which is an invariant quantity.

5.1.2 Relaxation Matrix for Magnetisation Modes

Having defined the magnetisation modes and their expectation values, we now turn to the relaxation matrix in the magnetisation mode basis, Γ, assuming that the three dipolar interactions (AM, AX, MX) are the only mechanism of relaxation. The calculation of the relaxation matrix elements is tedious and we shall limit ourselves here to providing a brief, general discussion of their properties and to defining some selected, important expressions. For details, the reader is referred to the review of Werbelow and Grant[4] and the original references quoted there. The first important property of the relaxation matrix is that if only the dipolar interactions are considered, then the antisymmetric and the symmetric manifolds are not connected by the relaxation matrix. We can thus write:

$$\frac{d}{dt}\begin{pmatrix} {}^{a}v(t) \\ {}^{s}v(t) \end{pmatrix} = -\begin{pmatrix} {}^{a}\Gamma & 0 \\ 0 & {}^{s}\Gamma \end{pmatrix}\begin{pmatrix} {}^{a}v(t) - {}^{a}v^{eq} \\ {}^{s}v(t) - {}^{s}v^{eq} \end{pmatrix} \tag{5.13}$$

The second important property is that the relaxation matrix itself is a symmetric matrix, $\Gamma_{ij} = \Gamma_{ji}$.

The diagonal elements, Γ_{ii}, corresponding to the three first antisymmetric magnetisation modes are important. These are obtained from a straightforward generalisation of the expression for $\rho_S = T_1^{-1}$ in Eq. (3.31a). In this basis, ${}^{a}\Gamma_{11}$, for example, becomes:

$${}^{a}\Gamma_{11} = \rho_{AM} + \rho_{AX} \tag{5.14a}$$

with

$$\rho_{AM} = \frac{\pi}{5} b_{AM}^2 \left[J_{AM,AM}(\omega_A - \omega_M) + 3J_{AM,AM}(\omega_A) + 6J_{AM,AM}(\omega_A + \omega_M) \right] \tag{5.14b}$$

The subscript *AM,AM* on the spectral density explicitly indicates the autocorrelation of the *AM* dipolar interaction. Analogous expressions for ${}^{a}\Gamma_{22}$, ${}^{a}\Gamma_{33}$ can readily be obtained *mutatis mutandis*. Likewise, the off-diagonal elements ${}^{a}\Gamma_{12} = \sigma_{AM}$, ${}^{a}\Gamma_{13} = \sigma_{AX}$, ${}^{a}\Gamma_{23} = \sigma_{MX}$ (and their symmetry-related counterparts) are cross-relaxation rates, fully analogous to the expressions given in Eqs. (3.22c) and (3.29). These terms couple together the magnetisation modes given by the expectation values of the $\hat{I}_z$ operators. The elements ${}^{a}\Gamma_{14}$, ${}^{a}\Gamma_{24}$, ${}^{a}\Gamma_{34}$ are qualitatively different and interesting. As opposed to the other elements in the antisymmetric part of the relaxation matrix, which were seen to contain only the autocorrelated spectral densities, the matrix elements ${}^{a}\Gamma_{14}$, ${}^{a}\Gamma_{24}$, ${}^{a}\Gamma_{34}$, coupling together the $\hat{I}_z$ terms and the $4\hat{I}_z^A\hat{I}_z^M\hat{I}_z^X$, only depend on the cross-correlated spectral densities:

$${}^{a}\Gamma_{14} = \delta_{MAX} = \frac{6\pi}{5} b_{AM} b_{AX} J_{AM,AX}(\omega_A) = \frac{6\pi}{5} b_{AM} b_{AX} K_{MAX}(\omega_A) \tag{5.15}$$

For that reason, we are going to call these relaxation matrix elements the *cross-correlated relaxation rates* (CCRRs), which is the standard nomenclature. The notation *cross-correlated cross-relaxation rates* might be more appropriate but is judged a bit too clumsy. The cross-correlated dipolar spectral densities are denoted $J_{AM,AX}(\omega)$ or $K_{MAX}(\omega)$; both symbols are commonly used in the literature. We shall return to calculations of the cross-correlation functions and corresponding spectral densities in Chapter 6. The matrix elements for the part of the relaxation matrix corresponding to the symmetric modes are given as combinations of various ρ, σ and δ terms; *cf.* Werbelow and Grant.[4]

Thus, we find that the longitudinal relaxation in AMX systems is in general multiexponential, which indeed is similar to the two-spin case described by the Solomon equations. Starting with the Solomon equations, we were able to recover a single, well-defined T_1 under some limiting conditions. One was the case of *I* and *S* being identical spins. The three-spin counterparts of this situation, the A_3 and AX_2 spin systems, are more complicated. The A_3 system was treated by Hubbard as early as 1958[6] and both these systems, along with several other cases, were discussed in large detail by Werbelow and Grant[4]. We refer the reader to that work for a comprehensive presentation.

Another limiting situation, as discussed in Chapter 3, resulting in a single exponential longitudinal relaxation in a two-spin system is obtained by applying the experimental trick of spin decoupling, *i.e.* saturation of all transitions for one of the spins, *I*. The analogue of this experiment can easily be realised in a heteronuclear three-spin system, where the A-spin represents another nuclear species than M and X. A common case of experimental interest occurs when A = ^{13}C, while M and X are non-equivalent protons. As discussed by Werbelow and Grant[4] and by Brondeau and co-workers[7], the broadband decoupling of protons results in a single exponential carbon relaxation. The ^{13}C T_1^{-1} is given by Eq. (5.14), which implies that the relaxation contributions from the individual carbon-proton dipolar interactions are additive. The phenomenon of the nuclear Overhauser enhancement (NOE) arises also in the form of a simple linear combination of two separate enhancements characteristic of the two-spin system. In the simple case of a methylene carbon, carrying two magnetically non-equivalent protons with identical CH distances, the relaxation rate expression in Eq. (5.14a) gives the result that the equation for longitudinal relaxation in Eq. (3.31a) has only to be multiplied by 2 to be valid. Eq. (3.31b) for the NOE retains its validity also for this case. Similar results are obtained for a carbon-13 nucleus interacting with more complicated proton spin systems as long as the protons are non-equivalent. In cases where magnetic equivalence is present, the situation becomes more complicated and the cross-correlation effects may not be completely negligible, even under broadband irradiation of the proton spins. The reader is referred to Werbelow and Grant[4] for a general discussion and to Kowalewski *et al.*[8] for a careful case study.

Summarising the longitudinal relaxation in the AMX case, we have found that one should exercise a certain degree of care when selecting the basis set in which the relaxation matrix is to be evaluated. The set of magnetisation modes is often a good choice. Using this formulation, we recover a three-spin analogue of the Solomon equations if we can neglect the cross-correlation effects between pairs of dipolar interactions. We shall return to the experimental demonstration of the role of dipolar cross-correlations in Chapter 10. Spin decoupling in a heteronuclear spin system can simplify the picture, but caution is recommended in case of magnetic equivalence.

Before finishing this section, we would like to mention an interesting recent development of the relaxation theory for the system of three homonuclear and dipole-coupled spin 1/2 nuclei in an arbitrary geometry presented by Chang and Halle[9]. They used the Liouville-space formulation of the BWR theory and a basis set consisting of the irreducible spherical tensor operators for three spins (compare Subsection 4.2.1). One can see this approach as an extension of the early work by Hubbard[6]. For an A_3 system of three equivalent spin 1/2 nuclei positioned in the corners of an equilateral triangle and under motional conditions that do not break the equivalence of the three spins, the results of Hubbard[6] and of Werbelow and Grant[4] are reproduced. Under more general conditions with lower geometrical symmetry (allowing for two or three different dipolar couplings) and anisotropic motions, it is demonstrated that the longitudinal relaxation is described by a 10 × 10 relaxation matrix with the basis operators corresponding to the three magnetisations and seven zero-quantum coherences. We note in passing that the three-spin longitudinal order introduced in Eq. (5.11) can be treated as a zero-quantum coherence. An important new finding by Chang and Halle is that the odd-valued spectral density functions, $L_l(\omega) = -L_l(-\omega)$ (compare the discussion following Eq. (4.53)) can actually under certain conditions influence longitudinal relaxation in three-spin systems. Finally, the authors also investigated the case of non-isochronous (*i.e.* characterised by different chemical shifts) homonuclear three-spin systems and found that the shift differences could result in a non-monotonous dependence of effective relaxation rates on the B_0 magnetic field.

5.2 Anisotropic Chemical Shielding and Its Relaxation Effects

The dipole-dipole interaction is often the most important source of spin relaxation for $I=1/2$ spins, but not the only one. Another important interaction is the anisotropic chemical shift or the chemical shielding anisotropy (CSA). The CSA as a relaxation mechanism has been discussed thoroughly in several reviews, *e.g.* by Smith *et al.*[10–12] and Anet and O'Leary[13,14].

5.2.1 The Chemical Shielding Tensor

In the presence of chemical shielding, the local magnetic field at the nucleus is not equal to the external field of the NMR magnet but is given by:

$$\mathbf{B}_{loc} = \mathbf{B}_0(1-\sigma) = \mathbf{B}_0 - \mathbf{B}_{ind} \tag{5.16}$$

with the induced field $\mathbf{B}_{ind}=\sigma\mathbf{B}_0$. The shielding is in general anisotropic, meaning that it is orientation dependent. This means that for a powder sample, the shielding experienced by different crystallites in the sample is different, depending on their orientation with respect to the magnetic field. The chemical shielding is described as a tensor property of rank two and the symbol σ denotes this shielding tensor. We can express the shielding tensor in a Cartesian form:

$$\sigma = \begin{bmatrix} \sigma_{xx} & \sigma_{xy} & \sigma_{xz} \\ \sigma_{yx} & \sigma_{yy} & \sigma_{yz} \\ \sigma_{zx} & \sigma_{zy} & \sigma_{zz} \end{bmatrix} \tag{5.17}$$

The shielding tensor is a property of a particular nucleus in a molecule or a crystal and is therefore naturally expressed in the molecular coordinate system. The meaning of the symbol σ_{xy} can be understood by recognising that $\sigma_{xy}B_0$ is the induced field along the molecular x-axis caused by an external field B_0 oriented along the molecular y-axis. The necessity of describing the phenomenon of shielding with a shielding tensor, rather than a simpler object such as a scalar, reflects the fact that the induced field does not need to be parallel to the external field. The shielding tensor in the Cartesian form does not need to be symmetric: $\sigma_{xy}\neq\sigma_{yx}$, which is an obvious consequence of the meaning of the tensor component. We thus have a set of nine independent elements.

The Cartesian form of the shielding tensor is not necessarily the most useful one. We can re-write the Cartesian rank-two tensor as a sum of three terms:

$$\begin{aligned}
\begin{bmatrix} \sigma_{xx} & \sigma_{xy} & \sigma_{xz} \\ \sigma_{yx} & \sigma_{yy} & \sigma_{yz} \\ \sigma_{zx} & \sigma_{zy} & \sigma_{zz} \end{bmatrix} &= \sigma_{iso}\begin{bmatrix} 1 & 0 & 0 \\ 0 & 1 & 0 \\ 0 & 0 & 1 \end{bmatrix} + \frac{1}{2}\begin{bmatrix} 0 & \sigma_{xy}-\sigma_{yx} & \sigma_{xz}-\sigma_{zx} \\ -\sigma_{xy}+\sigma_{yx} & 0 & \sigma_{yz}-\sigma_{zy} \\ -\sigma_{xz}+\sigma_{zx} & -\sigma_{yz}+\sigma_{zy} & 0 \end{bmatrix} \\
&+ \begin{bmatrix} \sigma_{xx}-\sigma_{iso} & \frac{1}{2}(\sigma_{xy}+\sigma_{yx}) & \frac{1}{2}(\sigma_{xz}+\sigma_{zx}) \\ \frac{1}{2}(\sigma_{xy}+\sigma_{yx}) & \sigma_{yy}-\sigma_{iso} & \frac{1}{2}(\sigma_{yz}+\sigma_{zy}) \\ \frac{1}{2}(\sigma_{xz}+\sigma_{zx}) & \frac{1}{2}(\sigma_{yz}+\sigma_{zy}) & \sigma_{zz}-\sigma_{iso} \end{bmatrix} \\
&= \sigma^{(0)} + \sigma^{(1)} + \sigma^{(2)}
\end{aligned} \tag{5.18}$$

with $\sigma_{iso} = \frac{1}{3}(\sigma_{xx} + \sigma_{yy} + \sigma_{zz})$. The first term, $\sigma^{(0)}$, is a scalar multiplied by a unit tensor. The second term, $\sigma^{(1)}$, is an antisymmetric tensor and the third, $\sigma^{(2)}$, is a symmetric and traceless tensor.

The symmetric part of the shielding tensor may be transformed into a diagonal form, *i.e.* into a form in which only the diagonal elements are non-zero, by a suitable rotation of the coordinate system. We call this process the *diagonalisation* of the tensor. The new frame, in which the symmetric part of the shielding tensor is diagonal, is called the *principal frame* or the *principal axis system* (PAS) of the tensor. In the principal frame, the traceless, symmetric, rank-two tensor becomes:

$$\sigma^{(2)} = \begin{bmatrix} \sigma_{xx}^{PAS} - \sigma_{iso} & 0 & 0 \\ 0 & \sigma_{yy}^{PAS} - \sigma_{iso} & 0 \\ 0 & 0 & \sigma_{zz}^{PAS} - \sigma_{iso} \end{bmatrix} = \delta \begin{bmatrix} -\frac{1}{2}(1-\eta) & 0 & 0 \\ 0 & -\frac{1}{2}(1+\eta) & 0 \\ 0 & 0 & 1 \end{bmatrix} \tag{5.19}$$

where

$$\delta = \sigma_{zz}^{PAS} - \sigma_{iso} = \frac{2}{3}\left[\sigma_{zz}^{PAS} - \left(\sigma_{xx}^{PAS} + \sigma_{yy}^{PAS}\right)/2\right] = \frac{2}{3}\Delta\sigma \tag{5.20a}$$

and

$$\eta = \left(\sigma_{xx}^{PAS} - \sigma_{yy}^{PAS}\right)/\delta = 3\left(\sigma_{xx}^{PAS} - \sigma_{yy}^{PAS}\right)/2\Delta\sigma \tag{5.20b}$$

We should remember that the rotation of the coordinate system leaves the trace of the tensor unchanged. Thus, in agreement with the properties of a scalar, σ_{iso} does not change when we move from an arbitrary initial frame to the PAS. The tensor components σ_{xx}^{PAS}, σ_{yy}^{PAS} and σ_{zz}^{PAS} are denoted as principal components of the shielding tensor, and the labels *x*, *y* and *z* follow the convention:

$$\left|\sigma_{zz}^{PAS} - \sigma_{iso}\right| \geq \left|\sigma_{xx}^{PAS} - \sigma_{iso}\right| \geq \left|\sigma_{yy}^{PAS} - \sigma_{iso}\right| \tag{5.21}$$

$\Delta\sigma$ is called the *chemical shielding anisotropy* and η the *asymmetry parameter.* When the asymmetry parameter vanishes, $\sigma_{xx}^{PAS} = \sigma_{yy}^{PAS} = \sigma_{\perp}$, we say that the shielding tensor is axially symmetric (but still anisotropic; the shielding becomes isotropic only if $\sigma_{xx}^{PAS} = \sigma_{yy}^{PAS} = \sigma_{zz}^{PAS}$). The symmetry axis is the *z*-axis according to the convention (5.21), and one often uses the notation $\sigma_{zz}^{PAS} = \sigma_{\parallel}$. It is important to realise that the symmetry properties of the shielding tensor are closely related to the local symmetry of the molecular framework around the nucleus under consideration. Referring to Anet and O'Leary[14] for a more detailed presentation, we wish here to mention a single example of these relations. For example, the carbon-13 nucleus in chloroform resides on a three-fold symmetry axis of the molecule, which makes the shielding tensor axially symmetric and lets us identify the molecular symmetry axis with the *z*-axis of the tensor.

It is illustrative to relate the principal components of the shielding tensor to the spectrum of a solid powder sample, analogous to the dipolar powder pattern in Figure 3.3. Such spectra, for the axial and non-axial case, are shown in Figure 5.2. In the axial case, the shielding anisotropy simplifies to $\Delta\sigma = \sigma_{\parallel} - \sigma_{\perp}$ and corresponds to the width of the powder pattern. The powder pattern is a superposition of signals corresponding to different orientations of the molecule/microcrystallite with respect to the magnetic field. Each orientation corresponds, in turn, to a slightly different local field felt by the

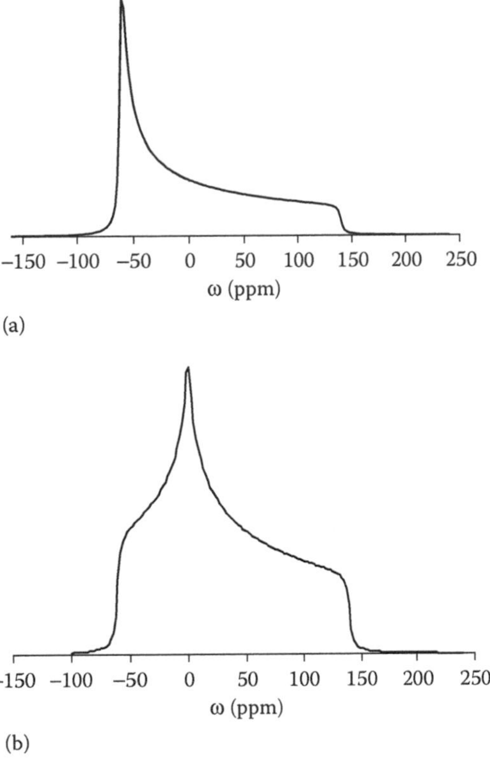

FIGURE 5.2 Illustration of the appearance of the powder pattern due to the CSA tensor for an axially symmetric case (a) and a non-axial case (b). Simulations were performed using $\sigma_{\perp} = -60$ ppm and $\sigma_{\parallel} = +140$ ppm (a) and $\sigma_{yy} = -60$ ppm, $\sigma_{xx} = 0$ ppm and $\sigma_{zz} = +140$ ppm (b).

nucleus. This means that the random tumbling of a molecule in solution therefore results in a randomly varying local field at the nucleus. This is reminiscent of the simple example considered in Chapter 2 and provides a qualitative explanation of the role of CSA as a relaxation mechanism.

Note that all of the discussion and terminology introduced following Eq. (5.19) refers to the symmetric part of the shielding tensor. Thus, the concepts of anisotropy of the tensor and its asymmetry parameter have nothing to do with the antisymmetric part, $\sigma^{(1)}$, of the shielding tensor. The choice of the axis system that makes the symmetric part diagonal *does not* make the antisymmetric part vanish. As discussed in the reviews by Anet and O'Leary[13,14] and in references quoted therein, the antisymmetric part of the shielding tensor always results in an induced field perpendicular to the external field with its magnitude dependent on the molecular orientation. The antisymmetric shielding tensor contributes to the solid-state powder pattern as a second-order effect. From the point of view of relaxation, it acts essentially as an independent source of relaxation (relaxation mechanism), usually much less important than the "ordinary" CSA mechanism connected to the symmetric rank-two tensor.

5.2.2 The CSA Hamiltonian

In order to formulate the theory of the CSA relaxation in quantitative terms, we need to formulate the chemical shielding Hamiltonian and, eventually, turn it into an expression in terms of irreducible tensors. The Hamiltonian for a single spin in the Cartesian coordinates is:

$$\hat{H}_I^{CS} = \gamma_I \hat{\mathbf{I}} \cdot \boldsymbol{\sigma} \cdot \mathbf{B} \tag{5.22}$$

We also introduce the irreducible spherical components of the shielding tensor[10] according to

$$\sigma_{0,0} = -\sqrt{3}\sigma_{iso} \tag{5.23a}$$

$$\sigma_{1,0} = -\frac{i}{\sqrt{2}}\left(\sigma_{xy} - \sigma_{yx}\right) \tag{5.23b}$$

$$\sigma_{1,\pm 1} = -\frac{1}{2}\left[\sigma_{zx} - \sigma_{xz} \pm i\left(\sigma_{zy} - \sigma_{yz}\right)\right] \tag{5.23c}$$

$$\sigma_{2,0} = \sqrt{\frac{3}{2}}\left(\sigma_{zz} - \sigma_{iso}\right) \tag{5.23d}$$

$$\sigma_{2,\pm 1} = \mp\frac{1}{2}\left[\sigma_{xz} + \sigma_{zx} \pm i\left(\sigma_{yz} + \sigma_{zy}\right)\right] \tag{5.23e}$$

$$\sigma_{2,\pm 2} = \frac{1}{2}\left[\sigma_{xx} - \sigma_{yy} \pm i\left(\sigma_{xy} + \sigma_{yx}\right)\right] \tag{5.23f}$$

We can easily recognise simple relations between the spherical components and the elements of $\sigma^{(0)}$, $\sigma^{(1)}$, $\sigma^{(2)}$ in Eq. (5.18). We concentrate for the moment on the rank-two spherical components, which have important implications for relaxation. These attain a simpler form when expressed in the principal axis system:

$$\sigma_{2,0}^{PAS} = \sqrt{\frac{3}{2}}\delta = \sqrt{\frac{2}{3}}\Delta\sigma \tag{5.24a}$$

$$\sigma_{2,\pm 1}^{PAS} = 0 \tag{5.24b}$$

$$\sigma_{2,\pm 2}^{PAS} = \frac{1}{2}\delta\eta \tag{5.24c}$$

Given the simplicity of these expressions, it is advantageous to use the PAS as a starting point for the discussion of the rank-two shielding tensor and the CSA relaxation. This discussion involves a series of coordinate frame transformations, which are conveniently carried out using the basic properties of the spherical tensors (see Brink and Satchler, *Further reading,* for more details). Using these properties, we can express the shielding tensor in the arbitrary axis system (AAS) as:

$$\sigma_{2,q}^{AAS} = \sum_{p=-2}^{2} D_{p,q}^{2}\left(\Omega_{PAS,AAS}\right)\sigma_{2,p}^{PAS} \tag{5.25}$$

The symbol $\Omega_{PAS,AAS}$ is a shorthand notation for the *three* Euler angles transforming the two frames into each other, here from the PAS to the AAS. This should not be confused with Ω_{IS}^{L}, introduced in Eq. (3.36), which denotes the set of *two* polar angles specifying the orientation of an axis (*IS*) in a given frame (*L*). To proceed, we need to convert the chemical shielding Hamiltonian in Eq. (5.22) to the general form discussed in the previous chapter and which was given by Eq. (4.37a). To do this, we need to express the spin and field factors as irreducible spherical tensors. This can be done formally by introducing a *spin-field tensor* $\hat{X}$, with Cartesian components $\hat{X}_{\alpha\beta} = \hat{I}_\alpha B_\beta$ (where α and β denote the Cartesian axes *x*, *y* and *z*). The irreducible spherical tensor form of the spin-field tensor operators in the AAS becomes:

$$\hat{X}_{2,0}^{AAS} = \frac{1}{\sqrt{6}}\left(2B_z\hat{I}_z - B_x\hat{I}_x - B_y\hat{I}_y\right) \tag{5.26a}$$

$$\hat{X}_{2,\pm1}^{AAS} = \mp\frac{1}{2}\left[B_z\hat{I}_x + B_x\hat{I}_z \pm i\left(B_z\hat{I}_y + B_y\hat{I}_z\right)\right] \tag{5.26b}$$

$$\hat{X}_{2,\pm2}^{AAS} = \frac{1}{2}\left[B_x\hat{I}_x - B_y\hat{I}_y \pm i\left(B_y\hat{I}_x + B_x\hat{I}_y\right)\right] \tag{5.26c}$$

If we choose to express the spin-field tensor in the laboratory frame, where $B_x^{LAB} = B_y^{LAB} = 0$ and $B_z^{LAB} = B_0$, these expressions simplify to:

$$\hat{X}_{2,0}^{LAB} = \sqrt{\frac{2}{3}}\hat{I}_z B_0 \tag{5.27a}$$

$$\hat{X}_{2,\pm1}^{LAB} = \mp\frac{1}{2}\hat{I}_\pm B_0 \tag{5.27b}$$

$$\hat{X}_{2,\pm2}^{LAB} = 0 \tag{5.27c}$$

We are going to incorporate the field strength, B_0, into the interaction strength ξ_{CSA} and define the operators $\hat{T}_{l,-q}$ of Eq. (4.37a) for the case of the CSA mechanism as:

$$\hat{T}_{2,0}^{CSA} = \sqrt{\frac{8}{3}}\hat{I}_z \tag{5.28a}$$

$$\check{T}_{2,\pm1}^{CSA} = \mp\check{I}_\pm \tag{5.28b}$$

$$\hat{T}_{2,\pm2}^{CSA} = 0 \tag{5.28c}$$

In order to make a contraction of the rank-two shielding tensor and rank-two spin-field tensor, we must express both tensors in a common frame. We choose the laboratory frame for this purpose, because this is the frame of preference for the dominant Zeeman interaction. To rotate the shielding tensor into that frame, we use Eq. (5.25) and note explicitly that the angle $\Omega_{PAS,LAB} \equiv \Omega_{I,L}$ rotating the PAS of the shielding tensor into the laboratory frame is time dependent because of molecular tumbling:

$$\sigma_{2,q}^{LAB} = \sum_{p=-2}^{2} D_{p,q}^2\left(\Omega_{I,L}(t)\right)\sigma_{2,p}^{PAS} \tag{5.29}$$

which results in

$$\hat{H}_1(t) = \xi'_{CSA}\sum_{q=-2}^{2}\sum_{p=-2}^{2}(-1)^q D_{p,q}^2\left(\Omega_{I,L}(t)\right)\sigma_{2,p}^{PAS}\hat{T}_{l,-q} \tag{5.30}$$

with the $\hat{T}$ -operators given in Eq. (5.28) and the σ^{PAS} components by Eq. (5.24). The symbol ξ'_{CSA} is a constant related to units and normalisation. We shall soon replace it by another interaction strength parameter, defined in the following paragraphs.

The situation simplifies significantly if the shielding tensor is axially symmetric: $\sigma_{xx}^{PAS} = \sigma_{yy}^{PAS}$ or $\eta = 0$. The only non-zero element of σ^{PAS} is then the $p = 0$ term and the summation over p reduces to a single term. Under these conditions, the only Wigner rotation matrix elements that need to be considered in Eq. (5.30) are $D_{0,q}^{2}$ and these are proportional to the spherical harmonics $Y_{2,q}$. Eq. (5.30) thus simplifies to:

$$\hat{H}_1(t) = \xi_{CSA} \sum_{q=-2}^{2} (-1)^q Y_{2,q}\left(\Omega_I^L(t)\right) \hat{T}_{1,-q} \tag{5.31}$$

With the spherical harmonics normalised as in Eq. (4.44) and with the operators $\hat{T}$ given in Eq. (5.28), the CSA interaction strength parameter to be used in Eq. (5.31) becomes:

$$\xi_{CSA} = -(2\pi/15)^{1/2} \gamma_I B_0 \Delta\sigma = (2\pi/15)^{1/2} \omega_I \Delta\sigma \tag{5.32}$$

In general, the chemical shielding tensor is not axially symmetric. However, as discussed by Goldman[15], an arbitrary shielding tensor can be decomposed into a sum of two axially symmetric tensors. We have thus the choice of either using the more complicated form of the CSA Hamiltonian given in Eq. (5.30) or working in terms of two axially symmetric shielding interactions on each nucleus. The Hamiltonian will then consist of two simpler terms, such as that of Eq. (5.31), and the calculations of the relaxation behaviour will require including the cross-correlation between the two CSA terms, in full analogy to the multiple dipolar interactions, treated in Section 5.1.

5.2.3 The CSA Relaxation Mechanism

Let us assume for simplicity that we deal with a single nucleus, characterised by an axial CSA. The unperturbed Zeeman Hamiltonian has the eigenstates $|1\rangle = |½, ½\rangle$ and $|2\rangle = |½, -½\rangle$. We here use this $|I,m\rangle$ notation to avoid confusing the common Greek letter symbols for these states with the indices in Eqs. (4.28) through (4.36) and in their analogues below. The density matrix for such a simple system has four elements. Two of them are populations, ρ_{11} and ρ_{22}, and two are coherences, ρ_{12} and ρ_{21}. With this simple system, we can easily exemplify the use of Redfield theory in detail. The evolution of the four density matrix elements is in principle described by a 4×4 relaxation supermatrix according to the formalism in Section 4.1. The situation simplifies, however, because of the secular approximation. The secular approximation removes couplings between populations and coherences, as well as between coherences with coherence orders +1 and –1. The full set of equations corresponding to Eq. (4.30) therefore takes the following form:

$$\frac{d}{dt}\begin{pmatrix} \rho_{11} \\ \rho_{22} \\ \rho_{12} \\ \rho_{21} \end{pmatrix} = \begin{pmatrix} R_{1111} & R_{1122} & 0 & 0 \\ R_{2211} & R_{2222} & 0 & 0 \\ 0 & 0 & R_{1212} & 0 \\ 0 & 0 & 0 & R_{2121} \end{pmatrix} \begin{pmatrix} \rho_{11} \\ \rho_{22} \\ \rho_{12} \\ \rho_{21} \end{pmatrix} \tag{5.33}$$

where we recognise the simplest possible form of the "Redfield kite" (*cf.* Figure 4.1). We compute the elements of the relaxation supermatrix using Eq. (4.28), which yields:

$$\begin{aligned} R_{1111} &= \frac{1}{2}\left(\mathcal{J}_{1111}(0) + \mathcal{J}_{1111}(0) - \mathcal{J}_{1111}(0) - \mathcal{J}_{2121}(\omega_0) - \mathcal{J}_{1111}(0) - \mathcal{J}_{2121}(\omega_0)\right) \\ &= -\mathcal{J}_{2121}(\omega_0) \end{aligned} \tag{5.34a}$$

$$R_{2222} = -\mathcal{J}_{1212}(\omega_0) \tag{5.34b}$$

$$R_{1122} = R_{2211} = \mathcal{J}_{1212}(\omega_0) \tag{5.34c}$$

$$R_{1212} = \frac{1}{2}\left(2\mathcal{J}_{1122}(0) - \mathcal{J}_{1111}(0) - \mathcal{J}_{2222}(0) - \mathcal{J}_{2121}(\omega_0) - \mathcal{J}_{1212}(\omega_0)\right) \tag{5.34d}$$

$$R_{2121} = \frac{1}{2}\left(2\mathcal{J}_{2211}(0) - \mathcal{J}_{1111}(0) - \mathcal{J}_{2222}(0) - \mathcal{J}_{2121}(\omega_0) - \mathcal{J}_{1212}(\omega_0)\right) \tag{5.34e}$$

In the first line of Eq. (5.34a), we explicitly list all six terms resulting from Eq. (4.28), where both $\gamma = 1$ and $\gamma = 2$ contribute. The spectral densities $\mathcal{J}_{\alpha\alpha'\beta\beta'}(\omega)$ are, in turn, calculated using Eq. (4.29). With the Hamiltonian of Eq. (5.31), we can re-write Eq. (4.29) as:

$$\begin{aligned}\mathcal{J}_{\alpha\alpha'\beta\beta'}(\omega) &= 2\xi_{CSA}^2 \sum_{q=-2}^{2} \int_0^{\infty} (-1)^{2q} \left\langle \alpha \middle| Y_{2,q}(t)\hat{T}_{2,-q} \middle| \alpha' \right\rangle \left\langle \beta' \middle| Y_{2,q}^*(t+\tau)\hat{T}_{2,-q}^{\dagger} \middle| \beta \right\rangle \exp(-i\omega\tau)d\tau \\ &= \xi_{CSA}^2 \sum_{q=-2}^{2} \left\langle \alpha \middle| \hat{T}_{2,-q} \middle| \alpha' \right\rangle \left\langle \beta' \middle| \hat{T}_{2,-q}^{\dagger} \middle| \beta \right\rangle 2\int_0^{\infty} Y_{2,q}(t) Y_{2,q}^*(t+\tau)\exp(-i\omega\tau)d\tau \\ &= \xi_{CSA}^2 J_l(\omega) \sum_{q=-2}^{2} \left\langle \alpha \middle| \hat{T}_{2,-q} \middle| \alpha' \right\rangle \left\langle \beta' \middle| \hat{T}_{2,-q}^{\dagger} \middle| \beta \right\rangle\end{aligned} \tag{5.35}$$

In the last line, we have implicitly assumed ensemble-averaging of the integral expression in the preceding line. The spectral densities $J_l(\omega)$ are strictly analogous to those occurring in the theory of the dipole-dipole relaxation, the only difference being that the axis whose reorientation we follow is the principal axis of the CSA rather than the dipole-dipole axis. To complete the derivation of the relaxation supermatrix, we need the matrix elements of the operators $\hat{T}_{l,-q}$, which can be obtained using the definitions of the step operators (Eq. (3.12)) and the Pauli spin matrices (Eq. (1.4)). We collect the results in Table 5.1. Note that the operators $\hat{T}_{2,\pm 2}$ and their matrix elements vanish.

Using Eq. (5.35), we obtain for the longitudinal (population-related) part of the relaxation matrix:

$$R_{1111} = R_{2222} = -\xi_{CSA}^2 J_2(\omega_0) \tag{5.36a}$$

$$R_{2211} = R_{1122} = \xi_{CSA}^2 J_2(\omega_0) \tag{5.36b}$$

TABLE 5.1 Matrix Elements, $\langle\alpha|\hat{T}_{l,-q}|\beta\rangle$, of the Operators $\hat{T}_{l,-q}$ in Eq. (5.28).

α	β	$\langle\alpha\|T_{2,-1}\|\beta\rangle$	$\langle\alpha\|T_{2,0}\|\beta\rangle$	$\langle\alpha\|T_{2,1}\|\beta\rangle$
1	1	0	$+\sqrt{\frac{2}{3}}$	0
1	2	0	0	-1
2	1	1	0	0
2	2	0	$-\sqrt{\frac{2}{3}}$	0

With this, the longitudinal part of the Redfield expression becomes:

$$\frac{d}{dt}\begin{pmatrix}\rho_{11}\\ \rho_{22}\end{pmatrix}=\begin{pmatrix}-\xi_{CSA}^2 J_2(\omega_0) & \xi_{CSA}^2 J_2(\omega_0)\\ \xi_{CSA}^2 J_2(\omega_0) & -\xi_{CSA}^2 J_2(\omega_0)\end{pmatrix}\begin{pmatrix}\rho_{11}\\ \rho_{22}\end{pmatrix} \tag{5.37}$$

As we did at the beginning of Section 2.1, we can choose to work in terms of a population difference (proportional to the expectation value of $\hat{I}_z$) and the sum of populations, which is constant in time. By doing so we obtain:

$$\frac{d}{dt}(\rho_{11}-\rho_{22})=-2\xi_{CSA}^2 J_2(\omega_0)(\rho_{11}-\rho_{22}) \tag{5.38a}$$

or

$$\frac{d}{dt}\langle\hat{I}_z\rangle=-2\xi_{CSA}^2 J_2(\omega_0)\langle\hat{I}_z\rangle \tag{5.38b}$$

After correcting for the finite temperature (compare Section 3.5), Eq. (5.38b) becomes analogous to the longitudinal Bloch equation, Eq. (1.22a), with:

$$\begin{aligned}\frac{1}{T_{1,CSA}}&=2\xi_{CSA}^2 J_2(\omega_0)=\frac{4\pi}{15}(\gamma_I B_0)^2(\Delta\sigma)^2 J_2(\omega_0)\\ &=\frac{2}{15}(\gamma_I B_0)^2(\Delta\sigma)^2\frac{\tau_c}{1+\omega_0^2\tau_c^2}\end{aligned} \tag{5.39}$$

In the second equality in Eq. (5.39), we have assumed that the molecule of interest tumbles as a rigid spherical object, with the correlation time τ_c, and used the normalised spectral density of the form $J_2(\omega_0)=(2\pi)^{-1}\tau_c/(1+\omega_0^2\tau_c^2)$. We will return to the derivation of model-dependent forms of the spectral density function in Chapter 6. The specific form of $J_2(\omega_0)$ in Eq. (5.39) will be derived in Section 6.1.

In the same way, we can calculate the decay rate constant for the two coherences, $R_{1212}=R_{2121}$ (the reader is encouraged to do that calculation!), and identify them with the transverse relaxation rate:

$$\begin{aligned}\frac{1}{T_{2,CSA}}&=\xi_{CSA}^2\left(\frac{4}{3}J_2(0)+J_2(\omega_0)\right)=\frac{2\pi}{15}(\gamma_I B_0)^2(\Delta\sigma)^2\left(\frac{4}{3}J_2(0)+J_2(\omega_0)\right)\\ &=\frac{1}{15}(\gamma_I B_0)^2(\Delta\sigma)^2\left(\frac{4}{3}\tau_c+\frac{\tau_c}{1+\omega_0^2\tau_c^2}\right)\end{aligned} \tag{5.40}$$

The effects of a non-axial CSA tensor are easy to include in the simple case of isotropic reorientation: the rate expressions in the second lines of Eqs. (5.39) and (5.40) are only multiplied by the factor $(1+\eta^2/3)$.

There are some interesting observations we can make concerning the relaxation rates resulting from the CSA mechanism as given by Eqs. (5.39) and (5.40). First, we note that the rates are proportional to the square of the external magnetic field. In the extreme narrowing regime, this field dependence can be used for separating the contributions from the CSA and DD relaxation mechanisms. By measuring the relaxation rate at several magnetic field strengths and plotting the measured rate against the square of the magnetic field, one obtains a straight line with the abscissa equal to the dipolar rate. Outside of the extreme narrowing range, this B_0^2 dependence will be offset by the term $\omega_0^2\tau_c^2$ in the denominators of

Eqs. (5.39) and (5.40). In the limit of $\omega_0^2\tau_c^2 >> 1$, the longitudinal rate will in fact become independent of the magnetic field. Apart from the magnetic field, the CSA relaxation depends on the magnitude of the shielding anisotropy. Large chemical shielding anisotropies are characteristic for many heteronuclei. In the case of ^{13}C, large $\Delta\sigma$ values (on the order of 200 ppm) are characteristic for carbons participating in double or triple bonds, for example, in carbonyl groups or in aromatic rings. The relaxation of such nuclei at the high magnetic fields of the present-day NMR spectrometers can be dominated by the CSA mechanism, in particular for carbon nuclei that are not directly bound to protons. Another nuclear species for which the CSA relaxation is important is ^{15}N in amide groups in peptides and proteins.

Another interesting property of the CSA relaxation is that the longitudinal and transverse relaxation rates are not equal to each other even in extreme narrowing, where the ratio becomes $T_1/T_2 = 7/6$. This is in contrast to dipole-dipole relaxation and most other relaxation mechanisms, and is the result of the properties of the CSA Hamiltonian under spatial rotation. We note, however, that at the limit of zero magnetic field, where the three Cartesian directions become equivalent in isotropic space, both relaxation rates vanish.

5.2.4 The DD/CSA Interference

Even though the elements $T_{2,\pm 2}$ are zero in the case of CSA, because of the nature of the spin-field tensor, it is important to keep in mind that the Hamiltonian in Eq. (5.31) is a scalar contraction of rank-two tensors. An important consequence of this fact is that the CSA Hamiltonian and the DD Hamiltonian in Eq. (3.13) have the same properties under rotation of coordinate systems and both depend on the spherical harmonics of the same rank. Because of this property, the CSA and dipolar interaction can interfere with each other, in full analogy with the case of multiple dipolar interactions discussed in Section 5.1. This type of cross-correlation effect can in the NMR context be traced back to the work of Shimizu in the early 1960s[16] and has been discussed in the review by Kumar *et al.*[1]. Consider a system consisting of two nuclear spins, I and S, both with the spin quantum number 1/2. We treat the two Zeeman interactions and the scalar spin-spin coupling as the unperturbed Hamiltonian, while the dipolar interaction and the two CSA interactions enter the perturbation Hamiltonian:

$$\begin{aligned}\hat{H}_1(t) = \xi_{IS}^{DD}\sum_{q=-2}^{2}(-1)^q Y_{2,q}\left(\Omega_{IS}^L(t)\right)\hat{T}_{2,-q}^{IS} + \xi_I^{CSA}\sum_{q=-2}^{2}(-1)^q Y_{2,q}\left(\Omega_I^L(t)\right)\hat{T}_{2,-q}^{I} \\ + \xi_S^{CSA}\sum_{q=-2}^{2}(-1)^q Y_{2,q}\left(\Omega_S^L(t)\right)\hat{T}_{2,-q}^{S}\end{aligned} \tag{5.41}$$

The interaction strength constants are defined in full analogy with Eqs. (3.13) and (5.31), including the definitions of the irreducible tensor operators of Eqs. (4.43) and (5.28). The set of angles $\Omega_{IS}^L(t)$ describes the orientation of the IS axis with respect to the laboratory frame, while the angles $\Omega_I^L(t)$ and $\Omega_S^L(t)$ refer to the orientation of the principal axis of the CSA for the I- and S-spins, respectively, in the same frame. Both shielding tensors are assumed to be axially symmetric. It is useful to transform the orientational part of the three terms in Eq. (5.41) in such a way that they all contain a spherical harmonic of the same time-dependent angle. We shall come back to the calculations of spectral densities, starting with this more complicated form of Hamiltonian, involving several coordinate frame transformations, in Chapter 6.

The result of applying the Redfield theory to the longitudinal part of this four-level system, analogous to that of Figure 3.5, is a relaxation equation, which is most conveniently expressed in the operator or magnetisation mode formalism as:

$$\frac{d}{dt}\begin{pmatrix}\langle \hat{I}_z\rangle \\ \langle \hat{S}_z\rangle \\ \langle 2\hat{I}_z\hat{S}_z\rangle\end{pmatrix} = -\begin{pmatrix}\Gamma_{11} & \Gamma_{12} & \Gamma_{13} \\ \Gamma_{21} & \Gamma_{22} & \Gamma_{23} \\ \Gamma_{31} & \Gamma_{32} & \Gamma_{33}\end{pmatrix}\begin{pmatrix}\langle \hat{I}_z\rangle - I_z^{eq} \\ \langle \hat{S}_z\rangle - S_z^{eq} \\ \langle 2\hat{I}_z\hat{S}_z\rangle\end{pmatrix} \tag{5.42}$$

The elements Γ_{11} and Γ_{22} of the relaxation matrix are sums of the corresponding elements occurring in the Solomon equations, Eqs. (3.20) (the dipolar ρ_I and ρ_S), and the CSA spin-lattice relaxation rates as given by Eq. (5.39). Γ_{12} and Γ_{21} are identical to the dipolar cross-relaxation rate, σ_{IS} and the CSA does not contribute to these elements. Γ_{33} is the decay rate of the longitudinal two-spin order, $\langle 2\hat{I}_z\hat{S}_z\rangle$. The elements connecting the expectation values of one-spin operators $\langle \hat{I}_z\rangle$ and $\langle \hat{S}_z\rangle$ with the two-spin operator $\langle 2\hat{I}_z\hat{S}_z\rangle$, Γ_{13}, Γ_{23}, Γ_{31} and Γ_{32}, are the result of interference or cross-correlation between the dipolar interaction and one of the CSA interactions.

It is interesting to note that these terms couple the expectation values of operators with opposite symmetries under spin inversion, contrary to the case of interferences between dipolar interactions, which couple expectation values of operators of the same symmetry. These terms arise through interference between the DD and CSA Hamiltonians. These CCRRs can be derived using the formulation of Section 4.1 (a detailed derivation can be found in the review by Smith *et al.*[17]) or using the operator approach of Section 4.2. The latter derivation can be found in the paper by Goldman[15]. The results are:

$$\Gamma_{13} = \Gamma_{31} = \delta_{I,IS} = \xi_{IS}^{DD}\xi_I^{CSA} J_{IS,I}(\omega_I) \tag{5.43a}$$

$$\Gamma_{23} = \Gamma_{32} = \delta_{S,IS} = \xi_{IS}^{DD}\xi_S^{CSA} J_{IS,S}(\omega_S) \tag{5.43b}$$

For the sake of completeness, we also provide an expression for Γ_{33}:

$$\begin{aligned}\Gamma_{33} = \rho_{IS} &= \frac{1}{8}\left(\xi_{IS}^{DD}\right)^2 J_{IS,IS}(\omega_I) + \frac{1}{8}\left(\xi_{IS}^{DD}\right)^2 J_{IS,IS}(\omega_S) \\ &\quad + 2\left(\xi_I^{CSA}\right)^2 J_{I,I}(\omega_I) + 2\left(\xi_S^{CSA}\right)^2 J_{S,S}(\omega_S)\end{aligned} \tag{5.43c}$$

$J_{IS,IS}(\omega)$ are autocorrelated dipolar spectral densities, identical to those used in Chapter 3 and in Section 5.1, while $J_{I,I}(\omega)$ and $J_{S,S}(\omega)$ are autocorrelated CSA spectral densities analogous to those appearing in Eq. (5.39). For a rigid, isotropic rotor with a rotational correlation time τ_c, the cross-correlated spectral densities take the form

$$J_{IS,I}(\omega) = \frac{1}{2}\left(3\cos^2\theta_{IS,I} - 1\right)\frac{1}{4\pi}\frac{2\tau_c}{1+\omega^2\tau_c^2} \tag{5.44a}$$

$$J_{IS,S}(\omega) = \frac{1}{2}\left(3\cos^2\theta_{IS,S} - 1\right)\frac{1}{4\pi}\frac{2\tau_c}{1+\omega^2\tau_c^2} \tag{5.44b}$$

One should notice that we use the same normalisation of the spectral density as in Eq. (3.16), *i.e.* we include the factor $1/4\pi$. The reader should be aware that different normalisations are used in the literature, the one omitting the $1/4\pi$ being rather popular. The factors $\frac{1}{2}(3\cos^2\theta_{IS,I} - 1)$ and $\frac{1}{2}(3\cos^2\theta_{IS,S} - 1)$ occur as a result of the time-independent coordinate transformations between the principal frames of the DD and CSA (for the *I*- and *S*-spin, respectively) interactions. The angles $\theta_{IS,I}$ and $\theta_{IS,S}$ are between the principal axes of the two CSA tensors and the DD internuclear vector. Replacing the ξs as strength constants with the dipolar coupling constant and shielding anisotropy and using the spectral densities in Eqs. (5.44), the CCRR of Eq. (5.43a) becomes:

$$\delta_{I,IS} = \frac{1}{5} b_{IS} (\omega_I \Delta\sigma)(3\cos^2\theta_{IS,I} - 1) \frac{\tau_c}{1+\omega_I^2\tau_c^2} \tag{5.45}$$

with an expression for $\delta_{IS,S}$ obtained *mutatis mutandis.*

An interesting feature of Eqs. (5.43) is that the CCRRs contain products of the two different interaction strength constants, in analogy with Eq. (5.15). The consequence of this fact is that a strong dipolar interaction can help to make a weak CSA visible. In other words, the DD/CSA CCRRs can be used to estimate the shielding anisotropy, $\Delta\sigma$, in solution, even if the anisotropy is so small that the contribution of the CSA mechanism to the spin-lattice relaxation rate (proportional to the square of $\Delta\sigma$) is negligible. As an example, the chemical shift anisotropies for the proton and carbon in chloroform, $\Delta\sigma_H$ and $\Delta\sigma_C$, were determined through measurements of longitudinal DD/CSA CCRRs[18] to be about 12 and 32 ppm, respectively. At the magnetic field of 9.4 T, this corresponds to $|\xi_H^{CSA}| = 20{\cdot}10^3 \text{rad s}^{-1}$ (3.1 kHz) and $|\xi_C^{CSA}| = 13{\cdot}10^3 \text{rad s}^{-1}$ (2.1 kHz), compared with $|\xi_{CH}^{DD}| = 51{\cdot}10^4 \text{rad s}^{-1}$ (81 kHz). It is also worth noticing that the proton CSA interaction strength constant is larger than its carbon counterpart, even though the $\Delta\sigma$ values expressed in parts per million show the opposite relation. This is an effect of the larger proton magnetogyric ratio. In analogy with the case of dipolar interference effects, the DD/CSA CCRRs give rise to asymmetries in multiplet relaxation. We shall return to this point in the context of experimental methods.

5.2.5 Transverse Relaxation Effects of the DD/CSA Interference

The DD/CSA interference effects influence also the transverse relaxation and the spectral lineshapes in the sense that they contribute to unequal line-broadening of the multiplet components. For a detailed discussion, we refer to the paper by Goldman[15]. Here, we just wish to summarise certain features of the results. The *J*-coupled spectrum is described very similarly to the case discussed at the end of Section 4.2. The *I*-spin spectrum is characterised in terms of two coherences, which we can express in several ways. One convenient possibility is to work with expectation values of $\langle \hat{I}_+ \rangle$ and $\langle 2\hat{I}_+\hat{S}_z \rangle$. The first of these expectation values corresponds to an in-phase doublet, split by the *J*-coupling to the *S*-spin, while the second is an antiphase doublet. Alternatively, we can formulate linear combinations of these operators, corresponding to individual transitions of the doublet. The evolution of the two expectation values follows a set of two coupled equations, reminiscent of Eq. (4.64), with the coupling terms containing the decay rates, the transverse cross-relaxation rates and the transverse cross-correlated relaxation rates. If the doublet components are well separated, the cross-relaxation effects can be neglected. Both lines are then expected to have Lorentzian lineshapes. In the absence of the CSA, the two doublet components have the same width. The presence of DD/CSA cross-correlation, however, has a peculiar effect on the two linewidths: it adds another term, η, to the linewidth of one of the doublet components, while the same term is subtracted from the linewidth of the second component. The phenomenon is referred to as *differential line broadening* (DLB). The difference in the linewidth, expressed in radians per second, is given by:

$$2\eta = \xi_{IS}^{DD} \xi_I^{CSA} \left[J_{IS,I}(\omega_I) + \frac{4}{3} J_{IS,I}(0) \right] \tag{5.46}$$

The DLB, a characteristic signature of the DD/CSA interference, is illustrated in Figure 5.3.

If we carefully inspect Eq. (5.41), we can conclude that the *IS* spin system might be influenced, at least in principle, by still another interference effect: that between the two CSA interactions. This interference

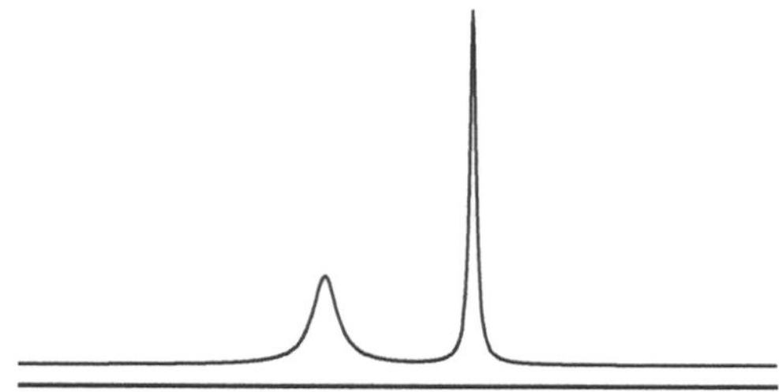

FIGURE 5.3 Differential line-broadening.

effect does not influence the longitudinal relaxation or the transverse relaxation of single quantum coherences. In a scalar-coupled *IS* two-spin system, one can, however, also excite and measure the relaxation of double-quantum (DQ) and zero-quantum (ZQ) coherences (we shall discuss how this can be done experimentally in Chapter 10). Each of these coherences is characterised by a decay rate constant, R_2^{DQ} and R_2^{ZQ}, related to the corresponding linewidth. The general properties of the linewidths of these multiple-quantum (MQ) coherences, as contrasted against the directly observable single-quantum coherences, were discussed in an important paper by Wokaun and Ernst[19]. The MQ relaxation rates are affected by the autocorrelated dipolar and CSA spectral densities, but not by the DD/CSA interference (an elegant symmetry argument for this was given by Werbelow[20]). On the other hand, as indicated in the same article by Werbelow and demonstrated experimentally by Pellecchia and co-workers[21], the DQ and ZQ coherences are affected by the CSA/CSA interference. In fact, the magnitude of this interference effect can be obtained by comparing the linewidth of these two transitions. The measured effects are, however, not strongly pronounced due to the fact that the coupling is between two interactions with relatively weak interaction strengths.

5.2.6 The Antisymmetric Component

We now return to the rank-one, antisymmetric component of the shielding tensor. The relaxation effects of this component were proposed for the first time by Blicharski[22] and were investigated thoroughly by Anet and O'Leary, who also described them in a couple of reviews[13,14]. We refer the reader to that work for details and choose to present here only the final results. The antisymmetric shielding contributes to relaxation in a rather simple way, adding another term to the total spin-lattice and spin-spin relaxation rates. The contributions have the form:

$$\frac{1}{T_{1,CSA}^{anti}} = \frac{1}{6}\gamma^2 B_0^2\left[\left(\sigma_{xy}-\sigma_{yx}\right)^2+\left(\sigma_{xz}-\sigma_{zx}\right)^2+\left(\sigma_{yz}-\sigma_{zy}\right)^2\right]\frac{\tau_1}{1+\omega_0^2\tau_1^2} \tag{5.47a}$$

and

$$\frac{1}{T_{2,CSA}^{anti}} = \frac{1}{12}\gamma^2 B_0^2\left[\left(\sigma_{xy}-\sigma_{yx}\right)^2+\left(\sigma_{xz}-\sigma_{zx}\right)^2+\left(\sigma_{yz}-\sigma_{zy}\right)^2\right]\frac{\tau_1}{1+\omega_0^2\tau_1^2} \tag{5.47b}$$

The correlation time τ_1 is not the same as the τ_c that we encountered in the case of the DD interaction or the symmetric CSA. Here, it is the characteristic correlation time for rank-one spherical harmonics, mentioned in Section 2.4. As will be shown in Section 6.1, we have the relation $\tau_1 = 3\tau_c$ for isotropic rotational diffusion. Two interesting observations can be made concerning the transverse relaxation rate contribution. First, in analogy with the symmetric CSA, it is not equal to the corresponding contribution to the longitudinal relaxation rate. Second, it has an unusual property for transverse relaxation, as it does not depend on the zero-frequency spectral density. Kowalewski and Werbelow[23] estimated the possible effect of the antisymmetric shielding anisotropy relaxation in peptides and proteins using the shielding tensor from *ab initio* calculations on formamide. The contributions of Eq. (5.47) were found to be negligible for the case of isotropic motion and moderate deviations from it. The relative importance of the antisymmetric shielding contributions might, however, increase dramatically in the case of strongly anisotropic motions (*cf.* Chapter 6). Conclusive experimental evidence for the occurrence of the antisymmetric CSA relaxation, along with a quantitative evaluation, was provided by Paquin and co-workers[24]. They measured the longitudinal spin-lattice relaxation rate (R_1) and heteronuclear NOE, along with the DD-CSA CCRRs, for the indole ^{15}N in tryptophan in solution. The measurements were carried out at four different magnetic fields over the temperature range $270 < T < 310$ K. The R_1 was assumed to contain contributions from the DD interaction with the directly bonded proton and from rank-two and rank-one CSA. The measurements of the NOE (cross-relaxation) provided an independent source of information on dipolar spectral densities, while the CCRRs depended on the DD interaction and the symmetric component of the CSA. Good fitting of the data to different motional models (see Chapter 6) was only possible after including the antisymmetric CSA, with the magnitude of about 10% of the total CSA effect.

5.3 Spin-Rotation Relaxation

The next source of relaxation that needs to be considered is the spin-rotation interaction. This relaxation mechanism is most important for spins in small symmetric molecules. The theory of relaxation through this interaction has been reviewed by McClung[25]. Detailed derivations of the spin-rotational relaxation can also be found in the paper by Hubbard[26] and in the book by McConnell (*Further reading*). The origin of the spin-rotation interaction is the fact that a rotating electronic charge distribution produces a magnetic field. The size of this magnetic field is proportional to the magnitude of the rotational angular momentum. The rotationally induced magnetic field can interact with the nuclear spins. The spin-rotation Hamiltonian can be expressed as:

$$\hat{H}_I^{SR} = \hat{\mathbf{I}} \cdot \mathbf{C} \cdot \hat{\mathbf{J}} \tag{5.48}$$

$\hat{\mathbf{I}}$ is the nuclear spin operator (a vector or a rank-one spherical tensor), $\hat{\mathbf{J}}$ is the angular momentum operator related to the rotation of the molecule (another vector or a rank-one tensorial quantity) and $\mathbf{C}$ is the rank-two spin-rotation coupling tensor. We assume, as elsewhere in this book, that the angular momentum operators are dimensionless and that the elements of the spin-rotation coupling tensor have the units rad s^{-1}. The Cartesian element C_{xy} is proportional to the magnitude of the magnetic field in the x-direction caused by the rotation around the molecular y-axis.

In analogy with the case of the CSA interaction, we have to be careful with coordinate frames. The spin-rotation coupling tensor is most conveniently expressed in its own principal axis system fixed in the molecule. We assume for simplicity that the tensor is axially symmetric, *i.e.* that it can in its principal frame be fully characterised by two components, one parallel to the axis, $C_{\parallel}$, and another perpendicular to the axis, $C_{\perp}$. The molecular frame is also appropriate for the rotational angular momentum. We therefore transform the spin operators, in irreducible spherical form (see Eq. (4.41)), into the molecular frame:

$$\hat{I}_k^{mol} = \sum_{q=-1}^{1} \hat{I}_{1,q} D_{q,k}^1 \left(\Omega_{L,mol}\right) \tag{5.49}$$

The Euler angles $\Omega_{L,mol}$ describe the transformation of the laboratory frame into the molecular frame. In analogy with the cases discussed earlier, the angles $\Omega_{L,mol}$ are time dependent through molecular tumbling. This is, however, not the only source of random time dependence in Eq. (5.48). The angular momentum, related to molecular rotation in condensed matter, also changes randomly. A common assumption is that the rotational motion of the molecule in a liquid can be treated classically and that the operator $\hat{J}_m$ (the mth component of the operator $\hat{\mathbf{J}}$ introduced in Eq. (5.48)) at time t can be replaced by $I_m\omega_m(t)$, where I_m is the mth principal component of the moment of inertia tensor and $\omega_m(t)$ is the corresponding angular velocity, with the stochastic time dependence now residing in the angular velocity. In a detailed analysis, one has to keep in mind that the principal axis system of the moment of inertia tensor and that of the spin-rotation tensor do not necessarily coincide. Re-formulating the spin-rotation Hamiltonian in the form of Eq. (4.38) is tedious and the reader is referred to the paper by Hubbard[26] or the book by McConnell (*Further reading*) for the full derivation. Here, it suffices to state that it is possible and that the resulting classical function $V_{l,q}(t)$ is time dependent through both the orientation of the molecular frame, with respect to the external magnetic field, and the angular velocity.

The calculation of the time-correlation functions, and spectral densities corresponding to these complicated quantities, is also tedious. Assuming that the orientation of the molecule at any instant and its angular velocity are uncorrelated, Hubbard obtained that the total time-correlation function can be expressed in terms of an angular velocity correlation function, $\langle\omega_m(t)\omega_m(t+\tau)\rangle$, and correlation functions for reorientation of rank-two spherical harmonics of the type encountered before in the context of the DD and CSA relaxation. Assuming that the molecule is a spherical top with the moment of inertia I, and that the molecular motion is isotropic rotational diffusion in small steps, the angular velocity correlation function becomes a decaying exponential with the time constant (correlation time) τ_J, which is simply related to the rotational correlation time for rank-two spherical harmonics, τ_c, through what is known as the *Hubbard relation*:

$$\tau_J\tau_c = \frac{I}{6k_BT} \tag{5.50}$$

The angular velocity (or angular momentum) correlation time is usually much shorter than the rotational correlation time. For the cases where spin-rotation relaxation may be important, for small symmetric molecules, both correlation times are short enough to guarantee the conditions of extreme narrowing. Under such conditions, the spin-rotation interaction gives rise to an exponential relaxation of both the longitudinal and the transverse component of the nuclear magnetisation, following the Bloch equations, with:

$$\frac{1}{T_{1,SR}} = \frac{1}{T_{2,SR}} = \left(\frac{2Ik_BT}{3\hbar^2}\right)\left(2C_\perp^2 + C_\parallel^2\right)\tau_J \tag{5.51}$$

We shall return to the spin-rotation relaxation in our discussion of relaxation in gas phase in Chapter 17.

5.4 Other Relaxation Mechanisms for Spin 1/2 Nuclei

In addition to the above-mentioned anisotropic interactions, which vary with time through processes related to molecular rotation, one more type of random modulation needs to be considered in the context of nuclear spin relaxation: chemical exchange. Chemical exchange means that the nuclear spins change their environment through a chemical process, which can be intramolecular (conformational exchange) or intermolecular. The exchange processes modulate anisotropic interactions, of the type mentioned in Sections 3.1 and 5.2, as well as isotropic interactions, such as chemical shift and J-coupling. Some of the exchange effects, which are important in the first place for transverse relaxation, can be described using Redfield theory, while others need to be dealt with in other ways. We choose to discuss all of these various aspects of chemical exchange in Chapter 13. In the case of dipolar or scalar IS coupling, a rapid relaxation of the S-spin through some other independent and efficient mechanism also provides a possible modulation of the two-spin interaction, which can contribute to the I-spin relaxation. We shall return to these effects in Chapters 14 and 15, discussing the quadrupolar nuclei and paramagnetic systems. In the case of indirect spin-spin coupling, modulated by either chemical exchange or S-spin relaxation, one usually speaks of *scalar relaxation*, of the first or second kind in the terminology of Abragam (*Further reading*).

The indirect spin-spin coupling interaction can contribute to relaxation in one more way. The interaction is, in fact, more complicated than indicated by Eq. (4.66). Besides the term $J_{IS}\hat{I}_z\hat{S}_z$ (and the non-secular terms $J_{IS}\hat{I}_x\hat{S}_x$ and $J_{IS}\hat{I}_y\hat{S}_y$) corresponding to the scalar product of the two spin operators, we should indeed take into account that the indirect coupling can in principle have an anisotropic, rank-two tensorial component similar to the dipolar coupling tensor. Actually, according to some authors (for example Levitt [*Further reading*]), the indirect spin-spin coupling is referred to as *indirect dipole-dipole interaction*. In the corresponding Hamiltonian, analogous to the dipolar Hamiltonian of Eq. (3.13), the J-tensor anisotropy expressed in rad s^{-1}, $2\pi\Delta J$, takes the place of the dipole-coupling constant. The anisotropy of the indirect coupling tensor is a very small correction to the direct dipolar coupling for light spin 1/2 nuclei, but the situation may turn out to be different for heavy nuclei, where the isotropic J-couplings may be very large and their anisotropies are unknown.

Summarising the chapter on relaxation mechanisms for spin 1/2 nuclei, we conclude that the simple additivity of contributions from different relaxation mechanisms to the total observed spin lattice-relaxation rate as usually presented:

$$\frac{1}{T_{1,obs}} = \frac{1}{T_{1,DD}} + \frac{1}{T_{1,CSA}} + \frac{1}{T_{1,CSA}^{anti}} + \frac{1}{T_{1,SR}} + \frac{1}{T_{1,SC}} \tag{5.52}$$

where $1/T_{1,DD}$ may include the J-coupling anisotropy and $1/T_{1,SC}$ refers to the scalar relaxation, is only valid under certain conditions. The conditions are that the cross-relaxation effects are suppressed through the experimental setup (which usually is the case for, *e.g.*, carbon-13 and nitrogen-15 relaxation under broadband decoupling of protons) and that the cross-correlation effects are either absent or suppressed. A similar statement is true for the transverse relaxation, where additional caution should be exercised in the presence of chemical exchange. Under more general conditions, one should work in terms of magnetisation modes (or a similar approach) and evaluate the contributions to different elements of the relaxation matrix, following the statement from the important paper by Szymanski *et al.*[27] that the concept of a relaxation mechanism should refer to a pair of (identical or different) interactions rather than to a single one.

Exercises for Chapter 5

1. Check, using relations in Eqs. (5.7–5.9), that Eq. (5.10) indeed is fulfilled.
2. Calculate the decay rate constant for the coherences, $R_{1212}=R_{2121}$, in analogy with Eq. (5.36).

References

1. Kumar, A.; Grace, R. C. R.; and Madhu, P. K., Cross-correlations in NMR. *Prog. NMR Spectr.* 2000, 37, 191–319.
2. Werbelow, L. G., *Relaxation Processes: Cross Correlation and Interference Terms*. Wiley: eMagRes, 2011.
3. Brutscher, B., Principles and applications of cross-correlated relaxation in biomolecules. *Concepts Magn. Reson.* 2000, 12, 207–229.
4. Werbelow, L. G.; and Grant, D. M., Intramolecular dipolar relaxation in multispin systems. *Adv. Magn. Reson.* 1977, 9, 189–299.
5. Canet, D., *Relaxation Mechanisms: Magnetisation Modes*. Wiley: eMagRes, 2007.

6. Hubbard, P. S., Nuclear magnetic relaxation of three and four spin molecules in a liquid. *Phys. Rev.* 1958, 109, 1153–1158.
7. Brondeau, J.; and Canet, D., Longitudinal magnetic relaxation of 13C (or 15N) interacting with a strongly irradiated proton system. *J. Chem. Phys.* 1977, 67, 3650–3654.
8. Kowalewski, J.; Effemey, M.; and Jokisaari, J., Dipole-dipole coupling constant for a directly bonded CH pair: A carbon-13 relaxation study. *J. Magn. Reson.* 2002, 157, 171–177.
9. Chang, Z. W.; and Halle, B., Longitudinal relaxation in dipole-coupled homonuclear three-spin systems: Distinct correlations and odd spectral densities. *J. Chem. Phys.* 2015, 143, 234201.
10. Smith, S. A.; Palke, W. E.; and Gerig, J. T., The Hamiltonians of NMR. Part I. *Concepts Magn. Reson.* 1992, 4, 107–144.
11. Smith, S. A.; Palke, W. E.; and Gerig, J. T.,The Hamiltonians of NMR. Part II. *Concepts Magn. Reson.* 1992, 4, 181–204.
12. Smith, S. A.; Palke, W. E.; and Gerig, J. T.,The Hamiltonians of NMR. Part III. *Concepts Magn. Reson.* 1993, 5, 151–177.
13. Anet, F. A. L.; and O'Leary, D. J., The shielding tensor. Part I: Understanding its symmetry properties. *Concepts Magn. Reson.* 1991, 3, 193–214.
14. Anet, F. A. L.; and O'Leary, D. J., The shielding tensor. Part II. Understanding its strange effects on relaxation. *Concepts Magn. Reson.* 1992, 4, 35–52.
15. Goldman, M., Interference effects in the relaxation of a pair of unlike spin-1/2 nuclei. *J. Magn. Reson.* 1984, 60, 437–452.
16. Shimizu, H., Theory of the dependence of nuclear magnetic relaxation on the absolute sign of spin-spin coupling constant. *J. Chem. Phys.* 1964, 40, 3357–3364.
17. Smith, S. A.; Palke, W. E.; and Gerig, J. T., The hamiltonians of NMR. Part IV: NMR relaxation. *Concepts Magn. Reson.* 1994, 6, 137–162.
18. Mäler, L.; and Kowalewski, J., Cross-correlation effects in the longitudinal relaxation of heteronuclear spin systems. *Chem. Phys. Lett.* 1992, 192, 595–600.
19. Wokaun, A.; and Ernst, R. R., The use of multiple quantum transitions for relaxation studies in coupled spin systems. *Mol. Phys.* 1978, 36, 317–341.
20. Werbelow, L. G., The use of interference terms to separate dipolar and chemical shift anisotropy contributions to nuclear spin relaxation. *J. Magn. Reson.* 1987, 71, 151–153.
21. Pellecchia, M.; Pang, Y. X.; Wang, L. C.; Kurochkin, A. V.; Kumar, A.; and Zuiderweg, E. R. P., Quantitative measurement of cross-correlations between N-15 and (CO)-C-13 chemical shift anisotropy relaxation mechanisms by multiple quantum NMR. *J. Am. Chem. Soc.* 1999, 121, 9165–9170.
22. Blicharski, J. S., Nuclear magnetic relaxation by anisotropy of the chemical shift. *Z. Naturforsch. A* 1972, 27, 1456–1458.
23. Kowalewski, J.; and Werbelow, L., Evaluation of spin relaxation induced by chemical shielding anisotropy: A comment on the importance of the antisymmetric component. *J. Magn. Reson.* 1997, 128, 144–148.
24. Paquin, R.; Pelupessy, P.; Duma, L.; Gervais, C.; and Bodenhausen, G., Determination of the antisymmetric part of the chemical shift anisotropy tensor via spin relaxation in nuclear magnetic resonance. *J. Chem. Phys.* 2010, 133, 034506.
25. McClung, R. E. D., *Spin-Rotation Relaxation Theory.* Wiley: eMagRes, 2007.
26. Hubbard, P. S., Theory of nuclear magnetic relaxation by spin-rotational interactions in liquids. *Phys. Rev.* 1963, 131, 1155–1165.
27. Szymanski, S.; Gryff-Keller, A. M.; and Binsch, G., A Liouville space formulation of Wangsness-Bloch-Redfield theory of nuclear spin relaxation suitable for machine computation. I. Fundamental aspects. *J. Magn. Reson.* 1986, 68, 399–432.

6

Spectral Densities and Molecular Dynamics

Spectral densities are a central concept in nuclear magnetic resonance (NMR) relaxation theory, as we saw already at the beginning of this book (Chapter 2). They provide the tool through which measurements of relaxation parameters convey information on molecular dynamics. So far, we have only introduced the spectral density function for the very simple model describing the motion of a rigid spherical object. In this chapter, we deal with the theory of time-correlation functions and spectral densities for different motional models, starting again with rigid, spherical objects and proceeding toward more complex systems. Throughout this chapter, we limit ourselves to dealing with ordinary, isotropic liquids, *i.e.* systems whose macroscopic properties do not depend on orientation. We shall very briefly discuss the case of anisotropic liquids in Chapter 17. In Sections 6.1 and 6.2, we concentrate on rigid molecular systems characterised by reasonably rapid motions, and in Section 6.3 we introduce the spectral density functions for molecules undergoing internal motion. Section 6.4 provides a brief presentation of spectral density functions based on distributions of correlation times. Experimental approaches to studies of molecular dynamics are presented in Chapter 11, in which the details of the physical information that can be obtained from spin relaxation are explored.

6.1 Time-Correlation Function for Isotropic Small-Step Rotational Diffusion

Several presentations on the subject of simple time-correlation functions exist in the literature, for example, the detailed description by Hennel and Klinowski (*Further reading*). A less detailed, but similar, presentation is also given in the book by Abragam (*Further reading*). The concepts behind the derivation can be traced back to Debye's classical work on dielectric relaxation[1] and the paper by Bloembergen, Pound and Purcell[2]. We shall start this chapter by dealing with a particularly simple system. The simplest possible case of reorientational motion in condensed matter is that of a rigid, spherical object undergoing a rotational diffusion process in small steps, as illustrated in Figure 2.1.

In the theory of relaxation through rank-two interactions, such as the dipole-dipole (DD) interaction and chemical shielding anisotropy (CSA) (as well as the quadrupolar interaction, to be discussed in Chapter 14), a time-correlation function for the normalised spherical harmonics $Y_{l,m}$ with $l=2$ is required. The case of $m=0$ was introduced already in Chapter 3 (compare Eq. (3.14)):

$$Y_{2,0} = \frac{1}{4}\sqrt{\frac{5}{\pi}}\left(3\cos^2\theta - 1\right) \tag{6.1}$$

Here, θ is the angle between a molecule-fixed vector and the external magnetic field. To fully specify the direction of the vector in the laboratory frame, we also need a second angle ϕ. We are going to use the symbol Ω (solid angle) for the full description of the direction of the vector, including both angles. The time-correlation function that we need to evaluate is:

$$G_2(\tau)=\left\langle Y_{2,0}(t)Y_{2,0}^*(t+\tau)\right\rangle=\iint Y_{2,0}(\Omega_0)Y_{2,0}^*(\Omega)P(\Omega_0)P(\Omega_0\,|\,\Omega,\tau)d\Omega_0 d\Omega \tag{6.2}$$

The subscript 2 on the time-correlation function refers to the fact that we are dealing with spherical harmonics with $l = 2$. As we shall see below, the correlation function is not dependent on the value of m. The symbol Ω_0 stands for the orientation of the axis of interest at time zero, Ω at time τ. The complex conjugation sign on $Y_{2,0}^*$ is, strictly speaking, unnecessary, as the function is real. We introduce it to make it easier to generalise the expression for future use in more complicated cases. $P(\Omega_0)$ is the probability density of the initial orientation specified by the solid angle Ω_0, and $P(\Omega_0|\Omega,\tau)$ is the conditional probability density that the axis will be oriented at the angle Ω at time τ, provided that it was Ω_0 at time zero.

We begin the calculation by looking at $P(\Omega_0)$. The problem of calculating this property becomes very simple if we assume that the medium in which the molecule reorients is *isotropic*. This assumption is fulfilled in all ordinary liquids and is much less stringent than the assumption of *isotropic motion*. In an isotropic medium, all angles Ω are equally probable:

$$P(\Omega)=P(\Omega_0)=\frac{1}{4\pi} \tag{6.3a}$$

This particular value for the uniform probability is consistent with the condition that the integral over all space of a probability density function must be equal to unity:

$$\int\limits_{\text{all space}} P(\Omega_0)d\Omega_0=\int_0^{2\pi}\int_0^{\pi}\frac{1}{4\pi}\sin\theta d\theta d\phi=\int_0^{\pi}\frac{1}{2}\sin\theta d\theta=1 \tag{6.3b}$$

By substituting Eq. (6.3a) into (6.2), we obtain:

$$G_2(\tau)=\frac{1}{4\pi}\iint Y_{2,0}(\Omega_0)Y_{2,0}^*(\Omega)P(\Omega_0\,|\,\Omega,\tau)d\Omega_0 d\Omega \tag{6.4}$$

Our next task is to find $P(\Omega_0|\Omega,t)$, the conditional probability density or the *propagator*. In order to do this, we need to examine the motion of our system. The physical model we deal with – small-step rotational diffusion, or diffusion on the surface of a sphere – is a special case of general three-dimensional diffusion. Fick's second law of diffusion describes the three-dimensional diffusion process in an isotropic medium:

$$\frac{\partial f(x,y,z)}{\partial t}=D\left(\frac{\partial^2 f(x,y,z)}{\partial x^2}+\frac{\partial^2 f(x,y,z)}{\partial y^2}+\frac{\partial^2 f(x,y,z)}{\partial z^2}\right) \tag{6.5a}$$

The function f is a measure of the concentration of diffusing particles at a certain point (x, y, z) and is a function of time. D is a diffusion coefficient.

Eq. (6.5a) can be transformed to spherical polar coordinates (r, θ, ϕ). This transformation is analogous to that performed in order to separate variables in the solution of the time-independent Schrödinger equation for the hydrogen atom, *cf.* Atkins and Friedman (*Further reading*). By concentrating on the angular part, and by assuming that r is constant and equal to the radius of the sphere, we obtain Fick's law for rotational diffusion:

$$\frac{\partial}{\partial\tau}f(\Omega,\tau)=D_R\hat{\Delta}_R f(\Omega,\tau) \tag{6.5b}$$

f is now a measure of the concentration of particles at a certain solid angle Ω, and D_R is the rotational diffusion coefficient. $\hat{\Delta}_R$ is an expression in θ and ϕ, called the *Legendre operator*; see Atkins and Friedman (*Further reading*). The Legendre operator is closely related to the total orbital angular momentum operator, $\hat{L}^2$. In analogy with the $\hat{L}^2$ operator, $\hat{\Delta}_R$ also has spherical harmonics as eigenfunctions, but with eigenvalues of the opposite sign, $-l(l+1)$:

$$\hat{\Delta}_R Y_{l,m} = -l(l+1)Y_{l,m} \tag{6.6}$$

In order to be useful as a propagator, $f(\Omega,\tau)$ has to fulfil two additional conditions. It has to be normalised in such a way that:

$$\int f(\Omega,\tau)d\Omega = 1 \tag{6.7}$$

and the solid angle Ω at $\tau = 0$ has to be identical to Ω_0:

$$f(\Omega,0) = \delta(\Omega - \Omega_0) \tag{6.8}$$

$\delta(x-a)$ is the Dirac delta function, with the following definition:

$$\int_{-\infty}^{\infty} f(x)\delta(x-a)dx = f(a) \tag{6.9}$$

We can say that when the δ-function is multiplied with another function and the product is integrated, the operation "projects" out the value of that other function at the point where the argument of the δ-function is zero. The properties of the Dirac delta function are discussed in the book by Hennel and Klinowski (*Further reading*) and here we will use the fact that the δ-function can be expanded in any complete and orthonormal set of functions:

$$\delta(x - x') = \sum_n \varphi_n(x)\varphi_n^*(x') \tag{6.10a}$$

The spherical harmonics form a complete and orthonormal set of functions of the solid angle:

$$\int Y_{l,m}^*(\Omega)Y_{l',m'}(\Omega)d\Omega = \delta_{ll'}\delta_{mm'} \tag{6.10b}$$

This allows us to expand the delta function of Eq. (6.8) as:

$$\delta(\Omega - \Omega_0) = \sum_{l,m} Y_{l,m}(\Omega)Y_{l,m}^*(\Omega_0) \tag{6.10c}$$

With the conditions specified by Eqs. (6.8) and (6.9), the solution to the rotational diffusion equation becomes acceptable as the propagator $P(\Omega_0|\Omega,t)$. To find this propagator, we are going to use the fact that the spherical harmonics form a complete set of eigenfunctions to $\hat{\Delta}_R$. We can expand $f(\Omega,\tau)$ at a certain time τ' in spherical harmonics, which gives:

$$f(\Omega,\tau') = \sum_{l=0}^{\infty}\sum_{m=-l}^{l} B_{l,m}Y_{l,m}(\Omega) \tag{6.11}$$

For other τ values, the coefficients $B_{l,m}$ are different, while $Y_{l,m}$ are the same (time independent). In general, we have:

$$f(\Omega,\tau)=\sum_{l=0}^{\infty}\sum_{m=-l}^{l}B_{l,m}(\tau)Y_{l,m}(\Omega) \tag{6.12}$$

and we only need to compute the time-dependent $B_{l,m}(\tau)$. By substituting Eq. (6.12) into Eq. (6.5b) and using Eq. (6.6), after some algebra the following is obtained:

$$\frac{\partial}{\partial\tau}B_{l,m}(\tau)+D_R l(l+1)B_{l,m}(\tau)=0 \tag{6.13}$$

which has a general solution:

$$B_{l,m}(\tau)=C_{l,m}\exp\left[-l(l+1)D_R\tau\right] \tag{6.14}$$

Combining Eqs. (6.12) and (6.14) gives the following expression for $f(\Omega,\tau)$:

$$f(\Omega,\tau)=\sum_{l=0}^{\infty}\sum_{m=-l}^{l}C_{l,m}Y_{l,m}(\Omega)\exp\left[-l(l+1)D_R\tau\right] \tag{6.15}$$

The constants $C_{l,m}$ can be determined from the initial condition, given in Eq. (6.8), and the expansion of the δ-function in spherical harmonics, according to Eq. (6.10c). This results in:

$$C_{l,m}=Y_{l,m}^{*}(\Omega_0) \tag{6.16}$$

and the propagator becomes:

$$P(\Omega_0\,|\,\Omega,\tau)=\sum_{l=0}^{\infty}\sum_{m=-l}^{l}Y_{l,m}^{*}(\Omega_0)Y_{l,m}(\Omega)\exp\left[-l(l+1)D_R\tau\right] \tag{6.17}$$

For the evaluation of $G(\tau)$, we now need to multiply Eq. (6.17) with $Y_{2,0}(\Omega_0)Y_{2,0}^{*}(\Omega)$ and to integrate over Ω_0 and Ω according to Eq. (6.4). The fact that the spherical harmonics form an orthonormal set, as specified in Eq. (6.10b), picks out one and only one term from the summation in Eq. (6.17), that with $l=2$, $m=0$. Thus, we finally obtain:

$$G_2(\tau)=\frac{1}{4\pi}\exp\left[-6D_R\tau\right] \tag{6.18}$$

which is identical to the exponential function we "guessed" in Section 2.3, if we identify

$$\tau_c=\frac{1}{6D_R} \tag{6.19}$$

The one-sided Fourier transformation of Eq. (6.18) results in:

$$J_2(\omega)=\frac{1}{4\pi}\frac{12D_R}{(6D_R)^2+\omega^2}=\frac{1}{4\pi}\frac{2\tau_c}{1+\omega^2\tau_c^2}=\frac{1}{2\pi}\frac{\tau_c}{1+\omega^2\tau_c^2} \tag{6.20}$$

where the last expression is identical to Eq. (3.16). The same result as in Eq. (6.18) would be obtained if we wanted a time-autocorrelation function for the spherical harmonic with $l=2$ and another m-value. The situation would be very different, however, if we tried to calculate a cross-correlation function involving two different spherical harmonics, *e.g.* $\langle Y_{2,0}(t)Y^*_{2,m}(t+\tau)\rangle$ with $m\neq 0$. Forming an expression analogous to Eq. (6.4) and using Eq. (6.17) for the propagator, we get the result that the cross-correlation function vanishes, because we cannot simultaneously fulfil two different conditions of the type specified in Eq. (6.10b). This explains why we obtain the symmetry relation that was presented in Chapter 4 (Eq. (4.52)) and that we can make use of its consequences. The same result, a vanishing cross-correlation function, would be the result if we tried to calculate $\langle Y_{l,m}(t)Y^*_{l',m'}(t+\tau)\rangle$ for $l\neq l'$. These important symmetry properties of the correlation functions relevant for NMR relaxation were derived by Hubbard[3].

Let us now turn our attention to cross-correlation functions (and cross-correlated spectral densities) involving different interactions, mentioned in Section 5.1 and, in larger detail, in Section 5.2. Taking the Hamiltonian of Eq. (5.41), which includes both the DD and CSA spin interactions, as a starting point, the relevant cross-correlation function will be of the form $\langle Y_{2,0}(\Omega^L_{IS}(0))Y^*_{2,0}(\Omega^L_I(\tau))\rangle$. In order to evaluate this function, we need to express $Y_{2,0}(\Omega^L_I(\tau))$ in terms of $Y_{2,0}(\Omega^L_{IS}(\tau))$. The transformation between the principal frames of the CSA and the DD interactions is time independent:

$$Y_{2,0}\left(\Omega_I^L\right)=\sum_{q=-2}^{2}D^2_{q,0}\left(\Omega_{IS,I}\right)Y_{2,q}\left(\Omega_{IS}^L\right) \tag{6.21a}$$

or, after complex conjugation:

$$Y^*_{2,0}\left(\Omega_I^L\right)=\sum_{q=-2}^{2}\left(D^2_{q,0}\left(\Omega_{IS,I}\right)\right)^*Y^*_{2,q}\left(\Omega_{IS}^L\right) \tag{6.21b}$$

The relevant spectral density becomes:

$$\begin{aligned}G_2^{DD-CSA}(\tau)&=\left\langle Y_{2,0}\left(\Omega_{IS}^L(0)\right)Y^*_{2,0}\left(\Omega_I^L(\tau)\right)\right\rangle\\
&=\sum_{q=-2}^{2}\left(D^2_{q,0}\left(\Omega_{IS,I}\right)\right)^*\left\langle Y_{2,0}\left(\Omega_{IS}^L(0)\right)Y^*_{2,q}\left(\Omega_{IS}^L(\tau)\right)\right\rangle\\
&=\left(D^2_{0,0}\left(\Omega_{IS,I}\right)\right)^*\left\langle Y_{2,0}\left(\Omega_{IS}^L(0)\right)Y^*_{2,0}\left(\Omega_{IS}^L(\tau)\right)\right\rangle\\
&=\frac{1}{2}\left(3\cos^2\theta_{IS,I}-1\right)\left\langle Y_{2,0}\left(\Omega_{IS}^L(0)\right)Y^*_{2,0}\left(\Omega_{IS}^L(\tau)\right)\right\rangle\\
&=\frac{1}{2}\left(3\cos^2\theta_{IS,I}-1\right)G_2(\tau)\end{aligned} \tag{6.21c}$$

Between the second and the third line above, we have used the fact that cross-correlation functions for $q\neq 0$ vanish. In the next to last line, we have introduced the explicit form of $D^2_{0,0}(\Omega_{IS,I})$. Clearly, for isotropic reorientation of rigid objects, the cross-correlation functions and the spectral densities only differ from the case of autocorrelation counterparts by a geometric factor, originating from a transformation between different molecule-fixed frames, the result anticipated in Eqs. (5.44).

Before leaving this section, we wish to comment on the case of autocorrelation functions for spherical harmonics with l-values other than $l=2$, *e.g.* rank-one interactions. The propagator of Eq. (6.17) is valid generally for the system following Eq. (6.5b). Multiplying Eq. (6.17) with $Y_{l,m}(\Omega_0)Y^*_{l,m}(\Omega)$, for an arbitrary l and m, and integrating over Ω_0 and Ω yields a generalisation of Eq. (6.18) to

$$G_l(\tau)=\frac{1}{4\pi}\exp\left[-l(l+1)D_R\tau\right] \tag{6.22a}$$

Clearly, the autocorrelation function remains a simple decaying exponential, but the correlation time depends on l:

$$\tau_l = \frac{1}{l(l+1)D_R} \tag{6.22b}$$

One should recall that we have encountered the case of $l=1$ in the context of the randomly reorienting field model (Chapter 2) and the antisymmetric part of the CSA in Section 5.2. Because of the result in Eq. (6.22b), the correlation time occurring in Eqs. (5.47) is three times longer than its more common $l=2$ counterpart.

6.2 More General Rigid Objects

The isotropic reorientation model discussed in the previous section applies, apart from spheres, also to other rigid, highly symmetric objects such as the octahedron, the cube, the icosahedron and some other related shapes. Even though it may be considered as a useful first approximation also for less symmetric objects, it is clear that it will certainly not apply to all molecules. Therefore, we need to introduce a model for the reorientation of non-symmetric molecules. In the first part of this section, we discuss the situation arising if we wish to loosen the assumption of isotropic reorientation, but still limit ourselves to rigid bodies undergoing small-step rotational diffusion. In the final part of the section, we connect the result of the small-step rotational diffusion approach to hydrodynamics and allow for certain complicating features in that description. Non-rigid molecules will be discussed in Section 6.3.

6.2.1 The Asymmetric and Symmetric Top

Perrin generalised Debye's treatment of rotational diffusion of a spherical top to an arbitrary rotational ellipsoid already in the 1930s[4]. In the NMR context, this was a subject that was treated early on by several authors [5–7]. Interesting recent reviews on rotational diffusion for non-symmetric molecules can be found in the NMR Encyclopedia (eMagRes), by Woessner[8] and by Grant and Brown[9]. We are going to treat a rigid object of arbitrary shape (*asymmetric top*) undergoing a small-step rotational diffusion. We denote this object as an *asymmetric diffusor*.

If the nuclear spin resides in an asymmetric diffusor, our coordinate transformation from the principal frame (principal axis system [PAS]) of the interaction governing the relaxation Hamiltonian (DD or the CSA interaction) to the laboratory frame needs to be done in two steps. First, we perform a time-independent transformation from the PAS to a coordinate system appropriate for the description of the anisotropic rotational diffusion (the *principal diffusion frame* [PDF]). Second, we transform from the PDF into the laboratory frame. This second transformation is time dependent through the time variation of the Euler angles between the two frames. Using these ideas, we can write the perturbing Hamiltonian in the following way:

$$\begin{aligned}\hat{H}_1(t) &= \xi_\eta \sum_{m=-2}^{2} (-1)^m V_{2,m}^{\eta,LAB} \hat{T}_{2,-m}^{\eta,LAB} = \xi_\eta \sum_{m=-2}^{2} (-1)^m \left(\sum_{q=-2}^{2} D_{q,m}^2(\Omega_{D,L}(t)) V_{2,q}^{\eta,PDF} \right) \hat{T}_{2,-m}^{\eta,LAB} \\ &= \xi_\eta \sum_{m=-2}^{2} (-1)^m \left(\sum_{p=-2}^{2} \sum_{q=-2}^{2} D_{p,q}^2(\Omega_{\eta,D}) D_{q,m}^2(\Omega_{D,L}(t)) V_{2,p}^{\eta,PAS} \right) \hat{T}_{2,-m}^{\eta,LAB}\end{aligned} \tag{6.23}$$

The symbol η designates the interaction under consideration, which can be DD or CSA. The Wigner rotation matrix elements are written as $D_{p,q}^2(\Omega_{old,new}) = D_{p,q}^2(\alpha, \beta, \gamma)$. The symbols α, β and γ are the three Euler angles (compare Figure 4.2), and Ω is a three-dimensional counterpart of the solid angle designated with the same symbol as in the preceding section. Since Ω is used as an argument of either spherical harmonics (functions of two angles) or Wigner rotation matrix elements (functions of three angles), we judge that there should be no confusion caused by this notation. The pairs of indices on the angles in Eq. (6.23) indicate explicitly the transformations: D,L is a shorthand notation for PDF,LAB (meaning the transformation *from* the PDF *to* LAB) and η,D for PAS(η),PDF. We note that the angle $\Omega_{\eta,D}$ transforms between two molecule-fixed frames and is independent of time.

In analogy with the symmetry relation in Eq. (4.52) and the discussion in the previous section, we are going to need the time-correlation function

$$
\begin{aligned}
G_{2,m}(\tau) &= \left\langle V_{2,m}^{\eta,LAB}(t)\left(V_{2,m}^{\eta,LAB}(t+\tau)\right)^*\right\rangle \\
&= \sum_{p=-2}^{2}\sum_{p'=-2}^{2}\sum_{q=-2}^{2}\sum_{q'=-2}^{2} D_{p,q}^2(\Omega_{\eta,D})\left(D_{p',q'}^2(\Omega_{\eta,D})\right)^* V_{2,p}^{\eta,PAS}\left(V_{2,p'}^{\eta,PAS}\right)^* \\
&\quad\times\left\langle D_{q,m}^2(\Omega_{D,L}(t))\left(D_{q',m}^2(\Omega_{D,L}(t+\tau))\right)^*\right\rangle
\end{aligned}
\tag{6.24}
$$

or, indeed, the spectral density function, which is the one-sided Fourier transform of this time-correlation function. In isotropic liquids, we saw in the previous section that the function $G_{2,m}(\tau)$ does not depend on m. For the sake of generality, and in order to conform to other presentations[7,9], we retain the m for the time being.

Instead of evaluating the time-correlation function from Eq. (6.2), we now need to derive the expression for the time-correlation function occurring in Eq. (6.24):

$$
\begin{aligned}
G_{q,q',m}^2(\tau) &= \frac{5}{4\pi}\left\langle D_{q,m}^2(t)\left(D_{q',m}^2(t+\tau)\right)^*\right\rangle \\
&= \frac{5}{4\pi}\iint D_{q,m}^2(\Omega_0)\left(D_{q',m}^2(\Omega)\right)^* P(\Omega_0)P(\Omega_0\,|\,\Omega,\tau)\,d\Omega_0 d\Omega
\end{aligned}
\tag{6.25}
$$

The factor $5/4\pi$ is inserted to account for the difference in normalisation of Wigner rotation matrices and spherical harmonics (we have, for example, $D_{0,0}^2 = \frac{1}{2}(3\cos^2\beta - 1) = \left(\frac{4\pi}{5}\right)^{1/2} Y_{2,0}$). In an isotropic medium, the probability density $P(\Omega_0)$ is in this three-dimensional case equal to:

$$
P(\Omega_0) = \frac{1}{8\pi^2} \tag{6.26}
$$

because the integration over the third angle contributes another factor 2π (compare with the two-dimensional case given in Eq. (6.3a)). Next, we need to find the propagator. For this purpose, we need a generalisation of Eq. (6.5b) describing the rotational diffusion. For a rigid body of arbitrary shape, one rotational diffusion coefficient is not sufficient, since diffusion is in general orientation dependent. Therefore, we instead introduce a (Cartesian) diffusion tensor, **D**, with the elements:

$$
D_{ij} = \frac{1}{2}\lim_{\Delta t\to 0}\frac{\theta_i\theta_j}{\Delta t} \tag{6.27}
$$

where θ_i and θ_j are the rotation angles about the i and j Cartesian axes, by which the body reorients during the time interval Δt. The limit of Δt approaching zero should not be taken literally: we need an interval short on the time scale of macroscopic events, but long on the typical molecular time scale. In terms of the diffusion tensor and the propagator, the rotational diffusion equation takes the form:

$$\frac{\partial P(\Omega_0\,|\,\Omega,\tau)}{\partial \tau} = -\sum_{i,j=x,y,z} \hat{L}_i D_{ij} \hat{L}_j P(\Omega_0|\Omega,\tau) \tag{6.28a}$$

The symbols $\hat{L}_i$ and $\hat{L}_j$ denote the operator for the i and j component of the angular momentum, respectively. The diffusion tensor is real and symmetric. We can convert it into diagonal form by a suitable rotation of the coordinate frame. In this principal diffusion frame, Eq. (6.28a) simplifies to:

$$\frac{\partial P(\Omega_0|\Omega,\tau)}{\partial \tau} = -\sum_{j=x,y,z} D_{jj} \hat{L}_j^2 P(\Omega_0|\Omega,\tau) = -\hat{R}^{PDF} P(\Omega_0|\Omega,\tau) \tag{6.28b}$$

$\hat{R}^{PDF}$ is a rotational diffusion operator in the principal diffusion frame. We note that Eq. (6.28b) is very similar to the time-dependent Schrödinger equation for rotational motion of an asymmetric rotor, which describes the rotation of an isolated molecule in gas under low pressure:

$$\frac{\partial \Psi(t)}{\partial t} = -i \sum_{j=x,y,z} \frac{\hbar^2}{2I_{jj}} \hat{L}_j^2 \Psi(t) \tag{6.29}$$

where I_{jj} denotes the jth principal component of the moment of inertia tensor in its principal frame.

The main difference between Eqs. (6.28b) and (6.29) is the presence of the imaginary number i in the latter. The physical consequence of this mathematical difference is that while Eq. (6.29) describes an oscillatory motion, Eq. (6.28b) characterises a damped dynamics. We can make use of Eq. (6.29) by using the complete set of eigenfunctions for the asymmetric rotor, $\Psi_j(\Omega)$, in an expansion similar to Eq. (6.12):

$$P(\Omega_0|\Omega,\tau) = \sum_{j=0}^{\infty} B_j(\tau)\Psi_j(\Omega) = \sum_{l}\sum_{\nu} B_{l,\nu}(\tau)\Psi_{l,\nu}(\Omega) \tag{6.30}$$

In the second equality above, we have used the fact that the total angular momentum quantum number l remains a good quantum number for the asymmetric rigid rotor problem. The explicit dependence of $P(\Omega_0|\Omega,\tau)$ on τ is now contained in the coefficients $B_j(\tau)$ or $B_{l,\nu}(\tau)$. Using an analogous procedure to that presented in Eqs. (6.13)–(6.16), we obtain the following expression for the propagator, which is similar to Eq. (6.17):

$$P(\Omega_0|\Omega,\tau) = \sum_{l,\nu} \Psi_{l,\nu}^*(\Omega_0)\Psi_{l,\nu}(\Omega)\exp(-b_{l,\nu}\tau) \tag{6.31}$$

The symbols $b_{l,\nu}$ denote the eigenvalues of the rotational diffusion operator $\hat{R}^{PDF}$, related to the asymmetric rotor Hamiltonian. The asymmetric rotor eigenfunctions can, in turn, be expanded in Wigner rotation matrices:

$$\Psi_{l,\nu}(\Omega) = \sum_{k,j} C_{\nu,k,j}^{l} \sqrt{\frac{2l+1}{8\pi^2}} D_{k,j}^{l}(\Omega) \tag{6.32}$$

We can treat the symbols $C_{\nu,k,j}^{l}$ as elements of the eigenvectors of the rotational diffusion operator. Using Eq. (6.32), we can re-write the propagator as an expansion in Wigner rotation matrices. The details of this calculation can be found in the paper by Grant and Brown[9]. To obtain the time-correlation function of Eq. (6.25), we use the orthogonality and normalisation of the Wigner rotation matrices and finally obtain:

$$G^2_{q,q',m}(\tau) = \frac{(-1)^{m-q'}}{4\pi}\sum_{v=-2}^{2} C^2_{v,q,m}C^2_{v,-q',m}\exp(-b_{2,v}\tau) \tag{6.33a}$$

In analogy with the case discussed below Eq. (6.20), the eigenvectors to the operator $\hat{R}^{PDF}$ do not depend on the index m, which can therefore be omitted in Eq. (6.33a), yielding:

$$G^2_{q,q'}(\tau) = \frac{(-1)^{q'}}{4\pi}\sum_{v=-2}^{2} C^2_{v,q}C^2_{v,-q'}\exp(-b_{2,v}\tau) \tag{6.33b}$$

The relevant functions that need to be used in Eq. (6.33b), the eigenvalues to $\hat{R}^{PDF}$ for $l=2$, $b_{2,v}$, and non-zero elements of the eigenvectors, are tabulated in Table 6.1, based on the review by Grant and Brown.[9]

Molecular symmetry simplifies the expressions in Table 6.1. In the case of an axially *symmetric diffusor* (or, briefly, symmetric diffusor), $D_{xx} = D_{yy} = D_{\perp}$ and $D_{zz} = D_{\parallel}$. The eigenvalues with $v=\pm k\,(k=1,2)$ become degenerate and take on simpler forms: $b_{2,0} = 6D_{\perp}$; $b_{2,1} = 5D_{\perp} + D_{\parallel}$; $b_{2,2} = 2D_{\perp} + 4D_{\parallel}$ (the reader is encouraged to show this!). For an even higher symmetry, $D_{\perp} = D_{\parallel} = D_R$, we obtain only one term in the sum over v, with the eigenvalue $6D_R$, *i.e.* we recover the solution of the problem with the spherical object discussed in the previous section.

The Fourier transform is a linear operation, *i.e.* it converts each exponential term in Eq. (6.33b) into a corresponding Lorentzian, which for the asymmetric diffusor results in spectral density-like objects, $M^2_{n,n'}(\omega)$, expressed as sums of five Lorentzians:

$$M^2_{q,q'}(\omega) = \frac{1}{2\pi}(-1)^{q'}\sum_{v=-2}^{2} C^2_{v,q}C^2_{v,-q'}\frac{b_{2,v}}{b^2_{2,v}+\omega^2} \tag{6.34}$$

Similar functions can be created for the symmetric diffusor case, where the summation simplifies to three terms, while a single Lorentzian is obtained for the spherical case in analogy with Eq. (6.20).

To obtain the spectral density $J_2(\omega)$, we need to perform the quadruple summation of Eq. (6.24). Assume that we deal with the case of an axially symmetric interaction, such as the DD interaction or the CSA with the asymmetry parameter $\eta=0$. The summations over p and p' then reduce to a single term, $p=p'=0$ and we obtain:

TABLE 6.1 Eigenvalues and Non-Zero Eigenvector Elements for Rotational Diffusion of an Asymmetric Diffusor

v	$b_{2,v}$	$C^2_{v,q}$
−2	$D_{xx}+D_{yy}+4D_{zz}$	$-C^2_{-2,-2} = C^2_{-2,+2} = \sqrt{1/2}$
−1	$D_{xx}+4D_{yy}+D_{zz}$	$-C^2_{-1,-1} = C^2_{-1,+1} = \sqrt{1/2}$
0	$6D+6\sqrt{D^2-R^2}$	$C^2_{0,-2} = C^2_{0,+2} = \sqrt{1/2}\cos(\chi/2)$; $C^2_{0,0} = \sin(\chi/2)$
1	$4D_{xx}+D_{yy}+D_{zz}$	$C^2_{1,-1} = C^2_{1,+1} = \sqrt{1/2}$
2	$6D-6\sqrt{D^2+R^2}$	$C^2_{2,-2} = C^2_{2,+2} = -\sqrt{1/2}\sin(\chi/2)$; $C^2_{2,0} = \cos(\chi/2)$

Note: In the expressions above, the following symbols have been used:

$$D = \frac{1}{3}(D_{xx}+D_{yy}+D_{zz})$$

$$R^2 = \frac{1}{3}(D_{xx}D_{yy}+D_{yy}D_{zz}+D_{zz}D_{xx})$$

$$\chi = \tan^{-1}\left[\frac{\sqrt{3}(D_{xx}-D_{yy})}{2D_{zz}-D_{xx}-D_{xx}}\right]$$

$$J_2(\omega)=\frac{1}{2\pi}\sum_{q=-2}^{2}\sum_{q'=-2}^{2}(-1)^{q'}D_{0,q}^2(\Omega_{\eta,D})\left(D_{0,q'}^2(\Omega_{\eta,D})\right)^*\sum_{v=-2}^{2}C_{v,q}^2C_{v,-q'}^2\frac{b_{2,v}}{b_{2,v}^2+\omega^2} \tag{6.35}$$

Thus, the only Wigner rotation matrix elements that we need are $D_{0,q}^2(\Omega_{\eta,D})=D_{0,q}^2(0,\beta_{\eta,D},\gamma_{\eta,D})$ and $(D_{0,q'}^2(\Omega_{\eta,D}))^*=(D_{0,q'}^2(0,\beta_{\eta,D},\gamma_{\eta,D}))^*=D_{0,-q'}^2(0,\beta_{\eta,D},\gamma_{\eta,D})$, which are proportional to the spherical harmonics $Y_{2,q}(\beta_{\eta,D},\gamma_{\eta,D})$ and $Y_{2,q'}^*(\beta_{\eta,D},\gamma_{\eta,D})$. The angles $\beta_{\eta,D}$ and $\gamma_{\eta,D}$ are the polar and azimuthal angles specifying the orientation of the unique *z*-axis of the PAS with respect to the PDF. The expression for $J_2(\omega)$ becomes quite complicated also in this case where the interaction is assumed to be axially symmetric, and even more unwieldy formulas are obtained for the non-axial CSA tensor. The reader can find full expressions in the articles of Grant and Brown[9] or Canet[10]. We limit the discussion here to describing the case of an axially symmetric interaction and an axially symmetric diffusion tensor (symmetric diffusor), which gives the following expression for the spectral density $J_2(\omega)$:

$$\begin{aligned}J_2(\omega)=\frac{1}{2\pi}\Bigg[&\frac{1}{4}\left(3\cos^2\beta-1\right)^2\frac{6D_\perp}{\left(6D_\perp\right)^2+\omega^2}+3\cos^2\beta\sin^2\beta\frac{5D_\perp+D_\parallel}{\left(5D_\perp+D_\parallel\right)^2+\omega^2}\\&+\frac{3}{4}\sin^4\beta\frac{2D_\perp+4D_\parallel}{\left(2D_\perp+4D_\parallel\right)^2+\omega^2}\Bigg]\end{aligned} \tag{6.36a}$$

or, introducing the correlation times, $\tau_{2,m}^{-1}=6D_\perp+m^2(D_\parallel-D_\perp)$:

$$\begin{aligned}J_2(\omega)=\frac{1}{2\pi}\Bigg[&\frac{1}{4}\left(3\cos^2\beta-1\right)^2\frac{\tau_{2,0}}{1+\omega^2\tau_{2,0}^2}+3\cos^2\beta\sin^2\beta\frac{\tau_{2,1}}{1+\omega^2\tau_{2,1}^2}\\&+\frac{3}{4}\sin^4\beta\frac{\tau_{2,2}}{1+\omega^2\tau_{2,2}^2}\Bigg]\end{aligned} \tag{6.36b}$$

Both expressions given in Eq. (6.36) simplify to Eq. (6.20) for the isotropic case where $D_\perp=D_\parallel$. The reader is advised to check that this indeed is the case. In the extreme narrowing range, when all the correlation times $\tau_{2,m}$ are much smaller than the inverse frequency, the expression for spectral density in Eq. (6.36b) becomes frequency independent and we can introduce an effective correlation time, analogous to that occurring in Eq. (3.32a). We shall return to this point in Chapter 11.

The analysis in the case of cross-correlation functions, involving different interactions η and η', is analogous, the only difference being that we need one more coordinate frame transformation to bring the PAS of the interaction η' into coincidence with that of η; compare Section 5.2. The reduced cross-correlation spectral density, analogous to Eq. (6.36b), then becomes:

$$\begin{aligned}J_2^{\eta,\eta'}(\omega)=\frac{1}{8\pi}\Big[&\left(3\cos^2\beta_\eta-1\right)\left(3\cos^2\beta_{\eta'}-1\right)\tau_{2,0}/\left(1+\omega^2\tau_{2,0}^2\right)\\&+12\cos\beta_\eta\cos\beta_{\eta'}\sin\beta_\eta\sin\beta_{\eta'}\cos\left(\phi_\eta-\phi_{\eta'}\right)\tau_{2,1}/\left(1+\omega^2\tau_{2,1}^2\right)\\&+3\sin^2\beta_\eta\sin^2\beta_{\eta'}\cos\left(2\phi_\eta-2\phi_{\eta'}\right)\tau_{2,2}/\left(1+\omega^2\tau_{2,2}^2\right)\end{aligned} \tag{6.37}$$

The symbols β_η and $\beta_{\eta'}$ are the polar angles that position the principal axes of the interactions η and η' with respect to the principal axis of the diffusion frame, while ϕ_η and ϕ_η' are the corresponding azimuthal angles.

6.2.2 Physical Interpretation of Correlation Times

In order for the rotational correlation times, or rotational diffusion coefficients, to have a physical meaning, they are often interpreted in terms of theories of liquids. In one such approach (see Huntress[7] and references quoted therein), related to the fluctuation-dissipation theorem connecting the transport coefficients to fluctuations and time-correlation functions (see Van Kampen, *Further reading*), one obtains the rotational diffusion coefficient as an integral over the angular velocity time-correlation function:

$$D_{ii} = \int_0^{\Delta t} \langle \omega_i(0)\omega_i(\tau) \rangle d\tau \tag{6.38}$$

where the upper limit of integration is the same as the Δt in Eq. (6.27), *i.e.* a time interval which is long compared with the molecular time scale, on which $\langle \omega_i(0)\omega_i(\tau) \rangle$ decays to zero. In other words, the upper limit in the integral in Eq. (6.38) can be extended to infinity. Using this approach, one can go on to derive the Hubbard relation, Eq. (5.50). In another approach[7,8], based on the Langevin equation, the diagonal elements of the rotational diffusion tensor are related to the principal elements of the rotational friction tensor, ξ:

$$D_{ii} = \frac{k_B T}{\xi_{ii}} \tag{6.39}$$

For a spherical reorienting object, we have only one friction coefficient, ξ, and one rotational diffusion coefficient, D_R. Following the Stokes–Einstein–Debye (SED) hydrodynamic approach, the friction coefficient can be related to the radius a of the sphere (or its volume, $V = 4\pi a^3/3$) and the viscosity of the medium, η, yielding:

$$D_R = \frac{k_B T}{8\pi\eta a^3} \tag{6.40}$$

which is fully equivalent to Eqs. (2.43). The hydrodynamic approach corresponds to treating the solvent as a continuous, viscous medium. The SED relation is derived under the *sticking boundary condition*, which means that the sphere carries along a layer of the hydrodynamic medium closest to it. An early modification of the SED approach was suggested by Gierer and Wirtz[11], who recognised the molecular nature of the solvent and introduced the concept of a microviscosity correction related to the finite values of the size ratio between the solute and the solvent molecules. For large molecules dissolved in water, the sticking boundary hydrodynamics seems to work very well.

Deviations from the sticking boundary hydrodynamics in small molecules have been an active field of research, mainly during the 1970s and 1980s, and have been reviewed by Boeré and Kidd[12], by Dote *et al.*[13] and by Woessner, among others[8]. In the notation of Dote and co-workers, the rank-two rotational correlation time is written as:

$$\tau_2 = \frac{V\eta f_{stick} C}{k_B T} + \tau_2^0 \tag{6.41}$$

where the term τ_2^0 is included to take into account deviations from the hydrodynamics and can be explained in terms of free (or inertially controlled) rotations[14] or certain cross-correlation effects[15], f_{stick} is a shape-dependent hydrodynamic friction factor, equal to unity for a sphere, and C is a dimensionless constant dependent on the prevailing boundary condition. For the sticking boundary condition, $C=1$, which combined with $\tau_2^0=0$ gives the SED equation. Another boundary condition for which the hydrodynamic equations can be solved is the *slipping boundary*, where the hydrodynamic medium slips along the reorienting particle, and where the relation to friction arises only through solvent displacement, caused by deviations from a spherical shape of the particle. The factors f_{stick} and C for spheroids reorienting under the slipping boundary condition have been calculated by Hu and Zwanzig[16]; here, we only note that for a sphere, one obtains in this case $C=0$.

We also wish to mention briefly hydrodynamic calculations using *bead models*, where a complicated molecule (for example a polypeptide or an oligosaccharide) is treated as a collection of (atomic) sites or beads, each bead being characterised by a hydrodynamic radius and a friction coefficient. Using hydrodynamics, the rotational and translational diffusion coefficient for a rigid object of a complicated shape can be calculated, including, if required, a solvent layer. This technique has been developed to calibrate the friction coefficients needed for dynamic simulations based on the Langevin equation (Langevin dynamics [LD])[17] but can (after this calibration is done) also be used to estimate the anisotropy of the rotational diffusion tensor[18].

Small values of the solute-to-solvent volume ratio and the slipping boundary conditions tend to reduce the rotational correlation time with respect to its Stokes–Einstein–Debye limit. An opposite effect, making the rotational correlation time longer than predicted by Eq. (2.43), is foreseen for a charged or polar solute, reorienting in a dielectric medium. This effect is known as *dielectric friction* and has been reviewed some time ago by Madden and Kivelson[19].

6.3 Dynamic Models for Non-Rigid Molecules

Real molecules are of course not rigid bodies but are subject to both small amplitude vibrations and more extensive internal motions. Nuclear spin relaxation properties reflect that fact and their measurements provide an excellent tool for studying molecular flexibility with atomic resolution. Studies of this type have become very popular, not least in the context of protein flexibility, and have been reviewed numerous times[20–27]. We shall start by looking at dipolar relaxation, since this is an important source of molecular dynamics information and is usually the dominating relaxation mechanism for proton-bearing carbon-13 or nitrogen-15 nuclei.

6.3.1 General

We are now at a point where we can examine dipolar relaxation in the presence of internal motions. In this case, autocorrelation functions (we shall come back to the cross-correlation functions later) similar to those of Eq. (3.36), which were introduced in connection with intermolecular dipolar relaxation, should be useful:

$$G_2^{nr}(\tau)=\left\langle \frac{Y_{2,0}\left[\Omega_{IS}^{L}(0)\right]}{r_{IS}^{3}(0)}\frac{Y_{2,0}^{*}\left[\Omega_{IS}^{L}(\tau)\right]}{r_{IS}^{3}(\tau)}\right\rangle \tag{6.42}$$

The superscript *nr* in $G_2^{nr}(\tau)$ stands for *non-rigid.* As earlier, $\Omega_{IS}^{L}(\tau)$ denotes the orientation of the *IS* internuclear axis with respect to the laboratory frame (L). We shall find it useful to express the spherical harmonics of that angle in terms of an orientation of the *IS* axis with respect to an arbitrary molecule-fixed frame (M) and the orientation of M in the L frame. Proceeding as in Eqs. (6.21), we obtain:

$$G_2^{nr}(\tau)=\sum_{n,n'=-2}^{2}\left\langle D_{0,n}^{2}(\Omega_{M,L}(0))\left(D_{0,n'}^{2}(\Omega_{M,L}(\tau))\right)^{*}\frac{Y_{2,n}\left[\Omega_{IS}^{M}(0)\right]}{r_{IS}^{3}(0)}\frac{Y_{2,n'}^{*}\left[\Omega_{IS}^{M}(\tau)\right]}{r_{IS}^{3}(\tau)}\right\rangle \tag{6.43}$$

We now make an important assumption implying that the global motion of M with respect to L and the intramolecular motion of the *IS* in the M frame are uncorrelated. This assumption is certainly reasonable for some kinds of internal motions, which do not cause any major change in the shape of the molecule, while caution may be required in the case of major conformational changes (we shall come back to this issue below). Fortunately, the latter processes often occur on a slower time scale than the overall reorientation, which simplifies the problem (see Section 6.3.4). With this "decorrelation assumption", the correlation function $G_2^{nr}(\tau)$ can be factorised into a sum of products of two correlation functions:

$$
\begin{aligned}
G_2^{nr}(\tau) &= \sum_{n,n'=-2}^{2} \left\langle D_{0,n}^2(\Omega_{M,L}(0))\left(D_{0,n'}^2(\Omega_{M,L}(\tau))\right)^* \right\rangle \left\langle \frac{Y_{2,n}\left[\Omega_{IS}^M(0)\right]}{r_{IS}^3(0)} \frac{Y_{2,n'}^*\left[\Omega_{IS}^M(\tau)\right]}{r_{IS}^3(\tau)} \right\rangle \\
&= \sum_{n=-2}^{2} \left\langle D_{0,n}^2(\Omega_{M,L}(0))\left(D_{0,n}^2(\Omega_{M,L}(\tau))\right)^* \right\rangle \left\langle \frac{Y_{2,n}\left[\Omega_{IS}^M(0)\right]}{r_{IS}^3(0)} \frac{Y_{2,n}^*\left[\Omega_{IS}^M(\tau)\right]}{r_{IS}^3(\tau)} \right\rangle \\
&= \left\langle D_{0,0}^2(\Omega_{M,L}(0))\left(D_{0,0}^2(\Omega_{M,L}(\tau))\right)^* \right\rangle \sum_{n=-2}^{2} \left\langle \frac{Y_{2,n}\left[\Omega_{IS}^M(0)\right]}{r_{IS}^3(0)} \frac{Y_{2,n}^*\left[\Omega_{IS}^M(\tau)\right]}{r_{IS}^3(\tau)} \right\rangle \\
&= \frac{4\pi}{5} \left\langle Y_{2,0}(\Omega_{M,L}(0)) Y_{2,0}^*(\Omega_{M,L}(\tau)) \right\rangle \sum_{n=-2}^{2} \left\langle \frac{Y_{2,n}\left[\Omega_{IS}^M(0)\right]}{r_{IS}^3(0)} \frac{Y_{2,n}^*\left[\Omega_{IS}^M(\tau)\right]}{r_{IS}^3(\tau)} \right\rangle
\end{aligned}
\tag{6.44}
$$

In the first line, the correlation function vanishes if the indices n and n' are not equal. This simplifies the expression in the second line to a single rather than a double sum. In proceeding from the second to the third line, we have used the fact that the correlation functions for Wigner rotation matrices do not, as mentioned in the previous section, depend on n and set $n=0$. In the last line, we have explicitly used the relation between Wigner rotation matrices and spherical harmonics.

The correlation functions of $Y_{2,n}(\Omega_{IS}^M)/r_{IS}^3$ deviate from a constant value (which, when combined with a constant value of b_{IS}, would correspond to the factor multiplying $G_2(\tau)$ in Eq. (6.21c)) as a result of intramolecular dynamics. We shall approach the intramolecular correlation functions in a few different ways. We should in this process keep in mind that we ultimately need spectral densities, evaluated at different combinations of Larmor frequencies of the nuclei involved or at the zero frequency.

Let us first discuss the effects of small-amplitude vibrational motions. Many such modes have high frequencies, far too high to provide frequency components at the relevant combinations of nuclear Larmor frequencies at which the spectral densities of importance for relaxation measurements are evaluated. It may be worthwhile to keep in mind that a harmonic vibration with a (fairly low) wavenumber of 200 cm^{-1} has an angular frequency of 38×10^{12} rad s^{-1}. For example, for protons to have such high Larmor frequency, a magnetic field of 14×10^4 Tesla would be required, almost four orders of magnitude larger than the NMR magnets available today. From the point of view of spin relaxation, we can consider the high-frequency vibrations of this type as extremely rapid ("ultrafast") motions, the only effect of which is to replace the internal motion correlation function by the average value of the sum of squares of absolute values of $Y_{2,n}[\Omega_{IS}^M(0)]/r_{IS}^3(0)$ over the vibrational motion. If the vibrationally averaged interaction tensor **D** (compare Eq. (3.6)) remains axially symmetric, we can simply replace the dipolar coupling constants in all expressions given in Chapter 3 by an effective, vibrationally averaged coupling constant, $\langle b_{IS}\rangle$. Vibrational effects of this type also occur in the solid state and, as discussed in an elegant paper by Henry and Szabo[28], the influence of vibrational motion on both solid-state lineshapes and relaxation in liquids is in fact very similar. The effects of vibrational averaging are normally to reduce the $\langle b_{IS}\rangle$ by a few percent with respect to the value expected for the equilibrium *IS* distance. Put in other words, the effective NMR *IS* distance, obtained from the dipolar coupling constant in either liquids or solids, is expected to be slightly longer than the distance obtained from X-ray or neutron diffraction measurements. An illustrative example of a relaxation study of these effects can be found in the paper by Kowalewski *et al.*[29].

6.3.2 Woessner's Models

Woessner studied the effect of internal rotation on the NMR spectral densities, using a simple model in which the overall reorientational motion is isotropic small-step rotational diffusion and the internal motion is either rotational diffusion of the *IS* axis around a single molecule-fixed axis or random jumps between equivalent sites[30]. The internal motion is in this case assumed not to influence the internuclear *IS* distance. An experimental situation where the Woessner model could be applied is carbon-13 relaxation in a reorienting methyl group attached to a spherically shaped molecule[31]. Using the assumption of isotropic overall reorientation and of invariant r_{IS}, we can write the time-correlation function of Eq. (6.44) as:

$$G_2^{nr}(\tau) = \frac{1}{5r_{IS}^6}\exp(-6D_R\tau)\sum_{n=-2}^{2}\left\langle Y_{2,n}\left[\Omega_{IS}^M(0)\right]Y_{2,n}^*\left[\Omega_{IS}^M(\tau)\right]\right\rangle$$
$$= \frac{4\pi}{5r_{IS}^6}G_2(\tau)G_2^i(\tau) \tag{6.45a}$$

where $G_2(\tau)$ is given by Eq. (6.18) and the internal motion (or intramolecular) correlation function $G_2^i(\tau)$ is:

$$G_2^i(\tau) = \sum_{n=-2}^{2}\left\langle Y_{2,n}\left[\Omega_{IS}^M(0)\right]Y_{2,n}^*\left[\Omega_{IS}^M(\tau)\right]\right\rangle \tag{6.45b}$$

The calculation of the intramolecular correlation functions for the internal rotational diffusion model follows a similar approach to that used in preceding sections for the rigid body motion. In Woessner's original work, the orientation was specified in terms of directional cosines rather than Euler angles, which were employed by Luginbühl and Wüthrich[21] and are used in this presentation. We first use the model based on rotational diffusion around the single axis. In order to derive the propagator for the internal motion, we start by an appropriate diffusion equation:

$$\frac{\partial P(\gamma,t)}{\partial t} = -D_i\hat{L}_i^2 P(\gamma,t) = D_i\frac{\partial^2}{\partial\gamma^2}P(\gamma,t) \tag{6.46}$$

The symbol γ is the rotation angle around the axis of internal motion, $\hat{L}_i^2$ is the angular momentum operator describing the rotation around that axis (equal to $-(\partial^2/\partial\gamma^2)$; see Atkins and Friedman, *Further reading*, for details) and D_i is the internal rotational diffusion coefficient.

The eigenfunctions and eigenvalues of $\hat{L}_i^2$ are given by:

$$\hat{L}_i^2\frac{1}{\sqrt{2\pi}}\exp(ik\gamma) = k^2\frac{1}{\sqrt{2\pi}}\exp(ik\gamma) \tag{6.47}$$

where k is an integer number. The probability density for the isotropic distribution of initial rotation angles is $P(\gamma_0) = 1/2\pi$. The relevant propagator $P(\gamma_0|\gamma,\tau)$ is expanded in the eigenfunctions of $\hat{L}_i^2$:

$$P(\gamma_0|\gamma,\tau) = \sum_k\frac{1}{2\pi}\exp(-ik\gamma_0)\exp(ik\gamma)\exp(-D_ik^2t) \tag{6.48}$$

In order to evaluate the correlation function of Eq. (6.45), we introduce one more coordinate frame transformation, from the *M*-frame to the internal rotation (*IR*)-frame, which has the following effect on the internal part of the correlation function:

$$\left\langle Y_{2,n}\left[\Omega_{IS}^{M}(0)\right]Y_{2,n}^{*}\left[\Omega_{IS}^{M}(\tau)\right]\right\rangle$$
$$= \sum_{a,a'=-2}^{2}\left\langle D_{n,a}^{2}\left(\Omega_{IR,M}(0)\right)\left(D_{n,a'}^{2}\left(\Omega_{IR,M}(\tau)\right)\right)^{*}\right\rangle Y_{2,a}\left(\Omega_{IS}^{IR}\right)Y_{2,a'}^{*}\left(\Omega_{IS}^{IR}\right) \tag{6.49}$$

The transformation between the M- and IR-frames is time dependent, because one of the Euler angles (γ) varies with time according to Eq. (6.46), while the other two Euler angles remain constant. By an appropriate choice of the IR frame, the orientation of the IS dipolar axis can be made time independent. After some manipulations (the reader is referred to the review by Luginbühl and Wüthrich[21]), one finally obtains:

$$G_{2}^{nr}(\tau) = \frac{1}{4\pi r_{IS}^{6}}\exp(-6D_{R}\tau)\left[\frac{1}{4}\left(3\cos^{2}\beta-1\right)^{2} + 3\sin^{2}\beta\cos^{2}\beta\exp(-D_{i}\tau) + \frac{3}{4}\sin^{4}\beta\exp(-4D_{i}\tau)\right] \tag{6.50}$$

The angle β is the angle between the IS axis and the axis of internal rotation, and can also be envisioned as a semi-angle of the cone on which the IS axis moves with respect to the rest of the molecule. This variety of the Woessner model, illustrated in Figure 6.1, is sometimes called *diffusion on cone*. The corresponding spectral density is:

$$J_{2}^{nr}(\omega) = \frac{1}{2\pi r_{IS}^{6}}\left[\frac{1}{4}\left(3\cos^{2}\beta-1\right)^{2}\frac{\tau_{2,0}}{1+\omega^{2}\tau_{2,0}^{2}} + 3\sin^{2}\beta\cos^{2}\beta\frac{\tau_{2,1}}{1+\omega^{2}\tau_{2,1}^{2}} + \frac{3}{4}\sin^{4}\beta\frac{\tau_{2,2}}{1+\omega^{2}\tau_{2,2}^{2}}\right] \tag{6.51a}$$

with

$$\tau_{2,m}^{-1} = 6D_{R} + m^{2}D_{i} \tag{6.51b}$$

Comparing Eq. (6.51) with Eq. (6.36b) for a rigid, symmetric diffusor, we see that the two spectral densities differ only by the constants in front of the square parentheses, if we identify D_R with $D_\perp$ and D_i with $D_\parallel - D_\perp$. The two motional models are indeed closely related. We notice also that in the absence of internal motion, $D_i = 0$, or with the IS vector parallel to the internal rotation axis, $\beta = 0$, Eq. (6.51a) reduces to the case of isotropic rigid body diffusion. In the limit of very fast internal motion, the correlation times

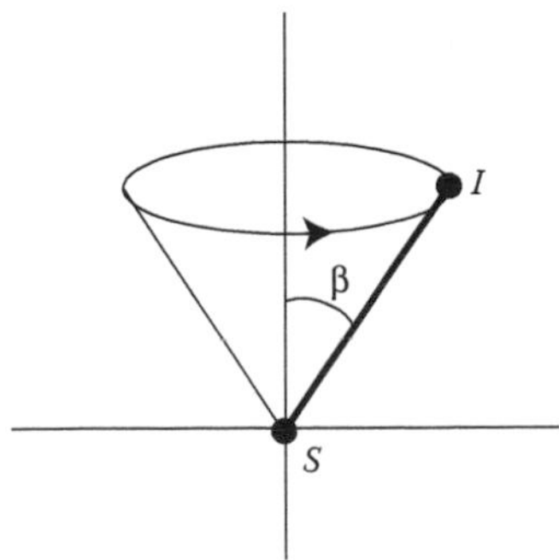

FIGURE 6.1 Diffusion on a cone. The IS axis moves on the cone, specified by the semiangle β.

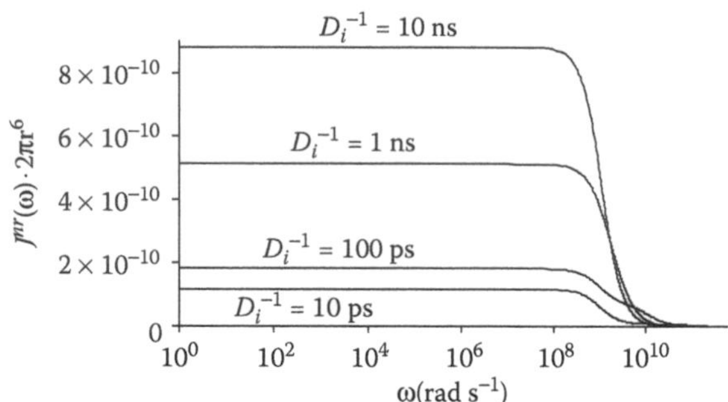

FIGURE 6.2 The Woessner spectral density function for a methyl group undergoing internal motion. The calculations, made with $\beta = 70.5°$ and $(6D_R)^{-1} = 1$ ns, show the spectral density function for different values of D_i.

$\tau_{2,1}$ and $\tau_{2,2}$ become very short compared with $\tau_{2,0}$ (which, in turn, is assumed not to be exceedingly long) and the last two terms in the spectral density can be neglected. We note that for a methyl group, assuming tetrahedral geometry, we obtain $\frac{1}{4}(3\cos^2\beta - 1)^2 = \frac{1}{9}$. If both motions are in the extreme narrowing regime, Eq. (3.32a) applies for the methyl carbon T_1^{-1}, with τ_c representing an *effective* correlation time, which becomes nine times shorter for an (infinitely) rapidly rotating methyl group than for a rigid methyl group. Plots of the spectral density of Eq. (6.51a) as a function of frequency for some D_i/D_R ratios are displayed in Figure 6.2.

The next question we wish to raise is the effect of internal motions on cross-correlation functions and corresponding spectral densities. For the case discussed in the previous paragraph, a methyl group undergoing internal rotation, the cross-correlation of dipolar CH (as well as HH) interactions can contribute to cross-correlated relaxation rates. The reader is referred to the review by Werbelow and Grant[32] for the details of the multispin relaxation effects in the A_3 and A_3X spin groupings. Here, we just note that a calculation analogous to that of Eq. (6.21c) is valid also in the presence of internal motion, provided that the transformation between the principal frames of the two interactions is indeed time independent. The resulting expression for the cross-correlated spectral density is in fact identical to Eq. (6.37), with the correlation times $\tau_{2,m}$ given in Eq. (6.51b). The reader can find a derivation in the review by Daragan and Mayo[20]. We can thus conclude that the close analogy between the rigid, symmetric top and the isotropic rotor with a single-axis internal rotation remains valid also for the cross-correlated spectral densities.

Another, conceptually different, description of the methyl group internal reorientation, also introduced by Woessner[30], is based on random jumps between three equivalent sites. Rather than using an internal diffusion equation, one here describes the internal dynamics by a master equation:

$$\frac{d}{dt}P_i(t) = \sum_{j=1}^{3} A_{ij}P_j(t) \tag{6.52}$$

where $P_i(t)$ denotes the population of site i and A_{ij} is the transition rate matrix:

$$\mathbf{A} = \begin{pmatrix} -2k_i & k_i & k_i \\ k_i & -2k_i & k_i \\ k_i & k_i & -2k_i \end{pmatrix} \tag{6.53}$$

The random jump model leads to a spectral density function of the same form as Eqs. (6.51), but with:

$$\tau_{2,0}^{-1} = 6D_R \tag{6.54a}$$

$$\tau_{2,1}^{-1} = \tau_{2,2}^{-1} = 6D_R + 3k_i \tag{6.54b}$$

Ericsson and co-workers compared the two Woessner models as tools for the interpretation of experimental variable-temperature carbon-13 relaxation data for methyl carbons in terpene solutions under extreme narrowing conditions and found them more or less equivalent[31].

There exists a large diversity of more complicated dynamic models for non-rigid molecules, allowing for the overall motion to be anisotropic and with the internal motion being described in a variety of ways, including possible modulation of the interaction strengths (*e.g.* the internuclear distance in the case of dipole-dipole interaction) with time. We refer the interested reader to selected reviews available in the literature[8,20-25].

6.3.3 The "Model-Free" Approach

In the next part of this section, we wish to discuss the "*model-free*" approaches, which have become very popular during the last decades, especially in the context of protein dynamics studies. We begin with the widely used Lipari–Szabo approach from the early 1980s[33] and proceed with its predecessors (the two-step model and the slowly relaxing local structures model) and modifications. A nice comparison of the Lipari–Szabo approach with other models can be found in the review by Luginbühl and Wüthrich[21].

The essence of the model-free approaches is that one makes an assumption about the functional form of the time-correlation function without formulating an equation for the propagator. The Lipari–Szabo approach can be derived for the case of isotropic overall reorientation, uncorrelated with rapid internal motions, thus taking Eq. (6.45) as the starting point. Independently of the details of the internal motions, the correlation function for internal motions, $G_2^i(\tau)$, can easily be predicted at the limits $\tau \to 0$ and $\tau \to \infty$. At $\tau \to 0$, we have $G_2^i(\tau) = 1$, while at $\tau \to \infty$ all the correlation can be assumed lost and the internal correlation function reaches the value:

$$\lim_{\tau\to\infty} G_2^i(\tau) = \sum_{n=-2}^{2} \langle Y_{2,n}(\Omega_{IS}^M)\rangle \langle Y_{2,n}^*(\Omega_{IS}^M)\rangle = \sum_{n=-2}^{2} \left|\langle Y_{2,n}(\Omega_{IS}^M)\rangle\right|^2 = S^2 \tag{6.55}$$

The average values $\langle Y_{2,n}(\Omega_{IS}^M)\rangle$ vanish for the case of completely unrestricted internal motions, where one has isotropic distribution of the orientations of the *IS* axis in the *M*-frame (compare Eq. (3.9)). If the local motion is restricted (locally anisotropic), the averages do not vanish. The quantity S is called the *generalised order parameter*, since it is reminiscent of the order parameter used in the description of incomplete averaging of anisotropic interactions in oriented systems. Thus, we can consider the generalised order parameter as a measure of the spatial restriction of the internal motion. In the absence of internal motion, we get $S^2 = 1$, while $S^2 = 0$ is obtained for completely unrestricted internal motion.

The internal correlation function is thus expected to decay with time from unity to S^2. In the Lipari–Szabo model, this decay is assumed to be monoexponential, introducing an approximate correlation function:

$$G_2^{i,A}(\tau) = S^2 + (1 - S^2)\exp(-\tau/\tau_{loc}) \tag{6.56}$$

The time constant τ_{loc} is called the *local correlation time* and can be defined in terms of an integral of the part of the $G_2^i(\tau)$ above the long time limit:

$$\tau_{loc} = \frac{1}{1-S^2}\int_0^\infty \left(G_2^{i,A}(t) - S^2\right)dt \tag{6.57}$$

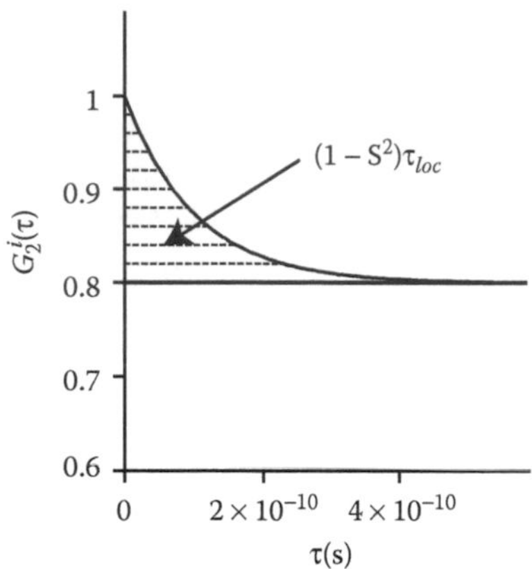

FIGURE 6.3 The Lipari–Szabo correlation function for internal motion, $G_2^i(\tau)$. The correlation function was generated with $S^2=0.8$ and $\tau_{loc}=100$ ps. The dashed area beneath $G_I(t)$ corresponds to the definition of the local correlation time.

This definition of τ_{loc} is illustrated in Figure 6.3. Obviously, the local correlation time defined in this way can easily accommodate more than one fast dynamic process. Substituting the approximate correlation function $G_2^{i,A}(\tau)$ from Eq. (6.56), instead of the correlation function $G_2^i(\tau)$, into Eq. (6.45a) yields the Lipari–Szabo time-correlation function, $G_2^{nr,LS}(\tau)$:

$$\begin{aligned} G_2^{nr,LS}(\tau) &= \frac{1}{4\pi r_{IS}^6}\exp(-6D_R\tau)\left[S^2+(1-S^2)\exp(-\tau/\tau_{loc})\right] \\ &= \frac{1}{4\pi r_{IS}^6}\left[S^2\exp(-\tau/\tau_R)+(1-S^2)\exp(-\tau/\tau_{eff})\right] \end{aligned} \tag{6.58a}$$

where the effective correlation time, τ_{eff}, and the overall rotational correlation time, τ_R, are defined by:

$$\tau_{eff}^{-1} = 6D_R + \tau_{loc}^{-1} = \tau_R^{-1} + \tau_{loc}^{-1} \tag{6.58b}$$

The corresponding spectral density becomes:

$$J_2^{nr,LS}(\omega) = \frac{1}{2\pi r_{IS}^6}\left[\frac{S^2\tau_R}{1+\omega^2\tau_R^2} + \frac{(1-S^2)\tau_{eff}}{1+\omega^2\tau_{eff}^2}\right] \tag{6.59a}$$

An interesting limiting situation arises when the effective correlation time is in the extreme narrowing regime at all relevant frequencies. Under such conditions the second term in the spectral density simplifies to:

$$J_2^{nr,LS}(\omega) = \frac{1}{2\pi r_{IS}^6}\left[\frac{S^2\tau_R}{1+\omega^2\tau_R^2} + (1-S^2)\tau_{eff}\right] \tag{6.59b}$$

In this case, the details of the short-time decay of the internal correlation function become immaterial, in the same way as the complicated details vanish in the description of the overall reorientation in extreme narrowing, yielding a single, effective correlation time (compare comments below Eq. (6.36) and Section 11.1). The Lipari–Szabo spectral density thus becomes exact for isotropic overall reorientation under these conditions.

Let us assume in addition that the *IS* distance (or any other interaction strength parameter) is constant, independent of the internal motion. We can then define a new Lipari–Szabo spectral density, where $1/r_{IS}^6$ is omitted:

$$J_2^{LS}(\omega) = \frac{1}{2\pi}\left[\frac{S^2\tau_R}{1+\omega^2\tau_R^2} + (1-S^2)\tau_{eff}\right] \tag{6.59c}$$

At this point, it is interesting to relate the Lipari–Szabo "model-free" approach to the diffusive model of Woessner. For a general case in which the local correlation time is outside of the extreme narrowing regime, a comparison only tells us that the Woessner spectral density consists of three Lorentzians, while only two Lorentzians build up the Lipari–Szabo spectral density, and that the two models are simply different. In the limit of very rapid internal motion, within the extreme narrowing regime for the correlation times $\tau_{2,1}^{-1}$ and $\tau_{2,2}^{-1}$ (see Eq. (6.51b)), we can identify:

$$S^2 = \frac{1}{4}(3\cos^2\beta - 1)^2 \tag{6.60a}$$

and

$$(1-S^2)\tau_{eff} = 3\sin^2\beta\cos^2\beta\tau_{2,1} + \frac{3}{4}\sin^4\beta\tau_{2,2} \tag{6.60b}$$

A similar comparison can be made between the Lipari–Szabo model and other models for internal motions. A general result of such comparisons is that S^2 can always be related to the spatial restrictions of the motion inherent in the model, while comparisons of the Lipari–Szabo local correlation time are impossible or complicated. In the applications of the Lipari–Szabo approach, a situation can arise in which the order parameter is quite high and the local correlation time rather short, which allows the second term in Eq. (6.59b) to be neglected. This leads to *the truncated Lipari–Szabo* spectral density:

$$J_2^{LS,trunc}(\omega) = \frac{1}{2\pi r_{IS}^6}\frac{S^2\tau_R}{1+\omega^2\tau_R^2} \tag{6.61}$$

We can easily see that Eq. (6.61) is equivalent to the result of infinitely fast vibrations, if we identify S^2/r_{IS}^6 with $(1/\langle r_{IS}^3\rangle)^2$. Full and truncated Lipari–Szabo spectral densities are compared for certain parameter sets in Figure 6.4.

Lipari and Szabo were actually not the first to propose biexponential time-correlation functions or double Lorentzian spectral densities. Expressions very similar to Eqs. (6.59) were earlier derived by Wennerström and co-workers[34–36] in the framework of the so-called *two-step model*. These authors were interested in describing surfactant dynamics in micelles and related systems rather than proteins, which may be an explanation for why their work is less frequently quoted. In their model, it was explicitly assumed that the averaging of the relaxation-causing interaction (quadrupolar, to which we shall return in Chapter 14, or dipolar) occurs in two steps. In the first step, the fast local motions within a micelle partially average out the interaction, down to the order of $S\times$interactions strength, where S is an order parameter, closely resembling the quantity denoted in the same way in analogous anisotropic systems. This residual interaction defines a local director whose orientation varies, in the second step, as a result of slower processes associated with the motions of the aggregate itself. The two-step model explicitly assumes that there is a three-fold (or higher) symmetry around the local director, which means that the residual interaction is axially symmetric. Polnaszek and Freed[37,38] also proposed a similar approach in the 1970s, denoted as the *slowly relaxing local structures* (SRLS) model, originally in the context of

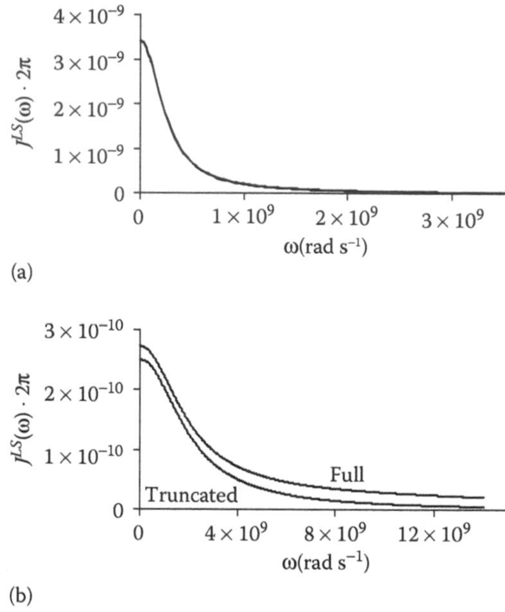

FIGURE 6.4 The Lipari–Szabo spectral density function and the truncated form for two different cases. (a) shows a typical situation for a small protein, with a global correlation time $\tau_R = 4$ ns, $S^2 = 0.85$ and $\tau_{loc} = 50$ ps. (b) shows the situation for a smaller molecule with $\tau_R = 0.5$ ns, $S^2 = 0.5$ and $\tau_{loc} = 50$ ps. In (a), no difference can be observed between the full Lipari–Szabo spectral density function and the truncated form, while in (b), a clear difference is seen.

electron spin relaxation in liquid crystals. We shall return to modern varieties of the SRLS model, allowing also for coupling between the local and global motions and applied to the nitrogen-15 relaxation in proteins, in the following subsection.

Two interesting questions arising in the context of the Lipari–Szabo (or similar) methods are those of the effects of slower internal motions and of distance variations. The issue of slow internal motions, much slower than the global reorientation, is simple. Such motions essentially do not contribute to relaxation through anisotropic interactions, since the faster global rotation dominates the dynamics. These motions can, on the other hand, influence relaxation measurements if they act as chemical (conformational) exchange. We shall return to this point in Chapter 13. Motions on the intermediate time scale, similar to the reorientational motion, are, however, not easily treated in the framework of the model-free approach. Another interesting problem is associated with variation in the internuclear distance. The internal motions can introduce variations in the internuclear distance (intergroup relaxation), leading to a complicated situation. This is particularly important in proton-proton cross-relaxation processes. This issue was dealt with in an important paper by Brüschweiler *et al.*[39]. They used a molecular dynamics (MD) simulation of the cyclic decapeptide antamanide in a chloroform solution in order to study fast angular and radial intramolecular dynamics (we may add, in passing, that it is quite common to use molecular dynamics simulation methods as tools for the analysis of NMR relaxation data in non-rigid molecules. A selection of references can be found in the listed reviews[21–25]). Analysing the data in a molecule-fixed reference frame, Brüschweiler *et al.*[39] were able to describe the spatial restrictions of the relative motion of two nuclei in terms of angular and radial order parameters. The angular order parameter for a proton pair *kl* (denoted $S_{\Omega,kl}$) was essentially calculated as the "normal" Lipari–Szabo order parameter from Eq. (6.55). In addition, radial order parameters ($S_{r,kl}$) were defined according to:

$$S_{r,kl}^2 = \left\langle \frac{1}{r_{kl}^3} \right\rangle^2 \Big/ \left\langle \frac{1}{r_{kl}^6} \right\rangle \tag{6.62}$$

so that the total order parameter squared was approximately expressed as a product:

$$S_{kl}^2 \approx S_{\Omega,kl}^2 S_{r,kl}^2 \tag{6.63}$$

An interesting observation, pointed out by Brüschweiler *et al.*[39], is that the rapid radial fluctuations lead to an *increased* cross-relaxation rate, since the distances shorter than average contribute to the dipolar interaction with a larger weight than the longer distances. By the same argument, we can state that $\langle r_{kl}^{-6}\rangle \geq \langle r_{rl}^{-3}\rangle^2$, *i.e.* the radial order parameter must be smaller than unity.

The Lipari–Szabo spectral densities are often chosen for fitting experimental relaxation data. Due to the complexity of the studied systems, it is not uncommon that this procedure is not fully successful. To deal with this situation for protein samples, Clore *et al.*[40] proposed an *ad hoc* extension of the simple formulation by assuming a biexponential form of the internal motion time correlation function with a slow and a fast internal motion:

$$G_2^{i,B}(\tau) = S_f^2 S_s^2 + \left(1 - S_f^2\right)\exp\left(-\tau/\tau_f\right) + S_f^2\left(1 - S_s^2\right)\exp\left(-\tau/\tau_s\right) \tag{6.64}$$

S_f and τ_f are the Lipari–Szabo parameters for the faster intramolecular local motion and S_s and τ_s are their slower counterparts. Both local correlation times must be shorter than the overall reorientation correlation time. Clearly, if the slower intramolecular motion is absent, then $S_s^2 = 1$ and Eq. (6.64) simplifies to Eq. (6.56). In the original work by Lipari and Szabo[33], an approximate expression to be used in the case of the overall motion being anisotropic was also given. This proposal was further extended and elaborated by Barbato and co-workers[41]. We shall return to the modifications of the Lipari–Szabo model-free approach of this kind in Section 11.3, dealing with determination of molecular dynamics from relaxation data.

Finally, we wish to mention the issue of the model-free approach to cross-correlation spectral densities, discussed in two reviews dealing with the cross-correlation function in non-rigid systems. Frueh[42] has described cross-correlation functions in a more general way, while Canet *et al.*[43] devoted their work specifically to the cross-correlation function within the model-free approaches of Wennerström and co-workers and Lipari and Szabo. Canet *et al.* derived an expression for the cross-correlated spectral density under several assumptions: (1) the global reorientation (in isotropic medium) is isotropic, (2) the two interactions are characterised by axially symmetric tensors, (3) the global and local motions are uncorrelated, with the latter much faster than the former, and (4) the local director has three-fold symmetry. The cross-correlation spectral density corresponding to Eq. (6.59c) for the autocorrelated spectral density then becomes:

$$J_2^{LS,cross}(\omega) = \frac{1}{2\pi}\left[\frac{SS'\tau_R}{1+\omega^2\tau_R^2} + \left(\frac{1}{2}\left(3\cos^2\theta - 1\right) - SS'\right)\tau_{eff}\right] \tag{6.65}$$

Here, θ is the angle between the principal frames of the two interactions (compare Eq. (6.21)), assumed constant in time, and the two order parameters, S and S', are defined through a cross-correlation analogue to Eq. (6.55):

$$\lim_{\tau\to\infty} G_2^{i,cross}(\tau) = \sum_{n=-2}^{2}\left\langle Y_{2,n}\left(\Omega_{IS}^M\right)\right\rangle\left\langle Y_{2,n}^*\left(\Omega_I^M\right)\right\rangle = SS' \tag{6.66}$$

We wish to stress that the product of the two order parameters does not need to be positive. The order parameters in Eqs. (6.65) and (6.66) can be related to models for molecular motions; interested readers can find this information in the review by Frueh[42].

6.3.4 Correlated Motions

The assumption that there is a lack of correlation between the overall and internal motions is central to the Woessner model as well as to the model-free and related approaches. This assumption is always fulfilled if the internal motions occur on a much faster time scale than the global motion (adiabatic approximation). In the case of complex molecular systems, such as, for example, proteins in solution, we may encounter situations when certain local motions occur with rates similar to the global reorientation and the two types of dynamics may be coupled or correlated. The SRLS method, the early version of which was mentioned in Section 6.3.3, has been developed to deal with this situation in proteins[44–46]. This present-day SRLS is a two-body model allowing for global motion of the whole molecule (body 1) and of the spin probe (body 2, *e.g.* a ^{15}NH vector). Both motions are described by rotational diffusion tensors. The motions of the two bodies occurs in a mutual ordering potential (potential of mean torque), which couples the two motions. The dynamics of the combined system is described by the two-body Smoluchowski equation:

$$\frac{\partial}{\partial t}P(X,t)=\hat{\Gamma}P(X,t) \tag{6.67}$$

where $P(X,t)$ is the probability density for the orientation of the probe and X is the set of two Euler angles, , $\Omega_{VF,OF}$ and $\Omega_{LF,VF}$ where LF denotes the laboratory frame (with the z-axis along the external magnetic field), VF refers to the local director frame (fixed in the protein) and OF is the principal frame of the local ordering tensor (fixed in the probe). The symbol $\hat{\Gamma}$ denotes the Smoluchowski operator (diffusion operator in the presence of a potential $U(\Omega_{VR,OF})$).

We can say that Eq. (6.67) plays the same role in the SRLS theory as Eq. (6.5b) does in the simple rotational diffusion model. Solving Eq. (6.67), by diagonalisation of the representation of the Smoluchowski operator, allows evaluation of the relevant spectral density functions.

Halle[47] presented some time ago a new derivation of the model-free equations, (6.59c) and (6.61), showing that their validity range (in terms of time scales and symmetries) is larger than stated in the original work. In addition, he criticised several aspects of the SRLS approach. In particular, the basic form of the two-body Smoluchowski equation, with its symmetric treatment of global and local motions, was challenged, based on analytical solution of the planar version of the SRLS model. Halle also discussed the mechanisms of dynamic coupling between the global and local motions, which can be either torque mediated or friction mediated (hydrodynamic), with only the former included in the SRLS model used for proteins. Moreover, it was argued that the internal motions in proteins that occur on the time scale of the overall rotational diffusion (and thus fulfil the condition for possible motional correlation) are slow because they are jump-like (involving barrier crossings) rather than characterised by slow rotational diffusion. In response, Meirovitch and co-workers[48] stated, among other things, that even if a model may not be physically reasonable for all possible systems and conditions, a major advantage of using a method based on a physical model lies in the ability to improve it by including features that are consistent with the properties of the system under investigation. As an example of such an improvement, we wish to mention a study of the dynamics of a small disaccharide molecule labelled with ^{13}C in one specific location by Zerbetto *et al.*[49]. They analysed multiple–magnetic field, variable-temperature relaxation data using a model conceptually close to the Smoluchowski-SRLS, with the local (probe) motion being the rotation of a $^{13}CH_2$ group. For the small system, it was however possible to obtain most of the relevant input through independent calculations. Thus, the diffusion tensor was obtained through a bead model hydrodynamic approach (including dynamic coupling) and the torsional potential for the $^{13}CH_2$ group through density functional theory (DFT) calculations. This left only one parameter, the HCH angle, which needed to be fitted to obtain very good agreement with a large set of experimental data.

A physical situation possibly involving correlated motions can be found in proteins with domain motions[24,25]. The domains are rather large units, which can attain different conformations with respect

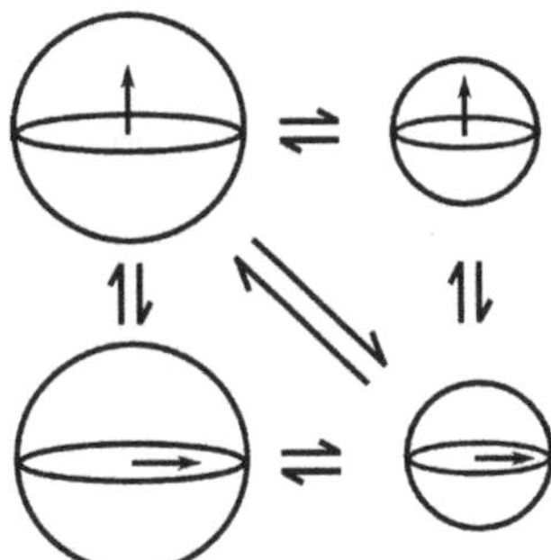

FIGURE 6.5 Simple model for local (vertical) and global (horizontal) conformational changes. The size of the circles corresponds to the magnitude of the global rotational correlation time ($1/6D_R$). The arrows indicate an orientation of a vector in the molecular frame. (Reprinted from Wong, V. *et al.*, *Proc. Natl. Acad. Sci. USA*, 106, 11016–11021, 2009. With permission.)

to the rest of the molecule. Changing the conformation of a domain can cause a significant modification of the global diffusion tensor. In the nomenclature introduced above, this corresponds to hydrodynamic coupling, not included in the usual SRLS approach. Theoretical methods able to deal with this situation are a topic of great current interest. Here, we wish to mention one approach, proposed a few years ago by Wong, Case and Szabo[50]. Their model involves jumps between discrete conformers with different overall diffusion tensors and different orientations of relevant vectors (such as the DD principal axis). In analogy with Woessner's jump model[30], the transitions between the conformers are described by a master equation, similar to Eqs. (6.52) and (6.53). The rates of the conformational changes can be similar to the overall diffusion rates, which allows for correlated motions. The work neglects, for simplicity, the possible presence of rapid local motions and treats two possible scenarios: (1) the diffusion tensors of the two conformers are both isotropic but are characterised by different diffusion coefficients; (2) the diffusion tensors of the two conformers are allowed to be anisotropic. The first of these scenarios follows the kinetic scheme of Figure 6.5. The vertical transitions represent the local motion, changing the orientation of a vector in the molecular frame. The horizontal transitions correspond to global changes of the rotational diffusion coefficient. By playing with rates of different kinetic processes, we can describe different physical situations. Retaining only one of the vertical transitions corresponds to a simple two-site jump model with a constant diffusion coefficient. Assigning a non-zero rate to the diagonal transitions only results in fully concerted local and global motions. For further details and discussions, the reader is referred to the original paper and to its generalisation[51].

6.4 Spectral Density Models for Low-Resolution Work

The material covered by Section 6.1 aims at deriving and explaining the simple case of exponentially decaying time-correlation functions. It is based on the Debye model[1], originally formulated for the time-correlation functions for the reorientation of the molecular electric dipole moment. Sections 6.2 and 6.3 cover improvements of the description of reorientational dynamics allowing for differences in local dynamics caused by anisotropic motions, quantifiable by atomic-resolution NMR measurements.

The Debye model can be made more general also in another sense, and much of this development has been carried out in the context of dielectric behaviour in time-dependent electromagnetic fields as described in the monograph by Böttcher and Bordewijk[52]. The complex dielectric constant of the Debye model is expressed as a function of frequency as follows:

$$\varepsilon(\omega) = \varepsilon_\infty + \frac{\varepsilon_0 - \varepsilon_\infty}{1 + i\omega\tau} \tag{6.68}$$

where ε_0 is the static dielectric constant, ε_∞ is the high-frequency limit of the dielectric constant and τ is the dielectric relaxation time.

In terms of the real ($\varepsilon'(\omega)$) and imaginary ($\varepsilon''(\omega)$) components of the dielectric constant, Eq. (6.68) takes the form:

$$\varepsilon'(\omega) = \varepsilon_\infty + \frac{\varepsilon_0 - \varepsilon_\infty}{1 + \omega^2\tau^2} \tag{6.69a}$$

$$\varepsilon''(\omega) = \frac{(\varepsilon_0 - \varepsilon_\infty)\omega\tau}{1 + \omega^2\tau^2} \tag{6.69b}$$

The real part, $\varepsilon'(\omega)$, is denoted the *frequency-dependent dielectric constant*, while the imaginary part, $\varepsilon''(\omega)$, is called the *loss factor*. The measurements of the complex dielectric constant as a function of frequency (dielectric spectroscopy) can in some cases be interpreted using Eqs. (6.69). However, one often needs to allow the functional form with more parameters. One such relation, of relevance also for NMR relaxation, is the Cole–Davidson (CD) equation[53–55]:

$$\varepsilon(\omega) = \varepsilon_\infty + \frac{\varepsilon_0 - \varepsilon_\infty}{(1 + i\omega\tau_0)^\beta} \tag{6.70a}$$

The parameter β can assume values between 0 and 1; in the latter limit, the Cole-Davidson equation simplifies to the Debye case. The real and imaginary components of the dielectric constants corresponding to Eq. (6.70a) take the form:

$$\varepsilon'(\omega) = \varepsilon_\infty + (\varepsilon_0 - \varepsilon_\infty)(\cos\varphi)^\beta \cos\beta\varphi \tag{6.70b}$$

$$\varepsilon''(\omega) = (\varepsilon_0 - \varepsilon_\infty)(\cos\varphi)^\beta \sin\beta\varphi \tag{6.70c}$$

where the symbol $\varphi = \arctan\omega\tau_0$. τ_0 is a variable with dimension time, which becomes identical to the relaxation time τ for $\beta = 1$, corresponding to the Debye model. Both these functions with $\beta = 0.67$ are plotted against $\omega\tau_0$ and compared with the $\beta = 1$ case in Figure 6.6. We notice that the high-frequency part of both CD curves is falling off more slowly than the Debye counterparts. The shape of the Debye $\varepsilon'(\omega)$ is analogous to the diagram of $1/T_1$ versus frequency in Figure 2.5. Further, we see that the shape of the $\varepsilon''(\omega)$ versus ln $\omega\tau_0$ for $\beta < 1$ is asymmetric, with the maximum shifted toward higher frequency. It is important to realise that the Cole-Davidson equation is in principle empirical, *i.e.* it is not derived from a more fundamental dynamical model. However, it can be associated with a distribution of dielectric relaxation times[53].

The Cole-Davidson model for the dielectric relaxation can be translated into the formalism of rotational spectral densities. Kruk (*Further reading*) gives the following formulation (omitting the normalisation), meant to be used in the NMR context:

$$J^{CD}(\omega) = \frac{\sin\left[\beta\arctan(\omega\tau_2)\right]}{\omega\left(1 + \omega^2\tau_2^2\right)^{\beta/2}} \tag{6.71}$$

The Cole-Davidson spectral densities are often used in connection with measurements of proton spin-lattice relaxation over a broad Larmor frequency range employing a field-cycling apparatus (mentioned in Section 3.5; see also Chapter 8). Measurements of this kind are normally carried out under low-resolution conditions, *i.e.* the effective T_1 is measured for all protons in the sample, and the observed dynamics is often complicated, reflecting both global and internal motions of various dipolar axes. Deviations

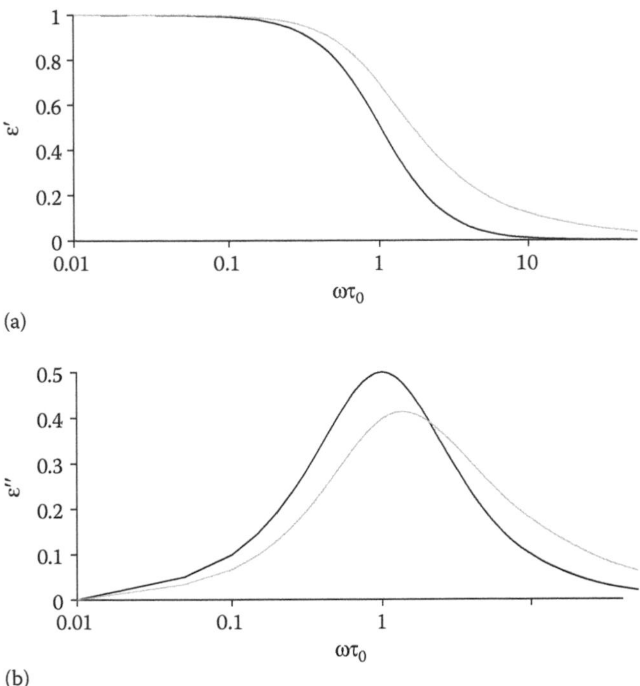

FIGURE 6.6 Frequency-dependent dielectric constant (a) and the loss factor (b) in arbitrary units plotted versus $\omega\tau_0$. Black lines: $\beta=1$ (Debye equation), grey lines: $\beta=0.67$ (Cole–Davidson equation).

from $\beta=1$ are typically observed for viscous and glassy liquids[54]. It is worth mentioning that the original work of Cole and Davidson[53] also dealt with dynamics of liquids of this type, such as glycerol, propylene glycol and *n*-propanol, at low temperatures. We shall return to the applications of relaxation measurements to this type of systems in Chapter 11.

Exercises for Chapter 6

1. Starting with Table 6.1, derive the eigenvalues of the rotational diffusion tensor for a symmetric diffusor.
2. Show that both expressions given in Eq. (6.36) simplify to Eq. (6.20) for the isotropic case, $D_{\perp} = D_{\parallel}$.

References

1. Debye, P., *Polar Molecules*. Chem. Catalog Co.: New York, 1929.
2. Bloembergen, N.; Purcell, E. M.; and Pound, R. V., Relaxation effects in nuclear magnetic resonance absorption. *Phys. Rev.* 1948, 73, 679–712.
3. Hubbard, P. S., Some properties of correlation functions of irreducible tensor operators. *Phys. Rev.* 1969, 180, 319–326.
4. Perrin, F., Mouvement Brownien d'un ellipsoide. I. Dispersion dielectrique pour des molecules ellipsoidales. *J. Phys. Radium* 1934, 5, 497–511.
5. Favro, L. D., Theory of the rotational brownian motion of a free rigid body. *Phys. Rev.* 1960, 119, 53–62.
6. Woessner, D. E., Nuclear spin relaxation in ellipsoids undergoing rotational brownian motion. *J. Chem. Phys.* 1962, 37, 647–654.

7. Huntress, W. T., The study of anisotropic rotation of molecules in liquids by NMR quadrupolar relaxation. *Adv. Magn. Reson.* 1970, 4, 1–37.
8. Woessner, D. E., *Brownian Motion and Correlation Times.* Wiley: eMagRes, 2007.
9. Grant, D. M.; and Brown, R. A., *Relaxation of Coupled Spins from Rotational Diffusion.* Wiley: eMagRes, 2007.
10. Canet, D., On the calculation of spectral density functions for spin interactions without axial symmetry. *Concepts Magn. Reson.* 1998, 10, 291–297.
11. Gierer, A.; and Wirtz, K., Molekulare Theorie der Mikroreibung. *Z. Naturforsch. A* 1953, 8, 532–538.
12. Boere, R. T.; and Kidd, R. G., Rotational correlation times in nuclear magnetic resonance. *Ann. Rep. NMR Spectr.* 1982, 13, 319–385.
13. Dote, J. L.; Kivelson, D.; and Schwartz, R. N., A molecular quasi-hydrodynamic free-space model for molecular rotational relaxation in liquids. *J. Phys. Chem.* 1981, 85, 2169–2180.
14. Alms, G. R.; Bauer, D. R.; Brauman, J. I.; and Pecora, R., Depolarized Rayleigh scattering and orientational relaxation of molecules in solution. I. Benzene, toluene and para-xylene. *J. Chem. Phys.* 1973, 58, 5570–5578.
15. Evans, G. T.; and Kivelson, D., The orientational-correlation time intercept in liquids. *J. Chem. Phys.* 1986, 84, 385–390.
16. Hu, C.-M.; and Zwanzig, R., Rotational friction coefficients for spheroids with the slipping boundary condition. *J. Chem. Phys.* 1974, 60, 4354–4357.
17. Venable, R. M.; and Pastor, R. W., Frictional models for stochastic simulations of proteins. *Biopolymers* 1988, 27, 1001–1014.
18. Rundlöf, T.; Venable, R. M.; Pastor, R. W.; Kowalewski, J.; and Widmalm, G., Distinguishing anisotropy and flexibility of the pentasaccharide LNF-1 in solution by carbon-13 NMR relaxation and hydrodynamic modeling. *J. Am. Chem. Soc.* 1999, 121, 11847–11854.
19. Madden, P.; and Kivelson, D., Dielectric friction and molecular reorientation. *J. Phys. Chem.* 1982, 86, 4244–4256.
20. Daragan, V. A.; and Mayo, K. H., Motional model analyses of protein and peptide dynamics using C-13 and N-15 NMR relaxation. *Prog. NMR Spectr.* 1997, 31, 63–105.
21. Luginbühl, P.; and Wüthrich, K., Semi-classical nuclear spin relaxation theory revisited for use with biological macromolecules. *Prog. NMR Spectr.* 2002, 40, 199–247.
22. Igumenova, T. I.; Frederick, K. K.; and Wand, A. J., Characterization of the fast dynamics of protein amino acid side chains using NMR relaxation in solution. *Chem. Rev.* 2006, 106, 1672–1699.
23. Jarymowycz, V. A.; and Stone, M. J., Fast time scale dynamics of protein backbones: NMR relaxation methods, applications, and functional consequences. *Chem. Rev.* 2006, 106, 1624–1671.
24. Peng, J. W., Exposing the moving parts of proteins with NMR spectroscopy. *J. Phys. Chem. Lett.* 2012, 3, 1039–1051.
25. Shapiro, Y. E., NMR spectroscopy on domain dynamics in biomacromolecules. *Prog. Biophys. Mol. Biol.* 2013, 112, 58–117. Erratum: *ibid.* 2014, 114, 13–13.
26. Peng, J. W.; and Wagner, G., Investigation of protein motions via relaxation measurements. *Methods Enzymol.* 1994, 239, 563–596.
27. Palmer, A. G.; Kroenke, C. D.; and Loria, J. P., Nuclear magnetic resonance methods for quantifying microsecond-to-millisecond motions in biological macromolecules. *Methods Enzymol.* 2001, 339, 204–238.
28. Henry, E. R.; and Szabo, A., Influence of vibrational motion on solid state line shapes and NMR relaxation. *J. Chem. Phys.* 1985, 82, 4753–4761.
29. Kowalewski, J.; Effemey, M.; and Jokisaari, J., Dipole-dipole coupling constant for a directly bonded CH pair: A carbon-13 relaxation study. *J. Magn. Reson.* 2002, 157, 171–177.
30. Woessner, D. E., Spin relaxation processes in a two-proton system undergoing anisotropic reorientation. *J. Chem. Phys.* 1962, 36, 1–4.

31. Ericsson, A.; Kowalewski, J.; Liljefors, T.; and Stilbs, P., Internal rotation of methyl groups in terpenes. Variable-temperature carbon-13 spin-lattice relaxation time measurements and force-field calculations. *J. Magn. Reson.* 1980, 38, 9–22.
32. Werbelow, L. G.; and Grant, D. M., Intramolecular dipolar relaxation in multispin systems. *Adv. Magn. Reson.* 1977, 9, 189–299.
33. Lipari, G.; and Szabo, A., Model-free approach to the interpretation of nuclear magnetic resonance relaxation in macromolecules 1. Theory and range of validity. *J. Am. Chem. Soc.* 1982, 104, 4546–4559.
34. Wennerström, H.; Lindblom, G.; and Lindman, B., Theoretical aspects on the NMR of quadrupolar ionic nuclei in micellar solutions and amphiphilic liquid crystals. *Chem. Scr.* 1974, 6, 97–103.
35. Wennerström, H.; Lindman, B.; Söderman, O.; Drakenberg, T.; and Rosenholm, J. B., 13C magnetic relaxation in micellar solutions. Influence of aggregate motion on T1. *J. Am. Chem. Soc.* 1979, 101, 6860–6864.
36. Halle, B.; and Wennerström, H., Interpretation of magnetic resonance data from water nuclei in heterogeneous systems. *J. Chem. Phys.* 1981, 75, 1928–1943.
37. Polnaszek, C. F.; and Freed, J. H., Electron spin resonance studies of anisotropic ordering, spin relaxation, and slow tumbling in liquid crystalline solvents. *J. Phys. Chem.* 1975, 79, 2283–2292.
38. Freed, J. H., Stochastic-molecular theory of spin-relaxation for liquid crystals. *J. Chem. Phys.* 1977, 66, 4183–4199.
39. Brüschweiler, R.; Roux, B.; Blackledge, M.; Griesinger, C.; Karplus, M.; and Ernst, R. R., Influence of rapid intramolecular motion on NMR cross-relaxation rates: A molecular dynamics study of antamanide in solution. *J. Am. Chem. Soc.* 1992, 114, 2289–2302.
40. Clore, G. M.; Szabo, A.; Bax, A.; Kay, L. E.; Driscoll, P. C.; and Gronenborn, A. M., Deviations from the simple two-parameter model-free approach to the interpretation of nitrogen-15 nuclear magnetic relaxation of proteins. *J. Am. Chem. Soc.* 1990, 112, 4989–4991.
41. Barbato, G.; Ikura, M.; Kay, L. E.; Pastor, R. W.; and Bax, A., Backbone dynamics of calmodulin studied by 15N relaxation using inverse detected two-dimensional NMR spectroscopy: The central helix is flexible. *Biochemistry* 1992, 31, 5269–5278.
42. Frueh, D., Internal motions in proteins and interference effects in nuclear magnetic resonance. *Prog. NMR Spectr.* 2002, 41, 305–324.
43. Canet, D.; Bouguet-Bonnet, S.; and Mutzenhardt, P., On the calculation of cross-correlation spectral density functions within the model-free approach. *Concepts Magn. Reson. Part A* 2003, 19A, 65–70.
44. Tugarinov, V.; Liang, Z. C.; Shapiro, Y. E.; Freed, J. H.; and Meirovitch, E., A structural mode-coupling approach to N-15 NMR relaxation in proteins. *J. Am. Chem. Soc.* 2001, 123, 3055–3063.
45. Meirovitch, E.; Shapiro, Y. E.; Polimeno, A.; and Freed, J. H., Structural dynamics of bio-macromolecules by NMR: The slowly relaxing local structure approach. *Prog. NMR Spectr.* 2010, 56, 360–405. Erratum: *ibid.* 2010, 57, 343–343.
46. Meirovitch, E., Polimeno, A., and Freed J.H., *Protein Dynamics by NMR Spin Relaxation: The Slowly Relaxing Local Structure Perspective.* eMagRes, 2007.
47. Halle, B., The physical basis of model-free analysis of NMR relaxation data from proteins and complex fluids. *J. Chem. Phys.* 2009, 131, 224507.
48. Meirovitch, E.; Polimeno, A.; and Freed, J. H., Comment on "The physical basis of model-free analysis of NMR relaxation data from proteins and complex fluids" [*J. Chem. Phys.* 131, 224507 (2009)]. *J. Chem. Phys.* 2010, 132, 207101.
49. Zerbetto, M.; Polimeno, A.; Kotsyubynskyy, D.; Ghalebani, L.; Kowalewski, J.; Meirovitch, E.; Olsson, U.; and Widmalm, G., An integrated approach to NMR spin relaxation in flexible biomolecules: Application to beta-D-glucopyranosyl-(1-> 6)-alpha-D-mannopyranosyl-OMe. *J. Chem. Phys.* 2009, 131, 234501.

50. Wong, V.; Case, D. A.; and Szabo, A., Influence of the coupling of interdomain and overall motions on NMR relaxation. *Proc. Natl. Acad. Sci. USA*. 2009, 106, 11016–11021.
51. Ryabov, Y.; Clore, G. M.; and Schwieters, C. D., Coupling between internal dynamics and rotational diffusion in the presence of exchange between discrete molecular conformations. *J. Chem. Phys.* 2012, 136, 034108.
52. Böttcher, C. J. F.; and Bordewijk, P., *Theory of Electron Polarization, Vol. 2*. Elsevier: Amsterdam, 1973.
53. Davidson, D. W., and Cole R. H., Dielectric relaxation in glycerol, propylene glycol and n-propanol. *J. Chem. Phys.* 1951, 19, 1484–1490.
54. Kruk, D.; Herrmann, A.; and Rössler, E. A., Field-cycling NMR relaxometry of viscous liquids and polymers. *Prog. NMR Spectr.* 2012, 63, 33–64.

7

NMR: The Toolbox

In this chapter, we will briefly go through the basics of a nuclear magnetic resonance (NMR) experiment. Naturally, since this is not the topic of this book, the presentation will be brief. For readers interested in details concerning NMR experiments and methodology, we refer to *Further reading* (see, for instance, the books by Keeler, Levitt, Ernst, Bodenhausen and Wokaun, or Cavanagh *et al.*). Pulsed NMR offers the possibility for the experimenter to create and monitor complicated non-equilibrium states. In the following, we shall go through the tools available in NMR spectroscopy with an emphasis on relaxation experiments.

A simple relaxation experiment can schematically be decomposed into three parts: a preparation period, a relaxation period and a detection period. This scheme bears a clear resemblance to the general scheme (preparation, evolution, mixing, detection) used to describe a two-dimensional experiment, although one should be aware of the difference. Since the first pulsed NMR experiments were introduced and the Fourier transform NMR methods were proposed (more than 50 years ago), enormous developments have occurred in NMR methodology, and two-dimensional methods for determining relaxation parameters are readily available as standard procedures. During the *preparation period* the system is prepared in a desired non-equilibrium state and this part of the experiment can be composed of many different radiofrequency (r.f.) pulses and time delays. In a two-dimensional approach, the evolution period for frequency-labelling in the indirectly detected dimension can be incorporated in the preparation period of the generic relaxation experiment. During the *relaxation period*, the non-equilibrium state is allowed to develop (relax). In practical implementations, the relaxation period is usually a variable delay, τ, which is arrayed, and a series of spectra with different values of τ is recorded. The relaxation rate can be obtained from a fit of the evolution of integrated intensities as a function of τ. After the relaxation period, some manipulation is often necessary for the creation of observable single-quantum coherences. This is usually referred to as the *mixing period*. In our terminology, this manipulation will be considered as a part of the *detection period*. The scheme may readily be extended to also contain the two-dimensional evolution, in which the indirect dimension usually involves the heteronuclear species for which relaxation is to be measured, and the detection is done on the more sensitive protons.

In a pulsed NMR experiment the spin system is perturbed by applying r.f. pulses of specific duration, amplitude and phase, and by introducing specific delays. These have the effect of changing the magnetisation of the spin system through specific flip angles of the pulses and magnetisation transfer through different couplings. For a weakly coupled spin system, the product operator formalism described in Section 1.4 provides a convenient tool for the description of NMR experiments. The evolution of the density operator under radiofrequency pulses, chemical shifts and scalar couplings will briefly be discussed in Section 7.1 and specific details concerning two-dimensional NMR and coherence selection will be discussed in Sections 7.2 and 7.3.

7.1 Evolution of Product Operators

The product operator formalism is a very useful approach for describing pulsed NMR experiments involving coupled spins in the *weak coupling* limit, *i.e.* when the J-couplings are much smaller than the differences in Larmor frequencies (chemical shifts) expressed in the same units, where all terms in the Hamiltonian commute. The idea of product operators was outlined in Section 1.4. For more extensive presentations of the subject, the reader is referred to the books by Ernst, Bodenhausen and Wokaun, by Keeler and by Levitt (*Further reading*). For two weakly coupled $I=1/2$ spins, I and S, the basis operators needed for the expansion of the density operator can be defined and 16 *product operators* are needed to fully describe the system: $\frac{1}{2}1_{op}, \hat{I}_x, \hat{I}_y, \hat{I}_z, \hat{S}_x, \hat{S}_y, \hat{S}_z, 2\hat{I}_x\hat{S}_x, 2\hat{I}_x\hat{S}_y, 2\hat{I}_x\hat{S}_z, 2\hat{I}_y\hat{S}_x, 2\hat{I}_y\hat{S}_y, 2\hat{I}_y\hat{S}_z, 2\hat{I}_z\hat{S}_x, 2\hat{I}_z\hat{S}_y, 2\hat{I}_z\hat{S}_z$. These operators have different meanings and names. $\frac{1}{2}1_{op}$ is the unity operator; $\hat{I}_x$ and $\hat{I}_y$ are transverse in-phase one-spin operators for spin I. The transverse one-spin operators are closely related to the shift operators $\hat{I}_+$ and $\hat{I}_-$, introduced in Eq. (3.12). $2\hat{I}_x\hat{S}_z$ and $2\hat{I}_y\hat{S}_z$ are two-spin operators, anti-phase for the I-spin. Double- and zero-quantum coherences for spins I and S are represented by $2\hat{I}_x\hat{S}_x$, $2\hat{I}_x\hat{S}_y$, $2\hat{I}_y\hat{S}_x$ and $2\hat{I}_y\hat{S}_y$. The longitudinal operators for the I-spin are the one-spin order, $\hat{I}_z$, and the two-spin order, $2\hat{I}_z\hat{S}_z$. The operators for the S-spin are defined in the same way. The only observable terms in the NMR experiment correspond to single-quantum coherences, represented for the I-spin by the in-phase $\hat{I}_x$ and $\hat{I}_y$ operators, and the anti-phase two-spin operators. Extending the formalism to include three spins is rather straightforward, but the calculations become a little cumbersome.

The density operator, ρ, may at all times in the NMR experiment be described by a combination of several product operators. Simple rules for the transformation of product operators can be obtained using the fact that the three single-spin operators satisfy the following commutation relationship:

$$\left[\hat{I}_x, \hat{I}_y\right] = i\hat{I}_z \tag{7.1}$$

and its cyclic permutations. This relationship leads to:

$$\exp\left(-i\theta\,\hat{I}_x\right)\hat{I}_y \exp\left(i\theta\,\hat{I}_x\right) = \hat{I}_y\cos\theta + \hat{I}_z\sin\theta \tag{7.2a}$$

This rotation of the operator $\hat{I}_y$, as seen in the left-hand side of Eq. (7.2a), can be identified formally with the rotation superoperator, $\hat{\hat{R}}_x(\theta)$, acting on the operator $\hat{I}_y$. The concept of the superoperator was first mentioned in the last section of Chapter 1.

Through the cyclic commutation relationship, we also get:

$$\exp\left(-i\theta\,\hat{I}_y\right)\hat{I}_z \exp\left(i\theta\,\hat{I}_y\right) = \hat{I}_z\cos\theta + \hat{I}_x\sin\theta \tag{7.2b}$$

$$\exp\left(-i\theta\,\hat{I}_z\right)\hat{I}_x \exp\left(i\theta\,\hat{I}_z\right) = \hat{I}_x\cos\theta + \hat{I}_y\sin\theta \tag{7.2c}$$

One should note that by substituting $\hat{I}_x$ for $-\hat{I}_x$ in Eq. (7.2a), we obtain a rotation in the opposite direction. This leads to the expression

$$\exp\left(i\theta\,\hat{I}_x\right)\hat{I}_y \exp\left(-i\theta\,\hat{I}_x\right) = \hat{I}_y\cos\theta - \hat{I}_z\sin\theta \tag{7.3}$$

An arbitrary product operator containing a single spin can be rotated through an angle θ about an a-axis using:

$$\hat{I}_b \xrightarrow{\phi\hat{I}_a} \hat{I}_b\cos\theta + \hat{I}_c\sin\theta \tag{7.4}$$

where *a,b,c* = *x,y,z*, and its cyclic permutations (*y,z,x* or *z,x,y*). For a two-spin operator, the effect of the rotation is evaluated from the individual operators according to:

$$2\hat{I}_b\hat{S}_{b'} \xrightarrow{\theta\hat{I}_a+\theta'\hat{S}_{a'}} 2\left(\hat{I}_b\cos\theta+\hat{I}_c\sin\theta\right)\times\left(\hat{S}_{b'}\cos\theta'+\hat{S}_{c'}\sin\theta'\right) \tag{7.5}$$

The commutator relationship in Eq. (7.1), and its consequences for rotations, is not enough to describe the NMR experiments. We may think of situations in which, for instance, $\hat{I}_x$ is rotated by $\hat{I}_y$. This corresponds to *a,b,c* = *y,x,z* (and its cyclic permutations). In this case, the commutator in Eq. (7.1) can be written as $[\hat{I}_y,\hat{I}_x]=-i\hat{I}_z$ or $[-\hat{I}_y,\hat{I}_x]=i\hat{I}_z$, which in analogy with Eq. (7.3) gives:

$$\exp\left(-i\theta\hat{I}_y\right)\hat{I}_x\exp\left(i\theta\hat{I}_y\right)=\hat{I}_x\cos\theta-\hat{I}_z\sin\theta \tag{7.6a}$$

and through the cyclic commutation relationships we also get

$$\exp\left(-i\theta\hat{I}_x\right)\hat{I}_z\exp\left(i\theta\hat{I}_x\right)=\hat{I}_z\cos\theta-\hat{I}_y\sin\theta \tag{7.6b}$$

$$\exp\left(-i\theta\hat{I}_z\right)\hat{I}_y\exp\left(i\theta\hat{I}_z\right)=\hat{I}_y\cos\theta-\hat{I}_x\sin\theta \tag{7.6c}$$

7.1.1 Radiofrequency Pulses and Free Precession

The usefulness of the product operator approach to describe NMR experiments can be exemplified by considering the effect of radio-frequency pulses. Let us consider the effect of an r.f. pulse of phase *x* applied to the equilibrium magnetisation, proportional to $\hat{I}_z$. The Hamiltonian for the pulse is proportional to $\hat{I}_x$, and Eq. (7.6b) thus describes the effect of applying an *x* pulse with the angle $\theta=\gamma_I B_1\tau$. This was discussed already in Section 1.3, and as mentioned there, if we adjust the magnitude of the r.f. field, B_1, and/or the duration, τ, during which it is applied, we can vary the angle, θ. If $\theta=90°$, the outcome is a density operator proportional to $-\hat{I}_y$.

The effect of several transformations, relevant in a real NMR experiment consisting of many events, such as r.f. pulses and delays, may be evaluated in a cascade of transformations. Each component of the product operator, *i.e.* components of $\hat{I}$ and $\hat{S}$, is rotated independently. Symbolically, in an arrow notation, this will look as follows:

$$\rho(t)\xrightarrow{\Omega_I\tau\hat{I}_z}\xrightarrow{\Omega_S\tau\hat{S}_z}\ldots\xrightarrow{\pi J_{IS}\tau 2\hat{I}_z\hat{S}_z}\ldots\rho(t+\tau) \tag{7.7}$$

where each of these transformations, here chemical shifts and *J*-couplings, corresponds to a rotation in a three-dimensional operator subspace.

It is revealing to investigate the effect of the different events in an experiment separately. First, we shall consider the effect of radiofrequency pulses, as exemplified in the beginning of Section 7.1.1. In general, the Hamiltonian for a pulse can be written $\phi\hat{I}_\alpha$, where α denotes the phase of the pulse, *i.e.* the specific axis along which it is applied. A radiofrequency pulse is defined by its flip angle, ϕ, given by the duration, τ, and the magnitude of the r.f. field, B_1. The effect of a pulse of phase $\pm x$ on the *I*-spin can, according to Eq. (7.4), be described by the following transformations:

$$\hat{I}_z\xrightarrow{\phi\hat{I}_{\pm x}}\hat{I}_z\cos\phi\mp\hat{I}_z\sin\phi \tag{7.8a}$$

$$\hat{I}_y\xrightarrow{\phi\hat{I}_{\pm x}}\hat{I}_y\cos\phi\pm\hat{I}_z\sin\phi \tag{7.8b}$$

$$\hat{I}_x\xrightarrow{\phi\hat{I}_{\pm x}}\hat{I}_x \tag{7.8c}$$

Similar results can be obtained for pulses of phase $\pm y$:

$$\hat{I}_z \xrightarrow{\phi \hat{I}_{\pm y}} \hat{I}_z \cos\phi \pm \hat{I}_x \sin\phi \tag{7.9a}$$

$$\hat{I}_y \xrightarrow{\phi \hat{I}_{\pm y}} \hat{I}_y \tag{7.9b}$$

$$\hat{I}_x \xrightarrow{\phi I_{\pm y}} \hat{I}_x \cos\phi \mp \hat{I}_z \sin\phi \tag{7.9c}$$

Thus, a 90° pulse of phase x will transform $\hat{I}_z$ into $-\hat{I}_y$, while a 90° pulse of phase y will transform $\hat{I}_z$ into $\hat{I}_x$. For a simple one-pulse experiment, consisting of a 90° pulse followed by detection, this formalism may seem unnecessary, but for a more complicated experiment using several pulses, of different length and phase, perhaps acting on different spins, these relations become very useful.

The NMR experiment becomes even more complicated when taking into account the periods of free precession (delays). The effect of free precession during an interval τ can be evaluated by taking into account evolution under chemical shifts and scalar couplings. The Hamiltonian describing the chemical shift for the I-spin can be written $\Omega_I \hat{I}_z$ (where, in the rotating frame, Ω_I is the offset frequency, in angular frequency units, with respect to the r.f. transmitter), and the transformations during the period τ are described by the relations:

$$\hat{I}_z \xrightarrow{\Omega_I \hat{I}_z \tau} \hat{I}_z \tag{7.10a}$$

$$\hat{I}_x \xrightarrow{\Omega_I \hat{I}_z \tau} \hat{I}_x \cos(\Omega_I \tau) + \hat{I}_y \sin(\Omega_I \tau) \tag{7.10b}$$

$$\hat{I}_y \xrightarrow{\Omega_I \hat{I}_z \tau} \hat{I}_y \cos(\Omega_I \tau) - \hat{I}_x \sin(\Omega_I \tau) \tag{7.10c}$$

For the weakly coupled two-spin system described herein, the Hamiltonian for evolution under scalar couplings can be written as $2\pi J_{IS}\hat{I}_z\hat{S}_z$ (compare Eq. (4.66)) and the evolution of the I-spin magnetisation is given by:

$$\hat{I}_z \xrightarrow{2\pi J_{IS}\hat{I}_z\hat{S}_z\tau} \hat{I}_z \tag{7.11a}$$

$$\hat{I}_x \xrightarrow{2\pi J_{IS}\hat{I}_z\hat{S}_z\tau} \hat{I}_x \cos(\pi J_{IS}\tau) + 2\hat{I}_y\hat{S}_z \sin(\pi J_{IS}\tau) \tag{7.11b}$$

$$\hat{I}_y \xrightarrow{2\pi J_{IS}\hat{I}_z\hat{S}_z\tau} \hat{I}_y \cos(\pi J_{IS}\tau) - 2\hat{I}_x\hat{S}_z \sin(\pi J_{IS}\tau) \tag{7.11c}$$

The evolution of two-spin operators under J-couplings is given by:

$$2\hat{I}_z\hat{S}_z \xrightarrow{2\pi J_{IS}\hat{I}_z\hat{S}_z\tau} 2\hat{I}_z\hat{S}_z \tag{7.12a}$$

$$2\hat{I}_x\hat{S}_z \xrightarrow{2\pi J_{IS}\hat{I}_z\hat{S}_z\tau} 2\hat{I}_x\hat{S}_z \cos(\pi J_{IS}\tau) + \hat{I}_y \sin(\pi J_{IS}\tau) \tag{7.12b}$$

$$2\hat{I}_y\hat{S}_z \xrightarrow{2\pi J_{IS}\hat{I}_z\hat{S}_z\tau} 2\hat{I}_y\hat{S}_z \cos(\pi J_{IS}\tau) - \hat{I}_x \sin(\pi J_{IS}\tau) \tag{7.12c}$$

Multiple-quantum operators in a two-spin system, two-quantum, or double-quantum, coherence (DQ), and zero-quantum coherence (ZQ) are represented by two-spin operators, in which both spins have transverse components. It is convenient to express multiple-quantum coherence in terms of shift operators. For instance, the operator $2\hat{I}_x\hat{S}_y$ can be expressed as:

$$2\hat{I}_x\hat{S}_y = 2\frac{1}{2}\left(\hat{I}^+ + \hat{I}^-\right)\frac{1}{2i}\left(\hat{S}^+ - \hat{S}^-\right) = \\ = \frac{1}{2i}\left(\hat{I}^+\hat{S}^+ - \hat{I}^-\hat{S}^-\right) - \frac{1}{2i}\left(\hat{I}^+\hat{S}^- - \hat{I}^-\hat{S}^+\right) \tag{7.13}$$

The first term in the second line of Equation 7.13 represents a *double-quantum* coherence, and the second term is a *zero-quantum* coherence. In the product operator formalism, *double-quantum* coherence and *zero-quantum* coherence can be obtained by suitable combinations of two-spin product operators:

$$DQ_x = \frac{1}{2}\left(2\hat{I}_x\hat{S}_x - 2\hat{I}_y\hat{S}_y\right) = \frac{1}{2}\left(\hat{I}^+\hat{S}^+ + \hat{I}^-\hat{S}^-\right) \tag{7.14a}$$

$$DQ_y = \frac{1}{2}\left(2\hat{I}_x\hat{S}_y + 2\hat{I}_y\hat{S}_x\right) = \frac{1}{2i}\left(\hat{I}^+\hat{S}^+ - \hat{I}^-\hat{S}^-\right) \tag{7.14b}$$

$$ZQ_x = \frac{1}{2}\left(2\hat{I}_x\hat{S}_x + 2\hat{I}_y\hat{S}_y\right) = \frac{1}{2}\left(\hat{I}^+\hat{S}^+ + \hat{I}^-\hat{S}^+\right) \tag{7.14c}$$

$$ZQ_y = \frac{1}{2}\left(2\hat{I}_y\hat{S}_x - 2\hat{I}_x\hat{S}_y\right) = \frac{1}{2i}\left(\hat{I}^+\hat{S}^- - \hat{I}^-\hat{S}^+\right) \tag{7.14d}$$

Rules for the transformation of *double-* and *zero-quantum* coherences can be derived as easily as for the single-quantum correlation, and it can be shown that a pure DQ coherence precesses at the sum of the chemical shifts of the two spins, while a pure ZQ coherence precesses at the difference of the chemical shifts. These transformation rules can be verified by evaluating the evolution of the two-spin operators in Eq. (7.14) by transforming the operators individually using Eq. (7.5). Furthermore, DQ and ZQ coherences do not evolve under scalar couplings (J-couplings). The transformation rules presented in this section can be applied to any pulse sequence and the outcome can readily be calculated. As already mentioned, the different transformations under shifts and couplings commute with each other and the evolution period can be divided into a cascade of transformations as described in Eq. (7.7).

7.1.2 A Simple Example: Insensitive Nuclei Enhanced by Polarisation Transfer (INEPT)

Pulses and delays can be combined in numerous ways to produce pulse sequences that allow the creation and detection of different non-equilibrium states. One of the more useful tools, which serves as a nice illustration of the use of product operators, is provided through the INEPT experiment (Figure 7.1), proposed originally by Morris and Freeman[1]. The experiment is designed to transfer magnetisation from a sensitive nucleus (usually protons, which have a high magnetogyric ratio; *cf.* Table 1.1) to a more insensitive nucleus. The result is a gain in signal intensity for the heteronucleus and this method is frequently used in NMR of carbon-13 or nitrogen-15, which have lower magnetogyric ratios, and where sensitivity enhancement is desired. The experimental scheme has perhaps gained even more attention in its two-dimensional variety, heteronuclear single quantum correlation (HSQC) spectroscopy[2].

The experiment starts with a $90°_x$ pulse on the I-spins (protons), which will produce the following transformation of the equilibrium state:

$$a\hat{I}_z + b\hat{S}_z \xrightarrow{90°\hat{I}_x} -a\hat{I}_y + b\hat{S}_z \tag{7.15}$$

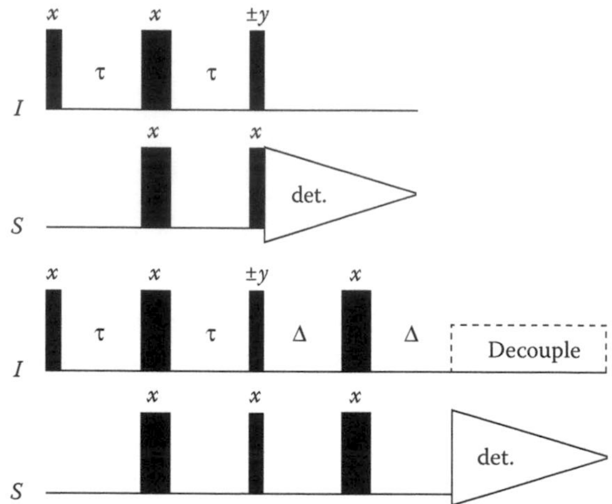

FIGURE 7.1 The INEPT pulse sequence. The top panel shows the conventional INEPT pulse sequence and the bottom panels the refocused INEPT. For an *IS* spin system, $\Delta=\tau$ is set to $1/4J_{IS}$ for maximum INEPT polarisation transfer. The dashed box in the lower panel indicates optional decoupling of the *I*-spins. The receiver phase is set to *x*, −*x*.

Here, *a* and *b* represent the relative magnitudes of the magnetisation associated with the *I*- and *S*-spin, respectively. If, for instance, *I* is a proton and *S* is a carbon-13 nucleus in a *J*-coupled *IS* spin system, $a \approx 4b$. The next step in the pulse sequence, $\tau-(180°_x)_{I,S}-\tau$, is known as a *spin-echo*, and it can be shown that the scalar *J*-coupling interaction evolves over the duration of the entire spin-echo, while the chemical shift is refocused during the echo. Taking into account evolution under *J*-coupling and the effect of the $180°_x$ pulses on both spins, we obtain:

$$\begin{aligned}
&-a\hat{I}_y + b\hat{S}_z \xrightarrow{2\pi J_{IS}\hat{I}_z\hat{S}_z\tau} -a\hat{I}_y\cos(\pi J_{IS}\tau) + a2\hat{I}_x\hat{S}_z\sin(\pi J_{IS}\tau) + b\hat{S}_z \\
&\xrightarrow{180°(\hat{I}_x+\hat{S}_x)} a\hat{I}_y\cos(\pi J_{IS}\tau) - a2\hat{I}_x\hat{S}_z\sin(\pi J_{IS}\tau) - b\hat{S}_z \\
&\xrightarrow{2\pi J_{IS}\hat{I}_z\hat{S}_z\tau} a\hat{I}_y\cos^2(\pi J_{IS}\tau) - a2\hat{I}_x\hat{S}_z\cos(\pi J_{IS}\tau)\sin(\pi J_{IS}\tau) \\
&-a2\hat{I}_x\hat{S}_z\sin(\pi J_{IS}\tau)\cos(\pi J_{IS}\tau) - a\hat{I}_y\sin^2(\pi J_{IS}\tau) - b\hat{S}_z
\end{aligned} \tag{7.16a}$$

which is equal to:

$$a\hat{I}_y\cos(2\pi J_{IS}\tau) - a2\hat{I}_x\hat{S}_z\sin(2\pi J_{IS}\tau) - b\hat{S}_z \tag{7.16b}$$

At this point, we introduce a specific value for the delay, τ equal to $\frac{1}{4}J_{IS}$, and it becomes clear that the purpose of this delay is to generate a pure *I*-spin *antiphase* term, as the first term in Eq. (7.16b) vanishes. Had we instead chosen to set the delay to $\frac{1}{2}J_{IS}$, the result would have been an inverted in-phase term originating from the *I*-spin. The $(90°_y)_I$ and $(90°_x)_S$ pulses at the end of the sequence provide the final outcome of the INEPT experiment:

$$-a2\hat{I}_z\hat{S}_y + b\hat{S}_y \tag{7.17}$$

Both of these terms correspond to *S*-spin signals. The first is the antiphase *y* magnetisation with an intensity factor originating from the *I*-spin (the factor *a* shows this). The second is the in-phase *y* magnetisation, which has a much lower intensity (from the factor b). The appearance of the spectrum will be an overlay of the two terms, as shown in Figure 7.2. The asymmetry in the spectrum can be removed by phase cycling, a procedure that will be described in Section 7.3.1. However, it can easily be seen that by replacing the final $(90°_y)_I$ by a $(90°_{-y})_I$, the outcome will be $a2\hat{I}_z\hat{S}_y + b\hat{S}_y$. By subtracting this from the first experiment, a pure antiphase term, represented by $-2 \times a2\hat{I}_z\hat{S}_y$ remains. This reasoning is equivalent to the phase cycle indicated in Figure 7.1.

There is a logical extension to this pulse sequence. By adding yet another spin-echo $\Delta - (180°_x)_{I,S} - \Delta$ after the final pulses in the INEPT sequence, and by setting also $\Delta = \tau = 1/4J_{IS}$, it can be shown that the antiphase term is transformed into an in-phase doublet according to:

$$-a2\hat{I}_z\hat{S}_y \xrightarrow{\Delta-180°(\hat{I}_x+\hat{S}_x)-\Delta} a\hat{S}_x \tag{7.18}$$

This is very practical, since proton decoupling will now produce one line in the spectrum that is enhanced by a factor of $a/b = \gamma_I/\gamma_S$ by the INEPT sequence, as compared with a conventional decoupled, one-dimensional spectrum with the nuclear Overhauser enhancement (NOE) suppressed (see Chapter 9). While $\Delta = 1/4\,J_{IS}$ is optimal for an IS spin system, other multiplets involving several spins need other values for optimum enhancement. Figure 7.2 illustrates the appearance of the INEPT spectrum for a two-spin system with and without refocusing.

The experiments described in this section are generally performed on a spin system containing one proton and one heteronucleus, for instance, carbon-13 or nitrogen-15. The r.f. pulses naturally only affect one species at a time, leading to a simple way of conducting multinuclear multipulse experiments. It is, however, often desirable to also be able to perform, for instance, proton experiments in which only one proton is excited. Other such applications include selective excitation of the solvent resonance to suppress this signal. Therefore, there is a need for *band-selective pulses*, which act on only a narrow range of the total spectrum. The easiest way to obtain such a pulse is simply to reduce the amplitude of the r.f. field and, by making the appropriate lengthening of the pulse, to retain the desired flip angle. In practice, this is not a good way to achieve selective excitation, since the rectangular pulse has very unfavourable characteristics, such as phase variation with offset and variable excitation profile. Many pulses have been designed to overcome the shortcomings of the rectangular pulse, and selective pulses of various shapes are used routinely for different nuclei in many applications. There are certain requirements of a selective pulse. It should be kept short, to minimise relaxation effects and evolution during the pulse, and the excitation profile should be uniform over the desired frequency range. Specific pulses have been designed for different purposes, for instance, for excitation and inversion of magnetisation. The more common include the Gaussian-shaped pulse, although there has been a rapid development of the area

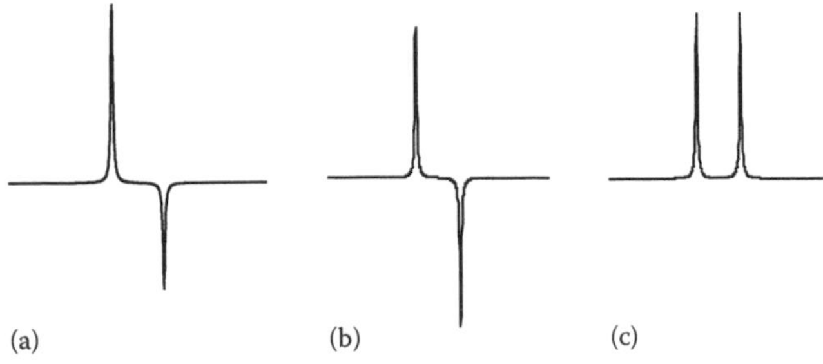

FIGURE 7.2 A schematic picture of the IS doublet. (a) shows the result of the INEPT sequence for a ^1H-^{13}C spin pair. (b) is the result of applying a phase cycle to the INEPT sequence to remove the non-transferred in-phase magnetisation corresponding to S_y. (c) shows the refocused INEPT without decoupling, and using a two-step phase cycle.

leading to a multitude of pulses being used. The issue of selective pulses has been reviewed by Freeman[3], and the reader should consult this work for a more detailed presentation.

7.2 Two-Dimensional NMR

The INEPT sequence described in Section 7.1 belongs to a class of experiments known as *polarisation transfer experiments*. They have proven extremely useful in a wide range of applications, especially as part of multidimensional NMR methods including heteronuclear correlations. Relaxation experiments, in particular, are very often run as two-dimensional X-^{1}H correlation experiments, where X is the species for which relaxation is to be measured, *e.g.* nitrogen-15 or carbon-13, and ^{1}H represents the directly bound proton(s). This is especially important for larger molecules, such as biological macromolecules, and spin-lattice, spin-spin and heteronuclear NOE relaxation measurements by two-dimensional techniques have become routine in protein NMR spectroscopy. A comprehensive discussion of two-dimensional NMR can be found in the text by Ernst, Bodenhausen and Wokaun, and detailed descriptions of the common experiments are beautifully provided in the book by Cavanagh *et al.* (*Further reading*).

A two-dimensional NMR experiment is in principle obtained by including yet another part in the general scheme of an NMR experiment – the indirect evolution period, usually denoted t_1. The evolution period is incremented in small steps to produce the indirectly detected free induction decay (FID) in the t_1 dimension. The directly detected free induction decay is recorded in the detection period as a function of a running time variable often denoted t_2. The analogous time for the indirectly detected FID, usually referred to as the *evolution period*, t_1, is incremented in a series of experiments. This evolution period consists of a delay, during which free precession occurs, sometimes complemented by pulses or some other form of r.f. irradiation. This leads to a directly detected FID for each incremented time point in the indirectly detected FID. The amplitude of the recorded free induction decay will thus depend on the evolution during t_1, and therefore the duration of t_1. A series of experiments recorded with increasing t_1 delays will give rise to an *amplitude-modulated* series of spectra after Fourier transformation, as we will see below. One should, however, note that the different segments (preparation, evolution, mixing and detection) in the two-dimensional scheme are not generally independent of each other, and may be combined in different ways. Nevertheless, for clarity, it is useful to break down the experiments into these parts.

In order to understand how the two-dimensional spectrum is achieved, we will briefly summarise the basics of Fourier transform NMR. A one-dimensional signal is usually obtained by so-called *quadrature detection*. Two orthogonal signals (recorded in the x and y directions in the rotating frame) are detected. These signals are cosine and sine functions of the offset frequency, or chemical shift (the difference between the Larmor frequency of the spin and the frequency at which the r.f. pulses are applied), Ω, and they decay with a rate, λ, equal in the homogeneous field to the transverse relaxation rate, $1/T_2$. The complex time-domain signal is given by:

$$s(t) = s_0[\cos\Omega t + i\sin\Omega t]\exp(-\lambda t) = \exp\{(i\Omega - \lambda)t\} \tag{7.19}$$

The Fourier transform of this signal leads to a *complex Lorentzian* frequency-domain signal given by:

$$S(\omega) = \int_0^\infty s(t)\exp(-i\omega t)dt = \frac{1}{i(\omega - \Omega) + \lambda} \tag{7.20}$$

The real part of the complex Lorentzian is called the *absorptive Lorentzian*, and the imaginary part the *dispersive Lorentzian*. The signal can alternatively be written as:

$$S(\omega) = A + iD \tag{7.21}$$

where

$$A=\frac{\lambda}{\lambda^2+(\omega-\Omega)^2} \tag{7.22a}$$

$$D=\frac{(\omega-\Omega)}{\lambda^2+(\omega-\Omega)^2} \tag{7.22b}$$

where A is the absorption part and D the dispersion part of the signal. If only one of the two components in Eq. (7.19) (the sine or cosine part) had been detected, it would not have been possible to determine the sense of the precession of the signal, or the sign of $(\omega-\Omega)$, which provides the signs of the frequencies in the NMR spectrum. This can easily be seen if one replaces the expression for the time-domain signal in Eq. (7.19) by either the cosine- or the sine-modulated part of the signal (and by noting that $\cos(\Omega t)=(1/2)[\exp(i\Omega t)+\exp(-i\Omega t)]$ and $\sin(\Omega t)=(1/2i)[\exp(i\Omega t)-\exp(-i\Omega t)]$) and by performing a Fourier transform (the reader is encouraged to do this).

In a two-dimensional experiment, the FID for the t_1 evolution is not detected directly, and consequently, quadrature detection in the conventional sense cannot be achieved. Instead, only one of the cosine or sine components is recorded, and one speaks of an *amplitude-modulated time-domain signal*, meaning that the amplitude of the recorded signal in t_2 is modulated by the evolution during t_1. The corresponding signal for a single peak in a two-dimensional spectrum will be of the form:

$$S_{\cos}(t_1,t_2)=s_0\cos(\Omega_1 t_1)\exp(-\lambda_1 t_1)\exp\{(i\Omega_2-\lambda_2)t_2\} \tag{7.23}$$

The sine-modulated signal can be obtained in an analogous way in a closely related experiment derived by simply shifting the phase of one (or several) pulses in the sequence by 90°:

$$S_{\sin}(t_1,t_2)=s_0\sin(\Omega_1 t_1)\exp(-\lambda_1 t_1)\exp\{(i\Omega_2-\lambda_2)t_2\} \tag{7.24}$$

The frequencies, Ω_1 and Ω_2, are the chemical shifts in the two dimensions. The terms λ_1 and λ_2 take into account relaxation during the evolution and detection and describe the decay of the signal or, in other words, the peak widths in the two dimensions. The two λs are, in general, not equal to each other.

Two problems are associated with the detection of two-dimensional data sets. First, it can easily be shown that the Fourier transform of the signal in either Eq. (7.23) or Eq. (7.24) will yield a real part of the spectrum that is a mixture of both absorptive and dispersive parts. The two-dimensional Fourier transform of Eq. (7.23) converts the signal into a *complex two-dimensional Lorentzian* spectrum:

$$S_{\cos}(\omega_1,\omega_2)=\int_0^\infty\int_0^\infty s_{\cos}(t_1,t_2)\exp\{(-i\omega_1)t_1\}\exp\{(-i\omega_2)t_2\}dt_1dt_2 \tag{7.25}$$

After performing the Fourier transform of the t_2 dimension we obtain:

$$\begin{aligned}S_{\cos}&=(t_1,\omega_2)=s_0\cos(\Omega_1 t_1)\exp(-\lambda t_1)[A_2+iD_2]\\&=\tfrac{1}{2}s_0\left[\exp(i\Omega_1 t_1)+\exp(-i\Omega_1 t_1)\right]\exp(-\lambda t_1)[A_2+iD_2]\end{aligned} \tag{7.26}$$

If we denote the Fourier transform of $\exp(\pm i\Omega_1 t_1)$ by $A_1^{\pm} + iD_1^{\pm}$ we obtain after the second transform that:

$$\begin{aligned} S(\omega_1,\omega_2) &= \tfrac{1}{2}s_0\left[\left(A_1^+ + iD_1^+\right)+\left(A_1^- + iD_1^-\right)\right]\left(A_2 + iD_2\right) \\ &= \tfrac{1}{2}s_0\left[A_1^+A_2 + A_1^-A_2 - D_1^+D_2 - D_1^-D_2\right] + \tfrac{i}{2}s_0\left[A_1^+D_2 + A_1^-D_2 + D_1^+A_2 + D_1^-A_2\right] \end{aligned} \quad (7.27)$$

where A_2, A_1^+, D_2 and $D_1^{\pm}$ denote the absorption and dispersion parts of the resulting signal in the ω_1 and ω_2 dimensions. The real part of this is readily seen to include both absorptive and dispersive components, and the spectrum will be a mixture of absorption and dispersion Lorentzians. The signal shape associated with this mixture is known as the *phase-twist* peak shape and will lead to spectra with poor properties concerning resolution. Similarly, the imaginary part of Eq. (7.26) is also a mixture of absorptive and dispersive parts, and will consequently also give rise to a peculiar peak shape.

The second problem in two-dimensional NMR is concerned with the discrimination between positive and negative frequencies. Fortunately, it is possible to solve both problems: to get rid of the dispersive part of the signal, and to obtain frequency discrimination by combining different data sets in a certain way. By performing two experiments, one in which the cosine-modulated signal, given by Eqs. (7.23) and (7.26), is recorded, and one in which the sine-modulated signal, given in Eq. (7.24), is obtained, we can combine the two according to:

$$\begin{aligned} S(t_1,\omega_2) &= \mathrm{Re}\left\{S_{\cos}(t_1,\omega_2)\right\} + i\,\mathrm{Re}\left\{S_{\sin}(t_1,\omega_2)\right\} \\ &= s_0\left[\cos(\Omega_1 t_1) + i\sin(\Omega_1 t_1)\right]\exp(-\lambda_1 t_1)A_2 \\ &= s_0\exp\left\{(i\Omega_1 - \lambda_1)t_1\right\}A_2 \end{aligned} \quad (7.28)$$

where the Fourier transformation with respect to t_2 has been performed, and the imaginary part has been deleted. Finally, Fourier transform with respect to t_1 provides the two-dimensional lineshape as

$$S(\omega_1,\omega_2) = \left[A_1^+ + iD_1^+\right]A_2 = A_1^+A_2^+ + iD_1^+A_2 \quad (7.29)$$

Taking the real part of this expression yields a spectrum in *pure absorption*. The method described above is known as the *hypercomplex*, or States, method[4] (after one of the inventors), but there are other methods, which will produce pure phase spectra in slightly different, although conceptually similar, ways. Among these, the time-proportional phase increment (TPPI) method[5], or a combination of the two, States-TPPI, should be mentioned. For a thorough comparison of the methods, the reader is referred to the paper by Keeler and Neuhaus[6]. More recently, the selection based on pulsed-field gradients has become popular.

For performing two-dimensional relaxation experiments, a method that correlates the frequency of the nuclei for which we wish to measure relaxation and their directly bound protons is needed. The HSQC sequence is most frequently used for these applications (Figure 7.3). This clever experiment, developed from an experiment originally proposed by Bodenhausen and Ruben[2], starts with an INEPT sequence that transfers magnetisation from protons to *J*-coupled heteronuclei in the way described for the INEPT experiment. At this stage, it is possible to introduce a suitable sequence of pulses and delays to create a desired non-equilibrium state for measuring relaxation (this will be discussed in subsequent sections covering the specific spectroscopic techniques). In the original HSQC experiment, the *S*-spin state, represented by the operator $2\hat{I}_z\hat{S}_y$ (see Eqs. (7.17) and (7.18)), is allowed to evolve during the t_1 evolution period to obtain frequency-labelling in the heteronuclear dimension. The 180° pulses refocus the *IS* *J*-coupling during the evolution period, which removes couplings from the lineshape in the indirect

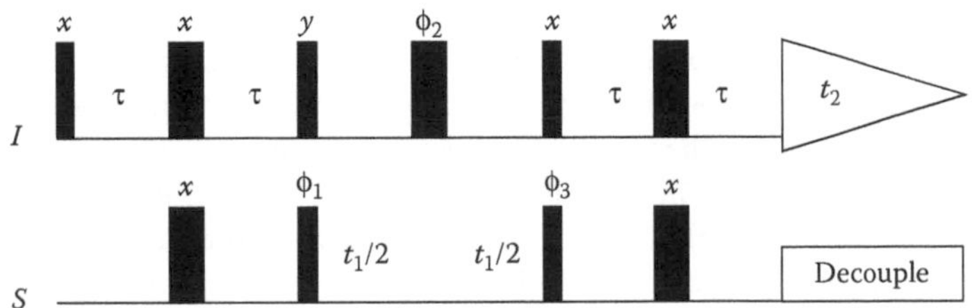

FIGURE 7.3 The HSQC pulse sequence. For an *IS* spin system, τ is set to $1/4J_{IS}$ for maximum INEPT polarisation transfer. The phase of each pulse is indicated. A suitable phase cycle for isotope editing, frequency discrimination and imperfections in the ^{1}H refocusing pulse is $\phi_1 = x, -x$, $\phi_2 = 4(y), 4(-y)$, $\phi_3 = 2(x), 2(-x)$ $\phi_{rec} = x, -x, -x, x$.

dimension. Finally, a *reverse* INEPT sequence is applied, during which the magnetisation is transferred back to the more sensitive proton nucleus. An appropriate phase cycle, as indicated in Figure 7.3, is needed in order to select for the desired coherence.

In order to determine a relaxation parameter by a two-dimensional method, an array of spectra with different relaxation delays must be recorded. This can be time-consuming both to record and to analyse. Nevertheless, this is the only way to achieve high enough resolution for larger molecules. Other important heteronuclear polarisation transfer techniques include the heteronuclear multiple-quantum coherence experiments (HMQC). The difference between the HMQC and the HSQC is basically that the transferred coherence evolves as a single-quantum coherence during the indirect evolution period in the HSQC, while it evolves as a *multiple-quantum* coherence in the HMQC. The HMQC experiment has the advantage of using fewer pulses, which in some applications can be important.

7.3 Coherence Order Selection

It is easily realised, as already seen from the INEPT example, that performing a certain pulse sequence most often produces many more states than the desired one. Thus, a technique is required to remove unwanted components of the resulting spectrum. There are currently two methods for doing this: phase cycling (which we have already touched upon) and the use of pulsed field gradients. The first method relies on the addition and subtraction of different spectra obtained using different phases of the pulses. A detailed presentation of the phase cycling methodology can be found in the books by Ernst, Bodenhausen and Wokaun, by Keeler and by Levitt (*Further reading*), and in the text by Keeler[7].

All product operators can be assigned a specific coherence order. For instance, the operators $\hat{I}_x, \hat{I}_y$ represent single-quantum one-spin operators, $p = \pm 1$ (the signed coherence order is associated with the shift operators, $p = +1$ with $\hat{I}_+$ and $p = -1$ with $\hat{I}_-$), while the antiphase term observed in the INEPT experiment, $2\hat{I}_z\hat{S}_y$, represents a single-quantum two-spin operator. The single-quantum coherence connects spin eigenstates differing by unity in the total *z*-component (the sum of *z*-components of all involved spins). In a two-spin system, there is also a possibility of creating *double-quantum* coherence, $p = \pm 2$, connecting the spin eigenstates with the *z*-components differing by two. The product operators $2\hat{I}_x S_x$, $2\hat{I}_y\hat{S}_y$, $2\hat{I}_x\hat{S}_y$, $2\hat{I}_y\hat{S}_x$ are *multiple-quantum* coherences, containing coherence orders $p = 0$ (connecting states with the same total *z*-component, *e.g.* $m_I = +½$, $m_S = -½$ with $m_I = -½$, $m_S = +½$) and $p = \pm 2$. The term $2\hat{I}_x\hat{S}_x - 2\hat{I}_y\hat{S}_y$ can easily be shown to represent pure *double-quantum* coherence (see Eq. (7.14)). Longitudinal magnetisation is represented by the operators $\hat{I}_z$, $\hat{S}_z$ and $2\hat{I}_z\hat{S}_z$, which correspond to populations rather than coherences but can formally be assigned the coherence order of zero. Only terms with $p = \pm 1$ are directly observable in an NMR experiment (by convention, the NMR receiver is usually stated to detect only $p = -1$), but others can be made observable by a suitable sequence of r.f. pulses and delays. Hence, it is crucial that we are able to be in control of what terms are retained in the experiment. The transformation of product operators in a pulse sequence is accompanied by changes in coherence orders. A $90_x°$ pulse changes the $\hat{I}_z$ operator, associated with a coherence order of $p=0$, into $-\hat{I}_y$, and a change in coherence order of $\Delta p=-1$ is obtained. In a coupled two-spin system, *double-quantum* coherence corresponds to a transition where two spins flip in the same sense, leading to a coherence order of $p=\pm 2$, whereas *zero-quantum* coherence corresponds to the two spins flipping in the opposite sense, leading to a coherence order of $p=0$. For the two-spin system, the coherence order can take on values of $-2<p<2$. A pulsed experiment can be described in terms of the pathway through various orders of coherence, and we can formally write the density operator as a sum of all possible coherence orders.

7.3.1 Phase Cycling

As we saw from the two-spin case, the range in coherence order for a system of M spins with $I=1/2$ is from $-M$ to $+M$. The change in coherence order is known as *coherence transfer* and is the basis for the phase cycling procedure. It is useful to identify the desired changes in coherence order in an experiment by drawing a *coherence order transfer pathway diagram*. To understand the usefulness of such diagrams, consider the two-dimensional 2Q-filtered correlation spectroscopy (COSY) experiment (Figure 7.4). The homonuclear two-dimensional COSY experiment suffers from the fact that the diagonal and cross-peaks cannot be absorptive at the same time. Usually this leads to a broad, dispersive diagonal, which tends to obscure the rest of the spectrum. One way of overcoming this is to select for 2Q-coherence at some point in the experiment. For our purposes it suffices to note that a 2Q- or *double-quantum* filter is needed. Single-quantum coherences are generated by the first 90° pulse, and are allowed to precess during the incremented delay, t_1. A 2Q-filter is applied, consisting of two pulses separated by a very short delay. During this delay, only 2Q-coherences should be allowed, and finally, coherence order $p=-1$ is selected for detection. Thus, the diagram in Figure 7.4 conveniently describes the desired coherences and indicates the necessary changes in coherence order. At each stage in a pulse sequence, we can indicate in the diagram the desired changes in coherence orders. We have a simple basic rule for how pulses and delays affect coherence orders: only r.f. pulses can cause changes in coherence orders, while the coherence order does not change during intervals of free precession. The coherence order diagram must always start with $p=0$, corresponding to thermal equilibrium, and end with $p=-1$, corresponding to single-quantum (order −1, by convention) coherence, in order to be detectable. The diagrams and the rules can be used to design phase cycles to select only the desired coherences. The way this works can be summarised in the following statements.

1. When the phase of a pulse or a group of pulses is shifted by ϕ, a coherence undergoing at that step a change in coherence order of Δp experiences a phase shift of $-\phi\Delta p$.
2. A phase cycle consisting of N steps, with 360°/N phase increments, selects coherence orders $\Delta p \pm nN$, where n is an integer.

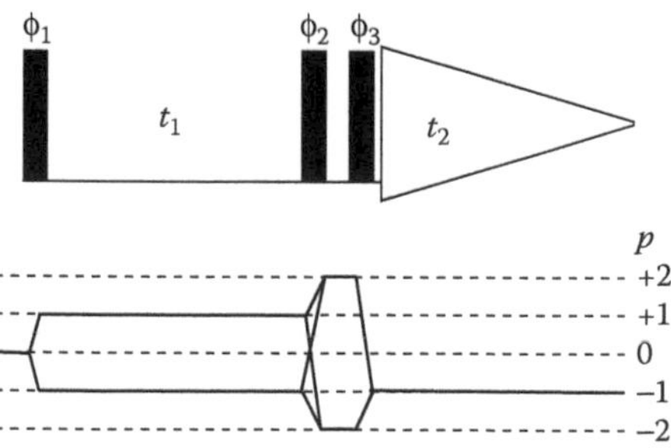

FIGURE 7.4 Coherence order diagram for the two-dimensional 2QF-COSY experiment. An appropriate phase cycle to select the desired changes in coherence levels is given by $\phi_1=\phi_2=x, y, -x, -y$, $\phi_3=x$ and $\phi_{rec}=2(x, -x)$.

3. By examining the coherence transfer pathway diagram, we can easily find out what the desired coherence order changes are. If the receiver phase is shifted by $-\phi_{rec}\Delta p$ following the phase of the desired coherence, the appropriate pathway is selected.

By co-adding several experiments in which the appropriate phases of the pulse(s), ϕ, and receiver phase, ϕ_{rec}, have been set, we can allow the desired coherence order to be selected, while others are deleted.

7.3.2 Pulsed Field Gradients

The phase cycling procedure can become time-consuming, as complicated pulse sequences require long phase cycles, implying that many transients have to be co-added to obtain the desired result. This is one of the reasons why the alternative procedure to select coherence transfer pathways – using pulsed field gradients – has become increasingly popular. Field gradients are used to create an inhomogeneous magnetic field in which transverse magnetisation dephases across the sample. The rate of dephasing is proportional to the coherence order and to the magnetogyric ratios of the nuclei involved. The phase shift experienced by a coherence of order p as a consequence of applying a static magnetic field of the strength B_G along the z-axis for a duration of τ is given by $p\gamma B_G\tau$. If the field varies over the NMR sample, *i.e.* a field gradient is applied across the sample, coherences in different regions of the sample will experience different fields. Thus, there will be a spread in the phase shift experienced by the different coherence orders. If the gradient pulse is strong enough and/or long enough, the applied field will completely dephase all non-zero coherence orders.

A second gradient pulse, of suitable strength and/or duration, can be applied after the r.f. pulse has changed the coherence order. This will refocus the phase of the desired coherence order, while the others remain dephased. This method allows coherence selection in only one transient, but in practice, several transients often have to be recorded to obtain sufficient signal-to-noise ratio. Readers interested in the principles of gradient-based coherence pathway selection techniques are referred to the excellent review by Keeler and co-workers[8].

Exercises for Chapter 7

1. Show that if only one of the two components in Eq. (7.19) (the sine or cosine part) is detected, it will not be possible to determine the sign of $(\omega-\Omega)$, which provides the signs of the frequencies in the NMR spectrum. Remember that $\cos(\Omega t) = (1/2)[\exp(i\Omega t) + \exp(-i\Omega t)]$ and $\sin(\Omega t) = (1/2i)[\exp(i\Omega t) - \exp(-i\Omega t)]$.

References

1. Morris, G. A.; and Freeman, R., Enhancement of nuclear magnetic resonance signals by polarization transfer. *J. Am. Chem. Soc.* 1979, 101, 760–762.
2. Bodenhausen, G.; and Ruben, D. J., Natural abundance nitrogen-15 NMR by enhanced heteronuclear spectroscopy. *Chem. Phys. Lett.* 1980, 69, 185–189.
3. Freeman, R., Shaped radiofrequency pulses in high resolution NMR. *Prog. NMR Spectr.* 1998, 32, 59–106.
4. States, D. J.; Haberkorn, R. A.; and Ruben, D. J., A two-dimensional nuclear Overhauser experiment with pure absorption phase in four quadrants. *J. Magn. Reson.* 1982, 48, 286–292.
5. Marion, D.; and Wüthrich, K., Application of phase sensitive two-dimensional correlated spectroscopy (COSY) for measurements of 1H-1H spin-spin coupling constants in proteins. *Biochem. Biophys. Res. Commun.* 1983, 113, 967–974.

6. Keeler, J.; and Neuhaus, D., Comparison and evaluation of methods for two-dimensional spectra with absorption-mode lineshapes. *J. Magn. Reson.* 1985, 63, 454–472.
7. Keeler, J., Phase cycling procedures in multiple pulse NMR spectroscopy of liquids. In *NATO ASI Series C. 322 (Multinucl. Magn. Reson. Liq. Solids: Chem. Appl.)*, 1990; 103–129.
8. Keeler, J.; Clowes, R. T.; Davis, A. L.; and Laue, E. D., Pulsed-field gradients: Theory and practise. *Methods Enzymol.* 1994, 239, 145–207.

8

Measuring T_1 and T_2 Relaxation Rates

We are now in a position to examine the specific methods used to determine relaxation rates and we will begin with the simple, single-spin relaxation properties related to the Bloch equations. The measurement of nuclear spin relaxation times has been an important and active field of research since the early days of Fourier transform nuclear magnetic resonance. As in all fields of NMR, the methodological development has been enormous, from direct measurements on low abundant nuclei to indirect multi-dimensional techniques applied to biological macromolecules. A useful book by Bakhmutov, intended for chemists wanting to explore the use of spin relaxation, contains a summary of the basic experimental techniques and applications (*Further reading*). The measurement of spin-lattice, T_1, and spin-spin, T_2, relaxation time constants has found particular use in characterising the reorientational and local motion of molecules.

In addition to the common one-dimensional inversion-recovery experiment for measuring T_1 relaxation time constants and the spin-echo for measuring T_2 relaxation time constants, there are now standard two-dimensional techniques to measure ^{15}N and ^{13}C T_1 and T_2 also for biological macromolecules to obtain valuable information about molecular dynamics in these complex systems. On a different note, the relaxation properties of the molecules under investigation are important in optimally designing novel NMR experiments in order to minimise signal loss due to relaxation and to select appropriate recycle delay between repeated sequences.

8.1 Spin-Lattice Relaxation Rates

T_1 relaxation measurements are among the easiest multipulse experiments to perform and also to conceptually understand. In order to measure T_1 relaxation time constants, non-equilibrium longitudinal magnetisation must be generated. The simplest way to achieve this is to apply an inversion pulse (180°) to the nucleus of interest. Together with a recycle delay, this constitutes the preparation period of the three-segmented relaxation experiment. This preparation is used in the famous inversion-recovery experiment developed in the early days of Fourier transform NMR spectroscopy[1] (Figure 8.1).

8.1.1 Inversion-Recovery

In this simple experiment, the equilibrium magnetisation M_0, directed along the static magnetic field B_0, is inverted by a 180° pulse. The magnetisation starts to relax according to Eq. (1.23) during the relaxation period, τ. The 90° pulse at the beginning of the detection period creates observable magnetisation in the x,y-plane, corresponding to $p = -1$ coherence, after which it is detected. The experiment is repeated with different durations of the relaxation period, and the time evolution of the signal gives the

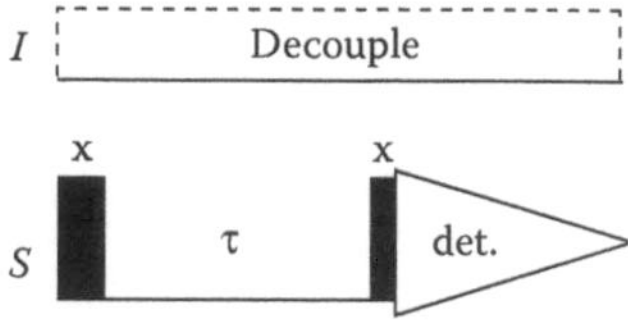

FIGURE 8.1 Inversion recovery pulse sequence for measuring T_1 relaxation of the S-spin while decoupling the I-spin.

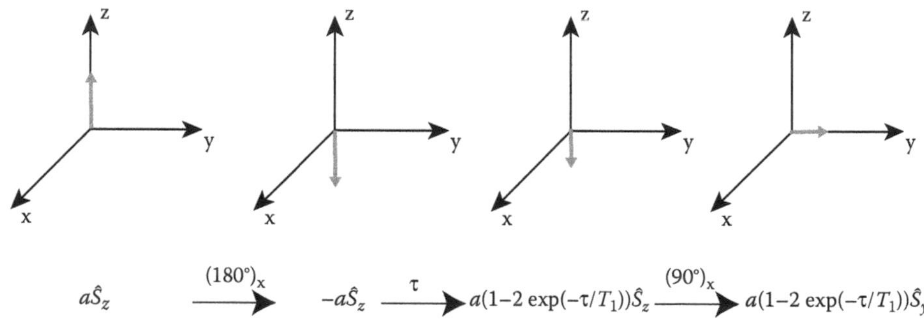

$$a\hat{S}_z \xrightarrow{(180^\circ)_x} -a\hat{S}_z \xrightarrow{\tau} a(1-2\exp(-\tau/T_1))\hat{S}_z \xrightarrow{(90^\circ)_x} a(1-2\exp(-\tau/T_1))\hat{S}_y$$

FIGURE 8.2 A vector representation of the inversion-recovery experiment, and the corresponding evolution of product operators.

relaxation time constant T_1. Figure 8.2 provides a vectorial model of how the experiment works. In the product operator formalism, the spin density operator undergoes the following transformations:

$$\begin{aligned} &a\hat{S}_z \xrightarrow{180^\circ \hat{S}_x} -a\hat{S}_z \xrightarrow{\tau} a\left[\left(1-2\exp\left(-\tau/T_1\right)\right)\right]\hat{S}_z \\ &\xrightarrow{90^\circ \hat{S}_x} -a\left[\left(1-2\exp\left(-\tau/T_1\right)\right)\right]\hat{S}_y \end{aligned} \tag{8.1}$$

where a is an amplitude factor characteristic of the equilibrium state of the S-spin. The observable magnetisation is a function of the duration of the relaxation period, τ. Measurement of peak amplitudes as a function of τ will thus result in an exponential recovery of magnetisation from the inverted state back to the equilibrium. At $\tau = 0$, the amplitude will be equal to $-a$; at some value for τ, the amplitude will be zero; and for very long τ -values ($\tau \gg T_1$), the equilibrium amplitude equal to a will be obtained. Typical results from a time-dependent experimental series using the inversion-recovery experiment are depicted in Figure 8.3.

The recycle delay in an inversion-recovery measurement should in principle be long enough (longer than about $5T_1$) to ensure that equilibrium conditions are reached between repetitions of the experiment. Assuming a perfect inversion, the T_1 relaxation time can then be obtained by a two-parameter exponential fitting of the following expression to the measured intensities, $S(\tau)$:

$$S(\tau) = S_0\left[1-2\exp\left(-\tau/T_1\right)\right] \tag{8.2}$$

The long recycle delay makes the inversion-recovery time-consuming, which is a serious problem in practical considerations.

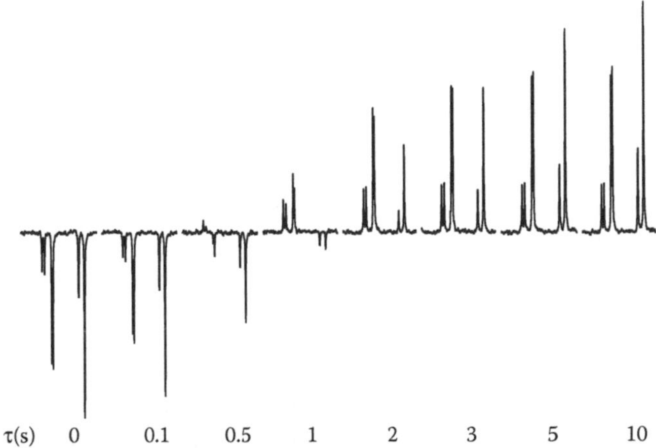

FIGURE 8.3 Carbon-13 T_1 relaxation behaviour for the phospholipid 1,2-dimyristoyl-sn-glycero-3-phosphocholine (DMPC) measured with the inversion-recovery experiment. Eight different values for the relaxation period, τ, are shown for part of the carbon-13 spectrum of the head-group region of DMPC.

8.1.2 Other Methods

In the interest of saving experiment time, the *fast inversion-recovery* experiment (FIR), proposed by Canet and co-workers[2], can instead be used. This simply means using a shorter recycle delay and instead of reaching equilibrium between transients, a steady-state condition is obtained by repeating the experiment a few times without storing the signal (this way of obtaining the steady state is commonly referred to as *dummy scans*). In this approach, the T_1 relaxation time constants are obtained from a three-parameter fit:

$$S(\tau) = S_0\left[1 - k\exp(-\tau/T_1)\right] \tag{8.3}$$

where k is a constant taking into account the relation between the recycle delay and the relaxation time, or a third parameter of the fit. The parameter k also absorbs possible imperfections of the 180° pulse. If the quality of the inverting pulse is judged to be important (for example in the case of spectra covering a large spectral region), one can replace the simple 180° pulse in the inversion-recovery (or the FIR) by a *composite* pulse, a pulse sandwich designed to compensate its own imperfections[3]. The reader is referred to the books by Ernst, Bodenhausen and Wokaun and by Canet (*Further reading*) for further discussion of composite pulses.

Another simple method for measuring the longitudinal relaxation time is the *saturation-recovery* experiment[4], in which the initial 180° pulse of the preparation period is replaced by a saturation sequence consisting of a 90° radiofrequency (r.f.) pulse followed by a field gradient pulse. The 90° pulse converts the longitudinal equilibrium (or steady-state) magnetisation into transverse magnetisation, which is dephased by the gradient pulse. The net result is that the magnetisation is destroyed (saturated) at $\tau = 0$ and recovers gradually towards equilibrium during the relaxation period. The saturation-recovery experiment can be of particular advantage if using a 180° pulse may cause problems with radiation damping (radiation damping is a phenomenon that arises for very strong signals when the NMR response current in the receiver coil is strong enough to create a radiofrequency magnetic field disturbing the experiment).

Alternative approaches to the measurements of spin-lattice relaxation times include the method known as the constant relaxation period (CREPE) experiment[5]. The 180° pulse of the inversion-recovery method is replaced by two 90° pulses, the first with a constant phase and the second with a variable phase. The phase variation leads to the creation of different amounts of longitudinal magnetisation at the beginning of the relaxation period, and the length of the relaxation period is kept constant throughout the experiment. The relationship between the initial longitudinal magnetisation and the resulting magnetisation after the constant relaxation period is linear, with a slope depending on T_1. The authors argue that their method increases the precision of the measurement by a factor of two compared with the FIR method with the same measurement time. In spite of that advantage, the CREPE method has not found very many users.

For rapid measurements, the super fast inversion-recovery (SUFIR) method was proposed, in which the entire relaxation experiment is recorded in one shot. This may be of use for systems with very long relaxation times[6]. Briefly, the experiment consists of the sequence $90°(S_1)–\tau–180°–\tau–90°(S_2)$, repeated until sufficient signal-to-noise ratio is obtained. A single value of the delay τ is used in the experiment. The free induction decays (FIDs), S_1 and S_2, are acquired after both 90° pulses and stored in separate memory blocks. The relaxation is assumed to be single exponential and the T_1 value for each resonance is obtained from the ratio of signal intensities in the spectra corresponding to S_1 and S_2:

$$T_1 = -\tau/\ln\left(1 - S_2/S_1\right) \tag{8.4}$$

Two groups have proposed another interesting time-saving modification of the longitudinal relaxation experiment[7,8]. The experiments are called *single-scan inversion-recovery* (SSIR). The idea is to apply a single 180° pulse and then measure the magnetisation corresponding to different τ values sequentially in different slices of the sample, selected by using selective r.f. pulses and field gradient pulses. The method can be considered a hybrid of the inversion-recovery and the echo-planar imaging protocol, and is thus related to Frydman and co-workers' method for multidimensional experiments[9]. As a result, it is possible to reduce the time necessary to determine the T_1 values by one or two orders of magnitude. The time gain occurs, however, at the expense of sensitivity and the method is likely to be most useful for slowly relaxing systems with good signal-to-noise ratio, conditions that apply, for example, for protons in small molecules. A more recent paper from the Frydman lab[10] describes a single-scan method designed for rapidly relaxing nuclei. In contrast to the SUFIR method, the SSIR technique allows the full longitudinal recovery curve to be followed and possible deviations from a simple exponential process to be detected.

An important aspect in longitudinal relaxation measurements concerns the fact that longitudinal magnetisation during the relaxation period can be transferred among the possible longitudinal spin states. To ensure that the relaxation is monoexponential in a two- or multi-spin system, such as the $I = {}^1H$, $S = X$ system, decoupling of the I-spin provides a way of rendering the X spin relaxation monoexponential (compare Section 3.2). The r.f. irradiation of the I-spins results in a pseudo-isolated S-spin system by making the cross-relaxation visible only in the form of a steady-state nuclear Overhauser enhancement (NOE) and by removing interference (cross-correlated relaxation) terms between the dipolar interaction and the S-spin chemical shielding anisotropy (CSA) (see Chapter 5). The multispin relaxation parameters will be further discussed in Chapters 9 and 10 and can be very useful in determining the local geometry of a molecule or in investigations of certain motional properties. In order to suppress the undesired effects of cross-relaxation and cross-correlation, proton broadband decoupling needs to be applied during the relaxation period.

Although decoupling during the relaxation period is sufficient to remove the complications of non-exponential relaxation, it is most often used throughout the experiment in order to benefit from the signal gain obtained from the NOE and from the collapse of the heteronuclear J-couplings. For small molecules, carbon-13 relaxation is often measured using direct carbon detection and a significant signal gain is obtained in this way: for carbon-13, the maximum NOE $(1 + \eta)$ is approximately 3. Outside of the extreme narrowing regime, the sensitivity gain for carbon-13 is smaller. In the case of nuclei with negative magnetogyric ratio, such as nitrogen-15, η is always negative and, under certain combinations of relaxation mechanisms and/or dynamic properties, the NOE phenomenon can actually lead to a signal loss.

The struggle for higher signal-to-noise in less time and for higher resolution, in particular in biological applications, has led to the development of other polarisation transfer schemes and of two-dimensional methods using indirect detection of the more sensitive proton nucleus. For molecules with a small NOE (or an unfavourable combination of negative magnetogyric ratio and dynamic properties), there is little signal intensity to be gained from applying broadband decoupling during the preparation period. Instead, the INEPT sequence can be employed to obtain signal enhancement (Figure 8.4, top panel). In this experiment, proposed by Kowalewski and Morris[11], the inversion pulse in the preparation period of the inversion-recovery method is simply replaced by the refocused INEPT sequence. The refocused INEPT is obtained by adding a spin-echo segment to the conventional INEPT experiment described in Section 7.1. The spin-echo has the effect of converting the spin operator corresponding to antiphase coherence $2\hat{I}_z\hat{S}_y$ into in-phase $\hat{S}_x$ coherence, as shown in Eq. (7.16). In this way, a transverse in-phase $\hat{S}_x$ coherence is generated, enhanced by the factor γ_I/γ_S, as described previously. For a spin pair consisting of $I = {}^1H$, $S = {}^{13}C$, the maximum signal gain will be $26.752 \times 10^7/6.728 \times 10^7 \approx 4$. Clearly, there is an advantage of using this experiment instead of NOE for signal enhancement and the gain becomes even more significant for nitrogen-15. The number of pulses and delays in the INEPT version might, however, cause problems. If the pulses are not calibrated properly, and the delays are not correctly set this will inevitably lead to signal loss during the sequence. Furthermore, one needs to consider that relaxation takes place throughout the sequence, leading to loss of intensity. Finally, the $90_y°$ pulse, after the refocused INEPT, will generate a non-equilibrium longitudinal state that is allowed to relax during the relaxation period in the same way as in the conventional inversion-recovery experiment.

From this point, it is straightforward to extend the pulse sequence by including a reverse INEPT to obtain a two-dimensional, proton-detected relaxation experiment (Figure 8.4, bottom panel). The two-dimensional inverse-detected measuring scheme will yield a two-dimensional spectrum for each relaxation delay and the cross-peak volumes are determined and fitted (in the usual way) to an exponentially decaying function. Figure 8.5 illustrates the typical appearance of a two-dimensional spectrum obtained with the pulse sequence in Figure 8.4. Today, there are variants that are designed to increase

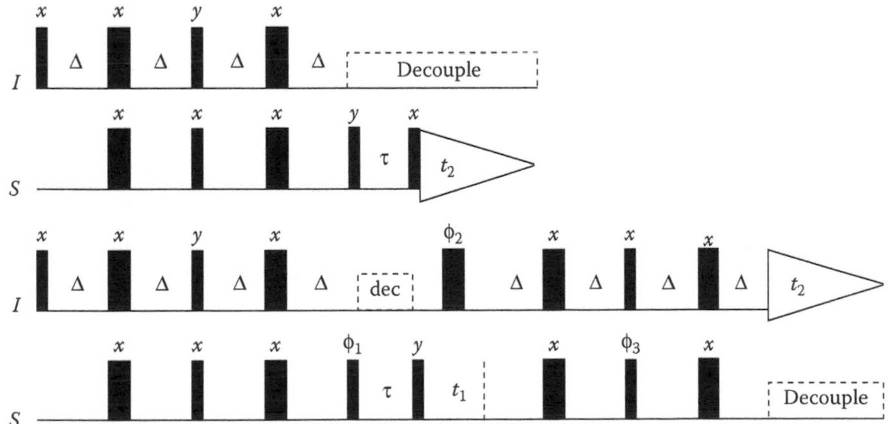

FIGURE 8.4 INEPT-enhanced T_1 experiment, one-dimensional (1D) and two-dimensional (2D) pulse sequences. The top panel shows the heteronuclear (S-spin) detected 1D experiment, and the bottom panel shows the inverse, or ^{1}H-detected, (I-spin) 2D experiment, where τ denotes the relaxation period, which is arrayed in a series of experiments. The phase cycle for the 2D experiment is $\phi_1 = 2(x, -x)$, $\phi_2 = 4(y), 4(-y)$, $\phi_3 = 2(y), 2(-y)$, and $\phi_{rec} = 2(x, -x, -x, x)$. The value for the delay Δ is set to $1/4J_{IS}$ for maximum INEPT polarisation transfer.

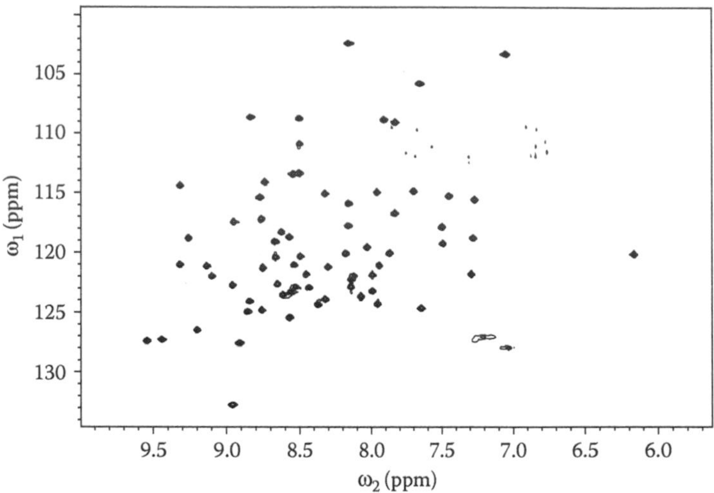

FIGURE 8.5 ^{1}H-^{15}N correlation spectrum for uniformly ^{15}N-labelled human ubiquitin recorded at 18.78 T (800 MHz ^{1}H Larmor frequency). In a T_1 relaxation experiment, the relaxation delay τ is arrayed and one two-dimensional correlation spectrum is recorded for each value of τ.

sensitivity even further, and that utilise pulse field gradients for coherence pathway selection. One such variant is named *preservation of equivalent pathways* (PEP)[12]. In the conventional heteronuclear single quantum correlation experiment, the term $2\hat{I}_z\hat{S}_y$ evolves during t_1 to produce a combination of $2\hat{I}_z\hat{S}_y$ and $2\hat{I}_z\hat{S}_x$ terms. The $2\hat{I}_z\hat{S}_x$ term does not contribute to the observed signal, and only half of the initial magnetisation produced in the INEPT sequence is observed. Without going into details, the PEP scheme allows the refocusing of both terms, leading to signal gain.

For situations where the increased resolution afforded by the two-dimensional spectrum is not needed and the proton spectrum offers a more favourable separation of signals, a one-dimensional indirect experiment can be used, in which the t_1 evolution period in the two-dimensional sequence is simply removed. This can be of interest for the measurement of T_1 for medium-sized organic molecules, for which the carbon-13 spectrum has unfortunate overlap.

In summary, the techniques outlined in this section have applications for different situations. Measurements of carbon-13 T_1 relaxation for small molecules are most easily made with the conventional fast inversion-recovery technique, using the NOE for signal enhancement. There are advantages with respect to both experimental time and simplicity in the acquisition and evaluation of data. The recycling delay does not have to be long enough for equilibrium to be reached, provided that the dummy scans leading to a steady state are inserted and a non-linear three-parameter exponential fit is used in the analysis. For large molecules, the only practical solution is to use two-dimensional inverse-detected methods for acquiring relaxation data.

8.1.3 Field-Cycling

The discussion in the previous section is concerned with T_1 experiments under high-resolution conditions at high magnetic fields. In studies of complex liquids (polymer solutions and melts, and liquid crystals), one is often interested in obtaining information on rather slow (micro- or nanosecond timescale) motions by measuring T_1 at low resolution and low magnetic field strengths using the technique of *fast field-cycling* (FFC). This technique was proposed (in a version very similar to that of the present day) by Anderson and Redfield a very long time ago[13], and we have alluded to it in Section 2.1. The outline of the experiment is shown in Figure 8.6. Briefly, the non-equilibrium state is created by

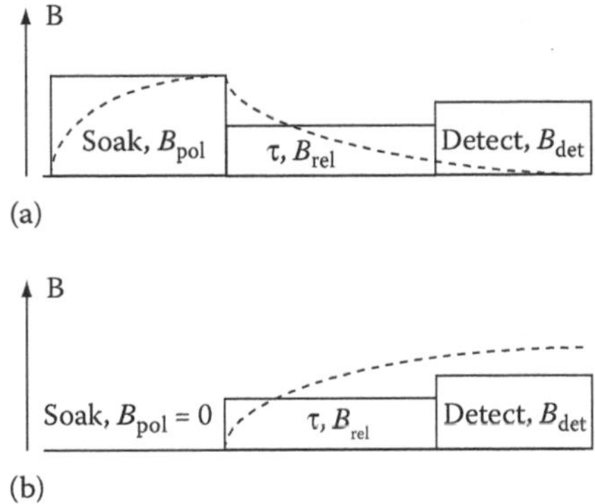

FIGURE 8.6 Field-cycling experiments. Figure (a) shows the original field-cycling experiment, in which the field is cycled in three steps. First, the sample is "soaked" at a high magnetic field strength $B_0 = B_{pol}$ to increase signal intensity. Second, the field is switched to a lower value, at which relaxation is measured. Finally, the field is set for detection. (b) shows a modified version of the experiment in which the soaking occurs at zero magnetic field strength. The field is switched to the value at which relaxation is measured and, finally, set to the appropriate field for detection. In both figures, the dashed lines indicate the build-up and decay of longitudinal magnetisation.

keeping (soaking) the sample at a certain moderately high polarising field B_0, and then rapidly switching B_0 to a lower value, at which we wish to measure T_1 (Figure 8.6a). After the variable relaxation delay τ, the field is switched again, this time to a value matching the frequency to which the detection circuit is tuned, and the signal is detected. If we wish to measure longitudinal relaxation at a moderate field, the variant of the experiment shown in Figure 8.6b can be useful. Here, we begin by soaking at zero-field and then follow the build-up of the magnetisation when the field is switched on at the value at which we wish to measure relaxation. The detection is carried out in the same way as in the previous variant. The fast field-cycling methods are described in detail by, for example, Kimmich and Anoardo[14]. Besides the FFC experiment, one speaks also about field-cycling in a slightly different context: high-resolution field-cycling, which was also proposed by Redfield[15,16]. He described a device, called a *shuttler*, designed to be used as an accessory to a standard commercial high-resolution, high-field spectrometer in a multi-user environment. The shuttler transports the NMR sample between the homogeneous zone of the high magnetic field, within the standard probe-head, and a prescribed location in the fringe field of the magnet. Thus, referring again to Figure 8.6, the soaking and detection occur at high field, while the relaxation takes place at the chosen value of the fringe field (which can additionally be manipulated to obtain very low B_{rel} values). The advantage compared with FFC is that the method works under high-resolution, high-sensitivity conditions and is compatible with a two-dimensional design of relaxation experiments. A disadvantage is that the mechanical shuttling is obviously slower than the electronic switching of the magnetic field in the FFC and does not allow measurements of very fast relaxation rates.

8.2 Spin-Spin Relaxation Rates

The spin-spin relaxation time constant, T_2, defined in the Bloch equation, is simply related to the width of the Lorentzian line at half height, $\Delta\upsilon_{1/2} = 1/(\pi T_2)$, as indicated in Figure 1.4. In principle, the linewidth can be used to measure T_2, but this seldom gives a reliable estimate of the relaxation time. In addition to the *homogeneous broadening* associated with transverse relaxation, there is *inhomogeneous broadening* caused by non-uniformity of the external magnetic field over the sample volume. In order to accurately measure T_2, a homogeneous static magnetic field is required. If the relaxation is rapid, leading to large linewidths as compared with those caused by the inhomogeneous broadening, determining T_2 by measuring linewidths is certainly fast and useful. This is the situation for quadrupolar nuclei, which have very efficient quadrupolar relaxation, leading to broad signals (*cf.* Chapter 14). For spin 1/2 nuclei, it is, however, often necessary to suppress the inhomogeneous broadening by using the *spin-echo* technique (Figure 8.7). Simply speaking, the spin-echo method allows the determination of the T_2 relaxation time constant, while the inhomogeneous broadening is suppressed.

8.2.1 Spin-Echo Methods

The simplest spin-echo experiment was proposed as early as 1950 by Erwin Hahn[17]. The (90°–τ/2–180°–τ/2-echo) experiment works as follows. The first $90_x°$ pulse creates M_y magnetisation (Figure 8.7). The magnetisation components start to precess in the *x*, *y*-plane with slightly different rates due to different chemical shifts, and possibly due to the inhomogeneous field causing different field strengths in different parts of the sample, according to:

$$\hat{S}_z \xrightarrow{90°\hat{S}_x} -\hat{S}_y \xrightarrow{\Omega_S\hat{S}_z\tau/2} \left(-\hat{S}_y\cos(\Omega_S\tau/2)+\hat{S}_x\sin(\Omega_S\tau/2)\right)\exp(-\tau/2T_2) \tag{8.5}$$

In Eq. (8.5), the scalar coupling has been ignored, since it can be shown that the spin-echo sequence refocuses heteronuclear couplings. Homonuclear scalar couplings are, however, not refocused, leading to a phenomenon known as J-modulation. Therefore, it has long been believed that T_2 cannot be determined in homonuclear, coupled systems using the spin-echo method. As demonstrated by the

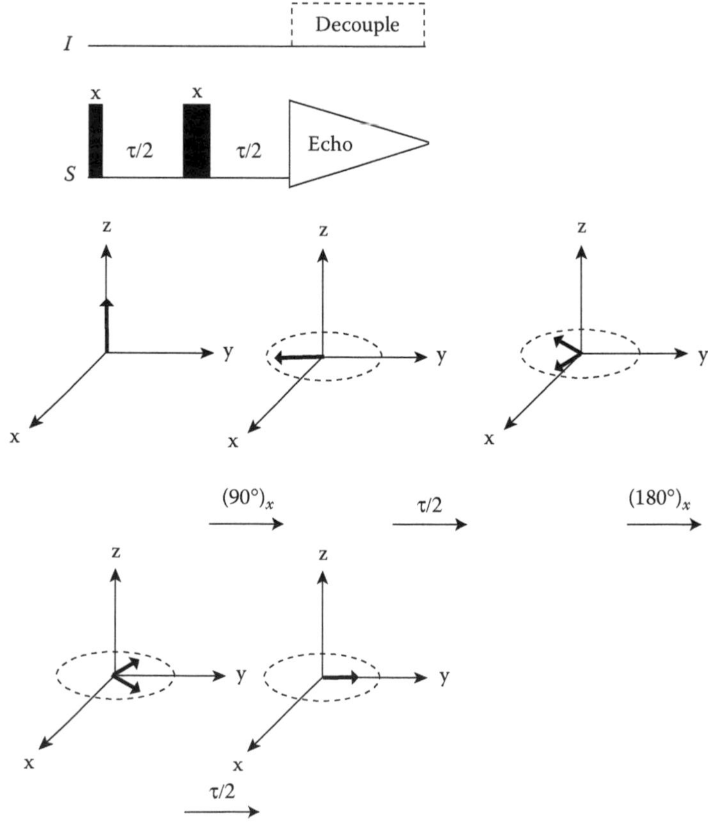

FIGURE 8.7 The spin-echo sequence. A vector representation of the effect of the different steps in the spin-echo pulse sequence. For more details see text.

Bodenhausen group, this is not quite right; it is possible to quench the echo modulation, at least under certain conditions[18]. We shall return to this point below.

The term $\exp(-\tau/2T_2)$ in Eq. (8.5) describes the transverse relaxation process during the delay. The $180_x°$ pulse reverses the sign of the $\hat{S}_y$ term, while the $\hat{S}_x$ is left unchanged (in the original work of Hahn, the second pulse was also 90°; its replacement by a 180° pulse was proposed a few years later by Carr and Purcell[19]). During the next interval, the precession is thus reversed, leading to a magnetisation vector at $-\hat{S}_y$. In other words, the components of the magnetisation meet to give rise to an echo. The result is:

$$\begin{aligned}
&\left(-\hat{S}_y\cos(\Omega_S\tau/2)+\hat{S}_x\sin(\Omega_I\tau/2)\right)\exp(-\tau/2T_2)\xrightarrow{180°\hat{S}_x}\\
&\left(\hat{S}_y\cos(\Omega_S\tau/2)+\hat{S}_x\sin(\Omega_S\tau/2)\right)\exp(-\tau/2T_2)\xrightarrow{\Omega_S\hat{S}_z\tau/2}\\
&\left(\hat{S}_y\cos^2(\Omega_S\tau/2)-\hat{S}_x\cos(\Omega_S\tau/2)\sin(\Omega_S\tau/2)\right.\\
&\left.+\hat{S}_x\sin(\Omega_S\tau/2)\cos(\Omega_S\tau/2)+\hat{S}_y\sin^2(\Omega_S\tau/2)\right)\exp(-\tau/T_2)=\hat{S}_y\exp(-\tau/T_2)
\end{aligned} \tag{8.6}$$

After the 180° pulse, an echo starts forming, and after a delay of exactly $\tau/2$ the echo is at its maximum. It is now possible to collect data and perform a Fourier transform to obtain the spectrum. The peak amplitudes in the NMR spectrum are therefore given by:

$$a(\tau)=S_0\exp(-\tau/T_2) \tag{8.7}$$

where S_0 refers to the amplitude of the equilibrium magnetisation. The experiment is repeated with different values for τ.

The simple procedure outlined here is prone to artefacts arising from *diffusion*. If the field is inhomogeneous, and if it changes over the duration of the experiment, the refocusing will be incomplete, resulting in loss of signal and an apparent T_2 decay time constant that is shorter than the true T_2. This may occur either if the field changes over time, or if the spins move into different regions of a sample experiencing an inhomogeneous field. This observation can in fact be turned into something useful. By deliberately disturbing the homogeneity of the magnetic field by applying a gradient pulse, it is possible to monitor the diffusion of the molecule within the sample. We shall come back to this issue in Chapter 16. Here, however, it is necessary to minimise these effects and to remove such artefacts in the T_2 measurements, Carr and Purcell[19] suggested using a short and constant delay τ, varying the relaxation period through repeating the number of echo cycles ($\tau/2$–180°–$\tau/2$) and making this number of repetitions the variable parameter of the relaxation experiment (Figure 8.8). In this way, the second half of the last echo is obtained after a number of echo cycles, and can be Fourier transformed and analysed analogously to the simple spin-echo procedure. The experiment, involving a large number of pulses, is likely to be sensitive to small miscalibration of the pulse durations. To reduce this problem, Meiboom and Gill modified the Carr–Purcell sequence by changing the phase of the 180° pulses from x to y[20]. The most common echo method, used today for T_2 studies, is widely known as the *CPMG sequence*, named after the initials of the authors who proposed the modifications of the basic spin-echo experiment (Carr, Purcell, Meiboom and Gill).

Until now, we have ignored scalar couplings between I and S. As stated above, the homonuclear couplings in a two-spin system are not refocused by ideal 180° pulses, affecting the two spins in the same way. If the r.f. in the CPMG experiment is of moderate power and the carrier frequency is set so that one spin (I) is on resonance, while the other one (S) is slightly off resonance, the echo modulation in the CPMG experiment may be suppressed. This was discovered by Dittmer and Bodenhausen[21] and a detailed explanation can be found in the review by Segawa and Bodenhausen[18]. Briefly, the off-resonance spin will rotate around the tilted effective r.f. field (compare Section 4.4) and the nominal 180° pulse will

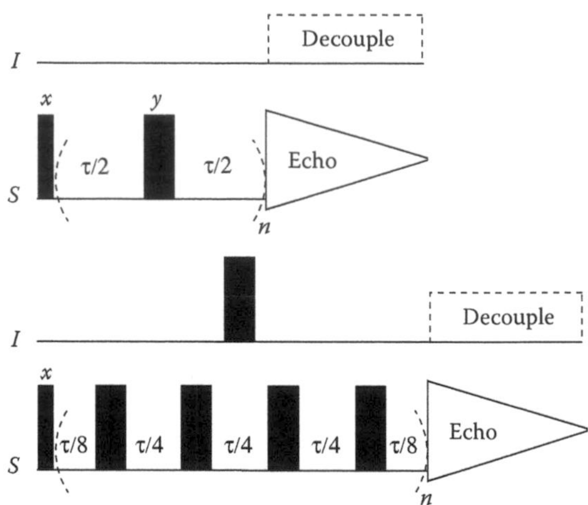

FIGURE 8.8 The Carr–Purcell–Meiboom–Gill experiment for measuring T_2. The echo cycle is repeated n times to produce a total relaxation delay of $n \times \tau$. The pulse sequence in the bottom panel includes the removal of cross-correlation effects during the echo cycle.

result in a rotation by a somewhat larger angle. This will lead to imperfect refocusing of the S-spin. When propagated by the CPMG sequence, the errors may lead to effective decoupling. The phenomenon is called *stabilisation by interconversion within a triad of coherences under multiple refocusing* (SITCOM). Tosner and co-workers[22] investigated the transverse magnetisation decay due to dipole-dipole relaxation under the SITCOM conditions and found that the effective decay rate, measured using different CPMG delays, τ, varied in a somewhat oscillatory manner between the value expected for a case of unlike spins without coupling (see Eq. (4.65a)) in the slow pulsing limit and a higher value, corresponding to identical spins (displaying a transverse analogue to the "3/2 effect", discussed in Section 3.4), in the fast pulsing limit.

Let us now turn to heteronuclear systems with J-coupling. Indeed, the couplings are now refocused by the spin-echo applied in the vicinity of the S-spin resonance frequency but result in both the $\hat{S}_x$, $\hat{S}_y$ operators and the antiphase $2\hat{I}_z\hat{S}_x$, $2\hat{I}_z\hat{S}_y$ terms being present during the τ delay. The measured apparent T_2 will therefore be a superposition of the relaxation rates for the in-phase and antiphase coherences. The measured transverse relaxation rate, $1/T_2 = R_2$, will be given by:

$$R_2 = 1/2\left\lfloor R_{S_y}\left(1+\sin\alpha/\alpha\right) + R_{2I_zS_y}\left(1-\sin\alpha/\alpha\right)\right\rfloor \tag{8.8}$$

where $\alpha = \pi J_{IS}\tau$. In order to reduce relaxation caused by the antiphase term, the second term has to be minimised by letting $\pi J_{IS}\tau \ll 1$. This means, in practice, that τ has to be kept sufficiently short not to let scalar couplings evolve. For carbon-13, assuming a scalar coupling of $J_{IS} = 145$ Hz, using a τ period of 1 ms will result in $1 - \sin\alpha/\alpha = 0.034$. This means that a systematic error of around 3% will be introduced. If τ is instead set to 0.5 ms, the term is reduced to less than 1%. Therefore, it is essential not to use too long values for τ.

Another consideration, concerning deviations from monoexponential transverse relaxation, is the possible presence of interference terms, or cross-correlation effects, between the S-spin CSA and IS dipolar interactions, in analogy with what was mentioned for T_1 measurements (see Chapter 5 for multispin effects in relaxation). Broadband irradiation of the I-spin (protons), which is commonly used in T_1 measurements, acts as an artificial T_2 relaxation mechanism and leads to artefacts in the T_2 measurement by the CPMG spin-echo sequence. This will result in too short T_2 values and is not a useful option. Instead, suitably spaced 180° I-spin pulses are employed during the CPMG sequence. A pulse sequence such as the ones depicted in Figure 8.8, bottom panel, will effectively remove the cross-correlation effects[23,24].

A problem with measuring T_2 can occur due to the fact that the *S*-spin radio-frequency pulses do not affect the entire spectrum equally. To ensure that *off-resonance* effects do not interfere with the measurement of T_2, the r.f. field strength must exceed the spectral width. This seldom holds for *e.g.* carbon-13, which has a relatively broad spectral range, and substantial errors can be obtained. The result of the spin-echo sequence will not produce magnetisation that is aligned along the *x*-axis, but rather, magnetisation that rotates about an effective axis in the *x*, *z*-plane, with the direction depending on the offset from the carrier frequency, the spin-echo delay τ, and the field strength of the pulse. Two ways of dealing with this problem include performing several measurements using different carrier frequencies, or correcting the measured relaxation parameters for off-resonance effects[25]. As a third possibility, one can consider using composite pulses, mentioned earlier in the context of T_1 measurements.

The R_2 rates are sensitive to chemical exchange phenomena, in which a spin exchanges between two states with different chemical shifts. This has led to the possibility of extending the timescale of molecular motions that can be studied by NMR methods from slow exchange phenomena on the second timescale to the millisecond–microsecond timescale. The relationship between chemical exchange and relaxation measurements will be discussed further in Chapter 13.

8.2.2 Spin-Lock Methods

An alternative approach to measuring transverse relaxation is to use a *spin-lock* sequence for measuring T_1 in the rotating frame, or $T_{1\rho}$. The experiment works in the following way. A $90°_x$ pulse on the *S*-spins creates $-\hat{S}_y$ magnetisation (Figure 8.9). A transverse spin-locking field is applied along the rotating *y*-axis, which locks the magnetisation in the rotating frame. The relaxation rate constant for the magnetisation along the direction of the field in the rotating-frame is called $T_{1\rho}$. The $T_{1\rho}$ relaxation parameter depends on the strength of the radio-frequency field, ω_1, as well as the resonance offset from the carrier frequency, $\Delta\omega = \omega_S - \omega_{rf}$. The theoretical background of the relaxation in the presence of r.f. fields was presented in Section 4.4. Here, we recapitulate briefly the main features. The Hamiltonian for a two-spin system in the presence of an r.f. spin-locking field on the *S*-spins is given by:

$$\hat{H}(t) = \hat{H}_Z + \hat{H}_{r.f.}(t) + \hat{H}_1(t) \tag{8.9}$$

in which $\hat{H}_1(t)$ is the relaxation Hamiltonian and $\hat{H}_{r.f.}(t)$ is:

$$\hat{H}_{r.f.}(t) = \omega_{1S}\left(\hat{S}_x \cos\left(\omega_{r.f.}\right)t + \hat{S}_y \sin\left(\omega_{r.f.}\right)t\right) \tag{8.10}$$

If the resonance offset, $\Delta = \omega_{1S} - \omega_{r.f.}$, is small compared with the r.f. field-strength, $\Delta << \omega_{1S}$, the spin will be locked in the *x*, *y*-plane by an effective field $\Omega = (\omega_{1S}^2 + \Delta^2)^{1/2}$. The tilt angle of the effective field with respect to the laboratory *z*-axis will then be given by $\tan\Theta = \omega_{1S}/\Delta$.

In a standard measurement, the offset is considered small compared with the amplitude of the applied field, ω_{1S}, and the tilt angle will be close to 90°, leading to an effective spin-lock of magnetisation along

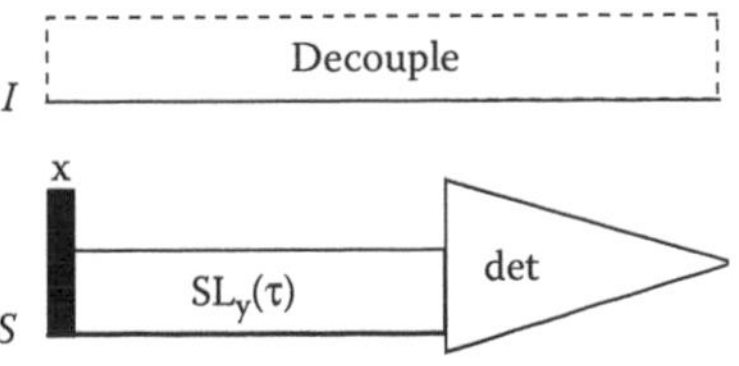

FIGURE 8.9 Spin-lock sequence for measuring $T_{1\rho}$.

the x-axis. If off-resonance effects are significant, the effective field will be directed in the x,z-plane, rotating around the z-axis. The expression for $R_{1\rho} = 1/T_{1\rho}$ then contains a combination of longitudinal relaxation and transverse relaxation according to:

$$R_{1,\rho} = R_1 \cos^2 \Theta + R_2 \sin^2 \Theta \tag{8.11}$$

This equation holds provided that the effective r.f. spin-lock field is much weaker than the external magnetic field, and if the condition $\omega_1^2 \tau_c^2 << 1$ is satisfied. For liquids undergoing rotational tumbling, this is always the case. However, if exchange processes are present for which the latter condition (with τ_c referring to the exchange lifetime) is not fulfilled, the situation becomes more complicated. We shall return to this case in Chapter 13.

In the same way as for T_1 measurements, heteronuclear T_2 relaxation is for large molecules most often measured by two-dimensional experiments, by combining either the spin-echo or spin-lock sequences with a polarisation transfer experiment. The experiments are commonly performed as inverse-detected experiments with the HSQC method as described in the previous section. Two-dimensional variants of the CPMG and spin-lock experiments are depicted in Figure 8.10. As an illustration of the advances in NMR methodology, one of the pulse sequences in Figure 8.10 ($T_{1\rho}$) is a sensitivity-enhanced experiment with gradients for coherence selection. For further details on determining relaxation rate constants by two-dimensional inverse-detected methods, the reader is referred to the reviews by Peng and Wagner[26] or by Palmer[27], as well as the book by Cavanagh *et al.* (*Further reading*).

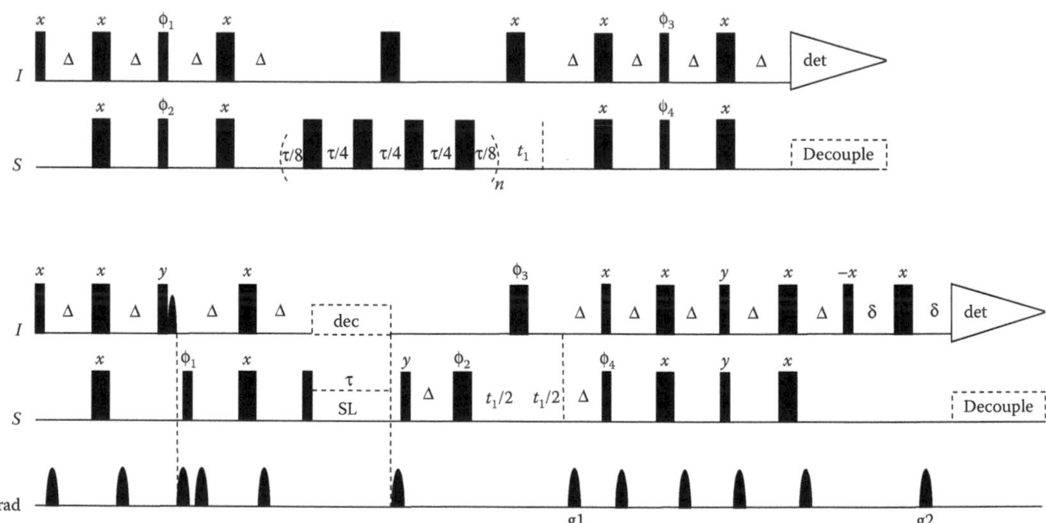

FIGURE 8.10 Two-dimensional pulse sequences for T_2 (top) and $T_{1\rho}$ (bottom). The pulse sequence for measuring $T_{1\rho}$ differs in that it uses gradients for coherence selection, and a pulse scheme for sensitivity enhancement. The delays are $\Delta = 1/4J_{IS}$; τ is the relaxation time delay and δ is set to longer than the duration of g2. The phase cycle used for the top pulse sequence is $\phi_1 = y, -y$; $\phi_2 = 2(x), 2(-x)$; $\phi_3 = 4(x), 4(-x)$; $\phi_4 = 8(x), 8(-x)$ and $\phi_{rec} = x, -x, -x, x, -x, x, x, -x, -x, x, x, -x, x, -x, -x, x$. The pulse train during the CPMG delay is designed to remove the dipole-dipole/CSA cross-correlation effects. In the enhanced sensitivity pulse sequence (bottom), two FIDs are collected with the ϕ_4 phase inverted and with the sign of gradient g1 inverted. The signals are stored separately and manipulated to obtain pure absorption data. Pure phase two-dimensional Fourier transform is subsequently performed by the States method. The phase cycle in the $T_{1\rho}$ experiment is $\phi_1 = x, -x$; $\phi_2 = 2(x), 2(y), 2(-x), 2(-y)$; $\phi_3 = 2(x), 2(-x)$; $\phi_4 = x$ and $\phi_{rec} = x, -x, x, -x, -x, x, -x, x$. Decoupling of the protons during the spin-lock period is achieved by a train of 180° pulses.

References

1. Vold, R. L.; Waugh, J. S.; Klein, M. P.; and Phelps, D. E., Measurement of spin relaxation in complex systems. *J. Chem. Phys.* 1968, 48, 3831–3832.
2. Canet, D.; Levy, G. C.; and Peat, I. R., Time saving in 13C spin-lattice relaxation measurements by inversion-recovery. *J. Magn. Reson.* 1975, 18, 199–204.
3. Levitt, M. H.; and Freeman, R., NMR population-inversion using a composite pulse. *J. Magn. Reson.* 1979, 33, 473–476.
4. Markley, J. L.; Horsley, W. J.; and Klein, M. P., Spin-lattice relaxation measurements in slowly relaxing complex spectra. *J. Chem. Phys.* 1971, 55, 3604–3605.
5. Zhao, H.; Westler, W. M.; and Markley, J. L., Precise determination of T-1 relaxation values by a method in which pairs of nonequilibrium magnetisations are measured across a constant relaxation period. *J. Magn. Reson. Ser. A.* 1995, 112, 139–143.
6. Canet, D.; Mutzenhardt, P.; and Robert, J. B., The Super Fast Inversion Recovery (SUFIR) experiment. In *Methods for Structure Elucidation by High-Resolution NMR*, Batta, G.; Kövér, K. E.; Szántay, C., Eds. Elsevier: Amsterdam, 1997; 317–323.
7. Loening, N. M.; Thrippleton, M. J.; Keeler, J.; and Griffin, R. G., Single-scan longitudinal relaxation measurements in high-resolution NMR spectroscopy. *J. Magn. Reson.* 2003, 164, 321–328.
8. Bhattacharyya, R.; and Kumar, A., A fast method for the measurement of long spin-lattice relaxation times by single scan inversion recovery experiment. *Chem. Phys. Lett.* 2004, 383, 99–103.
9. Frydman, L.; Scherf, T.; and Lupulescu, A., The acquisition of multidimensional NMR spectra within a single scan. *Proc. Natl. Acad. Sci. USA*. 2002, 99, 15858–15862.
10. Smith, P. E. S.; Donovan, K. J.; Szekely, O.; Baias, M.; and Frydman, L., Ultrafast NMR T-1 relaxation measurements: Probing molecular properties in real time. *ChemPhysChem.* 2013, 14, 3138–3145.
11. Kowalewski, J.; and Morris, G. A., A rapid method for spin-lattice time measurements on low magnetogyric ratio nuclei: INEPT signal enhancement. *J. Magn. Reson.* 1982, 47, 331–338.
12. Cavanagh, J., and Rance, M., Sensitivity-enhanced NMR techniques for the study of biomolecules. *Annu. Rep. NMR Spectrosc.* 1993, 27, 1–58.
13. Anderson, A. G.; and Redfield, A. G., Nuclear spin-lattice relaxation in metals. *Phys. Rev.* 1959, 116, 583–591.
14. Kimmich, R.; and Anoardo, E., Field-cycling NMR relaxometry. *Prog. NMR Spectr.* 2004, 44, 257–320.
15. Redfield, A. G., Shuttling device for high-resolution measurements of relaxation and related phenomena in solution at low field, using a shared commercial 500 MHz NMR instrument. *Magn. Reson. Chem.* 2003, 41, 753–768.
16. Redfield, A. G., High-resolution NMR field-cycling device for full-range relaxation and structural studies of biopolymers on a shared commercial instrument. *J. Biomol. NMR*. 2012, 52, 159–177.
17. Hahn, E. L., Spin echoes. *Phys. Rev.* 1950, 80, 580–594.
18. Segawa, T. F., and Bodenhausen, G., Modulation of spin echoes in liquids. In *eMagRes,* vol. 2, Wiley: 2013; 245–252.
19. Carr, H. Y.; and Purcell, E. M., Effects of diffusion on free precession in nuclear magnetic resonance experiments. *Phys. Rev.* 1954, 94, 630–638.
20. Meiboom, S.; and Gill, D., Modified spin-echo method for measuring nuclear relaxation times. *Rev. Sci. Instrum.* 1958, 29, 688–691.
21. Dittmer, J.; and Bodenhausen, G., Quenching echo modulations in NMR spectroscopy. *Chemphyschem* 2006, 7, 831–836.
22. Tosner, Z.; Skoch, A.; and Kowalewski, J., Behavior of two almost identical spins during the CPMG pulse sequence. *Chemphyschem*. 2010, 11, 638–645.

23. Palmer, A. G.; Skelton, N. J.; Chazin, W. J.; Wright, P. E.; and Rance, M., Suppression of the effects of cross-correlation between dipolar and anisotropic chemical shift relaxation mechanisms in the measurement of spin spin relaxation rates. *Mol. Phys.* 1992, 75, 699–711.
24. Kay, L. E.; Nicholson, L. K.; Delaglio, F.; Bax, A.; and Torchia, D. A., Pulse sequences for removal of the effects of cross correlation between dipolar and chemical-shift anisotropy relaxation mechanism on the measurement of heteronuclear T1 and T2 values in proteins. *J. Magn. Reson.* 1992, 97, 359–375.
25. Korzhnev, D. M.; Tischenko, E. V.; and Arseniev, A. S., Off-resonance effects in 15N T2 CPMG measurements. *J. Biomol. NMR*. 2000, 17, 231–237.
26. Peng, J. W.; and Wagner, G., Investigation of protein motions via relaxation measurements. *Methods Enzymol.* 1994, 239, 563–596.
27. Palmer, A. G.; Kroenke, C. D.; and Loria, J. P., Nuclear magnetic resonance methods for quantifying microsecond-to-millisecond motions in biological macromolecules. *Methods Enzymol.* 2001, 339, 204–238.

9

Cross-Relaxation Measurements

In this chapter, cross-relaxation phenomena will be investigated. For spin 1/2 nuclei, such as ^{1}H, ^{13}C and ^{15}N in organic molecules, relaxation seldom involves only one spin, since the dipole-dipole interaction with other nearby spins is the dominating relaxation mechanism. Thus, multispin contributions to relaxation become important. In these cases, the nuclear Overhauser effect (NOE), which was earlier seen to be related to the cross-relaxation rate, provides a measure of a third relaxation parameter, often as a complement to T_1 and T_2 measurements for heteronuclei. Important applications of relaxation are thus dynamical investigations, which are carried out by combining measurements of heteronuclear T_1, T_2 and heteronuclear NOE relaxation parameters. Perhaps the most widely studied relaxation phenomenon, however, is homonuclear proton-proton dipole-dipole cross-relaxation in organic and biological macromolecules, which contains an important source of information. Cross-relaxation studies offer possibilities to investigate the geometry of a molecule through the internuclear distance dependence of the dipole-dipole interaction. In Section 9.1, the NOE measurements for a two-spin system of unlike spins will be discussed. Homonuclear proton cross-relaxation applications will be discussed in Sections 9.2–9.4.

9.1 Heteronuclear NOE Measurements

The heteronuclear NOE experiment is conceptually very simple, but is for most applications one of the most difficult relaxation experiments to perform in practice. In its simplest form, proposed by Freeman and co-workers[1], the S-{^{1}H} *steady-state NOE factor* can be obtained through two separate experiments (Figure 9.1). In the first experiment, the NOE is built up in a delay during which protons are continuously irradiated prior to the actual S-spin pulse. The delay has to be long enough to ensure that the maximum NOE is obtained. The second experiment is performed without the proton irradiation, and the ratio of signal intensities obtained in the two experiments provides the NOE.

The Solomon equations (Eqs. (3.19)–(3.21)) provide a relationship between the time dependence of the signal intensities and the rate parameters, ρ_S, ρ_I, and the cross-relaxation rate, σ_{IS}. The transfer of magnetisation from the I-spin (protons) to the S-spin is governed by the cross-relaxation rate, σ_{IS}. If we saturate the I-spins by applying an r.f. field, *i.e.* $\langle \hat{I}_z \rangle = 0$, the evolution for $\hat{S}_z$ under steady-state conditions is given by Eq. (3.27), with the steady-state solution given by Eq. (3.28). Hence, the intensity of the S-spin signal will be multiplied by the factor $1+\eta$, usually referred to as the nuclear Overhauser enhancement factor. It should be emphasised that the equations given in Section 3.2 are only valid for an isolated two-spin system. This is approximately the case for naturally occurring carbon-13 spins bound to one proton or for proteins labelled with nitrogen-15 in the backbone. For a homonuclear proton spin system, in which many protons in the molecule contribute to the relaxation, this is seldom the case.

There are several practical aspects that need to be considered. First, the recycle delay between transients needs to be sufficiently long to ensure that the (unpolarised) equilibrium state is reached after each cycle of the pulse sequence. Usually, broadband decoupling of protons is used during acquisition

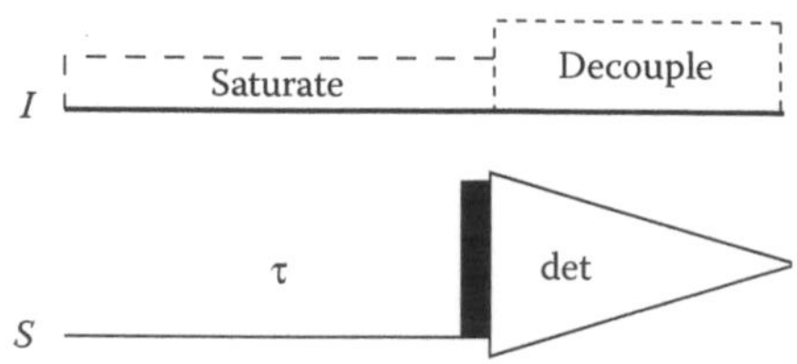

FIGURE 9.1 Steady-state NOE sequence for measuring heteronuclear NOE factors. Two experiments are required: one with a constant I-spin (^{1}H) irradiation during τ and the other without irradiation. In order to minimise heating, different power levels are used during the τ period and the detection.

and it is important to attain complete relaxation of the heteronuclear magnetisation between scans in order to obtain the correct intensities, in particular for the case without saturation of protons. This is a severe time-consuming factor in the NOE experiment; typically, a delay of $t_{rec} > 8T_1$ is used. Second, the long duration of the proton irradiation ($> 5T_1$) can lead to systematic errors due to sample heating. Off-resonance irradiation can be employed to compensate for heating in the NOE experiment. Heating will, however, affect measurements of different relaxation parameters to a varying extent, and there can be discrepancies between the temperature attained in measurements of T_1, T_2 and NOE relaxation parameters.

The determination of heteronuclear NOE in proteins, in analogy with the determination of T_1 and T_2 relaxation time constants, demands the highest possible sensitivity and resolution. Before going into the special methods conforming to these requirements, we begin by going through some problems that must be solved. The most common relaxation studies on protein samples involve nitrogen-15. The first difficulty is that the magnetogyric ratios for protons and nitrogen-15 have opposite signs, which means that the term η in Eq. (3.28) is negative and, as opposed to the case of carbon-13 with its positive γ, can cause signal loss rather than enhancement. For small, flexible peptides, and flexible regions in proteins, the term $1 + \eta$ can become slightly negative or close to zero, while for large, rigid molecules this term is typically close to unity, implying a small heteronuclear NOE. The transfer of magnetisation from solvent protons to exchangeable protons in the molecule is another problem that can lead to complications. This is typically an issue in determining backbone amide ^{15}N NOE factors in proteins in aqueous solution. This transfer is either due to chemical exchange or through direct NOE effects between solvent and the protein. The H_2O proton T_1 relaxation is relatively long and this has implications for the length of the recycle delay. This should be set sufficiently long to also ensure that the water proton magnetisation has reached equilibrium between scans. If not, the partial saturation of the water signal will reduce the intensity of the signals in the spectrum that are supposed to be free of NOE, and will lead to an overestimate of the NOE factors. An additional problem is caused by the need to suppress the water signal intensity in the spectrum. In the simplest way, this is done by applying a weak r.f. field selectively on the ^1H_2O signal prior to the pulse sequence. Perturbation of the strong ^1H_2O signal will, however, create magnetisation, which can be transferred to the dissolved molecule and interfere with the direct cross-relaxation mechanism between the S nucleus and the directly bound proton. In the NOE experiment, minimal perturbation of the water signal must be used. No irradiation on the water signal prior to the actual pulse sequence can be used, and the water magnetisation must be returned to the equilibrium state along the positive z-axis between transients[2,3].

The steady-state NOE measurements can be carried out as two-dimensional measurements in analogy with what has been discussed for T_1 and T_2 measurements (Figure 9.2). The overall scheme begins with a delay, with or without proton irradiation present. A 90° pulse on the S-spins creates transverse magnetisation, which is allowed to evolve during the t_1 evolution period. A reverse INEPT is applied to transfer magnetisation back to the protons for detection. The intensities carry information about the NOE factors and two separate measurements are performed, with and without proton irradiation. A way to minimise artefacts due to sample heating and different conditions in the two experiments is

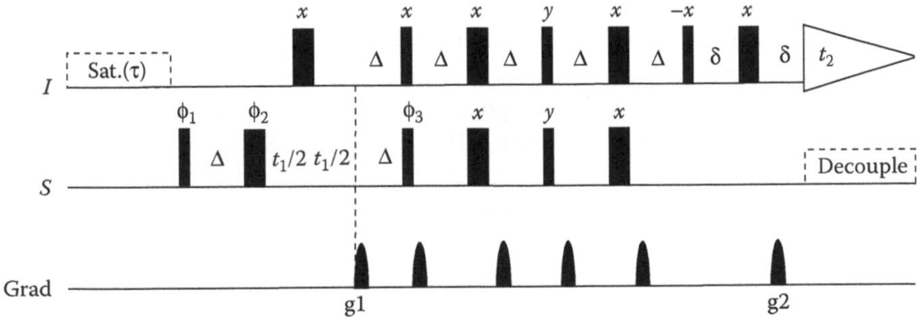

FIGURE 9.2 Enhanced sensitivity pulse sequence for measurement of heteronuclear NOE factors. Two experiments are recorded: one with a constant irradiation of the I-spins (^{1}H) during τ and one without. $\Delta = 1/4J_{IS}$, and δ is set to longer than the duration of g2. For each increment in t_1, two free induction decays are collected with the ϕ_3 phase inverted and with the sign of gradient g1 inverted. The signals are stored separately and manipulated to obtain pure absorption data. Pure phase two-dimensional Fourier transform is subsequently performed by the States method. The phase cycle is $\phi_1 = y$; $\phi_2 = x, y, -x, -y$; $\phi_3 = x$ and $\phi_{rec} = x, -x$.

to record the two experiments interleaved, *i.e.* to accumulate the spectra with and without the NOE in alternate scans. Artefacts can also arise for other reasons, *e.g.* because of a non-optimal saturation scheme. This issue has been discussed in depth by Ferrage and co-workers[4,5].

The heteronuclear NOE experiment can also be carried out as a *transient* rather than a steady-state measurement. A simple one-dimensional approach is obtained through a pulse sequence analogous to a selective inversion-recovery experiment[6] (Figure 9.3). A 180° inversion pulse is applied to the I-spins (^{1}H), and the cross-relaxation is monitored during a variable delay, τ. The initial non-equilibrium state after the inversion pulse is given by $\Delta\hat{I}_z(0) = -2I_{eq}$ and $\Delta\hat{S}_z(0) = 0$. In the initial rate regime, the time evolution of $\hat{S}_z(t)$ is linear with the rate $2\sigma_{IS}$. Thus, the build-up provides information on the cross-relaxation rate. The two-dimensional version of this experiment, suitable for ^{13}C in small molecules, is known as heteronuclear Overhauser spectroscopy (HOESY) and was proposed originally by Rinaldi, Yu and Levy[7,8]. The technique is similar to the homonuclear nuclear Overhauser enhancement spectroscopy (NOESY) method (to be discussed in the next section). The

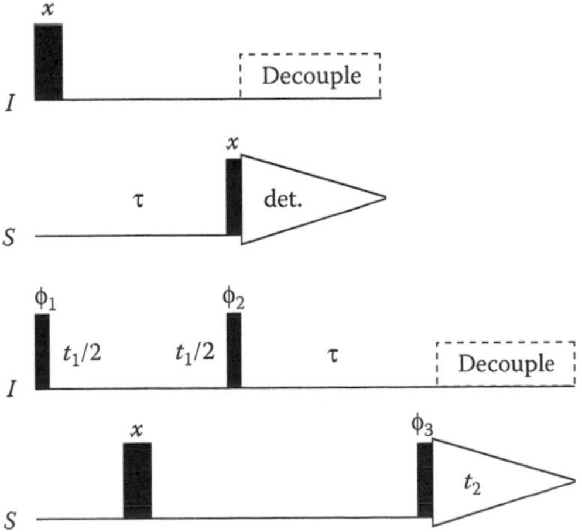

FIGURE 9.3 Experiments for measuring transient heteronuclear NOE factors. The one-dimensional pulse sequence is analogous to the selective inversion recovery, and the two-dimensional HOESY is analogous to a selective NOESY experiment. The phase cycle for the HOESY is $\phi_1 = x$; $\phi_2 = x, -x$; $\phi_3 = x, x, -x, -x, y, y, -y, -y$ and $\phi_{rec} = x, -x, -x, x, y, -y, -y, y$.

basic HOESY scheme is shown in Figure 9.3. The experiment begins with a long recycle delay. The two proton 90° pulses are separated by the evolution period, t_1, of the two-dimensional spectroscopy, which provides the frequency-labelling of the proton resonances and generates the proton longitudinal magnetisation (other proton magnetisation components are also generated but can be removed by phase cycling; see the next section). The carbon-13 180° pulse in the middle of the evolution period removes (refocuses) the heteronuclear J-coupling. During the mixing (or relaxation) period τ, the heteronuclear cross-relaxation process takes place, which transfers proton longitudinal magnetisation to carbon. The final carbon 90° pulse yields observable carbon-13 single-quantum coherence. Carbon detection, under proton broadband decoupling, has the advantage of generating a very simple spectrum and of removing the effects of the protons not interacting with the carbon-13 nuclei (for instance a residual water signal). By performing the experiment with a single mixing time, one can obtain a clear-cut picture of through-space connectivities associated with cross-relaxation. By repeating the experiment with different mixing times, one can follow the build-up of the heteronuclear NOE. The analysis is very similar to the case of NOESY (see next section). Several modifications of the basic HOESY scheme have been proposed and a partial bibliography can be found in the paper by Walker et al[9].

9.2 Homonuclear Cross-Relaxation: NOESY

Measurements of proton T_1 and T_2 are generally not very informative, since they are prone to artefacts arising from the assumption of an isolated set of spins, and not very common, although measurements of proton R_1 rates are included in the spectral density mapping approach (see Section 11.4). Proton relaxation is often difficult to quantify and interpret due to the fact that the spin systems easily become quite large in organic and biological molecules. However, proton-proton cross-relaxation rates are indeed one of the most important NMR parameters. The reason for this is that the σ_{IS} is proportional to the square of the dipolar interaction strength constant, which in turn contains a dependence on the inter-proton distance, r_{IS}. This has been used extensively to estimate inter-proton distances and constitutes the foundation of solution structure determination by NMR. In this section, we shall discuss in some detail the famous *NOESY experiment*. The focus will be on this experiment, since it is one of the best known homonuclear relaxation experiments. In subsequent sections, other techniques will also be presented, including cross-relaxation in the rotating frame (rotating-frame Overhauser effect spectroscopy [ROESY]).

9.2.1 General

One of the most common, and conceptually the simplest, two-dimensional proton NMR experiments is the homonuclear two-dimensional NOESY experiment[10,11]. The beauty of the experiment lies in the fact that, unlike other two-dimensional NMR experiments, magnetisation is transferred not through scalar couplings but instead, through cross-relaxation pathways. The expressions for dipolar cross-relaxation are given in Chapter 3, and for the applications described in this chapter, we need to revisit the Solomon equations for two spins, given by Eqs. (3.19)–(3.21) and (3.23). A general solution to the vector-matrix representation of Eq. (3.23) can be obtained for a two-spin system and is given by:

$$\mathbf{V} = \exp(-\mathbf{R}t)\left(\mathbf{V} - \mathbf{V}_{eq}\right) = \mathbf{X}^{-1}\exp(-\mathbf{D}t)\mathbf{X}\left(\mathbf{V} - \mathbf{V}_{eq}\right) \tag{9.1}$$

The expression $\exp(-\mathbf{R}t)$ (exponential of a matrix) is, in analogy with the exponential of an operator, defined by the power series expansion. To obtain a convenient representation of the problem, it is useful to begin by diagonalising the matrix $\mathbf{R}$; $\mathbf{D}$ in Eq. (9.1) is the diagonal matrix of eigenvalues of $\mathbf{R}$ and $\mathbf{X}$ is a unitary matrix of eigenvectors of $\mathbf{R}$, such that $\mathbf{D} = \mathbf{XRX}^{-1}$. The eigenvalues are given by:

$$\mathbf{D} = \begin{pmatrix} \lambda_+ & 0 \\ 0 & \lambda_- \end{pmatrix}$$
$$\lambda_\pm = \frac{1}{2}\left\{(\rho_I + \rho_S) \pm \left[\left((\rho_I - \rho_S)^2 + 4\sigma_{IS}^2\right)^{1/2}\right]\right\} \quad (9.2)$$

The solution to the Solomon equations can thus be written:

$$\begin{pmatrix} \Delta\hat{I}_z \\ \Delta\hat{S}_z \end{pmatrix} = \begin{pmatrix} a_{II}(t) & a_{IS}(t) \\ a_{IS}(t) & a_{SS}(t) \end{pmatrix} \begin{pmatrix} \Delta\hat{I}_z(0) \\ \Delta\hat{S}_z(0) \end{pmatrix} \quad (9.3)$$

where $\Delta\hat{I}_z$ and $\Delta\hat{S}_z$ represent the deviation from equilibrium magnetisation for the *I*- and *S*-spins. The time-dependent coefficients, $a_{II}(t) = [\exp(-\mathbf{R}t)]_{II}$ and $a_{IS}(t) = [\exp(-\mathbf{R}t)]_{IS}$, can be computed from the diagonal matrix of eigenvalues and will be given on the next page.

Turning to the two-dimensional NOESY experiment, we shall see how the signal intensities of the diagonal- (auto) and cross-peaks can be given by these coefficients. In its original form[10], the pulse sequence contains three pulses, separated by an evolution period, t_1, and a period during which dipolar relaxation is active (the relaxation or mixing period). The sequence is shown in Figure 9.4. Let us consider what will happen to one of the proton spins, *I*, during the experiment. The first 90° *x* pulse creates transverse $-\hat{I}_y$ magnetisation, which evolves and undergoes frequency-labelling during the evolution period t_1. The second pulse creates an array of terms, of which some are desirable in the NOESY experiment:

$$\hat{I}_z \xrightarrow{90^\circ \hat{I}_x - t_1 - 90^\circ \hat{I}_x} -\hat{I}_z \cos(\Omega_I t_1)\cos(\pi J_{IS} t_1) - 2\hat{I}_x\hat{S}_y \cos(\Omega_I t_1)\sin(\pi J_{IS} t_1)$$
$$+ \hat{I}_x \sin(\Omega_I t_1)\cos(\pi J_{IS} t_1) - 2\hat{I}_z\hat{S}_y \sin(\Omega_I t_1)\sin(\pi J_{IS} t_1) \quad (9.4)$$

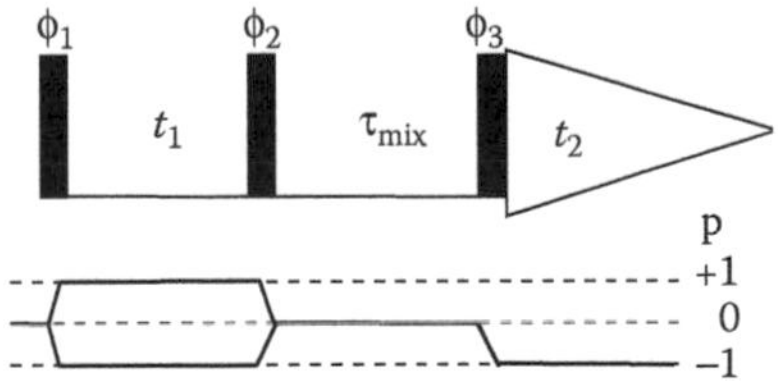

ϕ_1	ϕ_2	ϕ_3	ϕ_{rec}
x	x	x	x
x	x	y	y
x	x	$-x$	$-x$
x	x	$-y$	$-y$
$-x$	$-x$	x	$-x$
$-x$	$-x$	y	$-y$
$-x$	$-x$	$-x$	x
$-x$	$-x$	$-y$	y

FIGURE 9.4 The NOESY pulse sequence. The coherence order diagram shows the coherence orders that need to be selected for during the different delays in the pulse sequence. The minimal phase cycle for the NOESY experiment, with suppression of peaks at zero frequency in the ω_1 domain (the axial peaks). The symbols ϕ_1, ϕ_2 and ϕ_3 refer to the phases of the first, second and third 90° pulse, as indicated in the top panel, and ϕ_{rec} denotes the receiver phase. (Adapted from Keeler, J., *Understanding NMR Spectroscopy*, 2nd ed, Wiley, Chichester, 2010.)

For now, let us concentrate on the evolution of the $\hat{I}_z$ term during the mixing time, τ_m; most of the other terms can be rejected by a suitable phase cycle (see Section 9.2.2). Furthermore, the phase cycling procedure will remove the contribution from the equilibrium magnetisation, I_z^{eq}. The $\hat{I}_z$ magnetisation relaxes during τ_m, as governed by the Solomon equation. The evolution of $\hat{I}_z$ will thus become:

$$\begin{aligned}&-\hat{I}_z\cos(\Omega_I t_1)\cos(\pi J_{IS}t_1)\xrightarrow{\tau_m}-\hat{I}_z a_{II}(\tau_m)\cos(\Omega_I t_1)\cos(\pi J_{IS}t_1)\\&-\hat{S}_z a_{IS}(\tau_m)\cos(\Omega_I t_1)\cos(\pi J_{IS}t_1)\end{aligned} \tag{9.5}$$

where a_{II} and a_{IS} are coefficients depending on the relaxation matrix elements, as was discussed earlier in this section. The coefficients are given by:

$$\begin{aligned}a_{II}&=\frac{1}{2}\left[\left(1-\frac{\rho_I-\rho_S}{\lambda_+-\lambda_-}\right)\exp(-\lambda_-\tau_m)+\left(1+\frac{\rho_I-\rho_S}{\lambda_+-\lambda_-}\right)\exp(-\lambda_+\tau_m)\right]\\a_{SS}&=\frac{1}{2}\left[\left(1+\frac{\rho_I-\rho_S}{\lambda_+-\lambda_-}\right)\exp(-\lambda_-\tau_m)+\left(1-\frac{\rho_I-\rho_S}{\lambda_+-\lambda_-}\right)\exp(-\lambda_+\tau_m)\right]\\a_{IS}&=a_{SI}=\left(\frac{-\sigma_{IS}}{\lambda_+-\lambda_-}\right)\left[\exp(-\lambda_-\tau_m)-\exp(-\lambda_+\tau_m)\right]\end{aligned} \tag{9.6}$$

These equations are sometimes written in a slightly different way. The mixing coefficients can be conveniently expressed by two rate constants: a *cross-rate constant*, R_C, and a *leakage rate constant*, R_L. By defining these as:

$$\begin{aligned}R_C&=\lambda_+-\lambda_-=\left[(\rho_I-\rho_S)^2+4\sigma_{IS}^2\right]^{1/2}\\R_L&=\lambda_-=\frac{1}{2}(\rho_I+\rho_S)-\frac{1}{2}R_C\end{aligned} \tag{9.7}$$

we see that the coefficients now become:

$$\begin{aligned}a_{II}&=\frac{1}{2}\left[\left(1-\frac{\rho_I-\rho_S}{R_C}\right)+\left(1+\frac{\rho_I-\rho_S}{R_C}\right)\exp(-R_C\tau_m)\right]\exp(-R_L\tau_m)\\a_{SS}&=\frac{1}{2}\left[\left(1+\frac{\rho_I-\rho_S}{R_C}\right)+\left(1-\frac{\rho_I-\rho_S}{R_C}\right)\exp(-R_C\tau_m)\right]\exp(-R_L\tau_m)\\a_{IS}&=a_{SI}=\left(\frac{-\sigma_{IS}}{R_C}\right)\left[1-\exp(-R_C\tau_m)\right]\exp(-R_L\tau_m)\end{aligned} \tag{9.8}$$

This formulation can be convenient when describing a two-spin proton system. The cross-rate (R_C) is, for two protons with similar $1/T_1$, dominated by the cross-relaxation rate, σ_{IS}. If the protons have the same longitudinal relaxation rate constants, $\rho_I=\rho_S=\rho$, Equations (9.6)–(9.8) simplify to:

$$a_{II}=a_{SS}=\frac{1}{2}\left[1+\exp(-2\sigma_{IS}\tau_m)\right]\exp\left(-(\rho-\sigma_{IS})\tau_m\right) \tag{9.9a}$$

$$a_{IS}=a_{SI}=-\frac{1}{2}\left[1-\exp(-2\sigma_{IS}\tau_m)\right]\exp\left(-(\rho-\sigma_{IS})\tau_m\right) \tag{9.9b}$$

Finally, the last $90_x°$ pulse will convert the longitudinal terms into y-operators:

$$-\hat{I}_z a_{II}(\tau_m)\cos(\Omega_I t_1)\cos(\pi J_{IS} t_1) - \hat{S}_z a_{IS}(\tau_m)\cos(\Omega_I t_1)\cos(\pi J_{IS} t_1) \xrightarrow{90°(\hat{I}_x+\hat{S}_x)}$$
$$\hat{I}_y a_{II}(\tau_m)\cos(\Omega_I t_1)\cos(\pi J_{IS} t_1) + \hat{S}_y a_{IS}(\tau_m)\cos(\Omega_I t_1)\cos(\pi J_{IS} t_1) \quad (9.10)$$

After detection during t_2, the spectrum will thus consist of a diagonal peak, centred at (Ω_I, Ω_I), and a cross-peak centred at (Ω_I, Ω_S) (see Figure 9.5), with the time-dependent amplitude coefficients described earlier in this section. If the cross-relaxation rate between I and S is zero, $\sigma_{IS} = 0$, the amplitude coefficients for the cross-peaks will vanish (Eq. (9.9b)) and we can thus see that only diagonal peaks will appear in the spectrum, with the intensity given by the factor $\exp(-\rho\tau_m)$. If $\sigma_{IS} \neq 0$, off-diagonal cross-peaks will appear in the spectrum, and the amplitude of the off-diagonal peaks will in the initial-rate regime, *i.e.* for short mixing times, be given by $-\sigma_{IS}\tau_m$. For longer mixing times, the relationship between cross-peak intensity and cross-relaxation becomes more complicated, as seen in Eq. (9.9). It should also be noted that the isolated two-spin system approximation seldom holds, implying that for longer mixing times, *spin diffusion* involving several spins becomes important (spin diffusion is discussed briefly in the following sections).

9.2.2 Interpretation and Practical Aspects

For an isolated pair of protons, the cross-relaxation rate is given by:

$$\sigma_{HH} = \frac{\pi}{5} b_{IS}^2 \{6J(2\omega_H) - J(0)\} \quad (9.11)$$

which is obtained from Eqs. (3.22c) and (3.29) and where we have neglected the difference between the chemical shifts of the two protons. The fact that the cross-relaxation rate depends on the dipolar interaction constant b_{IS}^2, and thus on $1/r_{IS}^6$, has been widely used for estimating distances from cross-relaxation

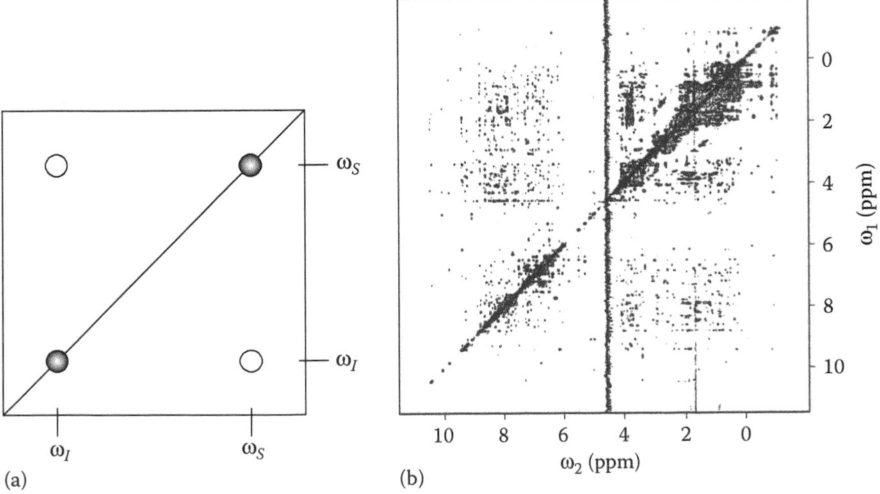

FIGURE 9.5 NOESY spectrum, schematic and real. (a) shows a schematic NOESY spectrum for an IS-spin system without resolved scalar couplings. The diagonal peaks (filled circles) have intensities proportional to a_{II} and a_{SS}, while the off-diagonal cross-peaks have intensities proportional to a_{IS} and a_{SI}. (b) shows a NOESY spectrum for 1 mM reduced horse heart cytochrome c acquired with $\tau_{mix} = 100$ ms.

measurements. Cross-peak volumes are integrated and, by normalising against known distances in a molecule, semi-quantitative (at best) values for distances can be obtained. This constitutes the basis for solution structure determination by NMR. Using the isotropic diffusor spectral density function presented in Section 6.1, the expression for σ_{HH} becomes:

$$\sigma_{HH} = \frac{1}{10} b_{IS}^2 \tau_c \left\{ \frac{6}{1 + 4\omega_H^2 \tau_c^2} - 1 \right\} \tag{9.12}$$

In the extreme-narrowing limit, the cross-relaxation rate becomes $\sigma_{HH} = \frac{1}{2} b_{IS}^2 \tau_c$. By examining Eq. (9.12), one can note that the value for σ_{HH} becomes zero if $\omega_H \tau_c = 1.12$ and, furthermore, that σ_{HH} changes sign at that point. In the slow tumbling limit ($\omega_H \tau_c \gg 1$), $\sigma_{HH} = -\frac{1}{10} b_{IS}^2 \tau_c$. It may thus happen that for certain molecular sizes, giving rise to the condition $\omega_H \tau_c = 1.12$, the cross-peaks vanish in the NOESY spectrum. From examining Eq. (9.9), we can thus predict that for small, rapidly tumbling molecules ($\omega_H \tau_c < 1.12$), the relaxation properties lead to weak negative cross-peaks. For large macromolecules in the slow tumbling limit, $\omega_H \tau_c \gg 1$, the cross-peaks will instead be positive and strong. The correlation-time dependence of the diagonal and cross-peak intensities for a few examples is shown in Figure 9.6.

Cross-relaxation for large macromolecules in the slow tumbling limit leads also to a phenomenon known as *spin-diffusion*, which relates to the fact that magnetisation is efficiently transferred among a group of spins. The spin diffusion can cause magnetisation to be transferred between two spins, which may be located quite far apart, via one or several intervening spins. In principle, this can be taken into account by including more than the two spins in the analysis of relaxation data, *i.e.* extending the size of the spin system. The analysis, known as the *full relaxation matrix* treatment [12], becomes significantly more complicated than the isolated spin pair model discussed above. If spin diffusion is not properly handled, difficulties in correctly assessing the relationship between the distance and the cross-peak volume will result. Although keeping the mixing time short leads to weaker cross-peak signals, on the other hand, this helps in minimising the spin-diffusion effect.

As was outlined in Section 7.3, a suitable phase cycling scheme can be devised to retain only terms of a particular coherence order. In the case of the NOESY experiment, we wish to retain the magnetisation with $p = 0$ during the mixing period. The minimum phase cycle that will have this effect as well as

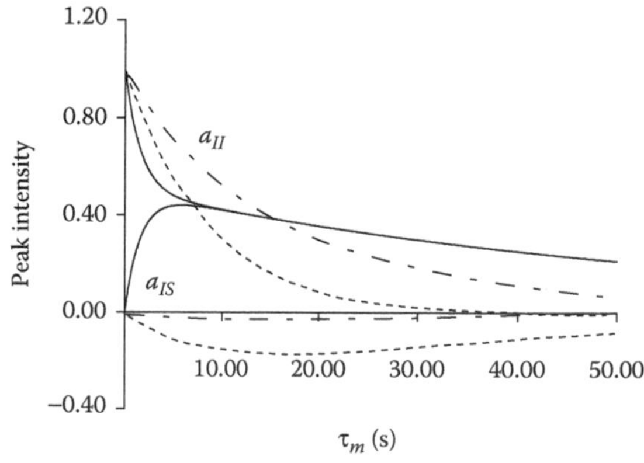

FIGURE 9.6 Dependence of the diagonal and cross-peak time-dependent amplitude factors, a_{II} and a_{IS}, on the mixing time, τ_m, in the two-dimensional ^{1}H-^{1}H NOESY spectrum shown for different values of the correlation time, τ_c. The simulations were done using Eq. (9.9) for $\tau_c = 0.1$ ns (- - - -), $\tau_c = 0.4$ ns (- — -) and $\tau_c = 4$ ns (——). The auto-relaxation rates and cross-relaxation rates were calculated for a Larmor frequency of 400 MHz using the isotropic rotor spectral density function assuming $b_{HH} = 27 \times 10^3$ rad s^{-1} ($b_{HH}/2\pi = 4.3$ kHz) (corresponding to $r_{HH} = 3$ Å).

suppression of the axial peaks, together with the corresponding coherence transfer pathway diagram, is shown in Figure 9.4.

Several things need to be achieved with the phase cycle. First, so-called axial peaks, arising due to the fact that $p = \pm 1$ coherences relax back to equilibrium longitudinal magnetisation during the evolution period, t_1, must be suppressed. The resulting terms will not be frequency-labelled and will give rise to artefacts along $\omega_1 = 0$ in the spectrum. This is a general problem in two-dimensional NMR. Second, to obtain a spectrum with pure phase, both $p = +1$ and $p = -1$ must be retained during t_1, while $p = 0$ is suppressed. Finally, the phase cycle must retain $p = 0$ during the mixing time, while all coherences must be suppressed. The phase cycle in Figure 9.4 will do the job and will reject the third and fourth terms in Eq. (9.4), since these are single-quantum coherences. The second term in Eq. (9.4) is more difficult, as it represents a mixture of zero-quantum (ZQ) coherence and double-quantum coherence: $2\hat{I}_x\hat{S}_y = 2(DQ_y - ZQ_y)$. The double-quantum term ($p = 2$) is suppressed by the phase cycle, but the zero-quantum coherence ($p = 0$) is similar to populations and can persist in the mixing time, τ_m, without being affected by phase cycling. The zero-quantum coherence $ZQ_y = \frac{1}{2}(2\hat{I}_y\hat{S}_x - 2\hat{I}_x\hat{S}_y)$ will precess according to the difference in chemical shift of I and S, and the following terms will be generated at the end of the mixing time:

$$\begin{aligned}
&\frac{1}{2}\left(2\hat{I}_y\hat{S}_x - 2\hat{I}_x\hat{S}_y\right)\cos(\Omega_I t_1)\sin(\pi J_{IS} t_1) \xrightarrow{\tau_m} \\
&\frac{1}{2}\left(2\hat{I}_y\hat{S}_x - 2\hat{I}_x S_y\right)\cos(\Omega_I t_1)\sin(\pi J_{IS} t_1)\cos\left[(\Omega_I - \Omega_S)\tau_m\right] \\
&-\frac{1}{2}\left(2\hat{I}_y\hat{S}_y + 2\hat{I}_x\hat{S}_x\right)\cos(\Omega_I t_1)\sin(\pi J_{IS} t_1)\sin\left[(\Omega_I - \Omega_S)\tau_m\right]
\end{aligned} \tag{9.13}$$

The zero-quantum coherence will result in observable terms after the final 90° pulse is applied:

$$\begin{aligned}
&\frac{1}{2}\left(2\hat{I}_y\hat{S}_x - 2\hat{I}_x\hat{S}_y\right)\cos(\Omega_I t_1)\sin(\pi J_{IS} t_1)\cos\left[(\Omega_I - \Omega_S)\tau_m\right] \\
&-\frac{1}{2}\left(2\hat{I}_y\hat{S}_y + 2\hat{I}_x\hat{S}_x\right)\cos(\Omega_I t_1)\sin(\pi J_{IS} t_1)\sin\left[(\Omega_I - \Omega_S)\tau_m\right] \xrightarrow{90^\circ\left(\hat{I}_x + \hat{S}_x\right)} \\
&\frac{1}{2}\left(2\hat{I}_x\hat{S}_z - 2\hat{I}_z\hat{S}_x\right)\cos(\Omega_I t_1)\sin(\pi J_{IS} t_1)\cos\left[(\Omega_I - \Omega_S)\tau_m\right] \\
&-\frac{1}{2}\left(2\hat{I}_z\hat{S}_z + 2\hat{I}_x\hat{S}_x\right)\cos(\Omega_I t_1)\sin(\pi J_{IS} t_1)\sin\left[(\Omega_I - \Omega_S)\tau_m\right]
\end{aligned} \tag{9.14}$$

The first term represents antiphase observable single-quantum coherence for both the I- and the S-spin and will for each spin give rise to two zero-quantum peaks centred at (Ω_I, Ω_I) and (Ω_I, Ω_S). They will be antiphase in both dimensions and dispersive with respect to the NOE cross-peaks, which may lead to difficulties in accurately measuring NOE-cross-peak intensities. This problem can partly be overcome by randomly varying the mixing time slightly between different t_1 increments. The cosine dependence of the ZQ-peak amplitudes will lead to a cancellation of the ZQ-peaks when the experiments are co-added. The amplitude of the ZQ-peaks will diminish due to transverse relaxation and the artefacts are therefore more pronounced at shorter mixing times. To avoid the ZQ artefacts, we thus need longer mixing times, more seriously affected by spin diffusion, and a trade-off between the two is therefore necessary. A thorough discussion concerning the outcome of the NOESY experiment and the appropriate phase cycle is given in text by Ernst, Bodenhausen and Wokaun, by Keeler and by Cavanagh and co-workers (*Further reading*).

9.3 Cross-Relaxation in the Rotating Frame: ROESY

The measurement of NOE by the two-dimensional NOESY experiment has its counterpart in the measurement of rotating-frame Overhauser enhancement (ROE). This is the two-dimensional *ROESY experiment*, in which cross-relaxation is measured in a tilted rotating frame, analogous to the discussion concerning measurement of $T_{1\rho}$ relaxation (see Chapter 8). A detailed description of ROESY and variants has been presented by Bull[13,14] and in a more recent review by Bax and Grzesiek[15]. The original name for the experiment is cross relaxation appropriate for minimolecules emulated by locked spins (CAMELSPIN)[16]. The name hints at the fact that the experiments work well for small and medium-sized molecules. In this frame, the auto-relaxation rate is given by Eq. (8.11), in which a dependence on the tilt angle of the r.f. field is introduced. The cross-relaxation rate constant can also be defined in the rotating frame in analogy with the $R_{1\rho}$ time constant and is given by:

$$\sigma_{IS\rho} = \frac{\pi}{5} b_{IS}^2 \left\{ \cos\Theta_I \cos\Theta_S \left(6J(2\omega_H) - J(0)\right) + \sin\Theta_I \sin\Theta_S \left(3J(\omega_H) + 2J(0)\right) \right\} \tag{9.15}$$

The first term within the brackets corresponds to pure cross-relaxation in the laboratory frame for a tilt angle of zero degrees, *i.e.* the conventional longitudinal cross-relaxation rate σ_{IS}. The second term, obtained for a tilt angle of 90°, describes the cross-relaxation in the rotating frame, identical to the R_{2IS} of Eq. (4.86b):

$$\sigma_{IS}^{ROE} = \frac{\pi}{5} b_{IS}^2 \left(3J(\omega_H) + 2J(0)\right) \tag{9.16}$$

In analogy with the laboratory-frame cross-relaxation rate constant, using the isotropic rigid rotor spectral density function, the following expression is obtained:

$$\sigma_{IS}^{ROE} = \frac{1}{10} b_{IS}^2 \tau_c \left(\frac{3}{1 + \omega_H^2 \tau_c^2} + 2 \right) \tag{9.17}$$

The ROE cross-relaxation rate is here seen to be always positive for all correlation times, as opposed to the NOE cross-relaxation, which passes through zero for a critical correlation time, such that $\omega_H \tau_c = 1.12$. The ROE is therefore useful for measuring cross-relaxation for medium-sized molecules with correlation times near the critical value. The frequency dependence of σ_{IS}^{ROE} and σ_{IS}^{NOE} for different correlation times is depicted in Figure 9.7.

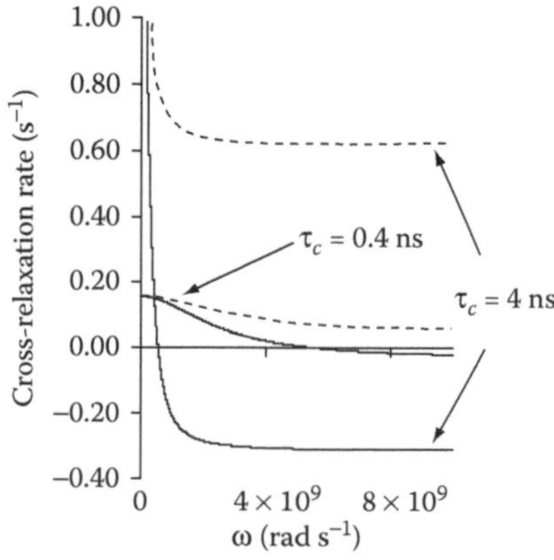

FIGURE 9.7 Cross-relaxation rates as a function of frequency for two different values of the correlation time. Solid lines show σ_{IS}^{NOE} while dashed lines show σ_{IS}^{ROE} relaxation rates. The calculations were performed assuming $b_{HH} = 27 \times 10^3$ rad s^{-1} ($b_{HH}/2\pi = 4.3$ kHz) (corresponding to $r_{HH} = 3$ Å).

The two-dimensional ROESY experiment consists in its simplest form of a $90^\circ_x - t_1 - \tau_m - t_2$ sequence in which a spin-lock is applied during τ_m (Figure 9.8). The 90° pulse generates transverse magnetisation, which is frequency-labelled during t_1, after which the following is obtained:

$$\begin{aligned}\hat{I}_z &\xrightarrow{90^\circ \hat{I}_x - t_1} -\hat{I}_y \cos(\Omega_I t_1)\cos(\pi J_{IS} t_1) + 2\hat{I}_x\hat{S}_y \cos(\Omega_I t_1)\sin(\pi J_{IS} t_1) \\ &+ \hat{I}_x \sin(\Omega_I t_1)\cos(\pi J_{IS} t_1) + 2\hat{I}_y\hat{S}_z \sin(\Omega_I t_1)\sin(\pi J_{IS} t_1)\end{aligned} \tag{9.18}$$

A strong r.f. field with a 90° tilt angle and with phase y is applied during the mixing time, τ_m, which locks the y-operators, while all terms containing the x-operator dephase due to r.f. inhomogeneity. Without the spin-lock, the transverse cross-relaxation will be "decoupled" by the spin precession taking place during the relaxation delay. In short, transverse magnetisation can be transferred between spins due to cross-relaxation, but the net effect will be that the transferred components will cancel due to free precession of the magnetisation (this statement is equivalent to saying that the cross-relaxation processes become non-secular). The spin-lock will, however, in effect, make all of the spins identical, acting as if they have the same precession frequency. Under these conditions, the cross-relaxation rate for the I- and S-spin system can be described by the transverse Solomon equations, fully analogous to Eq. (4.85):

$$\frac{d}{dt}\begin{pmatrix}\langle \hat{I}_y \rangle \\ \langle \hat{S}_y \rangle\end{pmatrix} = -\begin{pmatrix} R_{2I} & \sigma_{ROE} \\ \sigma_{ROE} & R_{2S}\end{pmatrix}\begin{pmatrix}\langle \hat{I}_y \rangle \\ \langle \hat{S}_y \rangle\end{pmatrix} \tag{9.19}$$

where $R_{2I} = 1/T_{2I}$, $R_{2S} = 1/T_{2S}$ and σ_{ROE} is the rotating frame cross-relaxation rate as given in Eqs. (9.16) and (9.17).

In a completely analogous fashion to the longitudinal cross-relaxation experiment (NOESY), the magnetisation will be transferred between spins due to this transverse cross-relaxation phenomenon. The result after the spin-lock period, τ_m, is therefore

$$\begin{aligned}-\hat{I}_y \cos(\Omega_I t_1)\cos(\pi J_{IS} t_1) &\xrightarrow{\tau_m} -\hat{I}_y a_{II}(\tau_m)\cos(\Omega_I t_1)\cos(\pi J_{IS} t_1) \\ &- \hat{S}_y a_{IS}(\tau_m)\cos(\Omega_I t_1)\cos(\pi J_{IS} t_1)\end{aligned} \tag{9.20}$$

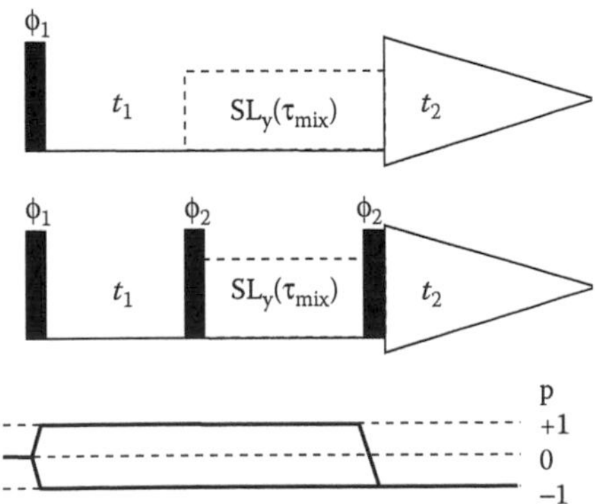

FIGURE 9.8 The ROESY pulse sequence. The top diagram shows the original ROESY pulse sequence, and the bottom the modifications suggested by Griesinger and Ernst to improve sensitivity. The appropriate coherence order diagram for the experiment is shown. In the original pulse sequence, $\phi_1 = x, -x$ and $\phi_{rec} = x, -x$, and in the modified pulse sequence, $\phi_1 = x, -x$, $\phi_2 = y$ and $\phi_{rec} = y, -y$.

where $a_{II}(\tau_m) = [\exp(-\mathbf{R}\tau_m)]_{II}$ and $a_{IS}(\tau_m) = [\exp(-\mathbf{R}\tau_m)]_{IS}$ and **R** is the matrix of rotating frame relaxation rates (Eq. (9.19)).

The antiphase term in Eq. (9.18) will also lead to observable terms at the end of the sequence. This is the antiphase S-spin coherence that has been generated by correlation between the two spins through the J-coupling, which is used in homonuclear correlation experiments, such as correlation spectroscopy (COSY). Distortions of the in-phase cross-peak pattern will therefore appear due to this COSY-like contribution to the signal, analogously to what was discussed for the zero-quantum interference in NOESY cross-peaks. There are two distinct mechanisms that cause artefacts due to J-couplings between spins in the ROESY experiments: the COSY magnetisation transfer and the total correlation spectroscopy (TOCSY)– or homonuclear Hartmann–Hahn cross-polarisation (HOHAHA)–like magnetisation transfer (the reader is referred to the texts by Levitt and by Ernst, Bodenhausen and Wokaun (*Further reading*) for discussion of the TOCSY and HOHAHA experiments). We wish to recommend the review by Schleucher and co-workers concerning possible artefacts from TOCSY in ROESY[17] and the review on ROESY by Bax and Grzesiek[15]. The net integrated contribution to the intensity is zero, due to the antiphase nature of the artefacts, but integrating the peak volumes can become difficult because of the distortions of lineshapes.

One way of minimising J-coupling effects in ROESY is to use a relatively weak spin-lock field strength, or to make sure that the two spins involved in the J-coupling experience different field strengths. The latter can be achieved by positioning the carrier at one edge of the spectrum. Both of these methods have implications for the effective tilt angle, which, on the other hand, may have severe consequences for the sensitivity of the experiment. From examining Eq. (9.15) we see that values for Θ deviating from 90° will decrease the signal intensity. The loss of intensity can be overcome if two 90° pulses, bracketing the spin-lock period, are inserted in the pulse sequence[18] (see Figure 9.8, bottom panel). The signal loss, given by $\sin\Theta_I \sin\Theta_S$, is in this case transferred to a phase error, generated by the resonance offset, and can be compensated for during processing of the data. In yet another modification of the ROESY experiment, aiming at suppressing the TOCSY effects and known as T-ROESY, the single frequency or continuous wave (CW) irradiation during the spin-lock period in the second diagram of Figure 9.8 is replaced by a multipulse sequence[19,20].

9.4 Homonuclear Cross-Relaxation: One-Dimensional Experiments

Instead of using the two-dimensional schemes discussed in Sections 9.2 and 9.3, homonuclear cross-relaxation rates can be measured by *selective one-dimensional* methods. By applying a selective inversion pulse on one of the protons, I, cross-relaxation to another proton, S, can be monitored during a mixing time, τ_m. The amount of magnetisation transferred to the S-spin via cross-relaxation is detected by applying a 90° read-pulse. This experiment is usually called the *selective transient NOE* experiment and

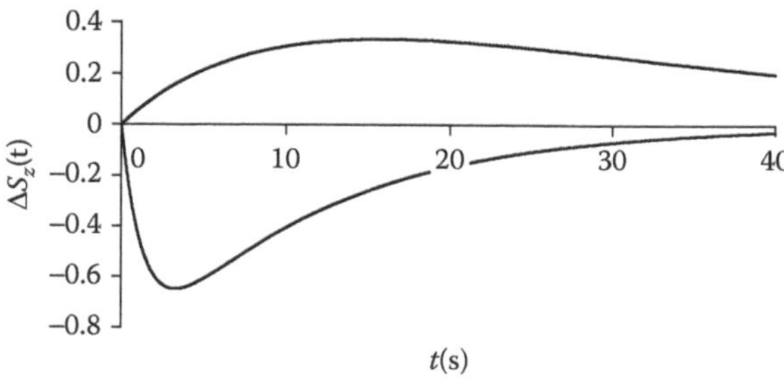

FIGURE 9.9 Plots of the evolution of $\Delta\hat{S}_z(t)$ after a selective inversion of the I-spin. The curves were calculated using Eq. (9.21), assuming $b_{HH} = 27 \times 10^3$ rad s^{-1} ($b_{HH}/2\pi = 4.3$ kHz) (corresponding to $r_{HH} = 3$ Å). The top curve is calculated for $\tau_c = 0.1$ ns and the bottom curve for $\tau_c = 4$ ns.

is essentially the same as outlined for a system of two unlike spins in Section 9.1. The cross-relaxation rate is most often obtained from a series of experiments using a variable mixing delay τ_m. The initial conditions in this experiment are given by: $\Delta\hat{I}_z(0) = \langle\hat{I}_z\rangle(0) - I_z^{eq} = -2I_z^{eq}$ and $\Delta\hat{S}_z(0) = 0$. The time evolution of the S-spin magnetisation during the mixing time is given by:

$$\Delta\hat{S}_z(t) = \exp\{-(\rho-\sigma)t\}\left[1-\exp(-2\sigma t)\right] \tag{9.21}$$

As can be seen from Eq. (9.21), the time evolution of the S-spin magnetisation is biexponential. Cross-relaxation build-up curves obtained in the selective transient NOE experiment for different values of ρ and σ (corresponding to different molecular sizes or correlation times) can be found in Figure 9.9. In the *initial rate* regime, valid for short mixing times, the slope of the build-up curve is given by:

$$\frac{d}{dt}\Delta\hat{S}_z(t) = 2\sigma \tag{9.22}$$

and the cross-relaxation rate can be evaluated from a linear fit. A selective inversion of one of the spins can easily be achieved by using a weak, shaped pulse with a narrow excitation range. The one-dimensional selective NOE experiment can be thought of as a way of recording selective strips of a two-dimensional NOESY. For small and medium-sized molecules, this approach is often more practical, as overlap in the spectrum is not such a big problem. In addition, it is feasible to obtain reliable cross-relaxation rate parameters from the initial regime of the build-up curve, which can be used to estimate distances more accurately. A sophisticated variety of the one-dimensional transient NOE experiments, known as the *double pulsed field gradient spin echo* (DPFGSE) experiment[21], uses a combination of selective r.f. pulses and field-gradient pulses in order to obtain very clean spectra, suitable for quantitative work.

In organic chemistry, the *NOE difference experiment* has often been employed to elucidate the structure of small organic compounds. Comparing NOE enhancement factors obtained in this way may help, for example, in distinguishing between *cis* and *trans* isomers in compounds with a double bond. Two spectra are recorded, one in which a certain resonance is irradiated selectively during a fixed delay prior to recording the spectrum. The second experiment is performed in the same way, but the irradiation is applied to a frequency far outside the spectral region and the spectrum is subtracted from the first. The resultant spectrum should in principle only contain signals due to NOE transfer of magnetisation from the selectively irradiated spin. It is, however, evident that the method is heavily dependent on the quality of subtraction. If the two spectra differ in properties not related to the NOE, such as those generated by instabilities in temperature and field homogeneity, severe artefacts comparable to small NOEs will arise. Nevertheless, this method has been widely used to estimate differences in inter-proton distances.

Finally, in this section, we wish to return briefly to the problem of spin diffusion. A few experiments have been designed to reduce the complications caused by spin diffusion in a large spin system, usually consisting of protons in organic and biological macromolecules. Failure to correctly assess the influence of spin diffusion will lead to errors in the determined cross-relaxation rates and internuclear distances. Protons that are distant may display apparent strong cross-relaxation, which is mediated through other spins.

If we wish to study cross-relaxation between two distant spins, A and X, in the presence of a third spin, K, which is close to both A and X, we need to somehow remove cross-relaxation pathways involving K. Figure 9.10 shows an example of a pulse sequence that will quench spin diffusion while retaining only the direct cross-relaxation pathway (quenching undesirable indirect external trouble in NOESY [QUIET-NOESY][22]). First, one of the spins, let us say A, is inverted by a selective inversion pulse. When a simultaneous inversion pulse is applied on both A and X in the middle of the mixing period, the sign of both A and X magnetisation changes, and thus the build-up of magnetisation on K changes. A second experiment is performed in which the initial inversion of the A spin is not performed, and the result is

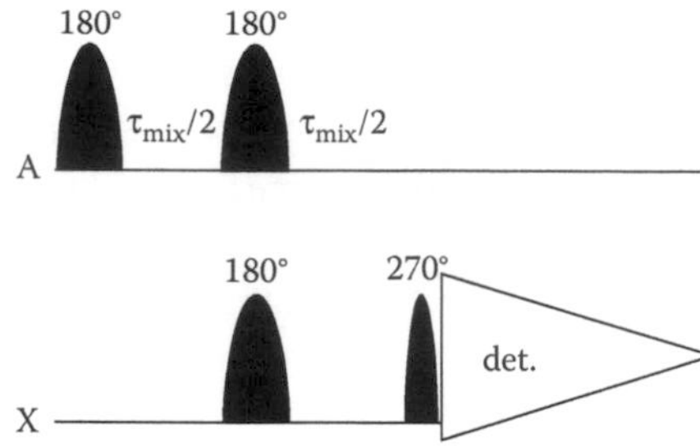

FIGURE 9.10 Pulse sequence for removal of spin diffusion, QUIET-NOESY. For clarity, the appropriate pulses applied to the A and X protons are shown on separate lines. The oval shape indicates a selective r.f. pulse. In practice, different selective shaped pulses are used to invert the A, and the A and X magnetisation, and for detecting the A-spin.

subtracted from the first experiment, which is equivalent to the difference spectroscopy technique. At the end of the mixing time, only cross-relaxation from A to X will therefore contribute to the X-spin signal. In this way the true cross-relaxation between A and X can be measured without the influence of the intervening spin K. Since the NOESY experiment and the one-dimensional variants are concerned with homonuclear proton-proton cross-relaxation, pulse schemes such as this must employ selective pulses, and typically different wave-form generated pulses are used for inversion pulses and excitation pulses. Difference spectroscopy is usually difficult to perform in practice, mainly due to large suppression artefacts, and therefore the original pulse sequence included a polarisation transfer step from the X-spin to a fourth "spy" nucleus for detection.

References

1. Freeman, R.; Hill, H. D. W.; and Kaptein, R., Proton-decoupled NMR spectra of carbon-13 with the nuclear Overhauser effect suppressed. *J. Magn. Reson.* 1972, 7, 327–329.
2. Clore, G. M.; Driscoll, P. C.; Wingfield, P. T.; and Gronenborn, A. M., Analysis of backbone dynamics of interleukin-1beta using two-dimensional inverse detected heteronuclear 15N-1H NMR spectroscopy. *Biochemistry* 1990, 29, 7387–7401.
3. Neuhaus, D.; and Van Mierlo, C. P. M., Measurement of heteronuclear NOE enhancements in biological macromolecules: A convenient pulse sequence for use with aqueous solutions. *J. Magn. Reson.* 1992, 100, 221–228.
4. Ferrage, F.; Piserchio, A.; Cowburn, D.; and Ghose, R., On the measurement of N-15-{H-1} nuclear Overhauser effects. *J. Magn. Reson.* 2008, 192, 302–313.
5. Ferrage, F.; Reichel, A.; Battacharya, S.; Cowburn, D.; and Ghose, R., On the measurement of N-15-{H-1} nuclear Overhauser effects. 2. Effects of the saturation scheme and water signal suppression. *J. Magn. Reson.* 2010, 207, 294–303.
6. Freeman, R.; Wittekoek, S.; and Ernst, R. R., High-resolution NMR study of relaxation mechanisms in a two-spin system. *J. Chem. Phys.* 1970, 52, 1529–1544.
7. Rinaldi, P. L., Heteronuclear 2D-NOE spectroscopy. *J. Am. Chem. Soc.* 1983, 105, 5167–5168.
8. Yu, C.; and Levy, G. C., Solvent and intramolecular proton dipolar relaxation in the three phosphates of ATP: a heteronuclear 2D NOE study. *J. Am. Chem. Soc.* 1983, 105, 6994–6996.
9. Walker, O.; Mutzenhardt, P.; and Canet, D., Heteronuclear Overhauser experiments for symmetric molecules. *Magn. Reson. Chem.* 2003, 41, 776–781.
10. Jeener, J.; Meier, B. H.; Bachmann, P.; and Ernst, R. R., Investigation of exchange processes by two-dimensional NMR spectroscopy. *J. Chem. Phys.* 1979, 71, 4546–4553.
11. Williamson, M. P., *NOESY.* Wiley: eMagRes, 2009.

12. Borgias, B. A.; Gochin, M.; Kerwood, D. J.; and James, T. L., Relaxation matrix analysis of 2D NMR data. *Prog. NMR Spectr.* 1990, 22, 83–100.
13. Bull, T. E., ROESY relaxation theory. *J. Magn. Reson.* 1988, 80, 470–481.
14. Bull, T. E., Relaxation in the rotating frame in liquids. *Prog. NMR Spectr.* 1992, 24, 377–410.
15. Bax, A.; and Grzesiek, S., *ROESY*. Wiley: eMagRes, 2007.
16. Bothner-By, A. A.; Stephens, R. L.; and Lee, J. M., Structure determination of a tetrasaccharide: Transient nuclear Overhauser effects in the rotating frame. *J. Am. Chem. Soc.* 1984, 106, 811–813.
17. Schleucher, J.; Quant, J.; Glaser, S. J.; and Griesinger, C., *TOCSY in ROESY and ROESY in TOCSY*. Wiley: eMagRes, 2007.
18. Griesinger, C.; and Ernst, R. R., Frequency offset effects and their elimination in NMR rotating-frame cross-relaxation spectroscopy. *J. Magn. Reson.* 1987, 75, 261–271.
19. Hwang, T. L.; and Shaka, A. J., Cross relaxation without TOCSY: Transverse rotating-frame Overhauser effect spectroscopy. *J. Am. Chem. Soc.* 1992, 114, 3157–3159.
20. Hwang, T. L.; Kadkhodaei, M.; Mohebbi, A.; and Shaka, A. J., Coherent and incoherent magnetisation transfer in the rotating frame. *Magn. Reson. Chem.* 1992, 30, S24–S34.
21. Stott, K.; Stonehouse, J.; Keeler, J.; Hwang, T. L.; and Shaka, A. J., Excitation sculpting in high-resolution nuclear magnetic resonance spectroscopy: Application to selective NOE experiments. *J. Am. Chem. Soc.* 1995, 117, 4199–4200.
22. Zwahlen, C.; Vincent, S. J. F.; Di Bari, L.; Levitt, M. H.; and Bodenhausen, G., Quenching spin diffusion in selective measurements of transient Overhauser effects in nuclear magnetic resonance: Applications to oligonucleotides. *J. Am. Chem. Soc.* 1994, 116, 362–368.

10

Cross-Correlation and Multiple-Quantum Relaxation Measurements

In the previous chapter, we discussed cross-relaxation, an important multispin effect. In this chapter, we proceed with other multispin phenomena, which occur through interference of several relaxation mechanisms, involving the cross-correlation spectral densities. Cross-correlation effects can arise through interference between, for instance, the dipole-dipole (DD) interaction of a spin pair and the chemical shielding anisotropy (CSA) of one of the spins, or through interference between two dipole-dipole interactions. The theoretical basis for these effects was discussed in Chapter 5. Much like the cross-relaxation measurements, cross-correlation can provide information about molecular geometry, here through an angular relation between the principal axes of the two interactions. Furthermore, for spin 1/2 nuclei, measurement of cross-correlation between the dominating DD interaction and the CSA provides unique information about the interaction strength of the CSA and thus of the shielding anisotropy.

Cross-correlation effects have been known from the early days of nuclear magnetic resonance, and in high-resolution NMR they were primarily observed in double-resonance experiments. One of the earlier observations of cross-correlation is the non-exponential recovery observed for proton-proton dipolar interactions in methyl groups[1]. Also, several studies of carbon-13 relaxation in methyl groups have demonstrated the presence of non-exponential recovery due to cross-correlation[2]. Experimental aspects of multispin effects in nuclear spin relaxation have been reviewed in detail several times over the years[3–5]. A thorough theoretical discussion on cross-correlation effects in a system of two spin 1/2 nuclei can be found in the papers by Werbelow and Grant[2,6] and by Goldman[7].

The experiments described in Chapter 8 are aimed at measuring auto-relaxation rates, T_1 and T_2. These rates are for most spin 1/2 nuclei dominated by the intramolecular dipole-dipole interaction, but other relaxation mechanisms need also to be considered. The CSA is for certain carbon-13 and nitrogen-15 nuclei an important source of relaxation. In the presence of unpaired electrons, the electron spin has a strong coupling to the nuclear spin and can cause rapid relaxation of the nuclear spin, provided they are close enough in space (*cf.* Chapter 15). For nuclei with spin greater than 1/2 possessing a quadrupole moment, the quadrupolar relaxation usually becomes the dominant relaxation mechanism (cf. Chapter 14). As mentioned in Section 5.4, various relaxation rates can be expressed as sums over all contributing mechanisms, including the cross-terms. The cross-terms, known as *cross-correlation* or *interference effects*, may be of significant magnitude and are the exclusive origin of certain off-diagonal elements in the relaxation matrices in multispin systems.

Several approaches to measuring cross-correlation have been proposed, both for small molecules and for biological macromolecules. In the following, a few experiments for measuring longitudinal and transverse cross-correlation will be examined. A detailed analysis will only be given for cross-correlation between the DD and CSA relaxation mechanisms, while the methods for measuring other interference phenomena will be presented in a rather sketchy way. Techniques for measuring longitudinal and transverse cross-correlated relaxation rates will be covered in Sections 10.1 and 10.2. Section 10.3 deals with the application of the phenomenon of cross-correlated transverse relaxation as a general spectroscopic enhancement tool. It is also feasible to monitor the relaxation of coherences that are not directly observable, such as multiple-quantum (MQ) and zero-quantum (ZQ) coherences. The associated relaxation rates may contain other sources of molecular information, and corresponding experimental methods will be dealt with in Section 10.4.

10.1 Cross-Correlated Longitudinal Relaxation

In a two-spin *IS* system, and in weakly coupled systems of spin 1/2 nuclei in general, interference terms manifest themselves as asymmetries in multiplet relaxation behaviour. In transverse relaxation measurements of an *IS* doublet, the two lines will be broadened to different extents due to different transverse relaxation rates. In longitudinal relaxation, the perturbation response of the two components in the doublet will also be different (Figure 10.1).

The longitudinal relaxation for a simple isolated two-spin system is most conveniently described by the extended Solomon equations (compare Eqs. (5.42) and (5.43)):

$$\frac{d}{dt}\begin{pmatrix} \langle \hat{I}_z \rangle \\ \langle \hat{S}_z \rangle \\ \langle 2\hat{I}_z\hat{S}_z \rangle \end{pmatrix} = -\begin{pmatrix} \rho_I & \sigma_{IS} & \delta_{I,IS} \\ \sigma_{IS} & \rho_S & \delta_{S,IS} \\ \delta_{I,IS} & \delta_{S,IS} & \rho_{IS} \end{pmatrix}\begin{pmatrix} \langle \hat{I}_z \rangle - I_z^{eq} \\ \langle \hat{S}_z \rangle - S_z^{eq} \\ \langle 2\hat{I}_z\hat{S}_z \rangle \end{pmatrix} \tag{10.1}$$

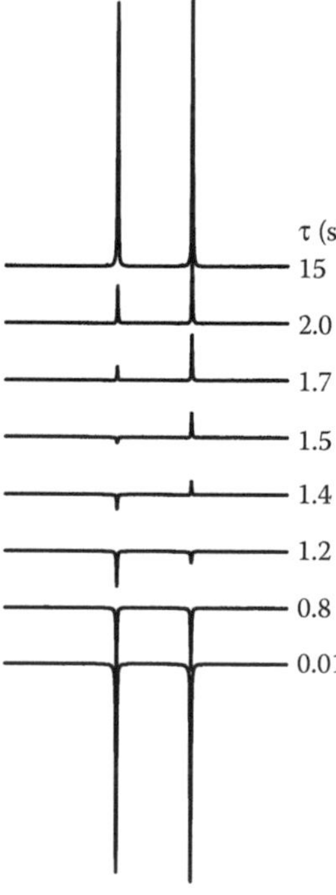

FIGURE 10.1 Different relaxation of the two components of the carbon-13 doublet in ^{13}C-labeled $CHCl_3$. The spectra were obtained using the conventional inversion-recovery pulse sequence, without proton decoupling, using values for the relaxation period as indicated in the figure.

The equation can be seen as an extension of Eq. (3.21) in which the two-spin order has been included. As introduced in Chapter 5, the relaxation rates responsible for converting one-spin order, $\hat{I}_z$, $\hat{S}_z$, into two-spin order, $2\hat{I}_z\hat{S}_z$, are denoted as cross-correlated relaxation rates (CCRRs), $\delta_{I,IS}$ and $\delta_{S,IS}$. The rate constants ρ_I, ρ_S and σ_{IS} were defined in Section 3.2 and $\rho_{IS}=1/T_{1IS}$ is the decay rate (auto-relaxation rate) of the two-spin order.

As discussed in Section 5.2, the cross-correlated relaxation rate is proportional to the interaction strength constants for both interactions involved in the interference. The longitudinal CCRR for the case of DD and CSA interactions was given in Eqs. (5.43)–(5.45). In a slightly modified formulation, the expression becomes:

$$\delta_{S,IS} = -\frac{4\pi}{5}\left(\frac{\mu_0}{4\pi}\right)\gamma_I\gamma_S^2\hbar B_0\Delta\sigma_S r_{IS}^{-3}\frac{1}{2}\left(3\cos^2\theta-1\right)J\left(\omega_S\right) \tag{10.2}$$

where θ is the angle between the principal axis of the CSA tensor and the dipolar internuclear vector of the length r_{IS}. $\Delta\sigma=\sigma_{\|}-\sigma_{\perp}$ is the shielding anisotropy, where $\sigma_{\|}$ and $\sigma_{\perp}$ are the principal components of the CSA tensor, which has been assumed to be axially symmetric.

The form of this relation suggests possible applications of the CCRRs: if $J(\omega)$, r_{IS} and θ are known, one can estimate $\Delta\sigma_S$, and conversely, with known $J(\omega)$, r_{IS} and $\Delta\sigma_S$, it is possible to use $\delta_{S,IS}$ to obtain θ.

10.1.1 Inversion-Recovery-Related Methods

A simple example of how longitudinal cross-correlation can be measured is related to the conventional inversion-recovery experiment (see Figure 8.1), with the important difference that no I-spin decoupling is employed during the experiment. A 180° pulse is applied to one of the spins, say the S-spins. In the case of a heteronuclear spin system, this is simply achieved in the standard way, while selective pulses need to be employed for measuring homonuclear proton cross-correlation. During the delay, τ, following the pulse, magnetisation is transferred from the created non-equilibrium state, $\langle\hat{S}_z\rangle(0)=-S_z^{eq}$, to various longitudinal operators according to the extended Solomon equations. The solution to these equations will lead to multiexponential relaxation behaviour. However, in the initial rate regime, the decay of the $\hat{S}_z$ state, the build-up of $\hat{I}_z$ and the build-up of the two-spin order, $2\hat{I}_z\hat{S}_z$, will be given by:

$$\left\langle\hat{I}_z\right\rangle(\tau)=I_z^{eq}-\sigma_{IS}\tau\left(\left\langle\hat{S}_z\right\rangle-S_z^{eq}\right) \tag{10.3a}$$

$$\left\langle\hat{S}_z\right\rangle(\tau)=\left\langle\hat{S}_z\right\rangle(0)-\rho_S\tau\left(\left\langle\hat{S}_z\right\rangle-S_z^{eq}\right) \tag{10.3b}$$

$$\left\langle2\hat{I}_z\hat{S}_z\right\rangle(\tau)=-\delta_{S,IS}\tau\left(\left\langle\hat{S}_z\right\rangle-S_z^{eq}\right) \tag{10.3c}$$

The two-spin order can be made observable by a selective 90° pulse on either the S- or the I-spin. The CCRR will have the effect of making the relaxation of the two components in the doublet different, as can be seen in Figure 10.1.

In order to quantify the CCRR, it is very useful to be able to exclusively detect only the build-up of two-spin order terms, due to cross-correlation effects. For that purpose, a double-quantum (DQ) filter can be used (Figure 10.2), as suggested by Jaccard *et al.*[8]. This works as follows. After the initial 180° pulse and the relaxation (mixing) period, various types of longitudinal order are present, with different weights. After a 90_y° pulse on the S-spins, the following density operator is obtained:

$$-a\hat{S}_z-b\hat{I}_z-c2\hat{I}_z\hat{S}_z\xrightarrow{90^\circ\hat{S}_y}-a\hat{S}_x-b\hat{I}_z-c2\hat{I}_z\hat{S}_x \tag{10.4}$$

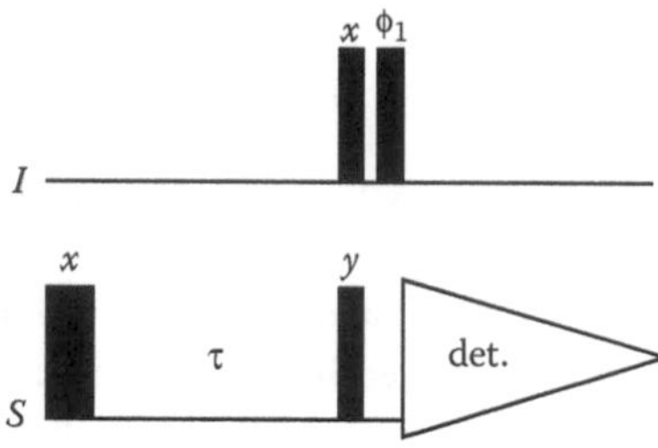

FIGURE 10.2 Heteronuclear multiple-quantum filtered inversion-recovery for measuring DD/CSA cross-correlation. Two experiments are performed, with $\phi_1 = x$ or $-x$. The sum gives the decay of the $\hat{S}_z$ magnetisation, and the difference gives the build-up and decay of longitudinal two-spin order, $2\hat{I}_z\hat{S}_z$.

If the S-spin magnetisation were detected at this stage, a superposition of the in-phase term and the antiphase term would be obtained. Instead of applying a 90°_y pulse on the S-spins (as was done in Figure 10.1), we can use a double-quantum filter, consisting of the sequence $((90^\circ_y)_S,(90^\circ_x)_I-\Delta-(90^\circ_\phi)_I)$, where ϕ denotes a two-step phase cycle. This allows us to discriminate between the in-phase and antiphase terms. The first 90°_x I pulse of the double-quantum filter yields:

$$-a\hat{S}_x - b\hat{I}_z - c2\hat{I}_z\hat{S}_x \xrightarrow{90^\circ \hat{I}_x} -a\hat{S}_x + b\hat{I}_y + c2\hat{I}_y\hat{S}_x \tag{10.5}$$

The last term represents a combination of double- and zero-quantum coherences and is not directly observable. The interval, Δ, is a short delay (typically only a few microseconds, short enough to safely neglect any evolution) inserted to allow for a change of the transmitter phase. By applying a second 90° pulse on the I-spins with the phase x, this last term in Eq. (10.5) is converted into the antiphase S-spin coherence:

$$-a\hat{S}_x + b\hat{I}_y + c2\hat{I}_y\hat{S}_x \xrightarrow{90^\circ \hat{I}_x} -a\hat{S}_x + b\hat{I}_z + c2\hat{I}_z\hat{S}_x \tag{10.6}$$

By the use of a simple phase cycling procedure, replacing the final 90°_x by a 90°_{-x} pulse in the double-quantum filter, and subtracting the outcome of this from the previous result, we obtain:

$$-2b\hat{I}_z - 2c2\hat{I}_z\hat{S}_x \tag{10.7}$$

The subtraction of large observable terms (such as the $\hat{S}_x$ in Eq. (10.6)) to observe small effects from the creation of two-spin order is sometimes known as *difference spectroscopy*. This experiment is also known as the *double-quantum filtered inversion-recovery*. The two-spin order, due to the cross-correlation effect, is in this case detected by observing exclusively the antiphase term in Eq. (10.6). In order to extract information about the cross-correlated relaxation rates, build-up curves for the creation of two-spin order from the $-\hat{S}_z$ state are measured in arrayed experiments using different (short) τ values and estimating the derivative with respect to τ at the limit $\tau \to 0$. If desired, the antiphase S-doublet can be converted into an in-phase doublet by a $1/4J_{IS}-180^\circ(I,S)-1/4J_{IS}$ sandwich in the same way as in refocused INEPT. This will allow the use of proton decoupling during the S-signal acquisition, leading to an improved signal-to-noise ratio[9].

By adding the spectra from the $(90^\circ_x-\Delta-90^\circ_x)$ and $(90^\circ_x-\Delta-90^\circ_{-x})$ sequences, one instead obtains $-2\hat{S}_x$, which can be used to derive the spin-lattice relaxation rate for the S-spin. It is worth noting at this point that in the standard S-spin (^{13}C or ^{15}N) inversion-recovery experiment, the cross-correlated relaxation will not lead to any observable terms as long as the I-spins (protons) are continuously irradiated during the τ delay. The decoupling effect, noticed in the context of the Solomon equations in Section 3.2, is thus equally valid in the case of systems described by the extended Solomon equations.

Homonuclear cross-correlation effects can be measured in a similar way by applying a selective inversion pulse on one of the spins involved in the interaction. The double-quantum filter does not need to be selective for either spin, as it can easily be shown that from the simple phase cycling, or difference spectroscopy, only the two-spin order term will be retained (the reader is encouraged to check this). Two-dimensional homonuclear relaxation experiments, such as the double-quantum filtered NOESY[10], function in an analogous way and can also be used.

A similar approach for retaining the antiphase term in the homonuclear experiment is to perform an inversion-recovery experiment with a selective inversion of one spin, followed by a relaxation delay, τ, and a single, hard detection pulse. If a hard read-pulse with a 90° flip angle is used, the two-spin order will be converted into non-observable double-quantum and zero-quantum coherence, $-2\hat{I}_x\hat{S}_x$. If a hard read-pulse with a smaller flip angle, say 45°, is applied instead, the antiphase term is retained[10], since only part of the two-spin order is converted into unobservable terms according to:

$$\begin{aligned} -a\hat{S}_z - b\hat{I}_z - c2\hat{I}_z\hat{S}_z \xrightarrow{45^\circ(\hat{I}_y+\hat{S}_y)} -\tfrac{1}{\sqrt{2}}a\hat{S}_z - \tfrac{1}{\sqrt{2}}a\hat{S}_x - \tfrac{1}{\sqrt{2}}b\hat{I}_z - \tfrac{1}{\sqrt{2}}b\hat{I}_x - \\ -\tfrac{1}{2}c2\hat{I}_z\hat{S}_z - \tfrac{1}{2}c2\hat{I}_x\hat{S}_z - \tfrac{1}{2}c2\hat{I}_z\hat{S}_x - \tfrac{1}{2}c2\hat{I}_x\hat{S}_x \end{aligned} \tag{10.8}$$

The observable terms in this expression will lead to an S-spin doublet in which the two components will relax differently due to the presence of the antiphase term. Thus, a series of experiments with increasing τ-values can be used to evaluate the cross-correlation rate. This technique was applied, for example, in a study by Mäler *et al.* (Figure 10.3) for a small organic molecule, *cis*-chloroacrylic acid[11]. A further comment concerning the homonuclear cross-correlation should be made. An important reflection was made in the work of Dalvit and Bodenhausen[10], who stated that although these effects have been known

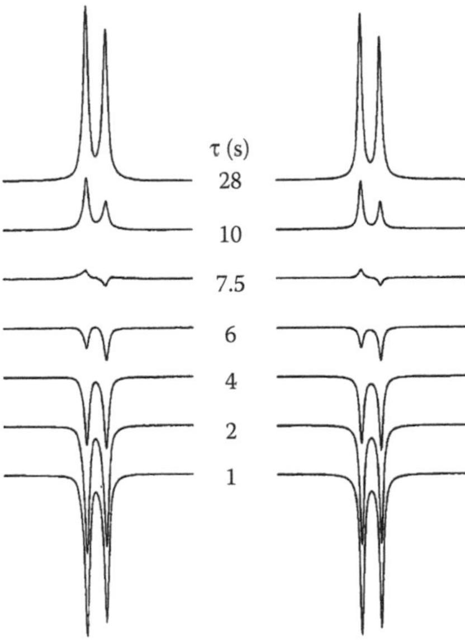

FIGURE 10.3 Different relaxation of the two components of one of the doublets in *cis*-chloroacrylic acid due to DD/CSA cross-correlation. The left-hand side shows experimental spectra, and the right-hand side shows simulated spectra after a selective inversion and a 45° monitoring pulse. (Reprinted with permission from Mäler, L. et al., *J. Magn. Reson. Ser. A*, 117, 220–227, 1995. Copyright (1995) Elsevier.)

to exist for a long time, the observations of their fingerprints have not been very common. The reason for this is simply the masking nature of the hard 90° pulse applied to the protons, which makes the cross-correlation effect invisible. Not until selective pulses, or observation pulses with different flip angles, are used do we observe the cross-correlation effects.

10.1.2 The Full Relaxation Matrix Approach

For a simple, isolated two-spin system in a small molecule, it is not necessary to make use of the initial rate approach. Instead, the full relaxation matrix method can be used[12]. In this approach, all of the longitudinal relaxation rates are measured at one particular τ_m value by a combination of selective or semi-selective two-dimensional NOESY experiments (Figure 10.4). Three different experiments are performed: (1) on the S-spin selectively, (2) on the I-spin selectively and (3) by applying the first two pulses on the I-spin, followed by the third pulse and the detection of the S-spin. The results of these experiments can be described, briefly, as selectively recording the two diagonal peaks and the cross-peak between I and S in a NOESY experiment (Figure 9.5). As mentioned previously, the hard pulses affecting all spins equally will render the terms resulting from cross-correlated relaxation invisible. When the semi-selective approach is used, asymmetries in the two-dimensional multiplet patterns will result due to cross-correlation effects.

It is useful, in this context, to talk about *magnetisation transfer modes*, which simply relate to the different transfer processes that can occur during the mixing time, τ_m, of the NOESY experiments. For instance, the transfer mode $\hat{S}_z \rightarrow \hat{S}_z$ is governed by the auto-relaxation rate ρ_S, while the transfer mode $\hat{S}_z \rightarrow 2\hat{I}_z\hat{S}_z$ is determined by the cross-correlated relaxation rate, $\delta_{S,IS}$, all according to Eq. (10.1). It is thus fairly straightforward to relate the contribution of the relaxation processes to the spectra obtained by the three different experiments. The intensities of the components of the "diagonal" peaks will thus contain information about the auto-relaxation rates (ρ_S, ρ_I) and about the cross-correlated relaxation rate ($\delta_{S,IS}$), while the "cross-peak" is influenced by the cross-relaxation rate, σ_{IS}, the cross-correlated relaxation rate and the two-spin order auto-relaxation rate. By the use of symmetry projection operators, corresponding to the different transfer modes, it is possible to obtain the total relaxation matrix in Eq. (10.1) from the multiplet component intensities. The method only works well for well-defined two-spin systems. Naturally, if the two spins are different species, the three NOESY experiments are simply

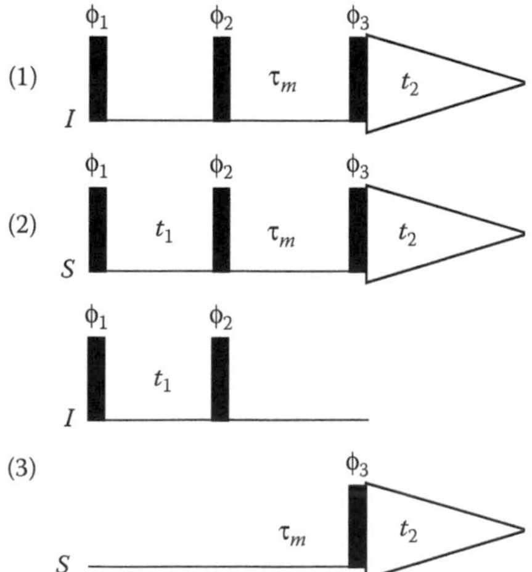

FIGURE 10.4 The NOESY/HOESY approach for measuring cross-correlation. The relaxation matrix for the two-spin system, IS, is obtained from transfer-mode analysis of normalised peak intensities from the three experiments. The phase cycle is the same as in the regular NOESY experiment (Figure 9.4).

performed as a heteronuclear double-resonance experiment[13], while selective pulses must be employed in the case of a proton spin system[12].

10.1.3 The Homogeneous Master Equation

A different approach was taken by Levitt and Di Bari[14,15], who demonstrated that by applying suitably spaced 180° pulses, it is possible to obtain a steady-state condition involving multispin longitudinal spin orders. These spin orders are created by cross-correlated relaxation, and experiments were designed to detect and evaluate cross-correlated relaxation parameters. The discussion below is based closely on the original work by Di Bari and Levitt[14,15]. To explain the observations, the *homogeneous master equation* (HME) approach was introduced. The master equation, or the Liouville–von Neumann equation, is given by

$$\frac{d}{dt}\hat{\rho} = -i\hat{\hat{L}}_0\hat{\rho} - \hat{\hat{\Gamma}}\left(\hat{\rho} - \hat{\rho}^{eq}\right) \tag{10.9}$$

in which $\hat{\hat{L}}_0$ is the Liouvillian (commutator with the Hamiltonian $\hat{H}_0$) describing the coherent interactions, the spin-spin and spin-field interactions. $\hat{\hat{\Gamma}}$ is the relaxation superoperator.

Eq. (10.9), in which the term $\hat{\rho}^{eq}$ has been added to account for the return to equilibrium, is inhomogeneous in the sense that the two superoperators act on different operators. The coherent part of the equation applies to the full density operator, $\hat{\rho}$, while the incoherent part, or relaxation, applies to the deviation from equilibrium. It is, however, possible to overcome this problem by replacing the relaxation superoperator by an "adjusted" superoperator, $\hat{\hat{\Upsilon}}$, suggested by Jeener[16], such that the homogeneous master equation can be written:

$$\frac{d}{dt}\hat{\rho} = -\left(i\hat{\hat{L}}_0 + \hat{\hat{\Upsilon}}\right)\hat{\rho} \tag{10.10}$$

in which $\hat{\hat{\Upsilon}} = \hat{\hat{\Gamma}} + \hat{\hat{\Theta}}$ under the high-temperature approximation. We now need to find an expression for the adjusted superoperator. Consider a spin system with eigenstates $|r\rangle$ and energy levels ω_r (in angular frequency units), defined so that the mean energy is zero. Next, we introduce the population operator (projector) $|P_r\rangle = |r\rangle\langle r|$ and a superoperator $\hat{\hat{P}} = |P_r\rangle\langle P_r|$. The matrix elements of the relaxation superoperator, $\langle P_r|\hat{\hat{\Gamma}}|P_s\rangle$, are the transition probabilities, W_{rs}. From recognising that the transition probability W_{rs} differs from W_{sr} by a small factor, $\exp\{(\omega_r - \omega_s)\tau_\Theta\}$, where $\tau_\Theta = \hbar/kT$, a new adjusted superoperator can be suggested as:

$$\hat{\hat{\Upsilon}} = \hat{\hat{\Gamma}}\exp\left(\hat{\hat{\omega}}\tau_\Theta\right) \tag{10.11}$$

where $\hat{\hat{\omega}} = \sum \omega_r \hat{\hat{P}}_r$. The thermal correction becomes $\hat{\hat{\Theta}} = \hat{\hat{\Gamma}}\hat{\hat{\omega}}\tau_\Theta$, an apparently simple expression, which, however, turns out to be less suitable for further analysis. After some additional manipulations[15], it can be shown that the effect of the addition of the correction term is to expand the equation of motion of the longitudinal magnetisation modes by adding the unity operator $\frac{1}{2}\hat{1}$. In a two-spin system, *IS*, the expanded equation of motion for the longitudinal spin orders becomes:

$$\frac{d}{dt}\begin{pmatrix} \langle \frac{1}{2}\hat{1} \rangle \\ \langle \hat{I}_z \rangle \\ \langle \hat{S}_z \rangle \\ \langle 2\hat{I}_z\hat{S}_z \rangle \end{pmatrix} = -\begin{pmatrix} 0 & 0 & 0 & 0 \\ \theta_I & \rho_I & \sigma_{IS} & \delta_{I,IS} \\ \theta_S & \sigma_{IS} & \rho_S & \delta_{S,IS} \\ \theta_{IS} & \delta_{I,IS} & \delta_{S,IS} & \rho_{IS} \end{pmatrix}\begin{pmatrix} \langle \frac{1}{2}\hat{1} \rangle \\ \langle \hat{I}_z \rangle \\ \langle \hat{S}_z \rangle \\ \langle 2\hat{I}_z\hat{S}_z \rangle \end{pmatrix} \tag{10.12}$$

The new elements of the relaxation matrix are:

$$\theta_I = -\tfrac{1}{2}\left(\rho_I\omega_I + \sigma_{IS}\omega_S\right)\tau_\theta \tag{10.13a}$$

$$\theta_S = -\tfrac{1}{2}\left(\rho_S\omega_S + \sigma_{IS}\omega_I\right)\tau_\theta \tag{10.13b}$$

$$\theta_{IS} = -\tfrac{1}{2}\left(\delta_{I,IS}\omega_I + \delta_{S,IS}\omega_S\right)\tau_\theta \tag{10.13c}$$

where ω_I and ω_S are the Larmor frequencies of the two spins. A schematic picture of the meaning of these terms is given in Figure 10.5. The three new terms represent the creation of spin order from the environment, represented by the unity operator. The different relaxation rates characterise transfers between the three longitudinal spin states (σ_{IS}, $\delta_{I,IS}$ and $\delta_{S,IS}$) or dissipation of spin order (ρ_I, ρ_S and ρ_{IS}).

The beauty of the approach becomes apparent if one considers the following experiment. Suitably separated 180° pulses are applied rapidly to both the I- and the S-spins. The spin operators transform under such a pulse sequence as $\frac{1}{2}\hat{1} \rightarrow \frac{1}{2}\hat{1}$, $\hat{I}_z \rightarrow -\hat{I}_z$, $\hat{S}_z \rightarrow -\hat{S}_z$ and $2\hat{I}_z\hat{S}_z \rightarrow 2\hat{I}_z\hat{S}_z$. The transformations can be divided into a *gerade* subspace, for operators that do not change sign, and an *ungerade* subspace, for operators that do change sign. This is equivalent to the discussion in Section 5.1 (see Eqs. (5.7)–(5.9)). The spin dynamics, due to the rapid pulsing, in the *gerade* subspace, containing the $\frac{1}{2}\hat{1} \rightarrow \frac{1}{2}\hat{1}$ and $2\hat{I}_z\hat{S}_z \rightarrow 2\hat{I}_z\hat{S}_z$ transfer modes, is given by the terms θ_{IS} and ρ_{IS} (see Figure 10.5). Hence, a steady state for the two-spin order is built up, governed by the two rates according to:

$$\frac{\left\langle 2\hat{I}_z\hat{S}_z \right\rangle}{\left\langle S_z^{eq} \right\rangle} = \frac{\delta_{I,IS}\omega_I + \delta_{S,IS}\omega_S}{\rho_{IS}\omega_S} \tag{10.14}$$

This remarkable result shows that it is possible to create a steady state of two-spin order resulting from cross-correlated relaxation through the relation in Equation 10.14. In the same manner, the spin dynamics in the *ungerade* subspace can be analysed and one finds, as expected, that the two one-spin orders,

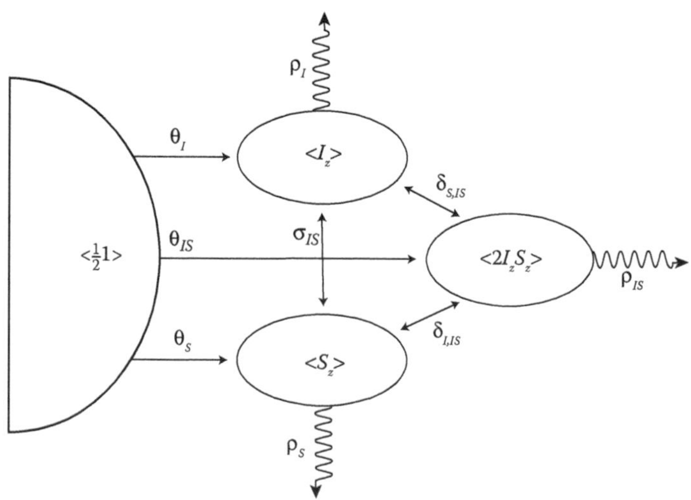

FIGURE 10.5 A physical interpretation of the different elements in the relaxation matrix for a two-spin system, *IS*. (Adapted from Levitt, M.H. and Di Bari, L., *Phys. Rev. Lett.*, 69, 3124–3127, 1992, and *Bull. Magn. Reson.*, 16, 94–114, 1994.)

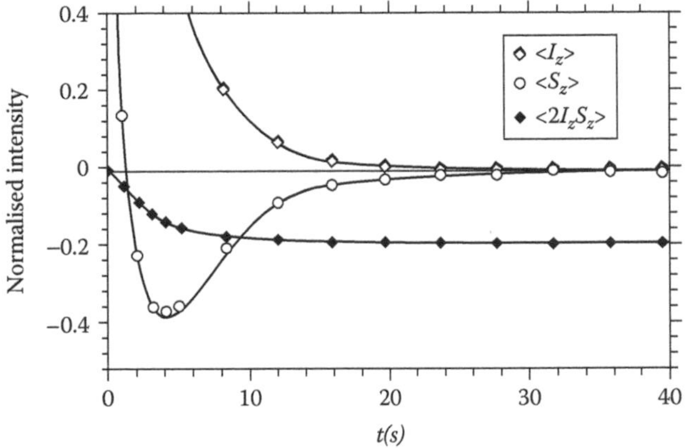

FIGURE 10.6 Trajectories of the two one-spin order terms $\langle \hat{I}_z \rangle$ and $\langle \hat{S}_z \rangle$ and the two-spin order $\langle 2\hat{I}_z\hat{S}_z \rangle$ as a function of time during a sequence of 180° pulses applied to both spins obtained for ^{13}C-labelled $CHCl_3$. (Reprinted with permission from Levitt, M. H. and Di Bari, L., *Phys. Rev. Lett.*, 69, 3124–3127, 1992. Copyright (1992) by the American Physical Society.)

$\langle \hat{I}_z \rangle$ and $\langle \hat{S}_z \rangle$, decay to zero with the rate constants ρ_I and ρ_S. Figure 10.6 shows results obtained for ^{13}C-labelled $CHCl_3$, which could be analysed to yield information about the various relaxation rate parameters, and in particular CSA values, from cross-correlated relaxation rates.

In summary, the measurements of longitudinal CCRRs require an asymmetric preparation of the spin system, followed by a relaxation period during which various transfers occur. In the subsequent detection period, the build-up of another component of the density operator is either measured selectively or can be isolated in the post-processing.

10.1.4 Cross-Correlated Dipolar Relaxation

The same philosophy can be applied to multispin dipolar relaxation phenomena, described in Section 5.1. Consider a heteronuclear *AMX* spin system, discussed in Section 5.1. An important example of such a system is the $^{13}CH_2$ methylene group, discussed in a pedagogic manner by Ghalebani *et al.*[17]. Let us assume that the two protons are characterised by equal (or very similar) *J*-couplings to the carbon-13 but are not magnetically equivalent. We can then identify the *A*-spin with the carbon-13, while *M* and *X* are mutually weakly coupled protons. An asymmetric preparation in this case might imply a carbon-13 inversion. In the absence of cross-correlation effects, the four lines (indeed, three in the 1:2:1 if the two $^1J_{CH}$ are equal) in the proton-coupled carbon-13 spectrum will then recover as a symmetric multiplet. In the presence of cross-correlated longitudinal relaxation, some multiplet asymmetries will evolve. For example, the build-up of the three-spin order, $4\hat{I}_z^A\hat{I}_z^M\hat{I}_z^X$, will lead to a different recovery rate of the outer lines of the multiplet compared with the inner lines. The results of a proton-coupled inversion-recovery experiment for the methylene group carbon-13 in the sugar trehalose are shown in Figure 10.7. We can see that the three lines in the triplet display slightly different relaxation behaviour. The central line relaxes faster towards a positive intensity after the inversion than the outer lines, while at the same time the two outer lines also show a certain degree of differential relaxation. The difference between the outer lines is, in the initial rate regime, a signature in the DD/CSA interference, while the deviation in the relaxation behaviour of the central line with respect to the mean of the outer lines indicates the interference of the two dipolar ^{13}C-^{1}H interactions. The theory of this phenomenon was presented in Section 5.1. Let us now for the moment concentrate on the dipolar interactions only and neglect the

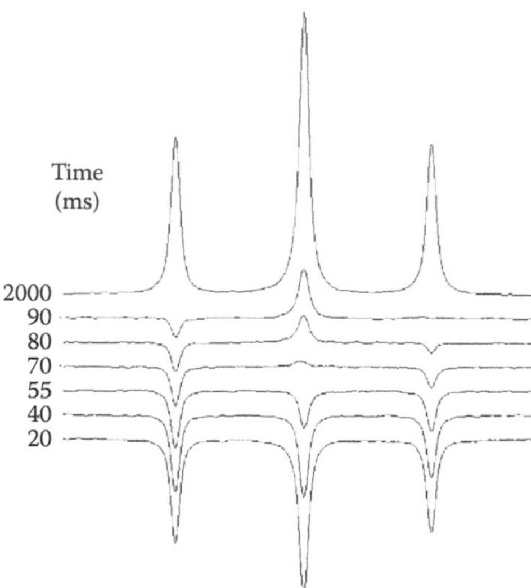

FIGURE 10.7 Selected traces from a proton-coupled carbon-13 inversion-recovery experiment on the methylene group in trehalose. The numbers on the left indicate the relaxation delay τ (compare Figure 8.2). (Reprinted with permission from Ghalebani, L., et al., *Concepts Magn. Reson.* 30A, 100–115, 2007. Copyright (2007) Wiley-VCH Verlag.)

DD/CSA cross-correlations. The relaxation matrix, expressed in the basis of the magnetisation modes defined in Eqs. (5.10) and (5.11), takes the form:

$$\frac{d}{dt}\begin{pmatrix} \langle \hat{A}_z \rangle \\ \langle \hat{M}_z \rangle \\ \langle \hat{X}_z \rangle \\ \langle 4\hat{A}_z\hat{M}_z\hat{X}_z \rangle \end{pmatrix} = -\begin{pmatrix} \rho_A & \sigma_{AM} & \sigma_{AX} & \delta_{A,AMX} \\ \sigma_{AM} & \rho_M & \sigma_{MX} & \delta_{M,AMX} \\ \sigma_{AX} & \sigma_{MX} & \rho_X & \delta_{X,AMX} \\ \delta_{A,AMX} & \delta_{M,AMX} & \delta_{X,AMX} & \rho_{AMX} \end{pmatrix}\begin{pmatrix} \langle \hat{A}_z \rangle - A_z^{eq} \\ \langle \hat{M}_z \rangle - M_z^{eq} \\ \langle \hat{X}_z \rangle - X_z^{eq} \\ \langle 4\hat{A}_z\hat{M}_z\hat{X}_z \rangle \end{pmatrix} \quad (10.15)$$

The symbols ρ and σ have the same meaning as in Section 10.1.3, while the cross-correlated relaxation rate $\delta_{A,AMX}$ is identical to ${}^a\Gamma_{14}$, given in Eq. (5.15), and the other δs are given by similar expressions. Ghalebani *et al.*[17] described a simple experimental protocol in which the CCRR was determined in the initial rate regime by measuring the relaxation rates for the three lines of the triplet in a proton-coupled inversion-recovery experiment:

$$\delta_{A,AMX} = \tfrac{1}{2}\left(\tfrac{1}{2}\left(W_o^+ + W_o^-\right) - W_i\right) \quad (10.16)$$

The symbols W_o^+ and W_o^- denote the relaxation rates for the two outer lines and W_i for the inner line. The authors recommend that the rates should be determined by exponential fitting of the line intensities after a judicious choice of the set of relaxation delays.

Various combinations of preparation and detection techniques, suitable for systems with magnetic equivalence, have been reviewed by Canet[4] and by Kumar and co-workers[5]. Other types of cross-correlation effects, related to quadrupolar or paramagnetic interactions, will be discussed in Chapters 14 and 15.

10.2 Cross-Correlated Transverse Relaxation

Let us now go back to the simple two-spin system. The transverse cross-correlated relaxation rates for the *S*-spin can also be measured in a straightforward way by in fact using the standard spin-echo pulse sequence for measuring R_2 relaxation rates. The relaxation period contains a refocusing 180° pulse on the *S*-spin that serves to refocus the evolution under *J*-couplings and chemical shift. This 180° *S* pulse does not affect, however, the fact that scalar couplings average the auto-relaxation rates for the in-phase and antiphase coherences present for the *S*-spin (compare Eq. (8.8)). In a similar way as described for longitudinal cross-correlation, cross-correlated relaxation (in this context, the rather clumsy notation *cross-correlated cross-relaxation* might, in fact, be useful!) between the in-phase $\hat{S}_y$ state and the antiphase $2\hat{I}_z\hat{S}_y$ state can occur.

10.2.1 General

Consider a case in which both the dipole-dipole and CSA relaxation mechanisms are important, as discussed for the longitudinal cross-correlation in the previous section. The corresponding equation for describing the time evolution of the density operator for the transverse coherences can thus be written as

$$\frac{d}{dt}\begin{pmatrix} \langle \hat{S}_y \rangle \\ \langle 2\hat{I}_z\hat{S}_y \rangle \end{pmatrix} = -\begin{pmatrix} \overline{R_2} & \eta_{S,IS} \\ \eta_{S,IS} & \overline{R_2} \end{pmatrix}\begin{pmatrix} \langle \hat{S}_y \rangle \\ \langle 2\hat{I}_z\hat{S}_y \rangle \end{pmatrix} \tag{10.17}$$

in which $\overline{R_2} = 1/2(R_{2S} + R_{2IS})$. R_{2S} is the transverse auto-relaxation rate of the in-phase magnetisation, equal to the sum of the dipolar and CSA terms (compare Eqs. (3.35) and (5.40)), while R_{2IS} is the auto-relaxation rate of the antiphase magnetisation:

$$\begin{aligned} R_{2IS} = {} & \frac{\pi}{10} b_{IS}^2 \left[4J(0) + J(\omega_I - \omega_S) + 3J(\omega_S) + 6J(\omega_I + \omega_S) \right] + \\ & + \frac{2\pi}{45} c_S^2 \left[4J(0) + 3J(\omega) \right] \end{aligned} \tag{10.18}$$

where b_{IS} is the dipole-dipole interaction strength constant, defined in Eq. (3.5), and c_S is the CSA interaction strength constant for the *S*-spin, equal to $\gamma_S B_0 \Delta\sigma_S$.

The average transverse relaxation rate should be compared to Eq. (8.8), in which it was seen that, indeed, a mixture of the two relaxation rates is obtained. In the Carr, Purcell, Meiboom, and Gill (CPMG) scheme, the pulsing can be made rapid enough to ensure that the measured $\overline{R_2}$ is dominated by R_{2S}.

The symbol $\eta_{S,IS}$ in Eq. (10.17) is the transverse cross-correlated relaxation rate, given by:

$$\eta_{S,IS} = -\frac{2\pi}{15}\left(\frac{\mu_0}{4\pi}\right)\gamma_I \gamma_S^2 \hbar B_0 \Delta\sigma_S r_{IS}^{-3} \frac{1}{2}\left(3\cos^2\theta - 1\right)\left[4J(0) + 3J(\omega_S)\right] \tag{10.19}$$

As mentioned in Section 8.2, it is often desirable to suppress the cross-correlation effects while measuring the transverse relaxation, in order to obtain "clean" R_{2S}. This can be done by inserting a set of suitably spaced *I*-spin 180° pulses into the CPMG sequence (*cf.* Figures 8.8 and 8.10).

Starting from Eq. (10.17), and following the discussion in Section 5.2, it can be seen that the transverse relaxation rates for the two components in a doublet are given by:

$$R_2^{\alpha} = \overline{R_2} - \eta_{S,IS} \tag{10.20a}$$

$$R_2^{\beta} = \overline{R_2} + \eta_{S,IS} \tag{10.20b}$$

in which α and β refer to the two components of the doublet. The difference in decay rates of the two components of the doublet is thus given by $2\eta_{S,IS}$ (compare Eq. (5.46)). Interestingly, if the cross-correlated relaxation rate in Eq. (10.20) is similar in size to the auto-relaxation rate, a significant line-narrowing effect is obtained for one of the doublet components, while the other component is severely broadened. This effect is utilised in the transverse relaxation-optimised spectroscopy (TROSY) experiment, proposed in the 1990s by Pervushin *et al.*[18], which was designed to increase resolution for large molecules with broad lines. This will be further discussed in the next section. Thus, as mentioned in Section 5.2, direct observations of transverse cross-correlation can be made by examining the differential line-broadening in doublet signals. The difference in the linewidth of the two doublet components can also be measured by the standard CPMG technique for a spin system consisting of two different spins, or by the use of selective excitation and inversion (or rather, refocusing) of protons in a homonuclear spin system. The same basic spin-echo technique as presented in Figure 8.7 will reveal asymmetry in the transverse relaxation of two components in a doublet, if no decoupling is applied. In principle, transverse cross-correlated relaxation is measured by generating a non-equilibrium transverse state, represented by $\hat{S}_y$ or $2\hat{I}_z\hat{S}_y$ and monitoring their inter-conversion in build-up curves from arrayed experiments.

Another important consequence of transverse cross-correlation is so-called *relaxation-allowed coherence transfer*. This phenomenon leads to the possibility of coherence transfer between different spins involved in cross-correlation during the evolution period in, for instance, correlation spectroscopy (COSY). These effects may be confused with the true cross-peaks based on *J*-couplings between spins[19,20].

10.2.2 Transverse Cross-Correlation in Proteins

Measurements of transverse cross-correlation effects in nitrogen-15-labelled proteins have become popular, since they can provide a means for quantifying the nitrogen-15 CSA values[21]. In addition, by combining R_1, R_2 and σ_{IS} with measurements of both longitudinal and transverse cross-correlated relaxation, one can obtain insights into the contribution of chemical exchange to R_2 relaxation rates[22]. A pulse sequence for measuring transverse CCRR for nitrogen-15-labelled proteins has been developed by Tjandra *et al.*[21]. The approach for measuring transverse cross-correlation is to generate a non-equilibrium density operator proportional to $2\hat{I}_z\hat{S}_y$, *i.e.* antiphase *S*-spin magnetisation, at the beginning of the relaxation period of the experiment. This is simply done by applying a (gradient-enhanced) INEPT sequence, as in the beginning of the pulse sequence shown in Figure 10.8. The relaxation period contains a nitrogen-15 180° pulse in the middle that serves to refocus evolution under chemical shifts and *IS J*-couplings, as discussed earlier in this section. Transfer of magnetisation from $2\hat{I}_z\hat{S}_y$ to $\hat{S}_y$ takes place during the relaxation period; at the end of the relaxation period, the density operator will be given by:

$$2\hat{I}_z\hat{S}_y\exp(-\overline{R_2}\tau)\cosh\left(\eta_{S,IS}\tau\right)+\hat{S}_y\exp(-\overline{R_2}\tau)\sinh\left(\eta_{S,IS}\tau\right) \tag{10.21}$$

It is straightforward to devise a route to create observable proton magnetisation from these terms by employing a reverse INEPT scheme. If we, however, consider carefully what it is we want to achieve, we realise that this will not lead to a method for determining $\eta_{IS,S}$, which is our goal. We need a final result that will depend only on τ and $\eta_{IS,S}$, which we will be able to obtain from an arrayed experiment as usual. If we make two experiments for each time point τ, one that rejects the first part of the expression in Eq. (10.21) and one that rejects the second part, this should lead to a way of directly measuring the cross-correlated relaxation. This can be done easily with the experimental scheme in Figure 10.8. After the relaxation delay, τ, the *S*-spin magnetisation is returned to the *z*-axis by applying a 90° pulse on the *S*-spins, yielding a combination of the two-spin order $2\hat{I}_z\hat{S}_z$ and $\hat{S}_z$.

The experiment can after this be executed in two ways, making it possible to detect either of the two terms in Eq. (10.21). In the first experiment, a 90° pulse on the *I*-spins is applied simultaneously with the *S*-spin pulse, followed by a gradient. The pulses create antiphase *I*-spin magnetisation, $2\hat{I}_y\hat{S}_z$, which

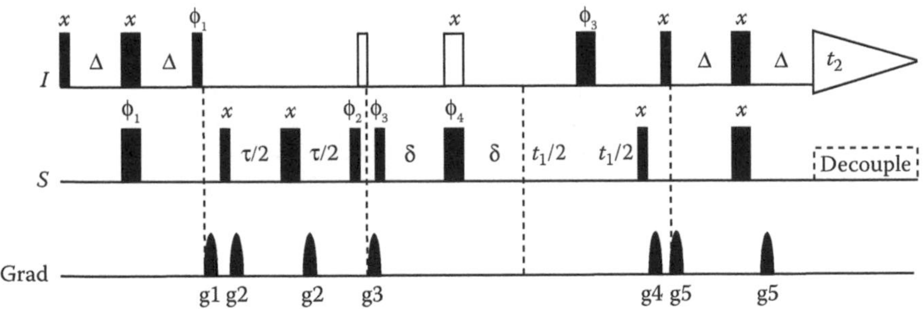

FIGURE 10.8 Two-dimensional correlation experiment for measuring transverse cross-correlated relaxation parameters. The delays are: $\Delta \leq 1/4J_{IS}$, τ is the arrayed relaxation delay, $\delta = 1/4J_{IS}$. The open bars indicate pulses that are employed in one of the two experiments carried out. The phase cycle is $\phi_1 = y, -y$; $\phi_2 = x, x, -x, -x$; $\phi_3 = x$; $\phi_4 = 4(x), 4(y), 4(-x), 4(-y)$; $\phi_5 = x$ and $\phi_{rec} = x, -x, -x, x, -x, x, x, -x$. Quadrature in the indirect dimension is achieved by incrementing ϕ_3 according to the States–time-proportional phase increment (TPPI) method. The duration of the gradients are g1 = 2.75 ms; g2 = 2 ms; g3 = 1 ms; g4 = 1.5 ms; and g5 = 0.4 ms, using the same amplitude for all gradients. (Adapted from Tjandra, N. et al., *J. Am. Chem. Soc.*, 118, 6986–6991, 1996).

is dephased by the gradient. The $\hat{S}_z$ term, on the other hand, remains intact, since the populations are not affected by the gradient. Next, $\hat{S}_z$ is converted into $\hat{S}_y$ magnetisation by yet another S-spin 90° pulse. The subsequent delay, 2δ, is chosen to maximise the conversion of $\hat{S}_z$ into the antiphase magnetisation through the J_{IS}-coupling between the two spins. Refocusing 180° pulses on both spins are used at the midpoint of the delay. Thus, the $\hat{S}_y$ evolves during 2δ as:

$$\hat{S}_y \xrightarrow{2\delta} \hat{S}_y \cos(\pi J_{IS} 2\delta) - 2\hat{I}_z\hat{S}_x \sin(\pi J_{IS} 2\delta) \tag{10.22}$$

and, if $2\delta = 1/2J_{IS}$, the $2\hat{I}_z\hat{S}_x$ term is maximised while the $\hat{S}_y$ term disappears. The rest of the pulse sequence is the same as the conventional heteronuclear single-quantum coherence (HSQC), just as described for the T_1 or T_2 relaxation measurements. In the second experiment, the 90° pulse on the I-spins is not applied. Remembering that it served to convert the two-spin order into antiphase I-spin coherence to be dephased by a gradient, we now see that the two-spin order in the second experiment remains, unaffected by the gradient. The next 90° pulse on the S-spin creates in-phase, $\hat{S}_y$, and antiphase $2\hat{I}_z\hat{S}_y$ coherence. If the refocusing pulse on the I-spins is removed, the result after the delay 2δ is this time given by:

$$\hat{S}_y + 2\hat{I}_z\hat{S}_y \xrightarrow{2\delta} \hat{S}_y + 2\hat{I}_z\hat{S}_y \tag{10.23}$$

i.e. both chemical shift and J-couplings are refocused by this scheme. Using a refocusing I-spin pulse during the subsequent t_1 evolution period will result in only the antiphase S-spin coherence contributing to the detected signal.

In summary, the two schemes provide a way of separately measuring the transfer of magnetisation from $2\hat{I}_z\hat{S}_y$ to $\hat{S}_y$ (the first experiment) and from $2\hat{I}_z\hat{S}_y$ to $2\hat{I}_z\hat{S}_y$ (the second experiment). The integrated intensity of the cross-peaks is of course related to the relaxation, as described by Eq. (10.21), and the ratio of the intensities obtained from experiment one, I_A, and two, I_B, is therefore given by:

$$\frac{I_A}{I_B} = \frac{\langle S_y \rangle(\tau)}{\langle 2I_zS_y \rangle(\tau)} = \frac{\exp(-\overline{R_2}\tau)\sinh(\eta_{I,IS}\tau)}{\exp(-R_2\tau)\cosh(\eta_{I,IS}\tau)} = \tanh(\eta_{I,IS}\tau) \tag{10.24}$$

The time dependence of this intensity ratio thus gives us the transverse cross-correlation rate. An example obtained for uniformly ^{15}N-labelled ubiquitin is shown in Figure 10.9.

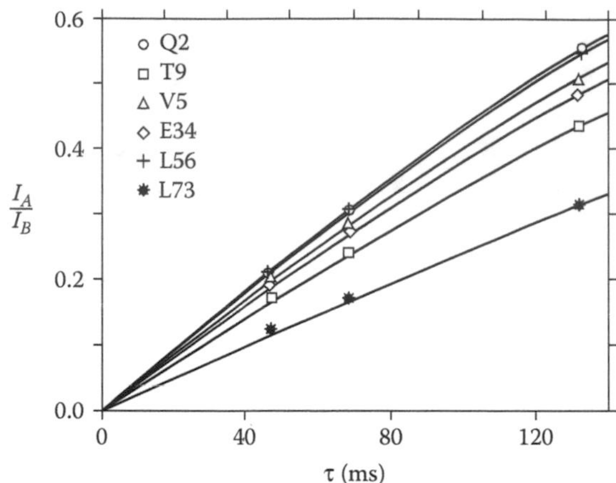

FIGURE 10.9 Plots of intensity ratios for different ^{15}N-^{1}H backbone spin pairs in ubiquitin obtained from the experiments recorded with the pulse sequence in Figure 10.7. The intensity ratios equal tanh(2τη), and the lines correspond to this relation. (Reprinted with permission from Tjandra, N. et al., *J. Am. Chem. Soc.*, 118, 6986–6991, 1996. Copyright (1996) American Chemical Society.)

In many applications, it is useful to measure both the longitudinal cross-correlated relaxation rate and the transverse CCRR. Comparing Eqs. (10.2) and (10.19), we notice that the two CCRRs differ by the presence of $J(0)$ in the transverse rate. Thus, a comparison of the two rates allows more accurate determination of the influence of slow dynamics on measured T_2 rates. In a very similar approach, the two longitudinal transfer rates corresponding to the $2\hat{I}_z\hat{S}_z \rightarrow \hat{S}_z$ (cross rate) and $2\hat{I}_z\hat{S}_z \rightarrow 2\hat{I}_z\hat{S}_z$ (auto rate) can be measured by two separate experiments[22] (the derivation of the experiment will not be given here). The cross-peak intensities are given by an expression completely analogous to Eq. (10.24) and can be used in an arrayed experiment to obtain the longitudinal cross-correlation rate.

The differential line-broadening within a multiplet can also be related to other types of cross-correlation effects, involving quadrupolar or paramagnetic interactions. We shall return to these issues in Chapters 14 and 15.

10.3 Transverse Relaxation-Optimised Spectroscopy: TROSY

Many biologically important molecular systems are too large to fall within the range of molecular sizes that can be studied by traditional NMR methods. Increasing the size limit is important for structural determination of molecules that are difficult to crystallise, such as membrane proteins, and molecular complexes and assemblies, and also for obtaining dynamic information through nuclear spin relaxation. There are two major problems associated with NMR of large macromolecules: resonance overlap, due to the increasing number of signals in the spectrum, and the fast transverse relaxation rates. The problems associated with resonance overlap can, at least in theory, be overcome by multidimensional NMR experiments, in combination with isotope-labelling schemes, through which the signals are dispersed in several dimensions. The fast T_2 relaxation is more problematic, since it leads to broadening of the signals, often beyond detection. A scheme to produce narrower lines, *i.e.* to decrease the transverse relaxation during the detection period, is needed. As discussed in previous chapters, the relaxation of heteronuclei such as carbon-13 and nitrogen-15 is dominated by dipolar relaxation caused by directly attached protons. A simple way to alleviate this difficulty is to replace the protons by deuterons. This will reduce the relaxation rates by a factor $\frac{3}{8}(\gamma_H/\gamma_D)^2 \approx 16$ (the factor 3/8 comes from the squares of the nuclear magnetic moments being proportional to $I(I+1)$), but will also reduce the sensitivity of the experiment.

An extremely important technique for investigating large biological macromolecules, in which cross-correlated relaxation plays a crucial role, has been developed by Pervushin and co-workers[18]. This method, TROSY, allows spectra with narrow lines to be recorded for very large molecules. At higher magnetic field strengths, the CSA relaxation mechanism becomes more important, as its interaction strength is proportional to the magnetic field. As a consequence, cross-correlation between the DD and CSA relaxation mechanisms becomes more pronounced at higher fields, leading to different transverse relaxation of the two multiplet components in a two-spin system, such as the ^{15}N-^{1}H or ^{13}C-^{1}H. The differential line-broadening effect can be quite substantial at high magnetic fields, leading to one narrow and one broad component. This is seldom observed in practice, since heteronuclear decoupling during acquisition of the proton spectrum is usually employed. The resulting spectrum is an average of the two components, which will be dominated by the fast relaxing components, resulting in a broad signal. In the TROSY experiment, only the narrow multiplet component is selected (Figure 10.10). The basic elements of the TROSY experiment are outlined in this section, and details as well as information about more recent developments can be found in a review by Wüthrich and Wider[23].

Consider again the *IS* two-spin system discussed in previous sections. The two-dimensional correlation spectrum for this system is a four-line multiplet (Figure 10.11). The evolution of the density operator in the slow tumbling limit, valid for the large macromolecules considered here, can be written retaining only terms including the spectral density at zero frequency. The relaxation rates for the individual components in the four-line multiplet spectrum can be written as:

$$R_{13}^{I} = \frac{\pi}{15}\left(\frac{3}{2}b_{IS}^{2} + 2f_{I}b_{IS}c_{I} + \frac{2}{3}c_{I}^{2}\right)4J(0) \tag{10.25a}$$

$$R_{24}^{I} = \frac{\pi}{15}\left(\frac{3}{2}b_{IS}^{2} - 2f_{I}b_{IS}c_{I} + \frac{2}{3}c_{I}^{2}\right)4J(0) \tag{10.25b}$$

$$R_{12}^{S} = \frac{\pi}{15}\left(\frac{3}{2}b_{IS}^{2} + 2f_{S}b_{IS}c_{S} + \frac{2}{3}c_{S}^{2}\right)4J(0) \tag{10.25c}$$

$$R_{34}^{S} = \frac{\pi}{15}\left(\frac{3}{2}b_{IS}^{2} - 2f_{S}b_{IS}c_{S} + \frac{2}{3}c_{S}^{2}\right)4J(0) \tag{10.25d}$$

Here, R_{12} and R_{34} denote the transitions $1 \longrightarrow 2$ and $3 \longrightarrow 4$ associated with the *S*-spin multiplet components at $\omega_S + \pi J_{IS}$ and $\omega_S - \pi J_{IS}$ and R_{13} and R_{24} denote the transitions involving the *I*-spin.

As in Section 10.2.1, b_{IS} is the dipole-dipole coupling constant, c_I is the CSA interaction strength for spin *I*, and c_S is the CSA interaction strength for spin *S*. f_I and f_S contain the angular dependence of the cross-correlated relaxation for the *I*- and *S*-spins, respectively, described earlier. From Eq. (10.25), it can readily be calculated that for a certain magnetic field strength, the relaxation rate of the I_{24} (in the *I*-spin dimension) and S_{34} (in the *S*-spin dimension) components (see Figure 10.11) will be near zero, leading to narrow signals in the spectrum. On the other hand, the two other components will have rapid transverse relaxation leading to broadened signals. If $I = {}^{1}$H and $S = {}^{15}$N in a protein backbone amide, we obtain that the minimum relaxation rate for ^{15}N is at, approximately, a magnetic field strength corresponding to 900 MHz ^{1}H frequency, while the minimum for the proton relaxation occurs at around 1000 MHz. It would appear that by choosing the appropriate magnetic field strength corresponding to a complete quenching in transverse relaxation for one of the components in a multiplet, an infinitely narrow signal would be the result. In practice, there is always a certain amount of "leakage" relaxation present. The most important source of relaxation is by remote protons in the molecule. This effect is reduced in deuterated proteins, where the only available efficient source of dipolar interactions is between distant amide protons.

The TROSY experiment (pulse sequence shown in Figure 10.12) selects, by the use of a combination of gradient pulses and phase cycling, only the slowly relaxing component in the multiplet. The advantage of using the TROSY measuring scheme over a conventional heteronuclear correlation method is clearly seen by inspecting the differences in lineshapes for the components of a multiplet, as illustrated

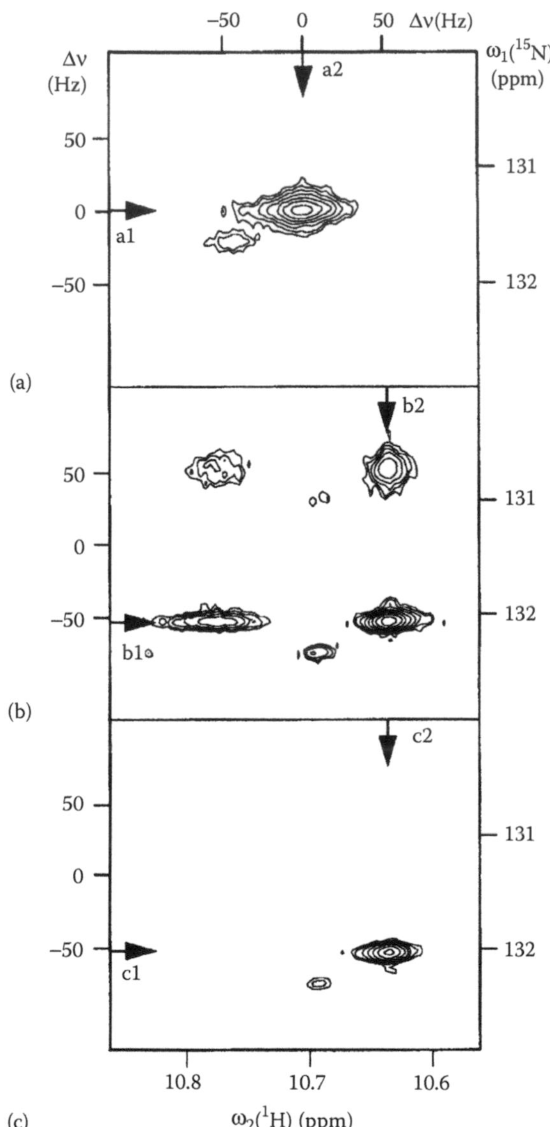

FIGURE 10.10 Multiplet patterns for the indole ^{15}N-^1H spin-pair in a Trp residue in the ftz homeodomain complexed to a 14 base-pair DNA duplex (17 kDa) obtained at a magnetic field strength of 17.6 T (750 MHz ^1H frequency). (a) shows the decoupled cross-peak pattern obtained from a conventional correlation experiment. (b) shows the coupled cross-peak from a conventional correlation spectrum. (c) shows the cross-peak pattern obtained with the TROSY experiment. (Reprinted with permission from Pervushin, K. et al., *Proc. Natl. Acad. Sci. USA*, 94, 12366–12371, 1997. Copyright (1997) National Academy of Sciences, U.S.A.)

in Figure 10.11. In principle, the TROSY scheme can be implemented in any NMR experiment, including the more common correlation techniques, such as those afforded by triple-resonance experiments on ^{15}N,^{13}C-labelled proteins, as well as in relaxation measurements.

Coping with fast transverse relaxation in large systems is not only limited to the indirect evolution and detection periods. For the whole duration of the experiment, magnetisation will be lost through rapid relaxation of the coherences present in the various steps in the pulse sequence. Typically, magnetisation is transferred between ^1H and ^{15}N or ^{13}C by INEPT steps. Rapid relaxation during the INEPT sequence

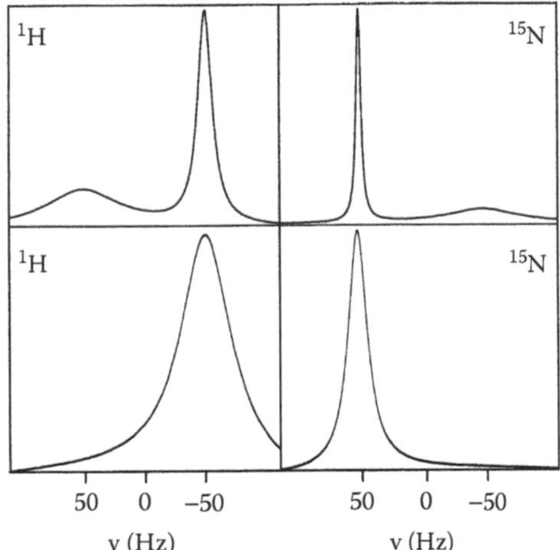

FIGURE 10.11 Dependence of multiplet peak widths on molecular size. The two top figures show cross-sections through the simulated multiplet pattern for a molecule corresponding to 150 kDa (corresponding to $\tau_c = 60$ ns). The two bottom figures show simulated peak widths for a molecule corresponding to 800 kDa (corresponding to $\tau_c = 320$ ns). The calculations were performed at a magnetic field strength of 17.6 T (750 MHz ^{1}H frequency). (Reprinted with permission from Pervushin, K. et al., *Proc. Natl. Acad. Sci. USA*, 94, 12366–12371, 1997. Copyright (1997) National Academy of Sciences, U.S.A.)

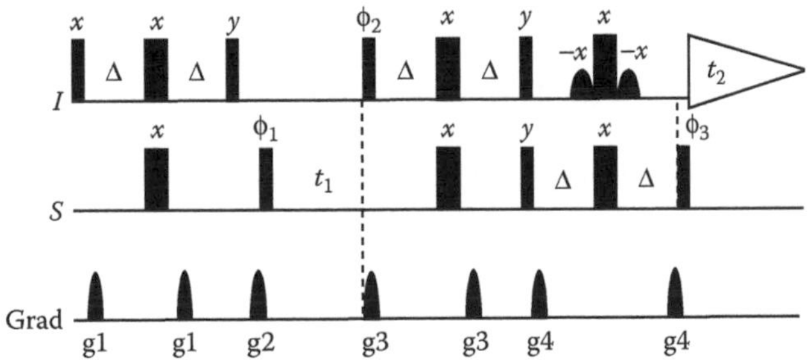

FIGURE 10.12 The TROSY pulse sequence. The two curved-shaped ^{1}H pulses are part of the Watergate H_2O suppression scheme. The delay Δ is set to $1/4J_{IS}$. The phase cycle is $\phi_1 = y, -y, -x, x$; $\phi_2 = 4(y), 4(-y)$; $\phi_3 = 4(x), 4(-x)$, and $\phi_{rec} = x, -x, -y, y, x, -x, y, -y$. Quadrature detection in the indirect dimension is achieved by incrementing ϕ_1 according to the States method. The relative gradient strengths are 12, −60, 20, and 28.8 for g1, g2, g3, and g4, respectively. (Adapted from Pervushin, K. et al., *Proc. Natl. Acad. Sci. USA*, 94, 12366–12371, 1997).

becomes a limiting factor in the experiment, as there often is no signal left to detect. Polarisation transfer methods based on cross-correlated relaxation have therefore been developed[24,25]. These methods have been named *cross-correlated relaxation-induced polarisation transfer* (CRIPT) and *cross-correlated relaxation-enhanced polarisation transfer* (CRINEPT). It can be shown that for large macromolecules with correlation times $\tau_c > 100$ ns, the CRIPT transfer of magnetisation becomes more efficient than can be achieved with the INEPT method.

10.4 Multiple- and Zero-Quantum Relaxation

So far, the discussion concerning multispin effect in this chapter has been centred on cross-correlated relaxation related to two-spin order and to transfer between single-quantum coherences. In this section, we discuss certain aspects of multiple-quantum relaxation, *i.e.* relaxation of coherences of orders other than one, which turns out to carry information about dynamics not possible to measure by other means. Differential relaxation of zero- and double-quantum coherence for a two-spin system has received considerable attention, since it can provide a measure of chemical exchange processes outside the fast exchange limit. More details on the influence of exchange on relaxation, and about methods to study exchange, can be found in Chapter 13.

In systems of two or more $I = \frac{1}{2}$ spins, the maximum possible order of coherence is given by the number of participating spins. In a two-spin system, we can deliberately create and study multiple-quantum coherences. As stated in Eqs. (7.13) and (7.14), terms such as $2\hat{I}_x\hat{S}_y$ are combinations of zero-quantum coherence and double-quantum coherence. Once created, these coherences have their characteristic decay rates – we remind the reader that the relaxation processes cannot change the order of coherence (see the Redfield kite; Figure 4.1).

The simplest possible two-dimensional experiment, which allows measurements of ZQ and DQ relaxation in a homonuclear system of two spin 1/2 nuclei, proposed in the seminal paper by Wokaun and Ernst[26], is shown in Figure 10.13. The first two 90° pulses, with a suitable delay in between, create multiple quantum coherences. It is possible (and usually necessary) to select a particular order of coherence by an appropriate phase cycling procedure. The MQ coherences are then allowed to evolve, with their characteristic frequencies, during the t_1-period. The 180° pulse in the middle of the evolution period serves to suppress chemical shift and field inhomogeneity effects (note that the linewidth corresponding to the DQ coherence is twice as sensitive to the inhomogeneity as a single-quantum linewidth, while the ZQ coherence does not broaden in an inhomogeneous field). The final 90° pulse converts the MQ coherence into observable single-quantum coherence, which is acquired during the t_2-period. The width of resulting line in the ω_1 domain yields a direct measure of the relaxation rate of the MQ coherence.

In the case where the relaxation rates for the two MQ coherences are different, $R_{DQ} \neq R_{ZQ}$, one speaks of *differential relaxation*. The difference between DQ and ZQ relaxation is another signature of cross-correlation and is due to a combination of dipole-dipole and/or CSA-CSA cross-correlation effects[27,28] in addition to cross-correlation rate between two exchange processes[29,30]. It is straightforward to show that the difference between DQ and ZQ relaxation rates is given by:

$$\Delta R_{MQ} = R_{DQ} - R_{ZQ} = R_{DQ}^0 - R_{ZQ}^0 + 4p_A p_B \gamma_I \gamma_S \Delta\omega_I \Delta\omega_S \tau_{ex} \tag{10.26}$$

where R_{DQ}^0 and R_{ZQ}^0 denote the relaxation rates in the absence of exchange. The last term in Eq. (10.26) may be present if there is a chemical exchange that alters the chemical shift for both the I- and the S-spin and if the two chemical exchange processes are cross-correlated. In this case, the cross-correlation occurs between the chemical exchange for the two spins, I and S, characterised by the chemical shift differences, $\Delta\omega_I$ and $\Delta\omega_S$, between the two states, A and B, with populations p_A and p_B. More details concerning chemical exchange processes and their effect on relaxation will be provided in Chapter 13. The difference between R_{DQ}^0 and R_{ZQ}^0 is given by[27]:

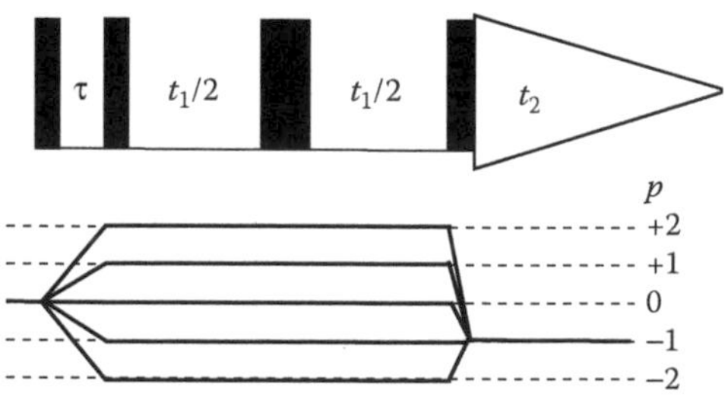

FIGURE 10.13 A simple two-dimensional pulse sequence for measuring MQ relaxation in a homonuclear spin system consisting of two non-equivalent spins.

$$R_{DQ}^0 - R_{ZQ}^0 = \eta_{cc} + \sigma_{IS} + \eta_{dd}^{remote} \tag{10.27}$$

in which η_{cc} is the I-spin CSA, S-spin CSA cross-correlation rate, σ_{IS} is the (dipolar) cross-relaxation rate and η_{dd}^{remote} is related to the dipole-dipole cross-correlations with other spins.

In practice, ΔR_{MQ} in large molecules is measured by recording two separate experiments, one in which the $2\hat{I}_y\hat{S}_y$ is present during a relaxation period and a second experiment in which the term represented by $2\hat{I}_x\hat{S}_x$ is present. By using relations such as those in Eqs. (7.13) and (7.14), we can see that the two operators contain different combinations of MQ coherences:

$$2\hat{I}_y\hat{S}_y = -\frac{1}{2}\left(\hat{I}^+\hat{S}^+ + \hat{I}^-\hat{S}^-\right) + \frac{1}{2}\left(\hat{I}^+\hat{S}^- + \hat{I}^-\hat{S}^+\right) \tag{10.28a}$$

$$2\hat{I}_x\hat{S}_x = \frac{1}{2}\left(\hat{I}^+\hat{S}^+ + \hat{I}^-\hat{S}^-\right) + \frac{1}{2}\left(\hat{I}^+\hat{S}^- + \hat{I}^-\hat{S}^+\right) \tag{10.28b}$$

in which $\frac{1}{2}(\hat{I}^+\hat{S}^+ + \hat{I}^-\hat{S}^-)$ is a DQ coherence and $\frac{1}{2}(\hat{I}^+\hat{S}^- + \hat{I}^-\hat{S}^+)$ is a ZQ coherence, each characterised by a relaxation rate, R_{DQ} and R_{ZQ}. The evolution of the two MQ coherences during the relaxation period τ is given by:

$$\begin{aligned}\left\langle 2\hat{I}_y\hat{S}_y\right\rangle(\tau) &= -\frac{1}{2}\exp\left(-R_{DQ}\tau\right)\left(\left\langle \hat{I}^+\hat{S}^+\right\rangle(0) + \left\langle \hat{I}^-\hat{S}^-\right\rangle(0)\right) \\ &+ \frac{1}{2}\exp\left(-R_{ZQ}\tau\right)\left(\left\langle \hat{I}^+\hat{S}^-\right\rangle(0) + \left\langle \hat{I}^-\hat{S}^+\right\rangle(0)\right)\end{aligned} \tag{10.29a}$$

$$\begin{aligned}\left\langle 2\hat{I}_x\hat{S}_x\right\rangle(\tau) &= \frac{1}{2}\exp\left(-R_{DQ}\tau\right)\left(\left\langle \hat{I}^+\hat{S}^+\right\rangle(0) + \left\langle \hat{I}^-\hat{S}^-\right\rangle(0)\right) \\ &+ \frac{1}{2}\exp\left(-R_{ZQ}\tau\right)\left(\left\langle \hat{I}^+\hat{S}^-\right\rangle(0) + \left\langle \hat{I}^-\hat{S}^+\right\rangle(0)\right)\end{aligned} \tag{10.29b}$$

and the time dependence of the intensity ratio from the two experiments can, consequently, be written as

$$\frac{\left\langle 2\hat{I}_y\hat{S}_y\right\rangle(\tau)}{\left\langle 2\hat{I}_x\hat{S}_x\right\rangle(\tau)} = \tanh\left(\Delta R_{MQ}\tau/2\right) \tag{10.30}$$

which means that it is possible to extract the differential relaxation from an arrayed experiment with different τ-values. This is a similar approach to the determination of the transverse (and longitudinal) cross-correlated relaxation rates discussed in Section 10.2 (see Eqs. (10.21) and (10.24)).

The renewed interest in multispin phenomena, and in differential MQ relaxation in particular, is largely due to the possibility of investigating chemical exchange, including outside the fast exchange limit[30]. Cross-correlation effects in MQ relaxation have also proved useful for determining angles between two dipole-dipole vectors[31,32]. This comes from the term η_{dd}^{remote} in Eq. (10.27), which in principle contains cross-correlation between all dipole-coupled spin pairs, but in many cases reduces to a dipole-dipole cross-correlation term between two spatially close vectors. Although such measurements are potentially of use for many different types of studies, one should be aware that dynamic investigations clearly require pre-determined information about the structural (distances, angles) as well as static parameters (CSA), while the determination of structural parameters, of course, requires knowledge about the molecular dynamics. Several attempts have been made to separate structure, dynamics and static parameters in various relaxation studies. We wish in particular to mention here the work of Vögeli and co-workers[33,34], who developed a protocol for measurements and analysis of that kind. The following points are important[34]:

1. Only dipolar CCRRs are used and measured with at least two methods each to estimate both systematic and random errors.
2. The measured MQ relaxation rates contain, besides the main terms of interest, also smaller contributions from less important interference terms. These are estimated by the full relaxation matrix calculations[35].
3. The orientations of the CH and NH bonds are obtained from residual dipolar couplings measured in a number of alignment media (which include the effects in internal dynamics over a large range of time scales, similarly to the CCRR spectral densities).
4. The calculations of the CCRRs are carried out allowing for anisotropic overall tumbling. The last two points are related to theoretical developments discussed in Section 6.3.

Further discussions concerning dynamic investigations from multispin phenomena can be found in Chapter 11, while structural aspects of such effects are discussed in Chapter 12. Examples of investigations of chemical exchange, including the use of cross-correlated relaxation, are found in Chapter 13.

Exercises for Chapter 10

1. Demonstrate that simple phase cycling, or difference spectroscopy, can be used to retain only the two-spin order term in a double-quantum filtered homonuclear inversion-recovery experiment using a non-selective double-quantum filter.

References

1. Hubbard, P. S., Nuclear magnetic relaxation of three and four spin molecules in a liquid. *Phys. Rev.* 1958, 109, 1153–1158.
2. Werbelow, L. G.; and Grant, D. M., Intramolecular dipolar relaxation in multispin systems. *Adv. Magn. Reson.* 1977, 9, 189–299.
3. Vold, R. L.; and Vold, R. R., Nuclear magnetic relaxation in coupled spin systems. *Prog. NMR Spectr.* 1978, 12, 79–133.
4. Canet, D., Construction, evolution and detection of magnetisation modes designed for treating longitudinal relaxation of weakly coupled spin 1/2 systems with magnetic equivalence. *Prog. NMR Spectr.* 1989, 21, 237–291.
5. Kumar, A.; Grace, R. C. R.; and Madhu, P. K., Cross-correlations in NMR. *Prog. NMR Spectr.* 2000, 37, 191–319.
6. Werbelow, L. G.; and Grant, D. M., Carbon-13 relaxation in multispin systems of the type AXn. *J. Chem. Phys.* 1975, 63, 544–556.
7. Goldman, M., Interference effects in the relaxation of a pair of unlike spin-1/2 nuclei. *J. Magn. Reson.* 1984, 60, 437–452.
8. Jaccard, G.; Wimperis, S.; and Bodenhausen, G., Observation of 2IzSz order in NMR relaxation studies for measuring cross-correlation of chemical shift anisotropy and dipolar interactions. *Chem. Phys. Lett.* 1987, 138, 601–606.
9. Batta, G.; Köver, K. E.; and Kowalewski, J., A comparison of 1D and 2D (unbiased) experimental methods for measuring CSA/DD cross-correlated relaxation. *J. Magn. Reson.* 1999, 136, 37–46.
10. Dalvit, C.; and Bodenhausen, G., Proton chemical shift anisotropy: Detection of cross-correlation with dipole-dipole interactions by double-quantum filtered two-dimensional exchange spectroscopy. *Chem. Phys. Lett.* 1989, 161, 554–560.

11. Mäler, L.; Mulder, F. A. A.; and Kowalewski, J., Paramagnetic cross-correlation effects in the longitudinal proton relaxation of cis-chloroacrylic acid in the presence of nickel(II) ions. *J. Magn. Reson. Ser. A.* 1995, 117, 220–227.
12. Di Bari, L.; Kowalewski, J.; and Bodenhausen, G., Magnetisation transfer modes in scalar-coupled spin systems investigated by selective 2-dimensional nuclear magnetic resonance exchange experiments. *J. Chem. Phys.* 1990, 93, 7698–7705.
13. Mäler, L.; and Kowalewski, J., Cross-correlation effects in the longitudinal relaxation of heteronuclear spin systems. *Chem. Phys. Lett.* 1992, 192, 595–600.
14. Levitt, M. H.; and Di Bari, L., Steady state in magnetic resonance pulse experiments. *Phys. Rev. Lett.* 1992, 69, 3124–3127.
15. Levitt, M. H.; and Di Bari, L., The homogeneous master equation and the manipulation of relaxation networks. *Bull. Magn. Reson.* 1994, 16, 94–114.
16. Jeener, J., Superoperators in magnetic resonance. *Adv. Magn. Reson.* 1982, 10, 1–51.
17. Ghalebani, L.; Bernatowicz, P.; Aski, S. N.; and Kowalewski, J., Cross-correlated and conventional dipolar carbon-13 relaxation in methylene groups in small, symmetric molecules. *Concepts Magn. Reson. A.* 2007, 30A, 100–115.
18. Pervushin, K.; Riek, R.; Wider, G.; and Wüthrich, K., Attenuated T2 relaxation by mutual cancellation of dipole-dipole coupling and chemical shift anisotropy indicates an avenue to NMR structures of very large biological macromolecules in solution. *Proc. Natl. Acad. Sci. USA* 1997, 94, 12366–12371.
19. Müller, N.; Bodenhausen, G.; and Ernst, R. R., The appearance of forbidden cross-peaks in two-dimensional nuclear magnetic resonance spectra due to multiexponential T2 relaxation. *J. Magn. Reson.* 1985, 65, 531–534.
20. Wimperis, S.; and Bodenhausen, G., Relaxation-allowed cross-peaks in two-dimensional NMR correlation spectroscopy. *Mol. Phys.* 1989, 66, 897–919.
21. Tjandra, N.; Szabo, A.; and Bax, A., Protein backbone dynamics and N-15 chemical shift anisotropy from quantitative measurement of relaxation interference effects. *J. Am. Chem. Soc.* 1996, 118, 6986–6991.
22. Kroenke, C. D.; Loria, J. P.; Lee, L. K.; Rance, M.; and Palmer, A. G., Longitudinal and transverse H-1-N-15 dipolar N-15 chemical shift anisotropy relaxation interference: Unambiguous determination of rotational diffusion tensors and chemical exchange effects in biological macromolecules. *J. Am. Chem. Soc.* 1998, 120, 7905–7915.
23. Wüthrich, K.; and Wider, G., *Transverse Relaxation-Optimized NMR Spectroscopy with Biomacromolecular Structures in Solution*. Wiley: eMagRes, 2007; pp. 468–477.
24. Riek, R.; Wider, G.; Pervushin, K.; and Wüthrich, K., Polarization transfer by cross-correlated relaxation in solution NMR with very large molecules. *Proc. Natl. Acad. Sci. USA* 1999, 96, 4918–4923.
25. Riek, R.; Fiaux, J.; Bertelsen, E. B.; Horwich, A. L.; and Wüthrich, K., Solution NMR techniques for large molecular and supramolecular structures. *J. Am. Chem. Soc.* 2002, 124, 12144–12153.
26. Wokaun, A.; and Ernst, R. R., The use of multiple quantum transitions for relaxation studies in coupled spin systems. *Mol. Phys.* 1978, 36, 317–341.
27. Konrat, R.; and Sterk, H., Cross-correlation effects in the transverse relaxation of multiple-quantum transitions of heteronuclear spin systems. *Chem. Phys. Lett.* 1993, 203, 75–80.
28. Tessari, M.; and Vuister, G. W., A novel experiment for quantitative measurement of CSA(1H(N))/CSA(15N) cross-correlated relaxation in 15N-labeled proteins. *J. Biomol. NMR* 2000, 16, 171–174.
29. Kloiber, K.; and Konrat, R., Peptide plane torsion angles in proteins through intraresidue H-1-N-15-C-13 ' dipole-CSA relaxation interference: Facile discrimination between type-I and type-II beta-turns. *J. Am. Chem. Soc.* 2000, 122, 12033–12034.
30. Wang, C.; and Palmer, A. G., Differential multiple quantum relaxation caused by chemical exchange outside the fast exchange limit. *J. Biomol. NMR* 2002, 24, 263–268.

31. Reif, B.; Hennig, M.; and Griesinger, C., Direct measurement of angles between bond vectors in high-resolution NMR. *Science.* 1997, 276, 1230–1233.
32. Schwalbe, H.; Carlomagno, T.; Hennig, M.; Junker, J.; Reif, B.; Richter, C.; and Griesinger, C., Cross-correlated relaxation for measurement of angles between tensorial interactions. *Methods Enzymol.* 2001, 338, 35–81.
33. Vögeli, B.; and Yao, L. S., Correlated dynamics between protein HN and HC bonds observed by NMR cross relaxation. *J. Am. Chem. Soc.* 2009, 131, 3668–3678.
34. Fenwick, R. B.; Schwieters, C. D.; and Vögeli, B., Direct investigation of slow correlated dynamics in proteins via dipolar interactions. *J. Am. Chem. Soc.* 2016, 138, 8412–8421.
35. Vögeli, B., Full relaxation matrix analysis of apparent cross-correlated relaxation rates in four-spin systems. *J. Magn. Reson.* 2013, 226, 52–63.

11

Relaxation and Molecular Dynamics

Relaxation measurements have found widespread use in many different areas within the fields of physics, chemistry, and biology. There are not many methods available by which dynamics on the timescale afforded by nuclear spin relaxation can be studied, and needless to say, many of the advances within the field of relaxation have therefore been focused on the reliable determination of molecular dynamic parameters. In the field of biomolecular NMR, the possibility of producing isotope-labelled compounds has opened the way for extensive studies of dynamics with atomic resolution in biological systems. In this way, NMR relaxation has found great use for investigating dynamics on a very wide timescale, ranging from fast local motions via molecular tumbling to slower motions related to exchange phenomena. In this chapter, we will present applications relating to studies of reorientational dynamics, while motion related to exchange will be explored in Chapter 13.

The first section of this chapter is an introduction to the others, briefly discussing the connection between molecular motion and relaxation, derived in Chapters 3 through 6. In the remaining sections, examples of applications of relaxation measurements to investigate dynamics for systems undergoing different degrees of complex motions will be presented and discussed. The text is not meant as a comprehensive review, and the selection of examples is a matter of the personal taste of the authors of this book as well as a matter of providing illustrative examples.

11.1 Molecular Motion

In this chapter, we will examine the effect of motions on relaxation and how this can be used to investigate molecular dynamics in systems of various levels of complexity. The theoretical background to relaxation, relaxation mechanisms, and molecular motion is given in Chapters 3 through 6, but the basic concepts are recapitulated here. Nuclear spin relaxation is caused by fluctuating interactions involving one or several nuclear spins. The time-dependent Hamiltonian, which is in general a sum of interactions containing the spin-lattice couplings giving rise to relaxation, can generally be written as a scalar contraction of two irreducible tensor operators, as given in Eqs. (3.13) and (4.37), or as a sum of such contractions; *cf.* Eq. (5.1) or (5.41). Various relaxation rates depend on three factors: the interaction strengths (which function as multiplicative factors), the matrix elements of spin operators (which provide a kind of selection rules for which interactions influence which relaxation matrix elements), and the spectral density functions. The latter quantities convey the dynamics information, central in this chapter.

The fundamental molecular dynamic quantities of primary interest for NMR are time correlation functions for rank-two spherical harmonics of the pair of angles specifying the direction of a given molecule-fixed axis with respect to the laboratory frame. The spectral density functions are twice one-sided Fourier transforms of the time-correlation functions and provide information on the distribution of the power available for causing spin transitions among different frequencies. As discussed in detail

TABLE 11.1 Selected Dynamic Models Used to Calculate Spectral Densities

Dynamic Model	Parameters Influencing the Spectral Densities	Number of Lorentzians	Comment
Isotropic rotational diffusion	Rotational diffusion coefficient, $D_R = 1/6\tau_R$	1	Useful as a first approximation
Rotational diffusion, symmetric top	Two rotational diffusion coefficients, $D_{\|\|}$ and $D_{\perp}$, the angle β between the symmetry axis and the internuclear axis	3	Rigid molecule, requires the knowledge of geometry
Rotational diffusion, asymmetric top	Three rotational diffusion coefficients, three Euler angles	5	Rigid molecule, rather complicated
Isotropic rotational diffusion with one internal degree of freedom	Rotational diffusion coefficient, D_R, internal motion rate parameter, angle between the internal rotation axis and the internuclear axis	3	Useful for *e.g.* methyl groups
"Model-free"	Global and local correlation times, generalised order parameter, S	2	Widely used for non-rigid molecules, *e.g.* biological macromolecules
Extended model-free	Global correlation time, local correlation times (fast and slow), two generalised order parameters, S_f and S_s (fast and slow)	3	Biological macromolecules with complex local motion
Anisotropic model-free	$D_{\|\|}$ and $D_{\perp}$, β, local correlation time, generalised order parameter, S	4	Non-spherical, non-rigid molecules

in Chapter 6, the time-correlation functions and the spectral densities can be derived using different motional models. In most cases, the spectral densities are linear combinations of Lorentzian functions:

$$J_2(\omega) = \sum_k a_k J_{2,k}(\omega) = \sum_k a_k \frac{2\tau_k}{1+\omega^2\tau_k^2} \tag{11.1}$$

where the coefficients, a_k, and the correlation times, τ_k, are model specific. Table 11.1 summarises the motional models that were discussed in Chapter 6 and in this chapter and provides a guide to where their use is appropriate.

11.2 Extreme Narrowing: Rapidly Reorienting Molecules

Some of the early applications of nuclear spin relaxation include the investigation of reorientational dynamics of small molecules in solution. Small molecules in low-viscosity solutions typically have rotational correlation times of a few tens of picoseconds or less, which means that the extreme narrowing conditions usually prevail. We remind the reader that the extreme narrowing regime implies that all the products $\tau_k^2\omega^2$ in Eq. (11.1) are much less than unity at all relevant frequencies, or that $J_2(\omega) = J_2(0)$. As a consequence, the interpretation of relaxation parameters in this range becomes particularly simple.

Let us consider, for example, a proton-carrying carbon-13 nucleus (S-spin) in an organic molecule, a convenient handle for studies of molecular dynamics. In small molecules at not too low concentrations, carbon-13 relaxation measurements can conveniently be carried out at natural isotopic abundance. The relaxation for proton-carrying carbons is usually dominated by the dipole-dipole (DD) interaction with the attached proton(s), and the dipolar relaxation under extreme narrowing conditions is independent of the magnetic field. Under proton (I-spin) decoupling conditions and in spin systems of low symmetry, the spin-lattice relaxation time, T_1, is well defined (compare Sections 3.2–3.4). The dipolar T_2 is equal to T_1 and therefore provides no extra information and need not be measured. Measurements of the nuclear Overhauser enhancement (NOE) factor allow the extent of the heteronuclear dipolar interaction to be determined by controlling whether Eq. (3.32b) is fulfilled. If other relaxation mechanisms are non-negligible, the measurement of the observed NOE factor, η_{obs}, allows the separation of the dipolar contribution to the total measured spin-lattice relaxation rate, $T_{1,obs}^{-1}$ (compare Eq. (5.52)) through the simple relation:

$$T_{1DD}^{-1} = T_{1,obs}^{-1}\frac{\eta_{obs}}{\eta_{max}} \tag{11.2}$$

where $\eta_{max}=\gamma_I/2\gamma_S$ (equal to 1.99 for S being carbon-13). For proton-carrying carbons, one usually neglects the relaxation contributions from dipolar interactions with more distant nuclei, and the dipolar spin-lattice relaxation rate is then given by Eq. (3.32a). The dipole-dipole interaction constant can in most cases be assumed to be known (which implies that the internuclear distance between the proton and the nucleus of interest is known), and therefore the dipolar relaxation is simply related to the effective correlation time, τ_c, of the internuclear vector. The site variation of τ_c within a molecule can provide information about different molecular properties.

11.2.1 Motional Anisotropy

Most molecules are not simple, symmetric objects, but rather, anisotropic. As already mentioned, different T_{1DD}, and thus τ_c, for different sites in a molecule can reveal information about anisotropy. From here on, interpretation of relaxation parameters is possible on several levels. First, a qualitative measure of motional anisotropy can be obtained directly by comparing carbon-13 relaxation times multiplied by the number of directly bonded protons, N_HT_1, for different sites in a molecule. The simple interpretation is that a larger value of N_HT_1 (slower relaxation) translates into shorter effective correlation time (faster reorientational motion). Studies of this type were common in the 1970s, and numerous examples can be found in the book by Levy *et al.*[1] and in the review by Levy and Kerwood in eMagRes[2]. Selected examples of small molecules that were studied early on are displayed in Figure 11.1. Due to the anisotropic shape of the diphenyldiacetylene molecule shown in Figure 11.1a, the motion around the long axis of the molecule occurs much faster than the tumbling of the long axis. The motion around the molecular axis does not influence the axis joining the carbon at position four of the phenyl ring with the proton bound to it. Thus, this carbon will experience a slower modulation of the dipolar coupling to its proton as compared

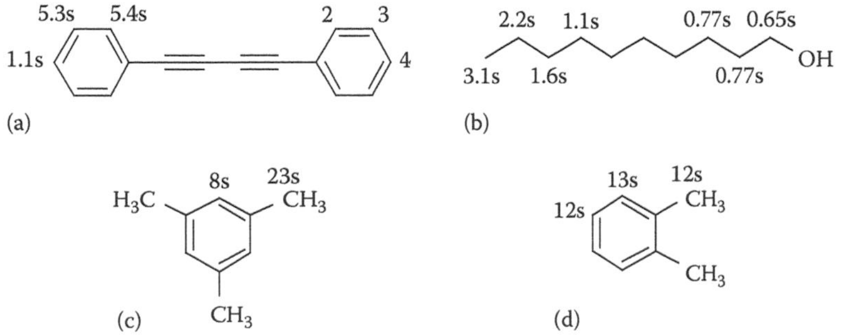

FIGURE 11.1 Examples of small organic molecules undergoing anisotropic reorientational motion. (a) diphenyldiacetylene; (b) 1-decanol; (c) 1,3,5-trimethylbenzene; (d) 1,2-dimethylbenzene (ortho-xylene). (The carbon-13 T_1 values shown in examples (a) and (b) are from Levy, G.C. *et al.*, *Carbon-13 Nuclear Magnetic Resonance Spectroscopy.* 2nd edn, Wiley, New York, 1980, and in examples (c) and (d) from Hore, P.J., *Nuclear Magnetic Resonance.* Oxford University Press, Oxford, 1995.)

with the carbons in positions two and three. This is clearly reflected in the T_1 relaxation times for the carbons in these positions.

The degree of intramolecular flexibility can be monitored in the same way. The carbons in the alkyl chains of 1-decanol, displayed in Figure 11.1b, have different T_1 relaxation times due to the variability in flexibility along the alkyl chain. The relaxation times indicate that the mobility increases along the chain from the position of the hydroxyl group, which acts as an anchoring site through hydrogen bonding. The fact that the carbon-13 relaxation in neat 1-decanol indeed is in the extreme narrowing regime was proved by measurements at two different magnetic fields[2]. The molecular motions in 1-decanol are obviously complicated and have more recently been investigated further in carbon-13-labelled material dissolved in various solvents by proton-coupled ^{13}C relaxation[3]. Other simple applications include investigations of hindered rotation in molecules with different substitution patterns (Figure 11.1c and d). The methyl carbon of 1,3,5-trimethylbenzene (mesitylene) has a much slower T_1 relaxation than the ring carbons, even though there are three protons attached to the methyl carbon. This is because of the rapid rotation of the methyl group. This motion is much faster than the relatively slow overall motion of the entire molecule, which is the only motion experienced by the ring carbons. In 1,2-dimethylbenzene (ortho-xylene) (*cf.* Figure 11.1d), the rapid methyl group rotation is slowed down by introducing a second substituent, which sterically hinders the rotation. The relaxation times for the methyl carbons are here comparable to the relaxation times of the ring carbons.

11.2.2 The Me_2THMN Example

Studies of motional anisotropy can be made more quantitative by connecting the effective correlation times to dynamic models. Let us start with molecules without too much internal mobility. A nice example of a system for which motional anisotropy was evaluated from relaxation data is the hydrocarbon 1,2,3,4-tetrahydro-5,6-dimethyl-1,4-methanonaphthalene (Me_2THMN), Figure 11.2, studied as a model system by Dölle and Bluhm in the mid-1980s[4]. The polycyclic structure of the molecule makes it reasonable to assume that it reorients essentially as a rigid body, ignoring for the moment the methyl groups. The molecule has no symmetry elements and is an asymmetric top. There are four CH carbons (two of which are aromatic) and three CH_2 carbons, providing a large number of non-parallel CH vectors. In a neat liquid at 309 K, all these proton-carrying carbons have, within experimental uncertainties, full NOE ($\eta = 1.99$) at a low magnetic field (2.1 T). Carbon-13 spin-lattice relaxation rates are non-uniform and can be interpreted in terms of the anisotropic rotational diffusion model described in Section 6.2.

The free rotation properties of an asymmetric top in gas phase are determined by the moment of inertia tensor for the molecule. This tensor and its principal frame can easily be calculated from the known atomic positions and masses in the molecule. From this geometry information, one can also derive dipolar coupling constants and the angles specifying the orientations of all the CH axes with respect to the moment of inertia principal frame. Since the hydrocarbons are not expected to interact strongly in the liquid state, one might assume as a first approximation that the principal frames of the rotational diffusion tensor coincide with that for the moment of inertia tensor. Under this assumption, the interpretation of carbon-13 T_1 in Me_2THMN simplifies to fitting the three principal elements of the diffusion tensor to the relaxation rates for the seven carbon sites in the molecule. As demonstrated by Dölle and Bluhm[4], this procedure turned out to indeed be possible. In addition, the authors demonstrated that an improved fit could be obtained by allowing the principal frames of the moment of inertia and rotational

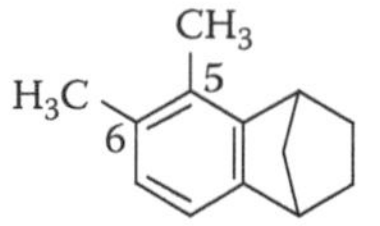

CH_3 5 H_3C 6

FIGURE 11.2 1,2,3,4-tetrahydro-5,6-dimethyl-1,4-methanonaphthalene (Me_2THMN).

diffusion tensor to deviate from each other. The fitting procedure included in that case six parameters: the three rotational diffusion coefficients and three Euler angles between the two frames.

One can take the analysis one step further by using the principal elements of the rotational diffusion tensor for Me_2THMN to validate the small-step diffusion model and to relate to the hydrodynamics. The first of these steps was reported in the same paper[4] by comparing the rotational correlation times, defined as $\tau_i = 1/6D_{ii}$, with the mean period a free-rotor (with a given moment of inertia tensor) needs to rotate one radian around each of the three principal axes, as calculated by Wallach and Huntress[5]:

$$\tau_i^{free} = \frac{3}{5}\left(\frac{I_i}{k_B T}\right)^{1/2} \tag{11.3}$$

Dölle and Bluhm reported the ratios τ_i/τ_i^{free} in the range 20–70 for the neat liquid at 309 K, which, according to Wallach and Huntress, confirmed the validity of the small-step diffusion model. The relation of the rotational diffusion coefficients in Me_2THMN, dissolved in various solvents, to the hydrodynamic theory has been analysed in another paper from the same group[6]. The interested reader is referred to that paper for further information. Dölle and co-workers have remained interested in carbon-13 relaxation in Me_2THMN and have also studied the neat liquid at higher fields and lower temperatures, where the extreme narrowing conditions do not apply[7]. The interpretation of the data in terms of the motion of a rigid body was found not to work very well, which may perhaps be related to the non-hydrodynamic conditions (the same molecular size of the solvent and solute) existing in the neat liquid.

11.2.3 Symmetric Tops and Methyl Group Rotation

In molecules with a unique, three-fold or higher, symmetry axis, the rotational diffusion follows the theory for symmetric tops as formulated in Eq. (6.36). In extreme narrowing, the spectral density reduces to an effective correlation time, which for an arbitrary axis, characterised by the angle β with respect to the symmetry axis, can be expressed as

$$\tau_c = \frac{1}{4}\left[\left(3\cos^2\beta - 1\right)^2 \tau_{2,0} + 12\cos^2\beta\sin^2\beta\tau_{2,1} + 3\sin^4\beta\tau_{2,2}\right] \tag{11.4a}$$

which for small-step rotational diffusion becomes:

$$\tau_c = \frac{1}{4}\left[\left(3\cos^2\beta - 1\right)^2 \frac{1}{6D_\perp} + 12\cos^2\beta\sin^2\beta\frac{1}{5D_\perp + D_\parallel}\right] + 3\sin^4\beta\frac{1}{2D_\perp + 4D_\parallel} \tag{11.4b}$$

An example of a molecule of this type, quinuclidine (1-azabicyclo[2.2.2]octane), is shown in Figure 11.3.[8] For the tertiary carbon, β = 0 and the relaxation of that carbon is influenced only by $\tau_{2,0}$ or by the rotational diffusion perpendicular to the axis. For the methylene carbons, the angle β is about 70° and both rotational processes contribute to the relaxation. Combining the measurements for all carbons, one can determine the rotational anisotropy expressed quantitatively as $D_\parallel/D_\perp$. This quantity turns out to be strongly dependent on the solvent, varying from about 1.3 in cyclohexane to about 2.5 in benzene and around 10 in chloroform and methanol. The anisotropy was also found to change, to a certain extent,

FIGURE 11.3 Quinuclidine (1-azabicyclo[2.2.2]octane).

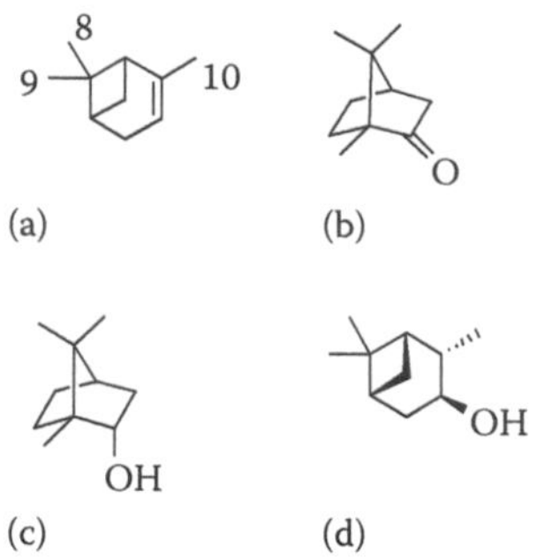

FIGURE 11.4 The terpenes studied in Ericsson *et al.*[10] (a) α-pinene; (b) camphor; (c) borneol; (d) isopinocampheol.

with temperature. The motional anisotropy is thus related to the strength of intermolecular interactions. The intriguing case of using benzene, which seems to hold an intermediate position between the inert cyclohexane and the hydrogen-bonding solvents, as a solvent was also subject to a molecular dynamics (MD) simulation[9], unfortunately without fully conclusive results.

The internal rotation of methyl groups is a particularly suitable motion for NMR relaxation studies. In the study of Me_2THMN by Dölle and Bluhm[4], the authors used a model allowing internal motion of the methyl groups superimposed on the asymmetric top rotational diffusion. They found that one of the methyl groups, that connected to C5 (Figure 11.2), characterised by a longer carbon-13 T_1, moves faster than the second one (the methyl group at C6; see Figure 11.2) and rationalised the finding by differences in steric hindrance. Ericsson and co-workers[10] studied the methyl group rotation in a series of terpenes (α-pinene, camphor and two alcohols) in chloroform solution (Figure 11.4). They used a simpler dynamic model, which assumed that the overall reorientation was isotropic and described the methyl group motion using both rotational diffusion and random three-site jumps, and reported measurements over a large temperature range. The results of both models for internal motion were in good agreement with each other. At each temperature, the relaxation of the tertiary and methylene carbons allowed the estimation of the average effective overall rotational correlation time. Combining this piece of information with the methyl carbon relaxation rates, the authors were able to evaluate the internal rate parameters, D_i or k_i of Eqs. (6.51) and (6.54). The temperature dependence of k_i for all the systems was analysed through the Arrhenius relation:

$$k_i = k_o \exp\left(-E_a/k_B T\right) \tag{11.5}$$

where E_a is the Arrhenius activation energy. As an example, the Arrhenius plots of ln k_i versus inverse temperature for the three methyl groups in α-pinene are shown in Figure 11.5. The E_a values for the methyl groups are related to the three-fold barriers hindering the internal motion[11]. Ericsson *et al.*[10] compared the experimentally determined activation energies with molecular mechanics calculations of the barrier heights. The effects of possible anisotropy of the overall motion were discussed and judged less important for α-pinene and camphor as compared with the alcohols.

11.2.4 Quadrupolar Interactions

Carbon-13 nuclei provide a convenient handle for studying molecular dynamics in solution, because a single dominant interaction can usually be identified and its strength assumed to be known. Quadrupolar nuclei, characterised by the spin quantum number $I \geq 1$, constitute another group of useful nuclear species for similar reasons: the quadrupolar interaction is normally completely dominant, and its strength

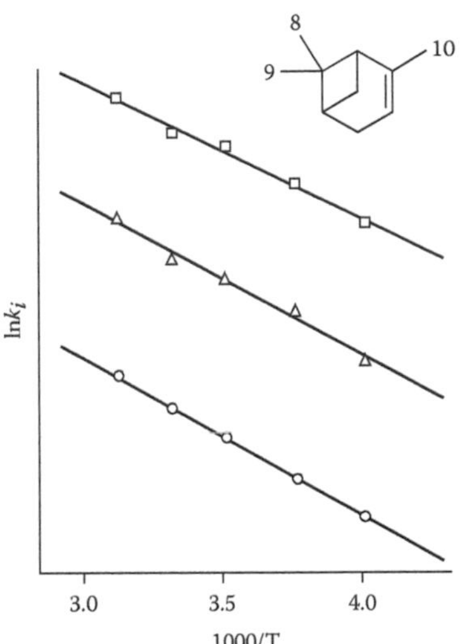

FIGURE 11.5 Arrhenius plots of ln k_i vs inverse temperature for the three methyl groups in α-pinene (shown in the inset). Circles indicate Me-8, triangles Me-9, and squares Me10. (Reprinted with permission from Ericsson, A. *et al.*, *J. Magn. Reson.*, 38, 9–22, 1980. Copyright (1980) Elsevier.)

is often known. We shall discuss the relaxation of this type of spin in more detail in Chapter 14. Here, we just need to state that in the extreme narrowing regime, the quadrupolar spin-lattice relaxation is a simple exponential process with the rate $T_{1,Q}^{-1}$, equal to the spin-spin relaxation rate and given by:

$$\frac{1}{T_{1,Q}} = \frac{1}{T_{2,Q}} = \frac{3}{40}\left(\frac{e^2qQ}{\hbar}\right)^2 \frac{2I+3}{I^2(2I-1)}\tau_c \tag{11.6}$$

where the symbol e is the elementary charge, Q is the nuclear quadrupole moment, a characteristic property of each nuclear species with $I \geq 1$ (see Table 1.1 for examples), q is the principal component of the electric field gradient tensor at the site of the nucleus, assumed to be axially symmetric. The quantity $(e^2qQ/\hbar)$, called the *quadrupolar coupling constant* (QCC), can be measured in the solid state and has the units of radians per second. Finally, τ_c is the effective rotational correlation time of the molecule-fixed principal axis of the field gradient tensor (the principal axis of the quadrupolar interaction). If we consider the particular case of a deuteron nucleus ($I = 1$) bound to a carbon in an organic molecule, the principal axis of the quadrupolar interaction coincides to an excellent approximation with the principal axis of the corresponding carbon-proton dipolar interaction in the analogous non-deuterated compound, and the assumption of the axial symmetry is fulfilled. Setting in $I = 1$ in Eq. (11.6) yields:

$$\frac{1}{T_{1,Q}} = \frac{3}{8}\left(\frac{e^2qQ}{\hbar}\right)^2 \tau_c \tag{11.7}$$

which is very similar to Eq. (3.32a) and where the effective correlation time has the same meaning as its carbon-proton dipolar counterpart in that equation.

We now wish to turn to some examples in which the information from dipolar relaxation of spin 1/2 nuclear species is combined with that from the quadrupolar interaction of $I \geq 1$ spins. A very simple such example is the so-called *dual spin probe technique*. An illustrative case where this technique has been applied is the aluminium complex *tris*(pentane-2,4-dionato)Al(III), or Al(*acac*)$_3$, shown in Figure 11.6. Dechter *et al.*[12] reported measurements of carbon-13 spin-lattice relaxation for the methine carbon and

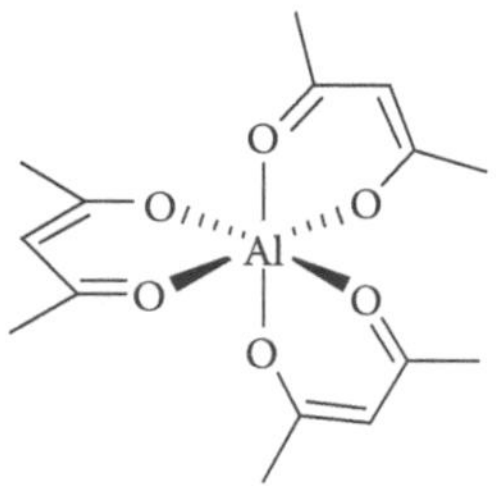

FIGURE 11.6 *Tris*(pentane-2,4-dionato)Al(III), or Al(*acac*)$_3$.

of the aluminium-27 linewidth in the complex dissolved in toluene. The carbon data allowed an estimation of the rotational correlation time for the reorientation of the CH axis.[25] Al is a quadrupolar nucleus ($I=5/2$) and its linewidth directly gives the $T_{2,Q}$. Assuming that the quadrupolar interaction is characterised by the same rotational correlation time, reasonable for the close-to-spherical molecular shape of Al(*acac*)$_3$, one can use Eq. (11.6) to estimate the effective QCC. The QCC in this particular case was also determined by solid-state NMR. The two measurements agreed very well with each other (in fact, there is a small error in the paper by Dechter *et al.*: the QCC measured from the linewidth should be a factor 2π larger than what is stated in the article).

More recently, Champmartin and Rubini[13] studied carbon-13 and oxygen-17 (another $I=5/2$ quadrupolar nucleus) relaxation in the same compound (as well as in the free pentane-2,4-dione) in solution. The carbon relaxation was measured as a function of the magnetic field. The methine carbon showed no field dependence, as expected, while the carbonyl carbon T_1^{-1} increased linearly with B_0^2. This indicates that the chemical shielding anisotropy (CSA) mechanism is important and allows an estimate of its interaction strength, which gives the anisotropy of the shielding tensor. Also, this quantity could be compared with the solid-state measurements on Al(*acac*)$_3$ and, again, the agreement was good. From the oxygen-17 linewidth, the authors also obtained the oxygen-17 QCC. The chemically interesting piece of information is the observation that the QCC changes only slightly between the free acid and the trivalent metal complex.

11.3 Field-Dependent Relaxation: Slower Motion

For medium-sized or large molecules undergoing rotational motion on a timescale outside the extreme narrowing regime, the spectral density functions become field dependent. The field dependence can be used to probe the spectral density function at different frequencies and thus, to test dynamic models. In this way, it is possible to extract information not only about a single effective rotational correlation time, as described for the extreme narrowing case, but also about fast local dynamics for each measured site within a molecule. Since proton relaxation is often complicated by the occurrence of several neighbouring protons, this is seldom used to probe molecular dynamics, and the main focus of this section is therefore on heteronuclear dipole-dipole relaxation. Proton relaxation rate is, however, one of the parameters used in the so-called spectral density mapping approach, which will be discussed in more detail in Section 11.4.2. It is also one of the cornerstones in determining structural parameters from nuclear Overhauser enhancement spectroscopy (NOESY)-based experiments, which will be further discussed in Chapter 12.

Here, relaxation for spin 1/2 S-nuclei, attached to one or several protons (I-spins), for which the dipole-dipole interaction with the directly bonded I-spins is the dominant relaxation mechanism will be discussed. The typical example of such a case is the relaxation of naturally occurring carbon-13 nuclei in organic or biological molecules. The equations describing the field-dependent behaviour of carbon-13 dipole-dipole T_1, T_2 and NOE relaxation were given in Section 3.4, where an introduction to carbon-13 relaxation was provided. The situation becomes a little more complicated if the molecule is uniformly labelled with carbon-13, in which case also the dipole-dipole interaction between neighbouring carbon-13 nuclei must be considered. For nitrogen-15, the relaxation through the chemical shift anisotropy mechanism must also be considered, since the CSA is generally much larger for nitrogen-15 than for carbon-13. As mentioned in the preceding section, the CSA is in certain cases important also for carbon-13 relaxation. Provided that the relaxation by the CSA mechanism can properly be taken into account when analysing the relaxation parameters, *i.e.* assuming that the CSA parameter is known with sufficient accuracy and that the bond length between the interacting nuclei is well characterised, the dipole-dipole relaxation can be used to elucidate overall rotational dynamics as well as providing information about local dynamics.

11.3.1 Applications in Organic Chemistry

An interesting example illustrating the possibilities and difficulties encountered when analysing carbon-13 relaxation outside of extreme narrowing is provided by the highly symmetric molecule hexamethylenetetramine (HMTA), studied by Kowalewski and co-workers[14] and shown in Figure 11.7. The T_d symmetry of the molecule guarantees that the overall reorientation has to be isotropic, and the polycyclic structure, without low-lying vibrational modes, makes the assumption of a rigid-body reorientation reasonable and attractive.

The molecule is so small that its motion is within the extreme narrowing regime in aqueous solution. In order to push the system outside of extreme narrowing, a cryosolvent consisting of a certain mixture of H_2O/dimethyl sulfoxide (DMSO) was used. The solution remains liquid at −30°C, but the carbon-13 relaxation is under these conditions dependent on the magnetic field (*cf.* Figure 11.7). The seven data points (three T_1^{-1}s, three NOEs and one T_2^{-1}) can be very well reproduced by Eqs. (3.31) and (3.35) combined with Eq. (3.16), assuming that the contributions of the two CH dipolar interactions within a methylene group to T_1^{-1} and T_2^{-1} are additive (this is an approximation, as discussed in the article, but without practical consequences for the case at hand). The field-dependent relaxation rates are thus described by only two parameters: the dipole-dipole coupling constant and the rotational correlation time. The correlation time from the fit is close to 1 ns (corresponding to $\omega_C^2\tau_c^2 \approx 0.4$ and $(\omega_C + \omega_H)^2\tau_c^2 \approx 10$ at the magnetic field of 9.4 T) and the dipole coupling constant $b_{CH} = -1.274 \cdot 10^5 \text{rads}^{-1}$. Using Eq. (3.5), this dipole-dipole coupling constant translates into r_{CH} of 114.2 pm, an unexpectedly long bond length.

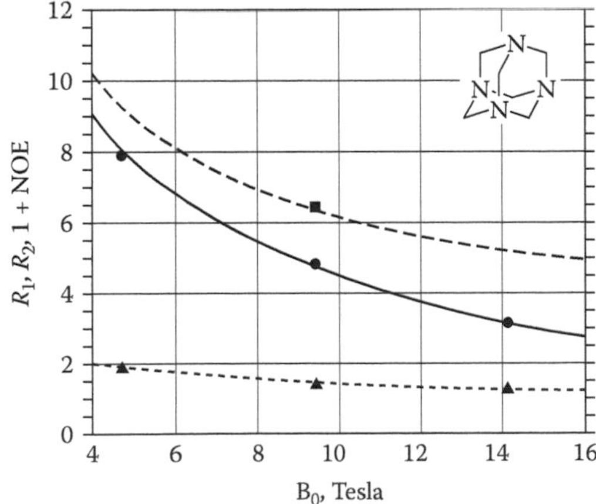

FIGURE 11.7 Relaxation data for hexamethylenetetramine (HMTA). Triangles show measured NOE factors, circles R_1, and squares R_2 relaxation rates. The curves are calculated relaxation rates from the fitted motional parameters using the truncated Lipari–Szabo spectral density function. (Reprinted with permission from Kowalewski, J. *et al. J. Magn. Reson.* 157, 171–177, 2002. Copyright (2002) Elsevier.)

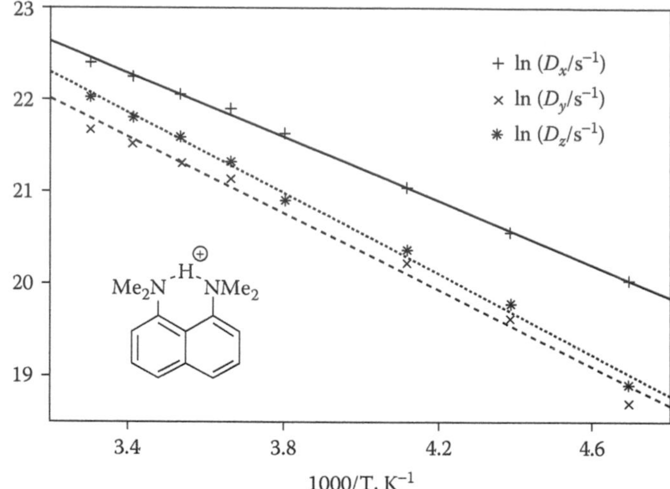

FIGURE 11.8 The rotational diffusion tensor elements for 1,8-bis(dimethyl amino)naphthalene (DMANH$^+$), shown in an Arrhenius-type plot. (Reprinted with permission from Bernatowicz, P. *et al.*, *J. Phys. Chem. A* 109, 57–63, 2005. Copyright (2005) American Chemical Society.)

The authors were able, at least in part, to explain the difference between this value and the neutron diffraction results of 107–110 pm by including a harmonic vibrational correction, resulting in a corrected dipole-dipole coupling constant of $1.389 \cdot 10^5$ rad s^{-1} or an internuclear distance of about 111 pm. The conclusion from this work is that care needs to be exercised when choosing the exact value of the dipolar coupling constant to be used in relaxation work, even in cases when the internuclear distance seems to be known from independent and reliable measurements.

In a somewhat related work, Bernatowicz and co-workers studied carbon-13 in solutions of the protonated form of 1,8-bis(dimethyl amino)naphthalene (DMANH$^+$) in dimethylformamide at three different magnetic fields[15]. By cooling the solution, they found the system to be far outside of the extreme narrowing conditions. The naphthalene part of the molecules is a reasonably rigid, asymmetric top, with sufficiently high molecular symmetry to determine the orientation of the principal axis of the rotational diffusion tensor (*cf.* Figure 11.8). The relaxation of the proton-bearing aromatic carbons was found to be dominated by the dipole-dipole interaction, while CSA relaxation was most important for the quaternary carbons. From the relaxation data together with CSA tensors from solid-state measurements and dipole coupling constants based on geometry from neutron scattering, it was possible to determine the magnitude of the three principal elements of the rotational diffusion tensor as a function of temperature. The tensor elements are plotted in an Arrhenius-type plot in Figure 11.8. The linearity of the plots is very good, with the activation energies of 14.5 ± 0.2, 17.4 ± 0.8 and 18.3 ± 0.5 kJ mol^{-1} for the rotational diffusion about the *x*-, *y*- and *z*-axis, respectively, which is a nice confirmation of the usefulness of the rigid-body, anisotropic tumbling model.

11.3.2 Carbohydrates

In organic chemistry, field-dependent relaxation in carbohydrates has been an area of intense investigation. One of the main questions has concerned the local motion about the glycosidic linkages. Sucrose (see Figure 11.9a) has served as a model compound for carbon-13 relaxation investigations, and the first attempt to characterise the dynamics in this molecule was presented in the late 1980s. Sucrose is a small molecule with reorientational dynamics within the extreme narrowing regime in aqueous solution at ambient and commonly used temperatures and available magnetic field strengths. Many of the

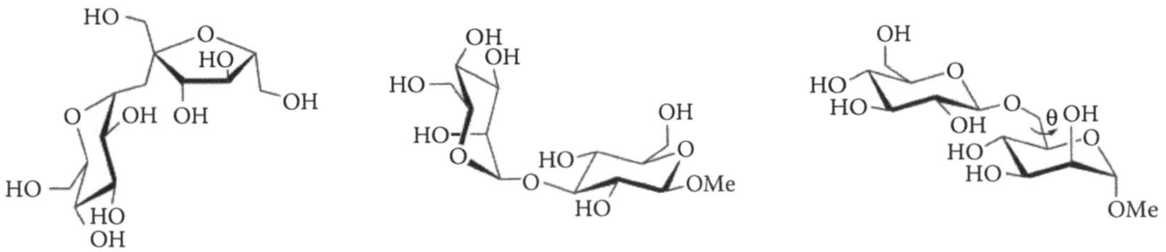

(a) (b) (c)

FIGURE 11.9 Three disaccharides for which relaxation is discussed in the text. (a) Sucrose; (b) methyl 3-O-α-D-mannopyranosyl-(1→3)-β-D-glucopyranoside; (c) β-D-Glc*p*-(1→6)-α-D-[6-^{13}C]-Man*p*-OMe. The θ angle is discussed in the text.

investigated carbohydrates have been, in fact, small di-, tri- or oligosaccharides, which are not necessarily outside the extreme narrowing regime in aqueous solution (Figure 11.9). In order to investigate the local dynamics, certain experimental tricks have been employed, such as decreasing the temperature and increasing viscosity by choosing solvent mixtures; for instance, the cryogenic water/DMSO or water/glycerol mixtures. In this way, the overall global reorientation is made sufficiently slow to render the relaxation field dependent. Under these conditions, the influence of local motion on the relaxation parameters can be probed. In these investigations, natural-abundance ^{13}C relaxation is often used to probe reorientational dynamics and in particular to study differences in local dynamics within a molecule. For proton-bearing carbons, the relaxation is almost completely dominated by intramolecular ^{13}C-^{1}H dipole-dipole relaxation, and relaxation by the CSA mechanism can safely be neglected. The R_1, R_2 and steady-state NOE factors are thus given by Eqs. (3.31) and (3.35).

Sucrose dynamics has been investigated in aqueous solution as well as in the cryogenic mixture of water/DMSO. McCain and Markley[16] studied the dynamics of sucrose in aqueous solution, where a weak field dependence of T_1 could be observed below room temperature. They showed that the dynamics could be accurately described by the introduction of an amplitude factor into the rigid-rotor spectral density function:

$$J(\omega) = \frac{1}{4\pi}\langle A(0)A(p)\rangle \frac{2\tau_c}{1+\omega^2\tau_c^2} \tag{11.8}$$

in which $\langle A(0)A(p)\rangle$ is the amplitude factor, evaluated at a time p, when the averaging by the internal motions is complete; that is, the very rapid initial decay in the correlation function caused by internal motion has already occurred. The amplitude factor can readily be identified as the Lipari–Szabo order parameter, S^2, in the truncated form of the model-free spectral density function given by:

$$J(\omega) = \frac{1}{4\pi} S^2 \frac{2\tau_c}{1+\omega^2\tau_c^2} \tag{11.9}$$

The study by McCain and Markley demonstrated that this spectral density function could be used successfully to describe internal motions in sucrose. They identified subtle differences in amplitude factors, which were explained as differences in internal motion. A subsequent study by Kovacs *et al.*[17] showed that in the cryogenic water/DMSO mixture, a more detailed picture of the internal motion could be obtained by pushing the system far outside the extreme narrowing regime.

The most common approach to elucidating dynamical properties from relaxation parameters is to model the spectral density function using the method often referred to as the *model-free approach*, originally developed by Lipari and Szabo[18]. A derivation of the relevant time-correlation functions and spectral densities is provided in Chapter 6, and a summary of the model is given in this section. The modulation of a local field produced by a molecule-fixed interaction (*e.g.* the DD interaction; *cf.* Chapter 3) is here produced by the overall rotational Brownian motion of a molecule as well as by local motions of nuclei within the molecule. The time-correlation function for isotropically tumbling molecules can be factored into two parts, providing the following equation:

$$G(t) = G_O(t)G_I(t) \tag{11.10}$$

in which $G_O(t)$ is the overall time correlation function for an isotropic reorientation of the molecule, characterised by a correlation time τ_M and given by Eq. (3.15) or (6.18). The symbol τ_M, commonly used in the literature, has the same meaning as τ_c or τ_R in earlier chapters.

The time-correlation function for the internal local motion, $G_I(t)$, can in general be written as a sum of exponentials, and the simplest, approximate internal correlation function is given by Eq. (6.56), repeated here for the convenience of the reader:

$$G_I(t) = S^2 + \left(1 - S^2\right)\exp\left(-t/\tau_e\right) \tag{11.11}$$

where S^2 is the square of the generalised order parameter that describes the spatial restriction of the internal motion (defined in Eq. (6.55)) and τ_e is the correlation time for the faster, local motion (the same as τ_{loc}, used in Section 6.3).

The corresponding spectral density function is obtained by performing a Fourier transformation and is given by:

$$J(\omega) = \frac{1}{2\pi}\left[S^2 \frac{\tau_M}{1+\omega^2\tau_M^2} + \left(1 - S^2\right)\frac{\tau}{1+\omega^2\tau^2}\right] \tag{11.12}$$

where $1/\tau = 1/\tau_M + 1/\tau_e$. A similar equation was formulated, using slightly different symbols, in Eq. (6.59c). The generalised order parameter takes on values satisfying $0 < S^2 < 1$. S^2 close to unity implies a highly restricted local mobility, and S^2 closer to zero a more unrestricted motion. This parameter has been found to be extremely useful for characterising local dynamics in molecules outside the extreme narrowing regime.

Although originally developed for studies of protein dynamics, this model has been widely used for describing molecular dynamics for organic compounds, such as carbohydrates. Generally, if a high degree of restriction of the local motion is present (high order parameter), the truncated form of the model-free spectral density function, with the second term in Eq. (11.12) omitted (compare Eq. (11.9)), can be used to model relaxation parameters for small and medium-sized molecules outside of extreme narrowing. Starting with an assumed value of the dipolar coupling constant, this treatment yields an overall global correlation time and an order parameter for each measured site. The actual magnitude of the dipolar coupling constant is not critically important, which is fortunate (compare the example of HMTA in Section 11.3.1). The possible errors in the coupling strength are absorbed in the order parameter, which does not matter too much if one wants to make comparisons between different sites in one molecule or between similar systems, provided that the internuclear distances can be assumed to be the same for all sites.

One of the interesting questions concerning dynamics in oligosaccharides is related to the possibility of local motion over the glycosidic linkage. Turning again to sucrose, NOE and rotating-frame Overhauser enhancement (ROE) results have indicated a high degree of local flexibility of the glycosidic linkage[19]. In a study of the disaccharide in Figure 11.9b[20], the question regarding the flexibility of the glycosidic linkage was again addressed. If this motion occurs on a timescale comparable to the overall motion, it is possible that this escapes carbon-13 relaxation investigations. Therefore, the carbon-13 relaxation data were complemented by measurements of proton cross-relaxation rates between nearby protons in different sugar units. It was found that the field and temperature dependence of both the intra-residue and the inter-residue ^{1}H-^{1}H cross-relaxation could be interpreted using the Lipari–Szabo dynamic parameters (S^2, τ_M) obtained from carbon-13 relaxation and fitting only the ^{1}H-^{1}H internuclear distance. The field dependence of the two cross-relaxation rates is displayed in Figure 11.10. It was possible to conclude that the inter-residue ^{1}H-^{1}H axis did not sense any additional mobility as compared with the intra-residue ^{13}C-^{1}H axes.

The combined carbon-13 and proton study of the disaccharide displayed in Figure 11.9b was followed by a molecular dynamics simulation[21], which yielded results in agreement with experiments. A question one can ask is why that disaccharide seems to behave differently from sucrose. This issue was studied by Effemey and co-workers[22]. A possible answer is that the protons in one of the sugar residues participating in the inter-ring cross-relaxation that was studied in sucrose reside in a hydroxymethyl group, which may have internal motions of its own, independently of the glycosidic linkage. It should, however, also be emphasised that the apparent rigidity of the glycosidic linkage is related to the timescale probed by nuclear spin relaxation[23].

The third disaccharide example is the compound displayed in Figure 11.9c. The spin system of principal interest is here the ^{13}C-labelled CH_2 group next to the glycosidic linkage. This case is more difficult in the sense that one can expect the torsional motion about the θ angle to be on a similar timescale to the global reorientation, while the motions involving the glycosidic linkage are quite slow. This molecule was studied by Zerbetto and co-workers[24], again using the same cryosolvent. The experimental data set

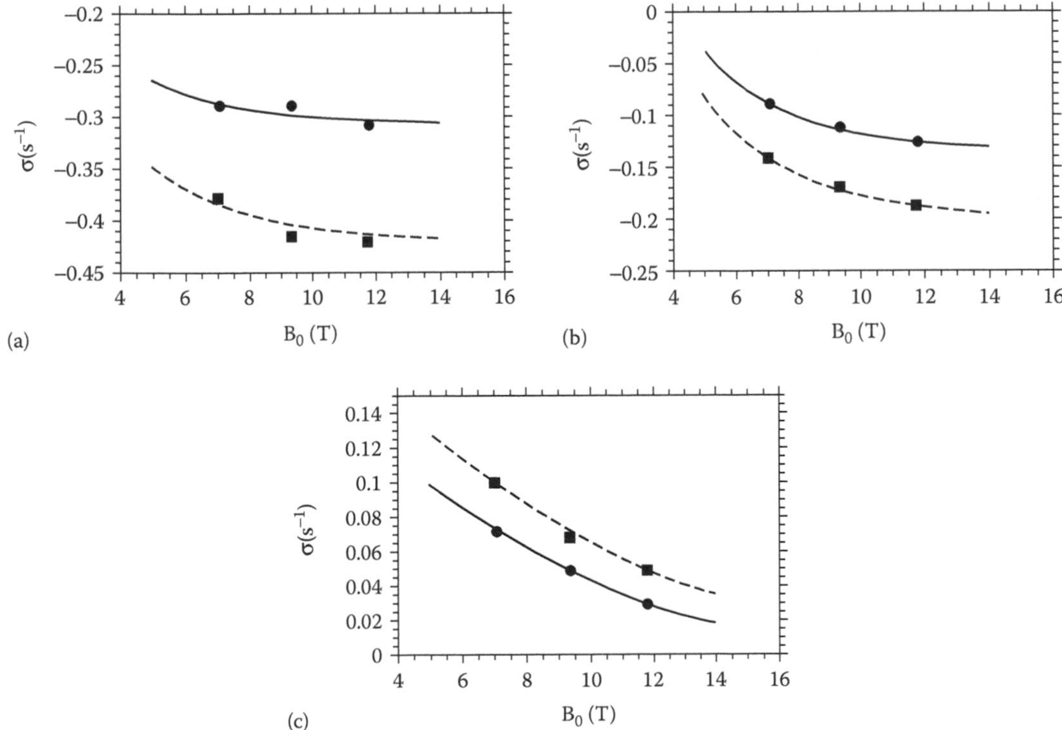

FIGURE 11.10 Field dependence of the cross-relaxation rates measured for the disaccharide shown in Figure 11.9b. Measurements were performed at three temperatures: (a) 268 K, (b) 282 K and (c) 323 K. Circles are the experimental points for the intra-residue cross-relaxation (H1′-H2′), and squares are the experimental points for the inter-residue cross-relaxation (H1′-H3′). (Reprinted with permission from Mäler, L. *et al.*, *J. Phys. Chem.*, 100, 17103–17110, 1996. Copyright (1996) American Chemical Society.)

consisted of five relaxation parameters relevant for the $^{13}CH_2$ group: carbon-13 T_1 and T_2, the heteronuclear NOE, and the cross-correlated dipolar relaxation rates (longitudinal and transverse) discussed in Chapter 10. The measurements were carried out at three temperatures and two magnetic fields (three fields at 293 K). The analysis was performed using an integrated mesoscopic/atomistic model, described in the same paper and mentioned in Chapter 6, designed in the spirit of the slowly relaxing local structures (SRLS) approach as a coupled two-body motion. One of the moving bodies was the molecule as a whole undergoing global tumbling, while the other one was the methylene group rotating around the θ angle. The dynamics was described by the Smoluchowski equation derived from a combination of atomistic bead model hydrodynamics (making use of the known viscosities at the three temperatures and providing the diffusive properties of the system) and a DFT calculation of the potential for the internal motion. The model thus contained only one parameter not assumed known *a priori*: the HCH angle, which was determined by fitting the data. The overall agreement between experiments and simulations was good. Related approaches have subsequently been applied to the data for γ-cyclodextrin[25] and linear oligosaccharides[26].

11.3.3 Guest-Host Complexes

Another class of chemical systems for which the question of internal mobility is of great interest are inclusion compounds or guest-host complexes. The "host" usually refers to a larger molecule (or crystal) with a cavity that can be occupied by a smaller "guest" molecule. Classical carbon-13 relaxation work concerned with the mobility of the guest inside the host cavity was published in the 1970s by Brevard *et al.*[27], who coined the concept of *dynamic coupling*, expressed as a ratio of the effective rotational correlation times for the host and the guest, obtained through carbon-13 relaxation in the extreme narrowing range (compare Section 11.1).

In some more recent studies, complexes between the host cryptophane E (*cf.* Figure 11.11) and chlorinated methanes (chloroform [$CHCl_3$] and dichloromethane [CH_2Cl_2] dissolved in tetrachloroethane) were investigated by carbon-13 relaxation. The solvent molecules are too big to enter the cryptophane cavity, while the chlorinated methanes exchange slowly, on both the carbon-13 and the proton timescale, between the bulk solution and the encaged site. The solution is viscous enough to bring the cryptophane complexes outside of extreme narrowing. The relaxation of host ^{13}C nuclei can be described by the Lipari–Szabo model, which yielded fairly high order parameters squared of about 0.7–0.8. In terms of the Lipari–Szabo model and its order parameter, the chloroform as a guest behaves as if it were an integral part of the molecule; its motion is characterised by an S^2 value similar to the methylene linkers in the host[28]. The situation is very different for the smaller dichloromethane guest[29]. The carbon-13 in this encaged molecule is characterised by a long T_1, similar to that of the free molecule in solution, with a weak but measurable field dependence, and an NOE factor that is a little less than full. The data

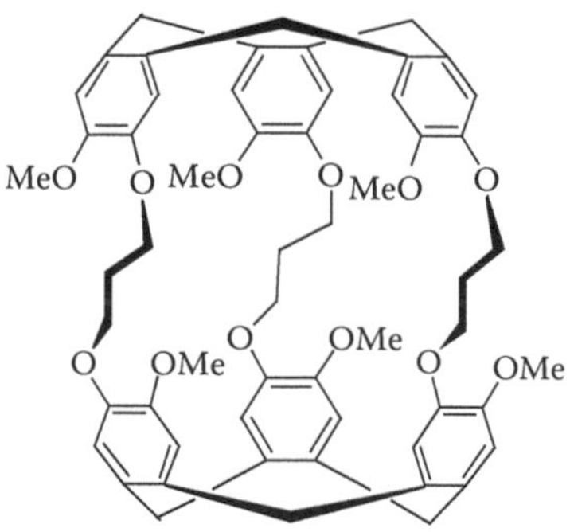

MeO O MeO O MeO O

O OMe O OMe O OMe

FIGURE 11.11 Structure of cryptophane E.

could be interpreted using the Lipari–Szabo spectral density with a very low order parameter, indicating an almost unrestrained motion of CH_2Cl_2 within the cryptophane E cavity. The great variation of the mobility of the two guests inside the cryptophane E cavity was nicely confirmed by the measurements of CH dipolar couplings in the solid inclusion compounds[30].

Similar work was also reported for a number of smaller cryptophanes and their complexes with chloroform and dichloromethanes; see, for example, papers by Nikkhou Aski and co-workers[31] and by Takacs *et al.*[32]. In the latter work, the authors described also an additional type of dynamic process: conformational exchange of the host. The mechanisms of these processes were discussed in terms of concepts introduced long ago in protein chemistry[33]: induced fit and conformational selection. In addition, structural aspects and energetics were discussed, based on a combination of DFT calculations and NOE measurements. We shall return to these problems in Chapters 12 and 13.

11.4 Molecular Dynamics in Proteins

NMR relaxation of proteins can provide a wealth of information concerning the dynamics on the nanosecond to picosecond timescale. Relaxation can be measured for backbone nuclei, such as the amide nitrogen-15, as well as for nitrogen-15 or carbon-13 nuclei in the side-chain. The limitation is usually set by the fact that the protein has to be uniformly labelled with nitrogen-15 or partially labelled with carbon-13 (and deuterium) for these types of measurements to be practically feasible. Dynamics on this timescale (complemented with measurements of chemical exchange; see Chapter 13) provide important information about biologically significant motions that complements information about protein structure.

Typically, several relaxation parameters (R_1, R_2, NOE) are measured for nitrogen-15 or carbon-13. The measurement of R_2 relaxation rates has proved to be important in determining whether slow motions, such as those described as chemical exchange, contribute to the relaxation. If this is the case, these slow motions must be accounted for properly when interpreting the relaxation data. Several ways of doing this exist, and the study of exchange phenomena by NMR relaxation will be dealt with in Chapter 13. The data are typically interpreted with the model-free approach, described in Chapter 6 and in the previous section, although extensions to this model as well as alternative methods exist.

11.4.1 Backbone Motion

Since the mid-1980s, spin relaxation studies have become increasingly popular within life science, where the interpretation of relaxation parameters in terms of restricted backbone flexibility in proteins has become an important application. R_1, R_2 and steady-state NOE factors for backbone amide nitrogen-15 nuclei (S-spin) are measured at one or, preferably, several magnetic field strengths. At this stage, chemical exchange is usually accounted for by adding a phenomenological term to the R_2 relaxation rate, R_{ex}, which becomes a parameter in the fitting of the relaxation data. With this in mind, the R_1, R_2 and NOE relaxation rates are given by Palmer[34]:

$$R_1 = \frac{\pi}{5} b_{NH}^2 \left[J(\omega_H - \omega_N) + 3J(\omega_N) + 6J(\omega_H + \omega_N) \right] + \frac{4\pi}{15} c_N^2 J(\omega_N) \tag{11.13a}$$

$$\begin{aligned} R_2 = {} & \frac{\pi}{10} b_{NH}^2 \left[4J(0) + J(\omega_H - \omega_N) + 3J(\omega_N) + 6J(\omega_H) + 6J(\omega_H + \omega_N) \right] \\ & + \frac{2\pi}{45} c_N^2 \left[4J(0) + 3J(\omega_N) \right] + R_{ex} \end{aligned} \tag{11.13b}$$

$$\sigma_{NH} = \frac{\pi}{5} b_{NH}^2 \left[6J(\omega_H + \omega_N) - J(\omega_H - \omega_N) \right] \tag{11.13c}$$

$$NOE = 1 + \frac{\gamma_H}{\gamma_N} \frac{\sigma_{NH}}{R_1} \tag{11.13d}$$

where the coefficient b_{NH} is the dipolar interaction strength constant (identical to the b_{IS} of Eq. (3.5)), and c_N is the interaction strength for the CSA relaxation mechanism, defined in Chapter 10.

The parameters c_N and b_{NH} are usually treated as constants, based on $r_{NH} = 102$ pm and $\Delta\sigma = -160$ ppm or similar values. If data are available at several magnetic field strengths, the contribution of chemical exchange to $J(0)$ in the equation for R_2 can be obtained from the expression[34,35]

$$\begin{aligned} Y &= R_2(B_0) - \frac{1}{2} R_1(B_0) - \frac{3\pi b_{NH}^2}{2} J(\omega_H) \\ &= \pi J(0) + \left(4\pi\gamma_X^2 \Delta\sigma_X^2 J(0)/9 + Q_{ex} \right) B_0^2 \end{aligned} \tag{11.14}$$

in which it is assumed that the exchange is dependent on the magnetic field strength as $R_{ex} = Q_{ex} B_0^2$. A linear plot of Y versus B_0^2, yields $J(0)$ directly from the intercept with the y-axis. A second way of determining whether exchange contributes to the measured R_2 relaxation rate is from cross-correlated relaxation, as will be further discussed in Chapter 13[36].

If sufficiently many relaxation parameters are measured, such as the three (R_1, R_2 and NOE) given in Eq. (11.13), perhaps at more than one field, Eq. (11.12) can be fitted to the data to yield S^2, τ_M and τ_e. If measurements are conducted at several fields, the most common way to account for chemical exchange is to include also R_{ex} as a parameter in the fit. The exchange contribution is assumed to be proportional to the square of the magnetic field. If the measurements are carried out for several sites within a molecule, say all of the backbone nitrogen-15 atoms, the degree of restriction of the local motion as a function of protein sequence can be elucidated. The model-free approach requires that one single overall rotational correlation time can be used to describe the global reorientation of the molecule. The overall correlation time is often estimated from the ratio of R_1 and R_2 relaxation rates for different sites in the molecule but is more reliably obtained by a simultaneous fit of all data to a global reorientational correlation time.

In the previous examples in this chapter, measurements were mainly performed on samples with natural abundance of the S-spins. The poor sensitivity and the relatively low concentration of protein in the sample make measurements on proteins with nitrogen-15 at natural abundance impossible even with today's instrumentation. All protein dynamic studies are therefore performed with isotopically enriched proteins.

One of the earlier examples of protein backbone dynamics studied by nitrogen-15 relaxation is staphylococcal nuclease (S. Nase)[37]. In this work, nitrogen-15 T_1, T_2 and NOE relaxation parameters were measured at several fields. The data were used to calculate sequence-dependent order parameters as well as linewidths (from T_2 values). The order parameters were found to be in the range of $S^2 = 0.86 \pm 0.04$ for well-ordered regions of secondary structure within the protein (Figure 11.12). Values within this range have today become the signature for structured regions in a protein, and the order parameter is a common way of describing differences in local dynamics within proteins. The T_2 values (or linewidths) were interpreted in terms of contributions from slower motions, although not as directly as the analysis indicated in Eqs. (11.13) and (11.14).

As a second example of protein dynamics, we will look at the backbone dynamics of the Ca^{2+}-binding protein calbindin D_{9k}, a small (75 amino acid residues) protein that is well characterised with respect to both structure[38–41] and dynamics[41–43]. The structure of this protein mainly consists of four helices that are connected by three loops. The helices come in pairs, held together by a Ca^{2+}-binding loop, and the two pairs are connected by a linker (Figure 11.13).

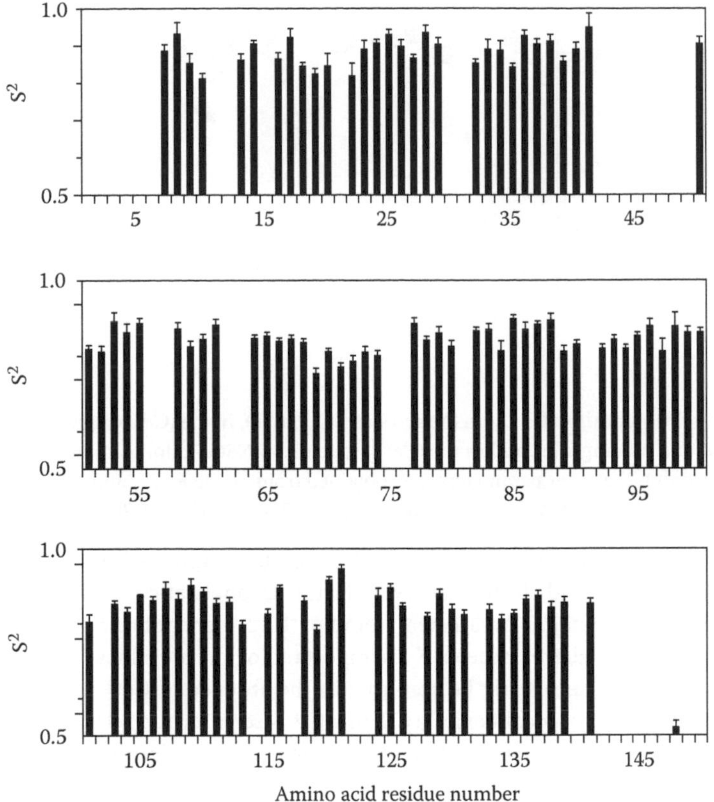

FIGURE 11.12 S^2 values for the backbone amide nitrogens of staphylococcal nuclease obtained from relaxation data recorded at 11.75 and 14.08 T. (Reprinted with permission from Kay, L.E. *et al.*, *Biochemistry*, 28, 8972–8979, 1989. Copyright (1989) American Chemical Society.)

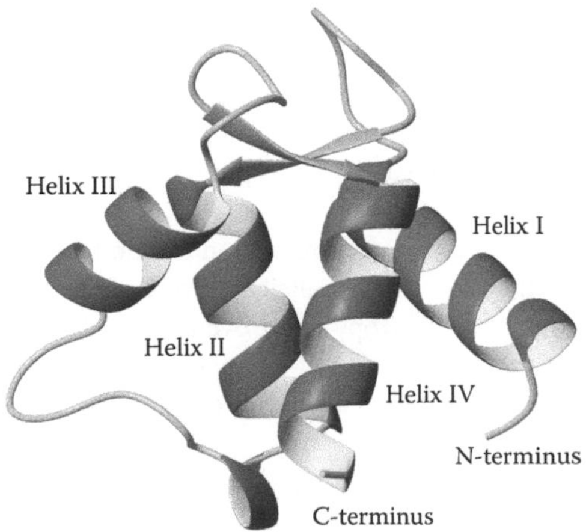

FIGURE 11.13 The structure of apo calbindin D_{9k}. The helices and sheets are displayed as ribbons, while the turns are represented as sticks. The coordinates were taken from the Protein Data Bank (www.rscb.org, accession code 1CLB). The figure was produced using Molmol (version 2.6).

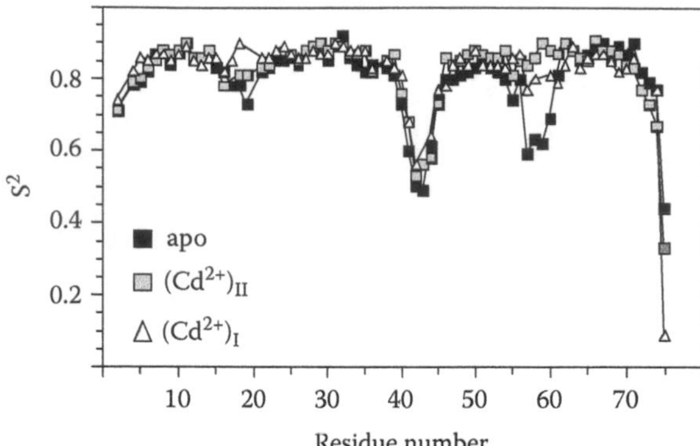

FIGURE 11.14 S^2 values for calbindin D_{9k} in the apo state (Ca^{2+}-free), in the $(Cd^{2+})_{II}$ state (modelling the state with a Ca^{2+} ion in the second binding site), and in the $(Ca^{2+})_I$ state (from N56A calbindin D_{9k} with a Ca^{2+} ion bound to the first binding site only). The order parameters were obtained from fitting R_1, R_2, and NOE relaxation parameters. (Reprinted with permission from Mäler, L. *et al.*, *Nature Struct. Biol.*, 7, 245–250, 2000.)

The dynamics of the protein, with and without Ca^{2+} ions bound in the binding sites, has been determined by measuring T_1, T_2, and steady-state NOE factors at one static magnetic field strength. From these data, the model-free dynamic parameters from Eq. (11.12) were obtained for nearly all amino acid residues. In particular, the order parameter, S^2, is important for the identification of differences in local mobility. In Figure 11.14, it can clearly be seen that in the absence of Ca^{2+} ions, the two ion-binding regions have a higher degree of local flexibility than what is seen for the relatively rigid helices, as evidenced by the values for the S^2. Upon addition of Ca^{2+} ions, the two binding regions become almost as motionally restricted as the rigid helices. This example illustrates the strength of the approach. One can relatively easily investigate changes in local flexibility that may be important in understanding underlying biological phenomena and that are not obvious from examining the structure alone. The intermediate states of calbindin D_{9k}, in which an ion was bound to only one binding site at a time, were also investigated by nitrogen-15 relaxation. In these investigations[41], data were recorded at two magnetic field strengths, which proved to be important, since regions with clear influences from chemical exchange on R_2 values were observed. This indicates that the two intermediate states are influenced by some sort of dynamic process on a relatively slow timescale. Furthermore, it was shown that for both intermediate states, the local mobility of both ion-binding loops was severely restricted (higher S^2 values) when a single ion was bound to either site (Figure 11.14).

In this example, the data could be used to analyse cooperativity in the ion-binding process, something that could not be done from the structures of the intermediates alone. The investigation of calbindin D_{9k} is an example of how dynamics revealed from spin relaxation can provide physical information that cannot be obtained by other means.

A folded, globular protein can to a first approximation be modelled as a not completely rigid but isotropically tumbling object, and this is often sufficient to draw at least qualitative conclusions about changes in local dynamics. However, anisotropy in the overall global motion may produce large errors, and even misinterpretations, in the resulting model-free parameters. The spectral density function for a molecule with an axially symmetric rotational diffusion tensor is given by[18,44]:

$$J(\omega) = \frac{1}{2\pi}\sum_{j=0}^{2} A_j\left[S^2\frac{\tau_j}{1+\omega^2\tau_j^2} + \left(1-S^2\right)\frac{\tau_j'}{1+\omega^2\tau_j'^2}\right] \tag{11.15}$$

in which $\tau_j^{-1} = 6D_\perp - j^2(D_\perp - D_\parallel)$, $D_\perp$ and $D_\parallel$ are principal components of the axially symmetric diffusion tensor, and the coefficients A_j are functions of the angle θ between the principal axes of the interaction tensors (the internuclear vector in the case of dipole-dipole interaction) and the unique axis of the diffusion tensor: $A_0 = \frac{1}{4}(3\cos^2\theta - 1)^2$, $A_1 = 3\sin^2\theta\cos^2\theta$, $A_2 = \frac{3}{4}\sin^4\theta$. Finally, $1/\tau_i' = 1/\tau_i + 1/\tau_e$, where τ_e is the correlation time for the internal motion of the internuclear vector in the molecular reference frame, in analogy with the spectral density function for isotropic overall tumbling.

For the limiting case of a rigid body ($S^2 = 1$), we recover the original function derived by Woessner, which was given in Eq. (6.36). For proteins, the parameters describing the anisotropy in the diffusion tensor, θ, $D_\perp$ and $D_\parallel$ are usually obtained from the structure of the molecule, determined by either NMR or X-ray crystallography, assuming that the diffusion tensor components are proportional to the corresponding elements of the moment of inertia tensor. If a structure is not available, these parameters can also be obtained from the fitting of relaxation parameters, keeping in mind that more experimental data are required, generally obtained from measurements at several fields. More complex expressions for the spectral density function are required for the fully anisotropic case, *i.e.* $D_{xx} \neq D_{yy} \neq D_{zz}$. A word of warning is appropriate in this context: while the Lipari–Szabo model with isotropic global motion can be derived rigorously, this is not the case for more complex overall dynamics.

It is often the case that the dynamics of a molecule, such as a protein, cannot be described sufficiently by one single internal motion. Therefore, an extended model-free spectral density function has been proposed by Clore and co-workers[45], in which two internal motional processes, one slow and one fast, are accounted for. This spectral density function, corresponding to the time-correlation function of Eq. (6.64), is given by:

$$J(\omega) = \frac{1}{2\pi}\left[S_f^2 S_s^2 \frac{\tau_M}{1+\omega^2\tau_M^2} + \left(S_f^2 - S_f^2 S_s^2\right)\frac{\tau_{s'}}{1+\omega^2\tau_{s'}^2} + \left(1 - S_f^2\right)\frac{\tau_{f'}}{1+\omega^2\tau_{f'}^2}\right] \tag{11.16}$$

in which S_s and S_f are the two order parameters, one associated with the slow internal motion with correlation time τ_s and the other with the fast motional process with correlation time τ_f. For the "primed" correlation times, we have $\tau_{s'} = \tau_M\tau_s/(\tau_M + \tau_s)$ and $\tau_{f'} = \tau_M\tau_f/(\tau_M + \tau_f)$. In the limit $S_f^2 \to 1$, the last term vanishes, and Eq. (11.16) simplifies to Eq. (11.12), with $S^2 = S_s^2$ and $\tau = \tau_{s'}$. The last term in Eq. (11.16) can also be neglected if τ_f is sufficiently small and a truncated form of the equation is obtained. It is independent of τ_f but still contains the two order parameters. In practice, this truncated form is often the one that is used for fitting protein relaxation data. The spectral density function in Eq. (11.16) is particularly important for molecules exhibiting internal motion of high amplitude, often characterised by several correlation times. This is commonly the case for flexible regions in proteins, which can display complex local dynamics.

It is worth noting that one of the examples provided by Clore *et al.*[45] as an illustration of the need to use the extended model was S. Nase, the protein investigated early on by Kay *et al.*[37] and discussed earlier in this section. For some of the residues, the extended model was used to better fit the data. More recently, the data were, however, again re-analysed by Jaremko and co-workers[46], who argued in favour of the simple model-free approach with a two-parameter description of the internal dynamics. This illustrates the difficulties associated with correctly interpreting relaxation in terms of a motional model.

In the calbindin D_{9k} example, described previously, different motional models were used to characterise the motional behaviour of different regions, and the need for an additional order parameter was seen for residues in the linker region. Molecules, such as peptides, that bind to larger objects, such as micelles or enzymes, sometimes require an additional local motion to describe the "local dynamics" of the entire peptide on or within the complex. A word of caution on using these motional models to interpret relaxation data is required, however. If more parameters are introduced into the model, the fitting will naturally be better. To distinguish between a significant improvement in the fit, related to introducing a better physical description, and an improvement caused by the mere fact of having more parameters, a careful statistical analysis of the results should be performed.

Another important question related to the protein backbone dynamics concerns the possibility of correlation between the motions of different residues. As mentioned in Chapter 10, measurements of multiple-quantum dipolar relaxation rates allow this type of information to be obtained. As an example, we wish to mention the work by Fenwick *et al.*[47] on the third immunoglobulin binding domain of protein G, GB3[48], comprising 56 residues. Fenwick *et al.* measured cross-correlated relaxation rates (CCRRs) involving dipolar ^{15}NH (proxy for the peptide planes) and ^{13}CH (proxy for side chains) interactions in neighbouring residues using the protocol described briefly in Section 10.4. The set of CCRRs was obtained at one magnetic field (14.1 T) and related to the residual dipolar couplings. The theoretical basis of the analysis is the observation, mentioned in Section 6.3.3, that the CCRRs are sensitive to motions on a timescale slower than the global rotational correlation time, *i.e.* slower than nanoseconds, in the backbone of GB3. The authors were able to reveal a complex dynamic network of slightly correlated motions. The origin of the correlations was in part in the local environment (exemplified by the side chain orientation either towards the solvent or buried in the core of the protein) and partly in the nature of the secondary structure elements (α-helices, β-sheets, loops).

11.4.2 Spectral Density Mapping

A different approach to investigating molecular dynamics from relaxation data is by the so-called *spectral density mapping* method. This method allows the calculation of the spectral density function at different frequencies, without postulating a model, by measuring a sufficient number of relaxation parameters. Different relaxation parameters contain distinct linear combinations of the spectral density function taken at different frequencies, and the spectral density function at a certain frequency can therefore be obtained by a suitable linear combination of relaxation rates. For a two-spin system, such as the S-^{1}H (S often being ^{15}N or ^{13}C), various relaxation rates can be measured; for instance, the longitudinal and transverse relaxation of the S-spin and the S-^{1}H cross-relaxation rates. It is, however, useful to also create other non-equilibrium states in order to monitor their relaxation. For instance, the relaxation rate associated with the relaxation of the S-spin antiphase coherence, $2\hat{S}_x\hat{H}_z$, is given by a different linear combination of the spectral density function than either of the abovementioned rates.

In this way, by measuring a larger number of relaxation parameters, the spectral density function at different frequencies can be obtained without ambiguity. In the original method, proposed by Peng and Wagner[49,50], the relaxation rates in Eq. (11.17), containing a total of five different frequencies, were measured. The relationship between the relaxation rates and the spectral density function can be summarised as follows:

$$\begin{pmatrix} R_{1S}\left(\hat{S}_z\right) \\ R_{2S}\left(\hat{S}_{x,y}\right) \\ R_{2SH}\left(2\hat{S}_{x,y}\hat{H}_z\right) \\ R_{1SH}\left(2\hat{S}_z\hat{H}_z\right) \\ R_{1H}\left(\hat{H}_z\right) \\ \sigma_{1H}\left(\hat{H}_z \to \hat{S}_z\right) \end{pmatrix} = \frac{\pi}{15} \begin{pmatrix} 0 & 9b_{SH}^2+4c_S^2 & 3b_{SH}^2 & 0 & 18b_{SH}^2 & 0 \\ \dfrac{18b_{SH}^2+8c_S^2}{3} & \dfrac{9b_{SH}^2+4c_S^2}{2} & \dfrac{3b_{SH}^2}{2} & 9b_{SH}^2 & 9b_{SH}^2 & 0 \\ \dfrac{18b_{SH}^2+8c_S^2}{3} & \dfrac{9b_{SH}^2+4c_S^2}{2} & \dfrac{3b_{SH}^2}{2} & 0 & 9b_{SH}^2 & 1 \\ 0 & 9b_{SH}^2+4c_S^2 & 0 & 9b_{SH}^2 & 0 & 1 \\ 0 & 0 & 3b_{SH}^2 & 9b_{SH}^2 & 18b_{SH}^2 & 1 \\ 0 & 0 & -3b_{SH}^2 & 0 & 18b_{SH}^2 & 0 \end{pmatrix} \bullet \begin{pmatrix} J(0) \\ J(\omega_S) \\ J(\omega_H-\omega_S) \\ J(\omega_H) \\ J(\omega_H+\omega_S) \\ \rho_{HH'} \end{pmatrix} \qquad (11.17)$$

where the coefficients b_{SH} and c_S have the same meaning as in Eq. (11.13). Most of the rate coefficients were introduced earlier: R_{1S} is the longitudinal relaxation rate of the S-spin, R_{2S} is the transverse relaxation rate for the S-spin, R_{2SH} is the relaxation rate of the antiphase S-spin coherence, R_{1SH} is the relaxation rate of the longitudinal two-spin order, R_{1H} is the proton longitudinal relaxation rate and finally, σ_{SH} is the cross-relaxation rate between the proton and the S-spin. The term $\rho_{HH'}$ denotes a sum of all rates containing purely ^{1}H-^{1}H relaxation, due to nearby protons within the molecule. The system of linear equations can be solved to obtain the spectral density function at the different frequencies involved, and the relationship between the relaxation rates and the spectral density function can instead be formulated as:

$$\begin{pmatrix} J(0) \\ J(\omega_S) \\ J(\omega_H-\omega_S) \\ J(\omega_H) \\ J(\omega_H+\omega_S) \\ \rho_{HH'} \end{pmatrix} = \frac{15}{\pi}\begin{pmatrix} -\frac{3}{8(9b_{SH}^2+4c_S^2)} & \frac{3}{4(9b_{SH}^2+4c_S^2)} & \frac{3}{4(9b_{SH}^2+4c_S^2)} & -\frac{3}{8(9b_{SH}^2+4c_S^2)} & -\frac{3}{8(9b_{SH}^2+4c_S^2)} & 0 \\ \frac{1}{18b_{SH}^2+8c_S^2} & 0 & 0 & \frac{1}{18b_{SH}^2+8c_S^2} & -\frac{1}{18b_{SH}^2+8c_S^2} & 0 \\ \frac{1}{12b_{SH}^2} & 0 & 0 & -\frac{1}{12b_{SH}^2} & \frac{1}{12b_{SH}^2} & -\frac{1}{6b_{SH}^2} \\ -\frac{1}{36b_{SH}^2} & \frac{1}{18b_{SH}^2} & -\frac{1}{18b_{SH}^2} & \frac{1}{36b_{SH}^2} & \frac{1}{36b_{SH}^2} & 0 \\ \frac{1}{72b_{SH}^2} & 0 & 0 & -\frac{1}{72b_{SH}^2} & \frac{1}{72b_{SH}^2} & \frac{1}{36b_{SH}^2} \\ -\frac{1}{4} & -\frac{1}{2} & \frac{1}{2} & \frac{1}{4} & \frac{1}{4} & 0 \end{pmatrix} \bullet \begin{pmatrix} R_{1S}(\hat{S}_z) \\ R_{2S}(\hat{S}_{x,y}) \\ R_{2SH}(2\hat{S}_{x,y}\hat{H}_z) \\ R_{1SH}(2\hat{S}_z\hat{H}_z) \\ R_{1H}(\hat{H}_z) \\ \sigma_{SH}(\hat{H}_z \to \hat{S}_z) \end{pmatrix} \tag{11.18}$$

The spectral density function is thus obtained as a function of frequency, and no *a priori* model is needed to analyse the relaxation data. Spectral density mapping was carried out for the ^{15}N-labelled protein eglin c using heteronuclear relaxation[50,51], and a typical spectral density as a function of frequency can be seen in Figure 11.15a, showing the spectral density function for Arg22, situated in an α-helical structure within the protein. The spectral density function was obtained from relaxation data at three magnetic field strengths.

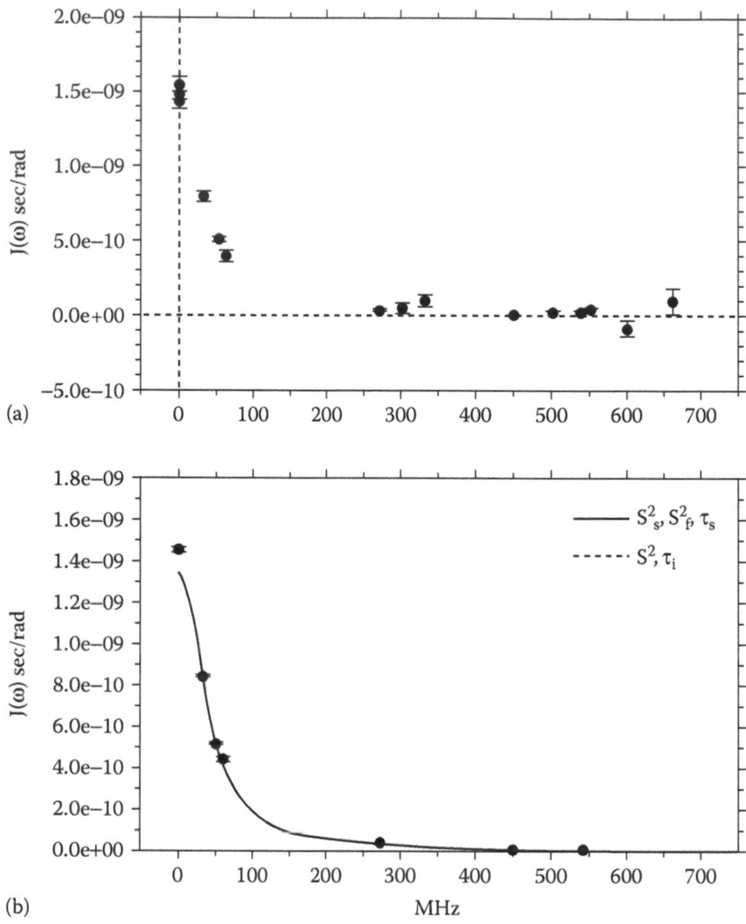

FIGURE 11.15 Spectral density mapping for Arg22 in eglin c. (a) shows full spectral density mapping of relaxation data recorded at three fields: 7.04, 11.74 and 14.1 T. (b) shows the reduced spectral density mapping of relaxation data obtained at the same fields and fits to the model-free spectral density function and the extended model-free spectral density function (assuming two internal motions). (Reprinted with permission from Peng, J.W. and Wagner, G., *Biochemistry*, 34, 16733–16752, 1995. Copyright (1995) American Chemical Society.)

A simplified spectral density mapping procedure, usually referred to as *reduced spectral density mapping*, was also proposed[51–54]. The approximation is valid under the condition that the spectral density function varies slowly with frequency around $\omega_H \pm \omega_S$, and that the value for the spectral density function at high frequencies is much lower than at zero frequency or ω_S. This is the case for a ^{15}N-^{1}H pair, with the proton frequency being much higher than the nitrogen-15 (compare $\gamma_H = 26.752 \times 10^7$ T^{-1} s^{-1} with $\gamma_N = -2.713 \times 10^7$ T^{-1} s^{-1}). When this is the case, the spectral density function taken at high frequencies involving the proton frequency, $J(\omega_H + \omega_N)$, $J(\omega_H)$ and $J(\omega_H - \omega_N)$ may be averaged to $J(\omega_h)$. Thus, only three independent measurements of relaxation parameters, which typically are the common R_1, R_2 and NOE relaxation parameters, are required to determine the spectral density function at the three frequencies according to:

$$J(\omega_h) = \frac{1}{\pi} \frac{\gamma_N}{\gamma_H} (NOE - 1) R_1 / b_{NH}^2 \tag{11.19a}$$

$$J(\omega_N)=\left[R_1-\frac{7\pi b_{NH}^2}{5}J(\omega_h)\right]\Big/\left(\frac{3\pi b_{NH}^2}{5}+\frac{4\pi c_N^2}{15}\right) \tag{11.19b}$$

$$J(0)=\left[R_2-\left(\frac{3\pi b_{NH}^2}{10}+\frac{6\pi c_N^2}{45}\right)J(\omega_N)-\frac{13\pi b_{NH}^2}{10}J(\omega_h)\right]\Big/\left(\frac{4\pi b_{NH}^2}{10}+\frac{8\pi c_N^2}{45}\right) \tag{11.19c}$$

An additional advantage with the reduced spectral density mapping approach is that the proton relaxation rate, which can be associated with experimental difficulties, does not have to be measured. Under the assumption that the spectral density function can be decomposed into two parts, one frequency-dependent component accounting for the overall (slow) motion, which can safely be modelled as $1/\omega^2$ at the high frequencies involved, and one frequency-independent part for the fast internal motion, the average frequency for different situations can be evaluated. From examining the expression for the heteronuclear cross-relaxation, an average frequency can be calculated from $6/(\omega_H+\omega_N)^2-1/(\omega_H-\omega_N)^2=5/\omega_h^2$, and the spectral density function taken at ω_h can be identified with the value $J(0.870\omega_H)$. Thus, the spectral density can be measured at three different frequencies: ω_0, ω_N, and $0.870\omega_H$. Under certain limiting conditions, which usually apply to proteins, this is in fact identical to the Lipari–Szabo approach, if the fast internal motions are within the extreme narrowing regime, and the overall motion in the slow tumbling limit. Figure 11.15b shows the reduced spectral density for eglin c, obtained using the same relaxation parameters as in 11.15a.

It should, however, be mentioned that the spectral density function that has been measured in this way is frequently interpreted in terms of a motional model, such as the model-free approach provided by the formalism described previously (Sections 6.3.3, 11.3.2 and 11.4.1). It is not always clear which method is to be preferred, since the spectral density mapping approach requires several relaxation rates to be measured, of which the proton longitudinal relaxation rate can be difficult to measure accurately, while the model-free approach with several parameters to fit often requires measurements at several fields. Fits of the reduced spectral density function for eglin c were performed in this way with the Lipari–Szabo model-free approach (Figure 11.15b), and order parameters for the different residues were thus obtained.

The spectral density mapping can also be applied to ^{13}C instead of ^{15}N. In fact, Eqs. (11.17) and (11.18) can be applied in exactly the same way for both these species. On the other hand, the reduced spectral density mapping is not really applicable for ^{13}C, because the differences between $(\omega_I-\omega_S)$, ω_I and $(\omega_I+\omega_S)$ are larger. Allard, Jarvet and co-workers applied the spectral density mapping to study ^{13}C-^1H vector dynamics in a selectively labelled, 22-residues-long peptide, using measurements at a single magnetic field[55] and at multiple magnetic fields[56]. In more recent work, Kaderavek *et al.*[57] investigated the utility of the spectral density mapping approach to ^{13}C relaxation data. A suite of protocols, including measurements of CCRRs and experiments at several magnetic fields, were developed and evaluated using samples of a small, selectively labelled disaccharide and a uniformly labelled RNA hairpin. The conclusion of the study was that sufficiently accurate values of auto- and cross-correlated spectral densities at zero and ^{13}C resonance frequency could be obtained using the data acquired at three magnetic fields for the latter case when the rotational diffusion tensor was only moderately anisotropic. For the case of highly anisotropic motions, exemplified by the disaccharide, analysis of auto-correlated relaxation rates at five magnetic fields was recommended.

11.4.3 Side-Chain Motion

Studies of protein dynamics have largely been focused on backbone ^{15}N-^1H motion as described in the previous subsection. However, many of the important contributions to the thermodynamics (see next subsection) in, for instance, folding processes and ion binding come from protein side-chain motions and other local motions, which are not accessible through measurements on the protein backbone atoms. In order to more fully understand protein dynamics, several methods for measurements on side-chain atoms in proteins have been developed. Measurements of side-chain ^{15}N-^{1}H relaxation in amino acids containing additional ^{15}N-^{1}H groups are straightforward and are essentially performed as described for backbone amide internuclear vectors. Measurements of carbon-13 relaxation parameters in side-chains as well as for backbone C^{α} and C' sites have also been conducted but suffer from the fact that partial isotope-labelling schemes must be employed to avoid complications due to ^{13}C-^{13}C dipolar interactions. Specific labelling of methyl groups has been employed, in which methyl-group dynamics could be measured. A few carbon-13 studies at natural abundance have been conducted but have the obvious disadvantage of poor sensitivity. The use of new methods and protein labelling schemes to determine side-chain motions has been reviewed several times by, for example, Wand and co-workers[58].

Recent advances in NMR methodology in general have made fractional deuteration of proteins important. The main goal of this is to reduce the number of dipolar interactions, but other possibilities can also be useful. The presence of ^{2}H in the protein opens up the possibility of ^{2}H relaxation measurements on side-chains. ^{2}H relaxation may in fact be more straightforward to analyse than relaxation for spin 1/2 nuclei, due to the fact that one single relaxation mechanism, the quadrupolar relaxation, dominates. Quadrupolar relaxation will be covered in more detail in Chapter 14, and we shall here only provide the necessary background to describe the methods behind side-chain motion investigations. Kay and co-workers have in a series of studies developed methods to measure ^{2}H relaxation in fractionally labelled protein side-chains[59–63]. The goal is to obtain longitudinal and transverse relaxation rates for a deuteron in methyl groups containing a ^{13}C (C), one ^{2}H spin (D) and two proton spins, I^i and I^j. The methyl group, $^{13}CD^{1}H_2$, can thus be represented as two three-spin systems, I^iCD and I^jCD. Two-dimensional ^{13}C-^{1}H correlation experiments have been designed to measure the decays of the $4\hat{I}_z\hat{C}_z\hat{D}_z(T_1)$ and $4\hat{I}_z\hat{C}_z\hat{D}_y$ $(T_{1\rho})$ terms, and a simple magnetisation transfer pathway for such a correlation experiment can be described as:

$$I \xrightarrow{J_{CH}} C(t_1) \xrightarrow{J_{CD}} D(\tau) \xrightarrow{J_{CD}} C \xrightarrow{J_{CH}} I(t_2) \tag{11.20}$$

Briefly, the experiment starts with proton magnetisation, which is transferred through the first step of a heteronuclear single quantum coherence (HSQC) experiment to the carbon-13. Carbon-13 frequency-labelling occurs during the evolution period, t_1, and the magnetisation is transferred to the deuterium spin through the heteronuclear carbon-deuteron coupling. Deuterium relaxation is monitored during the variable relaxation delay, τ, and the magnetisation is transferred back to the proton, via the carbon-13 spin, for detection. This is performed as a constant-time HSQC, in which the relaxation delay, τ, during which deuterium magnetisation is present and relaxes has been inserted.

The experiment can be designed to monitor the expectation value of either the $4\hat{I}_z\hat{C}_z\hat{D}_z$ or the $4\hat{I}_z\hat{C}_z\hat{D}_y$ terms during the relaxation delay, τ. The resulting spectrum is a ^{1}H-^{13}C correlation spectrum, in which the peak amplitudes depend on the relaxation of either of the triple-spin orders. The longitudinal relaxation for the three-spin order, $\langle 4\hat{I}_z^i\hat{C}_z\hat{D}_z\rangle$, is given by:

$$\frac{d}{dt}\langle 4\hat{I}_z^i\hat{C}_z\hat{D}_z\rangle = -\rho_i\langle 4\hat{I}_z^i\hat{C}_z\hat{D}_z\rangle - \sigma_{ij}\langle 4\hat{I}_z^j\hat{C}_z\hat{D}_z\rangle - \sum_{k\neq i,j}\sigma_{ik}\langle 4\hat{I}_z^k\hat{C}_z\hat{D}_z\rangle \tag{11.21}$$

where ρ_i is the auto-relaxation rate for the proton I^i, including, in principle, all interactions responsible for its relaxation, and σ_{ij} is the cross-relaxation between the two dipolar-coupled proton spins. A third term has been included to account for cross-relaxation between the I^i-spin and all other proton spins, k, in the vicinity of the methyl group. An analogous equation describing the relaxation for the $\langle 4\hat{I}_z^j\hat{C}_z\hat{D}_z\rangle$ state can also be formulated. Alternatively, by noting that the two proton spins, $\hat{I}^i$ and $\hat{I}^j$, are equivalent, we can write:

$$\frac{d}{dt}\langle 4\hat{I}_z\hat{C}_z\hat{D}_z\rangle = -R_{1ICD}\left(4\hat{I}_z\hat{C}_z\hat{D}_z\right) - \sum_{k\neq i,j}\left(\sigma_{ik}+\sigma_{jk}\right)\langle 4\hat{I}_z^k\hat{C}_z\hat{D}_z\rangle \tag{11.22}$$

in which $R_{1ICD} = \rho_i + \sigma_{ij}$ and $\hat{I}_z = \hat{I}_z^i + \hat{I}_z^j$. In order to obtain the relaxation of the pure deuterium spin order, it is necessary to measure the dipolar contributions to the relaxation in a separate experiment to provide R_{1IC}, the rate constant for the decay of the two-spin order, $\langle 2\hat{I}_z\hat{C}_z \rangle$. One can show that the relaxation of pure deuterium spin order can approximately be obtained from:

$$R_{1ICD} - R_{1IC} = R_1(D) \tag{11.23}$$

In a similar way, the equation describing the relaxation of the $\langle 4\hat{I}_z\hat{C}_z\hat{D}_y \rangle$ state can be written as

$$\frac{d}{dt}\langle 4\hat{I}_z\hat{C}_z\hat{D}_y \rangle = -R_{2ICD}\langle 4\hat{I}_z\hat{C}_z\hat{D}_y \rangle - \sum_{k\neq i,j}(\sigma_{ik}+\sigma_{jk})\langle 4\hat{I}_z^k\hat{C}_z\hat{D}_y \rangle \tag{11.24}$$

and we can obtain the transverse relaxation of the deuterium spin order as

$$R_{2ICD} - R_{1IC} = R_2(D) \tag{11.25}$$

By recording three separate experiments, yielding R_{1ICD}, R_{1IC}, R_{2ICD}, as described above, the two relaxation rate constants, $R_1(D)$ and $R_2(D)$, for the deuteron are thus obtained. These are given by:

$$R_1(D) = \frac{3\pi}{20}\left(\frac{e^2qQ}{\hbar}\right)^2\left[J(\omega_D)+4J(2\omega_D)\right] \tag{11.26a}$$

$$R_2(D) = \frac{\pi}{40}\left(\frac{e^2qQ}{\hbar}\right)^2\left[9J(0)+15J(\omega_D)+6J(2\omega_D)\right] \tag{11.26b}$$

We note that Eqs. (11.26a) and (11.26b) simplify under extreme narrowing conditions to Eqs. (11.6) and (11.7). Usually, the Lipari–Szabo spectral density function, described earlier, is used also here to analyse the relaxation data. The order parameter, S, is in this case given by $S = S_{axis}\frac{1}{2}(3\cos^2\theta - 1)$, where S_{axis} is the order parameter describing the local motion of the methyl group axis, and θ is the angle between this axis and the C-D bond. Assuming tetrahedral symmetry for a methyl group, this angle is 109.5°, and $S^2 = 0.11 \times S_{axis}^2$.

Muhandiram and co-workers[59] analysed deuterium relaxation data for methyl groups in the C-terminal SH2 domain from phospholipase $C_{\gamma 1}$. They found that the average S_{axis}^2 for methyl groups from different amino acid residues depends largely on the side-chain length. Interesting observations about differential side-chain dynamics, not related to side-chain length, were, however, also made, which could be important for understanding the folding processes, protein-ligand interactions, and other phenomena that are not characterised by backbone dynamic events. Schnell *et al.* examined the side-chain methyl dynamics in dihydrofolate reductase, an enzyme responsible for the production of tetrahydrofolate, which is an essential coenzyme in the biosynthesis of several amino acids[64]. The enzyme requires NADPH, and several differences in side-chain dynamics were found between the enzyme-folate and the enzyme-folate-$NADP^+$ complexes. Methyl-containing side-chains in the catalytic domain, as well as in the active site, were seen to become motionally restricted upon the co-factor binding, and the observations could in this case be correlated to changes in backbone dynamics as measured from ^{15}N relaxation.

11.4.4 Order Parameters and Thermodynamics

Attempts have been made to correlate differences in generalised order parameters with thermodynamic properties in proteins. For instance, differences in the order parameters of folded and unfolded states of proteins can be related to changes in free energy upon folding. Subsequent development has led to a method to relate changes in S^2 to conformational entropy changes[65,66].

The details of deriving the thermodynamic relations will not be given here (the reader is instead referred to the original works), but we note that both the generalised order parameter of a bond vector and the entropy depend on the distribution of orientations of the vector. If one assumes a diffusion-in-a-cone model for the motion of a bond vector, the relationship between the generalised order parameter and the conformational entropy s_p (we use here the lower-case symbol for entropy to avoid confusion with the order parameter) for a given site j can be written as[66]:

$$s_p(j)/k_B = \ln \pi \left[3 - \left(1 + 8S_j^b\right)^{1/2} \right] \tag{11.27}$$

where S_j is the generalised order parameter for the bond vector j. Consequently, the conformation entropy change associated with two states, such as an unfolded, a, and folded, b, state of a protein, can be obtained as:

$$\Delta s_p(j)/k_B = \ln \left\{ \left[3 - \left(1 + 8S_j^b\right)^{1/2} \right] \Big/ \left[3 - \left(1 + 8S_j^a\right)^{1/2} \right] \right\} \tag{11.28}$$

The beauty of this approach lies in the fact that conformational entropy contributions to the folding processes can be obtained per residue, or site, in the protein. This was first done for the folding of the *Escherichia coli* ribonuclease HI protein, in which it was possible to identify specific sites where the contribution of conformational entropy was important[65]. Similarly, the entropic contribution to Ca^{2+}-binding in calbindin D_{9k} was investigated by analysing the order parameters using Eq. (11.28)[41,65]. Here, it was shown that depending on the ion-binding pathway, different patterns in conformational entropy changes were observed (Figure 11.16). It was also possible to determine that one of the Ca^{2+}-binding sites, site II, made a significantly larger contribution to the conformational entropy than the other.

Although traditionally analyses have been performed on the protein backbone, it has become evident that the biologically relevant motions connected to conformational thermodynamics concern the protein side-chains[67,68]. Wand and co-workers have used a combination of measurements on side-chains to develop a way to investigate the conformational entropy of ligand binding to proteins[69]. In this way, it is possible to estimate the contribution of the conformational entropy to the free energy of binding of a target to the protein.

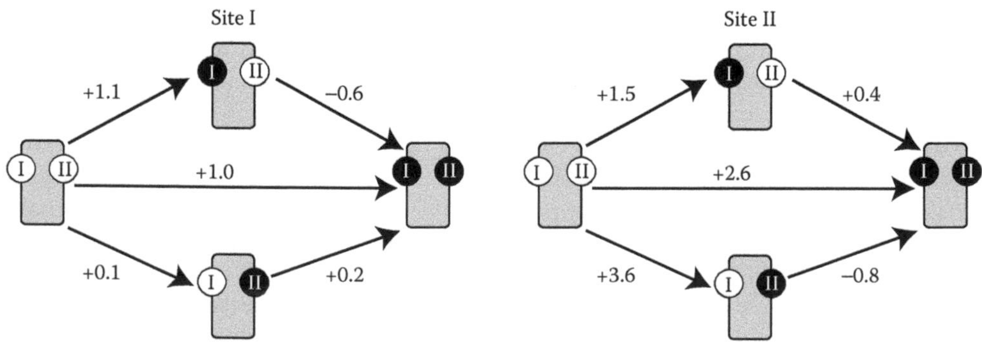

FIGURE 11.16 Ca^{2+}-binding pathway diagram for calbindin D_{9k} showing Ca^{2+}-induced changes in backbone conformational entropy. The displayed numbers are calculated as $-T\Delta s_{backbone}$ (kcal mol^{-1}) from the order parameters using Eq. (11.28). The left panel is for the first binding site (I) only, and the right panel is for the second binding site (II) only. (Adapted from Mäler et al., *Nature Struct. Biol.*, 7, 245–250, 2000.)

11.4.5 The Ubiquitin Example

Before concluding this section on molecular dynamics applications in proteins, we would like to mention that separating the static (interaction strengths, in particular the CSA) and dynamic information in proteins and other complex systems may lead to some complications. A nice illustration of the problem of separating dynamics from interaction strength, and of separating local motion from overall motional anisotropy, has been addressed in a series of relaxation studies on the small protein ubiquitin. Human ubiquitin is a globular protein, 76 amino acid residues long, that has been well studied by NMR since the mid-1980s. The structure of ubiquitin is well known, both from X-ray crystallographic studies and from solution-NMR (Figure 11.17). It is much used in methods development and serves as a model protein in many aspects.

One question that has been addressed is the dynamic behaviour of the protein. The first investigation of backbone dynamics by nitrogen-15 spin relaxation was performed in 1992 by Schneider and co-workers[70]. They measured conventional backbone nitrogen-15 R_1, R_2 and NOE relaxation parameters at two magnetic field strengths and interpreted the data in terms of the Lipari–Szabo "model-free" approach. They found that the dynamics could be described by a global correlation time for isotropic reorientation, while the internal motion for each backbone amide site was described by an order parameter (Figure 11.18). For most sites, the internal correlation time was found to be close to zero, implying that the truncated form of the Lipari-Szabo spectral density could be used (Eq. (6.61) or (11.9)). The backbone dynamics in ubiquitin was found to be fairly uniform throughout the sequence, with a higher degree of internal mobility at the two termini, especially at the C-terminus.

Proteins are, however, seldom perfect isotropic rotors, and a few years later, Tjandra *et al.*[71] addressed the issue of rotational diffusion anisotropy. The NMR data were interpreted in terms of axially symmetric diffusion and the anisotropy was found to be in agreement with hydrodynamic calculations based on the X-ray structure. From a combined molecular dynamics simulation and spin relaxation study by Lienin *et al.*[72], it was demonstrated that by combining both backbone ^{13}C′ carbonyl and ^{15}N amide relaxation data, a model accounting for anisotropic motion of entire molecular subunits, such as helices or β-sheets, could be constructed. The motional parameters were, however, found to depend on the magnitude and orientation of the interactions, the dipolar vectors and the CSA tensors, demonstrating the need for careful investigations of these interaction parameters.

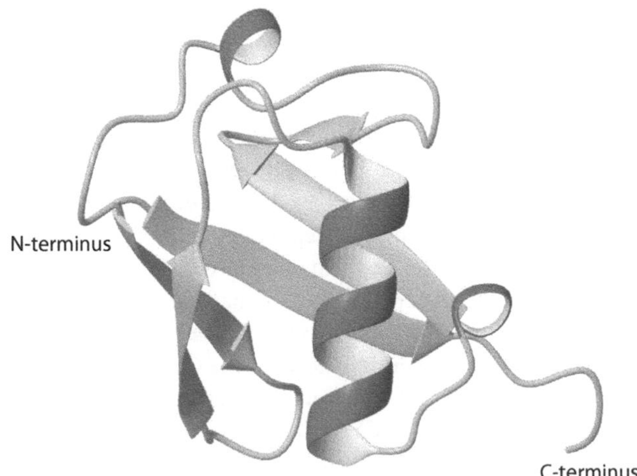

FIGURE 11.17 Structure of human ubiquitin. The helices and sheets are displayed as ribbons, while the turns are represented as sticks. The coordinates were taken from the Protein Data Bank (www.rscb.org, accession code 1D3Z). The figure was produced using Molmol (version 2.6).

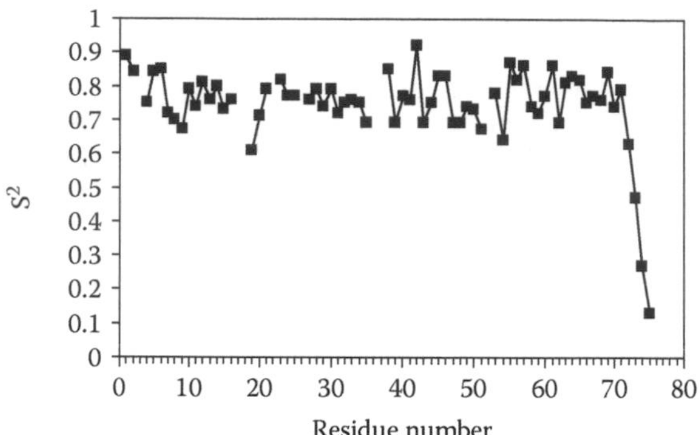

FIGURE 11.18 Order parameters for human ubiquitin obtained from fitting R_1 and NOE relaxation parameters recorded at 7.05 and 14.1 T. (The figure was produced with data taken from Table 1 of Schneider, D.M. *et al.*, *Biochemistry*, 31, 3645–3652, 1992.)

In the light of the previous results, subsequent studies on ubiquitin have included cross-correlated relaxation (CSA/DD) to also allow for the possibility of a variable CSA parameter, *i.e.* a different CSA for each backbone amide ^{15}N. CSA relaxation contributes to the overall relaxation for nitrogen-15 in a significant way, and it is therefore important to have accurate knowledge about the variability of the CSA parameter. Several groups have addressed this issue[73–78], but there is no conclusive evidence for residue-specific CSA values, or any correlation with structure. Tjandra *et al.*[73] devised an experiment to measure the transverse cross-correlation rate and demonstrated its applicability to ubiquitin. The experiment was described in Section 10.2.2, and the transverse cross-correlation rates can be obtained from this experiment by using Eq. (10.24). The cross-correlation rate was given in Eq. (5.46) and is repeated here in a slightly modified form:

$$\eta_{IS,S} = -\frac{2\pi}{15}\left(\frac{\mu_0}{4\pi}\right)\gamma_I\gamma_S^2\hbar B_0\Delta\sigma_S r_{IS}^{-3}\frac{1}{2}\left(3\cos^2\theta-1\right)\left[4J(0)+3J(\omega_S)\right] \tag{11.29}$$

From Eq. (11.29), it can readily be observed that the cross-correlation rate depends on several parameters. The CCRR depends explicitly on the CSA parameter $\Delta\sigma_S$, assuming an axially symmetric shielding tensor, on the internuclear distance r_{IS}, and on the angle, θ, between the two interactions. Furthermore, the dynamics, through the spectral density function $J(\omega)$, needs to be known. To a good approximation, one can in Eq. (11.29) use the dynamic parameters obtained from the dipole-dipole autocorrelation relaxation, *i.e.* order parameters and correlation times. If the internuclear distance is known, the cross-correlation rate can be used to determine the CSA parameter for each measured site in the protein.

Using this approach, Tjandra *et al.* did not find any significant differences in CSA parameters between different residues. Fushman and co-workers, on the other hand, used the same method to record multiple-field data, and found large residue-dependent variations in the CSA parameter[74,75]. The CSA parameter can also be included in the simultaneous fit of relaxation data, such as those used to obtain the Lipari–Szabo dynamic parameters. This was done in a study by Köver and Batta, who also included measurement of the longitudinal cross-correlation relaxation[76]. They found, as did Fushman and co-workers, substantial residue-dependent variation in the CSA parameter. Damberg and co-workers reported multiple-field longitudinal and transverse relaxation measurements, which revealed, on the other hand, only a modest variation in the CSA parameter[77]. In that study, the results were analysed jointly with the previously mentioned data, and the analysis revealed limited, but statistically significant, site-to-site variability in the CSA parameter. Although it is not clear what the differences in results are due to, it is important to emphasise the need for more accurate determination of the static molecular parameters that govern the relaxation rates. The example also highlights difficulties in accurately separating motional properties from the static molecular parameters using only relaxation data.

Loth and co-workers[78] reported measurements of a set of 14 CCRRs involving CSA interactions for ^{15}N, ^{13}C′ and ^{1}H^{N} in 61 residues of ubiquitin. The experimental data allowed determination of the individual components of the shielding tensors and their orientation in the molecular frame, assuming that one of the principal components is perpendicular to the peptide plane. The CSA parameters are to a certain extent dependent on the description of the internal dynamics, and the authors made use of several such models. In spite of these ambiguities, it was possible to identify some general trends. Thus, the ^{13}C′ principal shielding components showed very limited dispersion around the averages. The spreads of individual components for ^{15}N and ^{1}H^{N} were considerably larger.

Fenwick *et al.*[79] studied the motional backbone correlations between distant residues in the secondary structure elements of human ubiquitin using the CCRR-based methods mentioned earlier in Subsection 11.4.1. In the initial step of the analysis of the data, the authors determined the ensemble of structures describing the structural heterogeneity of the protein on the sub-millisecond timescale, making use of a large set of previously reported NOEs and residual dipolar couplings (we shall return to this type of analysis in Chapter 12). The ensemble of structures was found to be consistent with the measured CCRRs. In the next step, evidence for correlated motions was sought. For the β-strands, many relatively slow motions of H^{N}-N bond were found to be weakly correlated, which may be biologically important. A recently reported 1 ms MD simulation of ubiquitin[80] was in good agreement with the experimental findings of Fenwick *et al.*[79].

The last example of human ubiquitin dynamics that we wish to mention here is a study by Charlier and co-workers[81]. They used the sample shuttling technique, similar to that mentioned in Subsection 8.1.3, in addition to conventional ^{15}N relaxation measurements at three high fields. This resulted in a data set of ^{15}N T_1 obtained under high-resolution conditions at ten magnetic fields ranging from 0.5 to 22.3 T, together with usual NOE and T_2 results at three high fields. The extensive data sets were analysed using various versions of the model-free approach. In certain parts of the molecule, *e.g.* the β_1–β_2 turn region (residues 7–12), evidence for slow local motions with correlation times of around 2 ns, only slightly faster than the global correlation time of 4–5 ns, was found. One of the conclusions of the study is that the availability of this kind of data set, including relaxation at low magnetic fields, calls for development of new, more sophisticated theoretical models.

11.5 Low-Field Experiments: Slow Motions and Intermolecular Relaxation

In this section, we wish to mention a few selected applications of the low-resolution fast field-cycling measurements (FFC) of nuclear magnetic relaxation dispersion (NMRD) data for liquids displaying slow molecular motions. Such liquids can be of different types; for example, highly viscous molecular liquids, often with the ability to form a glassy state, and synthetic polymer melts. Studies of this kind have been subject to numerous reviews, of which we wish to refer to a few[82–84]. In addition to these types of systems, FFC-NMRD experiments are also common for paramagnetic systems, liquid crystal, and liquids contained in porous media. We shall return to these topics in Chapters 15 and 17. The FFC relaxation measurements are most usually performed for protons or deuterons.

The molecular system we wish to treat as an example is neat glycerol, a liquid characterised by a melting point of 291 K and a boiling point of 563 K, both at ambient pressure. Glycerol becomes very viscous when cooled and can easily be supercooled to form a glassy state. As a starting point for this discussion, we use the papers by Meier *et al.*[85] and Gainaru *et al.*[86] dealing with comparisons of proton NMRD measurements with dielectric spectroscopy (DS) data using glycerol (along with several other viscous liquids) as model systems. The DS data are usually presented as a plot of dielectric loss (the imaginary part of the susceptibility) versus frequency. The range of frequencies covered is typically several decades. Being able to cover a broad frequency range is also a feature of the FFC method. In order to facilitate the comparison between DS and FFC, the NMR spin-lattice relaxation data are often converted to the susceptibility representation by multiplying the relaxation rate ($1/T_1$) by the Larmor frequency:

$$\omega/T_1 = C\left[\chi''(\omega) + 2\chi''(2\omega)\right] \equiv C\chi''_{NMR}(\omega) \tag{11.30}$$

The constant C reflects the effective dipolar interaction strength (we are dealing with a system with many different pairwise interactions) and the susceptibilities $\chi''(\omega)$ are related to spectral densities: $\chi''(\omega) = \omega J(\omega)$. Thus, the first equality in Eq. (11.30) is related to the formulation in Eq. (3.33). In the last equality in Eq. (11.30), we neglect the factor of two difference between the frequencies in the form in the middle. This is in accordance with the idea of the looking at the gross features of the frequency variation over several orders of magnitude. If the spectral density is assumed to be Lorentzian (Debye limit), the low-frequency susceptibility follows the power law $\chi''(\omega) \propto \omega^1$.

In the work on viscous liquids, it appears common to assume that the spectral densities deviate from the Lorentzian form and can be described by the Cole–Davidson form as given in Eq. (6.71). Another assumption frequently used in this context is that the shape of the susceptibility is independent of temperature, *i.e.* that the amplitude of the curve and the quantity β do not change with temperature, while temperature influences τ_{rot} (compare Section 2.4), also commonly denoted as *structural relaxation time* τ_α. If this is the case, then one can introduce the frequency-temperature superposition and plot the relaxation rate in the susceptibility form, obtained at different temperatures against a dimensionless quantity $\omega\tau_{rot}$. This is in the spirit of introducing the susceptibility representation, and as discussed by Kruk (*Further reading*), a consequence of moving a relaxation expression, such as Eq. (3.33), to the susceptibility picture is that one can no longer talk about correlation time and frequency separately, but only about their product. The plot of suitably scaled susceptibility versus the product $\omega\tau_{rot}$ is known as the *master curve.*

The master curves for glycerol, as obtained from dielectric spectroscopy and FFC NMR, are shown in Figure 11.19. The data agree well with each other in the vicinity of the maximum, while deviations occur at lower and higher frequencies. The differences at high frequencies are explained by differences between dynamics of rank-one (DS) and rank-two (NMR) spectral densities (remember $\tau_1 = 3\tau_2$ in the small-step rotational diffusion limit as shown in Section 6.1). The same group showed convincingly and elegantly – by comparing results for different hydrogen isotope compositions[87] and by dilution experiments[88] – that the differences in the low-frequency range can be attributed to intermolecular relaxation. To get a consistent interpretation of the whole NMRD profile, they assumed that the relaxation rate at each point was a sum of the intra- and intermolecular contributions. The intramolecular contribution was expressed in terms of Cole–Davidson spectral densities. The intermolecular relaxation rate was in turn allowed to be a sum of two components: one purely translational term corresponding to the force-free diffusion model discussed in Section 3.5 and one term allowing for the fact that the spins do not

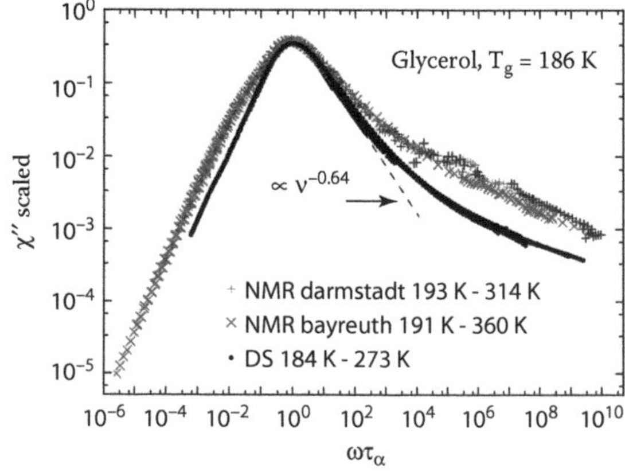

FIGURE 11.19 Master curves for glycerol obtained from dielectric spectroscopy (circles) and from NMR relaxation (crosses and plus signs). The Cole–Davidson line with $\beta = 0.64$ is shown as a dashed line. (Adapted with permission from Gainaru, C. *et al., J. Chem. Phys.*, 128, 174505, 2008.)

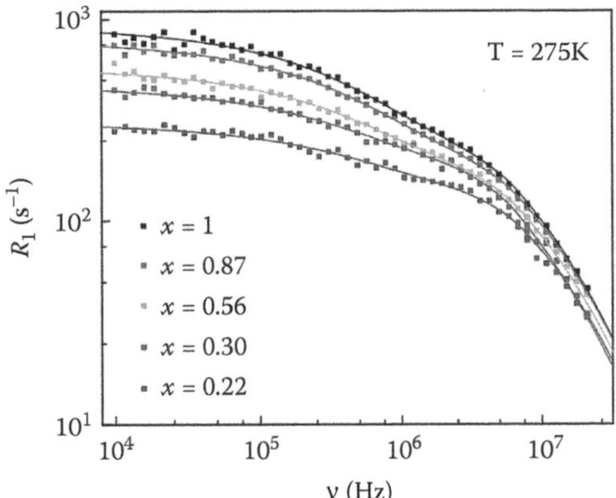

FIGURE 11.20 Proton relaxation dispersion data for different mole fractions x glycerol-d$_3$ (deuterated in hydroxyl groups) in glycerol-d$_8$ at 275 K. The top curve corresponds to $x=1$, and the relaxation rate decreases with decreasing mole fraction proton-containing species. (Adapted with permission from Meier, R. *et al.*, *J. Chem. Phys.*, 136, 034508, 2012.)

reside in the centres of spherical particles, described again by the Cole–Davidson function. The latter effect, known as the *eccentricity effect* (see Section 3.5), arises because of rotation-translation coupling. The quality of the fits is illustrated in Figure 11.20, showing the NMRD profiles obtained at 275 K for different molar fractions of glycerol-d$_3$ in fully deuterated glycerol.

The last example we wish to mention in this chapter relates to NMRD studies of synthetic polymer melts. As discussed by Kimmich and co-workers[82,83] and by Kruk *et al.*[84], measurements of this kind allow a description of different aspects of polymer dynamics. The situation here is more complicated, and depending on the length of polymer chain, several dynamic regimes can be identified. The shortest relevant timescale is that of reorientational motion of small segments (Kuhn segments) of the polymer, with a characteristic time constant denoted τ_s. This motion is more or less anisotropic and is responsible for the decay of a large fraction of the dipolar time-correlation function. The remainder of the correlation decays on a slower timescale due to more complicated collective chain dynamics. Discussing different models for the collective chain dynamics is beyond the scope of this book. We choose therefore only to illustrate the point by showing the "polymer dynamics" master curves, obtained after separating the τ_s process, for polybutadiene with different molecular weights (Figure 11.21).[89]

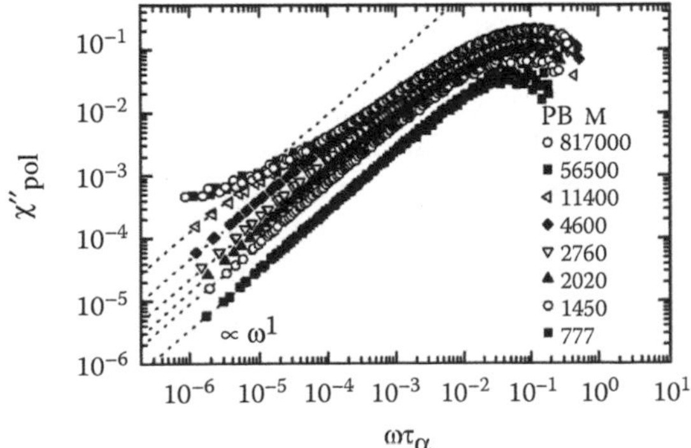

FIGURE 11.21 Master curves corresponding to the polymer dynamics after subtracting the counterpart corresponding to the fast τ_s (denoted here τ_α) process. At lowest frequencies, a linear behaviour in frequency is observed, except for the highest molecular weights. (Reprinted with permission from Kariyo, S. *et al.*, *Phys. Rev. Lett.*, 97, 207803, 2006. Copyright (2006) American Physical Society.)

References

1. Levy, G. C.; Lichter, R. L.; and Nelson, G. L., *Carbon-13 Nuclear Magnetic Resonance Spectroscopy*. 2nd edn; Wiley: New York, 1980.
2. Levy, G. C.; and Kerwood, D. J., *Carbon-13 Relaxation Measurements: Organic Chemistry Applications*. Wiley, eMagRes, 2007.
3. Liu, F.; Horton, W. J.; Mayne, C. L.; Xiang, T. X.; and Grant, D. M., Molecular dynamics of 1-decanol in solution studied by NMR coupled relaxation and stochastic dynamic simulations. *J. Am. Chem. Soc.* 1992, 114, 5281–5294.
4. Dölle, A.; and Bluhm, T., Motional dynamics in liquid 1,2,3,4-tetrahydro-5,6-dimethyl-1,4-methano-naphthalene II. Orientation of the rotational diffusion principal axis system by interpretation of 13C spin-lattice relaxation data. *Mol. Phys.* 1986, 59, 721–736.
5. Wallach, D.; and Huntress, W. T., Anisotropic molecular rotation in liquid N,N-dimethylformamide by NMR. *J. Chem. Phys.* 1969, 50, 1219–1227.
6. Klüner, R. P.; and Dölle, A., Friction coefficients and correlation times for anisotropic rotational diffusion of molecules in liquids obtained from hydrodynamic models and C-13 relaxation data. *J. Phys. Chem. A* 1997, 101, 1657–1661.
7. Dölle, A., Reorientational dynamics of the model compound 1,2,3,4-tetrahydro-5,6- dimethyl-1,4-methanonaphthalene in neat liquid from temperature-dependent C-13 nuclear magnetic relaxation data: Spectral densities and correlation functions. *J. Phys. Chem. A* 2002, 106, 11683–11694.
8. Maliniak, A.; Kowalewski, J.; and Panas, I., Nuclear spin relaxation study of hydrogen bonding in solutions of quinuclidine. *J. Phys. Chem.* 1984, 88, 5628–5631.
9. Maliniak, A.; Laaksonen, A.; Kowalewski, J.; and Stilbs, P., Molecular dynamics simulation of a weakly interacting system, quinuclidine-benzene II. *J. Chem. Phys.* 1988, 89, 6434–6441.
10. Ericsson, A.; Kowalewski, J.; Liljefors, T.; and Stilbs, P., Internal rotation of methyl groups in terpenes. Variable-temperature carbon-13 spin-lattice relaxation time measurements and force-field calculations. *J. Magn. Reson.* 1980, 38, 9–22.
11. Kowalewski, J.; and Liljefors, T., On the relationship between the potential barrier and the activation energy for the internal rotation of a methyl group. *Chem. Phys. Lett.* 1979, 64, 170–174.
12. Dechter, J. J.; Henriksson, U.; Kowalewski, J.; and Nilsson, A.-C., Metal nucleus quadrupole coupling constants in aluminum, gallium and indium acetylacetonates. *J. Magn. Reson.* 1982, 48, 503–511.
13. Champmartin, D.; and Rubini, P., Determination of the O-17 quadrupolar coupling constant and of the C-13 chemical shielding tensor anisotropy of the CO groups of pentane-2,4-dione and beta-diketonate complexes in solution. NMR relaxation study. *Inorg. Chem. Commun.* 1996, 35, 179–183.
14. Kowalewski, J.; Effemey, M.; and Jokisaari, J., Dipole-dipole coupling constant for a directly bonded CH pair – A carbon-13 relaxation study. *J. Magn. Reson.* 2002, 157, 171–177.
15. Bernatowicz, P.; Kowalewski, J.; and Sandström, D., NMR relaxation study of the protonated form of 1,8-bis(dimethylamino)naphthalene in isotropic solution: Anisotropic motion outside of extreme narrowing and ultrafast proton transfer. *J. Phys. Chem. A* 2005, 109, 57–63.
16. McCain, D. C.; and Markley, J. L., Rotational spectral density functions for aqueous sucrose: Experimental determination using 13C NMR. *J. Am. Chem. Soc.* 1986, 108, 4259–4264.
17. Kovacs, H.; Bagley, S.; and Kowalewski, J., Motional properties of two disaccharides in solutions as studied by carbon-13 relaxation and NOE outside of the extreme narrowing region. *J. Magn. Reson.* 1989, 85, 530–541.
18. Lipari, G.; and Szabo, A., Model-free approach to the interpretation of nuclear magnetic resonance relaxation in macromolecules 1. Theory and range of validity. *J. Am. Chem. Soc.* 1982, 104, 4546–4559.
19. Poppe, L.; and Van Halbeek, H., The rigidity of sucrose - just an illusion? *J. Am. Chem. Soc.* 1992, 114, 1092–1094.

20. Mäler, L.; Widmalm, G.; and Kowalewski, J., Dynamical behavior of carbohydrates as studied by carbon-13 and proton nuclear spin relaxation. *J. Phys. Chem.* 1996, 100, 17103–17110.
21. Vishnyakov, A.; Widmalm, G.; Kowalewski, J.; and Laaksonen, A., Molecular dynamics simulation of the alpha-D-Manp-(1 -> 3)-beta-D- Glcp-OMe disaccharide in water and water/DMSO solution. *J. Am. Chem. Soc.* 1999, 121, 5403–5412.
22. Effemey, M.; Lang, J.; and Kowalewski, J., Multiple-field carbon-13 and proton relaxation in sucrose in viscous solution. *Magn. Reson. Chem.* 2000, 38, 1012–1018.
23. Höög, C.; Landersjö, C.; and Widmalm, G., Oligosaccharides display both rigidity and high flexibility in water as determined by C-13 NMR relaxation and H-1,H-1 NOE spectroscopy: Evidence of anti-phi and anti-psi, torsions in the same glycosidic linkage. *Chem. Eur. J.* 2001, 7, 3069–3077.
24. Zerbetto, M.; Polimeno, A.; Kotsyubynskyy, D.; Ghalebani, L.; Kowalewski, J.; Meirovitch, E.; Olsson, U.; and Widmalm, G., An integrated approach to NMR spin relaxation in flexible biomolecules: Application to beta-D-glucopyranosyl-(1 -> 6)-alpha-D-mannopyranosyl-OMe. *J. Chem. Phys.* 2009, 131, 234501.
25. Zerbetto, M.; Kotsyubynskyy, D.; Kowalewski, J.; Widmalm, G.; and Polimeno, A., Stochastic modeling of flexible biomolecules applied to NMR relaxation. I. Internal dynamics of cyclodextrins: Gamma-cyclodextrin as a case study. *J. Phys. Chem. B* 2012, *116*, 13159–13171.
26. Kotsyubynskyy, D.; Zerbetto, M.; Soltesova, M.; Engström, O.; Pendrill, R.; Kowalewski, J.; Widmalm, G.; and Polimeno, A., Stochastic modeling of flexible biomolecules applied to NMR relaxation. 2. Interpretation of complex dynamics in linear oligosaccharides. *J. Phys. Chem. B* 2012, 116, 14541–14555.
27. Brevard, C.; Kintzinger, J. P.; and Lehn, J. M., Nuclear relaxation and molecular properties. VII. Molecular dynamics. Component analysis of local molecular motions. *Tetrahedron* 1972, 28, 2447–2460.
28. Lang, J.; Dechter, J. J.; Effemey, M.; and Kowalewski, J., Dynamics of an inclusion complex of chloroform and cryptophane-E: Evidence for a strongly anisotropic van der Waals bond. *J. Am. Chem. Soc.* 2001, 123, 7852–7858.
29. Tosner, Z.; Lang, J.; Sandström, D.; Petrov, O.; and Kowalewski, J., Dynamics of an inclusion complex of dichloromethane and cryptophane-E. *J. Phys. Chem. A* 2002, 106, 8870–8875.
30. Tosner, Z.; Petrov, O.; Dvinskikh, S. V.; Kowalewski, J.; and Sandström, D., A C-13 solid-state NMR study of cryptophane-E: Chloromethane inclusion complexes. *Chem. Phys. Lett.* 2004, 388, 208–211.
31. Nikkhou Aski, S.; Lo, A. Y. H.; Brotin, T.; Dutasta, J. P.; Edén, M.; and Kowalewski, J., Studies of inclusion complexes of dichloromethane in cryptophanes by exchange kinetics and C-13 NMR in solution and the solid state. *J. Phys. Chem. C* 2008, 112, 13873–13881.
32. Takacs, Z.; Steiner, E.; Kowalewski, J.; and Brotin, T., NMR investigation of chloromethane complexes of cryptophane-A and its analogue with butoxy groups. *J. Phys. Chem. B* 2014, 118, 2134–2146.
33. Koshland, D. E., Application of a theory of enzyme specificity to protein synthesis. *Proc. Natl. Acad. Sci. U. S. A.* 1958, 44, 98–104.
34. Palmer, A. G., NMR probes of molecular dynamics: Overview and comparison with other techniques. *Ann. Rev. Biophys. Biomol. Struc.* 2001, 30, 129–155.
35. Phan, I. Q. H.; Boyd, J.; and Campbell, I. D., Dynamic studies of fibronectin type I module pair at three frequencies: Anisotropic modelling and direct determination of conformational exchange. *J. Biomol. NMR* 1996, 8, 369–378.
36. Kroenke, C. D.; Loria, J. P.; Lee, L. K.; Rance, M.; and Palmer, A. G., Longitudinal and transverse H-1-N-15 dipolar N-15 chemical shift anisotropy relaxation interference: Unambiguous determination of rotational diffusion tensors and chemical exchange effects in biological macromolecules. *J. Am. Chem. Soc.* 1998, 120, 7905–7915.
37. Kay, L. E.; Torchia, D. A.; and Bax, A., Backbone dynamics of proteins as studied by 15N inverse detected heteronuclear NMR spectroscopy: Application to staphylococcal nuclease. *Biochemistry* 1989, 28, 8972–8979.

38. Kördel, J.; Skelton, N. J.; Akke, M.; and Chazin, W. J., High-resolution solution structure of calcium-loaded calbindin D9k. *J. Mol. Biol.* 1993, 231, 711–734.
39. Skelton, N. J.; Kördel, J.; and Chazin, W. J., Determination of the solution structure of apo calbindin D9k by NMR spectroscopy. *J. Mol. Biol.* 1995, 249, 441–462.
40. Akke, M.; Forsén, S.; and Chazin, W. J., Solutions structure of (Cd2+)1-calbindin D9k reveals details of the stepwise structural changes along the apo->(Ca2+)1II->(Ca2+)2I,II binding pathway. *J. Mol. Biol.* 1995, 252, 102–121.
41. Mäler, L.; Blankenship, J.; Rance, M.; and Chazin, W. J., Site-site communication in the EF-hand Ca2+-binding protein calbindin D9k. *Nature Struct. Biol.* 2000, 7, 245–250.
42. Kördel, J.; Skelton, N. J.; Akke, M.; Palmer, A. G.; and Chazin, W. J., Backbone dynamics of calcium-loaded calbindin D9k studied by two-dimensional proton-detected nitrogen-15 NMR spectroscopy. *Biochemistry* 1992, 31, 4856–4866.
43. Akke, M.; Skelton, N. J.; Kördel, J.; Palmer, A. G.; and Chazin, W. J., Effects of ion binding on the backbone dynamics of calbindin D9k determined by nitrogen-15 NMR relaxation. *Biochemistry* 1993, 32, 9832–9844.
44. Woessner, D. E., Nuclear spin relaxation in ellipsoids undergoing rotational brownian motion. *J. Chem. Phys.* 1962, 37, 647–654.
45. Clore, G. M.; Szabo, A.; Bax, A.; Kay, L. E.; Driscoll, P. C.; and Gronenborn, A. M., Deviations from the simple two-parameter model-free approach to the interpretation of nitrogen-15 nuclear magnetic relaxation of proteins. *J. Am. Chem. Soc.* 1990, 112, 4989–4991.
46. Jaremko, L.; Jaremko, M.; Nowakowski, M.; and Ejchart, A., The quest for simplicity: Remarks on the free-approach models. *J. Phys. Chem. B* 2015, 119, 11978–11987.
47. Fenwick, R. B.; Schwieters, C. D.; and Vögeli, B., Direct investigation of slow correlated dynamics in proteins via dipolar interactions. *J. Am. Chem. Soc.* 2016, 138, 8412–8421.
48. Gronenborn, A. M.; Filpula, D. R.; Essig, N. Z.; Achari, A.; Whitlow, M.; Wingfield, P. T.; and Clore, G. M., A novel, highly stable fold of the immunoglobulin N binding domain of streptococcal protein G. *Science* 1991, 253(5020), 657–661.
49. Peng, J. W.; and Wagner, G., Mapping of spectral density functions using heteronuclear NMR relaxation measurements. *J. Magn. Reson.* 1992, 98, 308–332.
50. Peng, J. W.; and Wagner, G., Mapping of the spectral densities of N-H bond motions in eglin c using heteronuclear relaxation experiments. *Biochemistry* 1992, 31, 8571–8586.
51. Peng, J. W.; and Wagner, G., Frequency spectrum of NH bonds in eglin c from spectral density mapping at multiple fields. *Biochemistry* 1995, 34, 16733–16752.
52. Farrow, N. A.; Zhang, O.; Forman-Kay, J. D.; and Kay, L. E., Comparison of backbone dynamics of a folded and an unfolded SH3 domain existing in equilibrium in aqueous buffer. *Biochemistry* 1995, 34, 868–878.
53. Farrow, N. A.; Zhang, O.; Szabo, A.; Torchia, D. A.; and Kay, L. E., Spectral density function mapping using 15N relaxation data exclusively. *J. Biomol. NMR* 1995, 6, 153–162.
54. Ishima, R.; and Nagayama, K., Quasi-spectral-density function analysis for nitrogen-15 nuclei in proteins. *J. Magn. Reson. Ser. B* 1995, 108, 73–76.
55. Allard, P.; Jarvet, J.; Ehrenberg, A.; and Gräslund, A., Mapping of the spectral density function of a Ca–Ha bond vector from NMR relaxation rates of a 13C-labelled α-carbon in motilin. *J. Biomol. NMR* 1995, 5, 133–146.
56. Jarvet, J.; Allard, P.; Ehrenberg, A.; and Gräslund, A., Spectral-density mapping of C-13(alpha)-H-1(alpha) vector dynamics using dipolar relaxation rates measured at several magnetic fields. *J. Magn. Reson. Ser. B* 1996, 111, 23–30.
57. Kaderavek, P.; Zapletal, V.; Fiala, R.; Srb, P.; Padrta, P.; Precechtelova, J. P.; Soltesova, M.; Kowalewski, J.; Widmalm, G.; Chmelik, J.; Sklenar, V.; and Zidek, L., Spectral density mapping at multiple magnetic fields suitable for C-13 NMR relaxation studies. *J. Magn. Reson.* 2016, 266, 23–40.

58. Igumenova, T. I; Frederick, K. K.; and Wand, A. J., Characterization of the fast dynamics of protein amino acid side chains using NMR relaxation in solution. *Chem. Rev.* 2006, 106, 1672–1699.
59. Muhandiram, D. R.; Yamazaki, T.; Sykes, B. D.; and Kay, L. E., Measurement of H-2 T-1 and T-1ρ relaxation times in uniformly C-13-labeled and fractionally H-2-labeled proteins in solution. *J. Am. Chem. Soc.* 1995, 117, 11536–11544.
60. Kay, L. E.; Muhandiram, D. R.; Farrow, N. A.; Aubin, Y.; and Forman-Kay, J. D., Correlation between dynamics and high affinity binding in an SH2 domain interaction. *Biochemistry* 1996, 35, 361–368.
61. Yang, D.; Mittermaier, A.; Mok, Y. K.; and Kay, L. E., A study of protein side-chain dynamics from new 2H auto-correlation and 13C cross-correlation experiments: Application to the N-terminal SH3 domain from drk. *J. Mol. Biol.* 1998, 276, 939–954.
62. Millet, O.; Muhandiram, D. R.; Skrynnikov, N. R.; and Kay, L. E., Deuterium spin probes of side-chain dynamics in proteins. 1. Measurement of five relaxation rates per deuteron in C-13-labeled and fractionally H-2-enriched proteins in solution. *J. Am. Chem. Soc.* 2002, 124, 6439–6448.
63. Skrynnikov, N. R.; Millet, O.; and Kay, L. E., Deuterium spin probes of side-chain dynamics in proteins. 2. Spectral density mapping and identification of nanosecond time-scale side-chain motions. *J. Am. Chem. Soc.* 2002, 124, 6449–6460.
64. Schnell, J. R.; Dyson, H. J.; and Wright, P. E., Effect of cofactor binding and loop conformation on side chain methyl dynamics in dihydrofolate reductase. *Biochemistry* 2004, 43, 374–383.
65. Akke, M.; Brüschweiler, R.; and Palmer, A. G., NMR order parameter and free energy: An analytical approach and its application to cooperative Ca(2+) binding by calbindin D9k. *J. Am. Chem. Soc.* 1993, 115, 9832–9833.
66. Yang, D.; and Kay, L. E., Contributions to conformational entropy arising from bond vector fluctuations measured from NMR-derived order parameter: Applications to protein folding. *J. Mol. Biol.* 1996, 263, 369–382.
67. Lee, A. L.; Kinnear, S. A.; and Wand, A. J., Redistribution and loss of side-chain entropy upon formation of a calmodulin–peptide complex. *Nature Struct. Biol.* 2000, 7, 72–77.
68. Wand, A.J. Dynamic activation of protein function: A view emerging from NMR spectroscopy. *Nature Struct. Biol.* 2001, 8, 926–931.
69. Marlow, M. S.; Dogan, J.; Frederick, K. K.; Valentine, K. G.; and Wand, A. J. The role of conformational entropy in molecular recognition by calmodulin. *Nature Chem. Biol.* 2010, 6, 352–358.
70. Schneider, D. M.; Dellwo, M. J.; and Wand, A. J., Fast internal main-chain dynamics of human ubiquitin. *Biochemistry* 1992, 31, 3645–3652.
71. Tjandra, N.; Feller, S. E.; Pastor, R. W.; and Bax, A., Rotational diffusion anisotropy of human ubiquitin from N-15 NMR relaxation. *J. Am. Chem. Soc.* 1995, 117, 12562–12566.
72. Lienin, S. F.; Bremi, T.; Brutscher, B.; Brüschweiler, R.; and Ernst, R. R., Anisotropic intramolecular backbone dynamics of ubiquitin characterized by NMR relaxation and MD computer simulation. *J. Am. Chem. Soc.* 1998, 120, 9870–9879.
73. Tjandra, N.; Szabo, A.; and Bax, A., Protein backbone dynamics and N-15 chemical shift anisotropy from quantitative measurement of relaxation interference effects. *J. Am. Chem. Soc.* 1996, 118, 6986–6991.
74. Fushman, D.; Tjandra, N.; and Cowburn, D., Direct measurement of N-15 chemical shift anisotropy in solution. *J. Am. Chem. Soc.* 1998, 120, 10947–10952.
75. Fushman, D.; Tjandra, N.; and Cowburn, D., An approach to direct determination of protein dynamics from N-15 NMR relaxation at multiple fields, independent of variable N-15 chemical shift anisotropy and chemical exchange contributions. *J. Am. Chem. Soc.* 1999, 121, 8577–8582.
76. Köver, K. E.; and Batta, G., Separating structure and dynamics in CSA/DD cross-correlated relaxation: A case study on trehalose and ubiquitin. *J. Magn. Reson.* 2001, 150, 137–146.

77. Damberg, P.; Jarvet, J.; and Gräslund, A., Limited variations in N-15 CSA magnitudes and orientations in ubiquitin are revealed by joint analysis of longitudinal and transverse NMR relaxation. *J. Am. Chem. Soc.* 2005, 127, 1995–2005.
78. Loth, K.; Pelupessy, P.; and Bodenhausen, G., Chemical shift anisotropy tensors of carbonyl, nitrogen, and amide proton nuclei in proteins through cross-correlated relaxation in NMR spectroscopy. *J. Am. Chem. Soc.* 2005, 127, 6062–6068.
79. Fenwick, R. B.; Esteban-Martin, S.; Richter, B.; Lee, D.; Walter, K. F. A.; Milovanovic, D.; Becker, S.; Lakomek, N. A.; Griesinger, C.; and Salvatella, X., Weak long-range correlated motions in a surface patch of ubiquitin involved in molecular recognition. *J. Am. Chem. Soc.* 2011, 133, 10336–10339.
80. Lindorff-Larsen, K.; Maragakis, P.; Piana, S.; and Shaw, D. E., Picosecond to millisecond structural dynamics in human ubiquitin. *J. Phys. Chem. B* 2016, 120, 8313–8320.
81. Charlier, C.; Khan, S. N.; Marquardsen, T.; Pelupessy, P.; Reiss, V.; Sakellariou, D.; Bodenhausen, G.; Engelke, F.; and Ferrage, F., Nanosecond time scale motions in proteins revealed by high-resolution NMR relaxometry. *J. Am. Chem. Soc.* 2013, 135, 18665–18672.
82. Kimmich, R.; and Anoardo, E., Field-cycling NMR relaxometry. *Progr. NMR Spectr.* 2004, 44, 257–320.
83. Kimmich, R.; and Fatkullin, N., Polymer chain dynamics and NMR. *Adv. Polym. Sci.* 2004, 170, 1–113.
84. Kruk, D.; Herrmann, A.; and Rössler, E. A., Field-cycling NMR relaxometry of viscous liquids and polymers. *Progr. NMR Spectr.* 2012, 63, 33–64.
85. Meier, R.; Kahlau, R.; Kruk, D.; and Rössler, E. A., Comparative studies of the dynamics in viscous liquids by means of dielectric spectroscopy and field cycling NMR. *J. Phys. Chem. A* 2010, 114, 7847–7855.
86. Gainaru, C.; Lips, O.; Troshagina, A.; Kahlau, R.; Brodin, A.; Fujara, F.; and Rössler, E. A., On the nature of the high-frequency relaxation in a molecular glass former: A joint study of glycerol by field cycling NMR, dielectric spectroscopy, and light scattering. *J. Chem. Phys.* 2008, 128, 174505.
87. Kruk, D.; Meier, R.; and Rössler, E. A., Translational and rotational diffusion of glycerol by means of field cycling H-1 NMR relaxometry. *J. Phys. Chem. B* 2011, 115, 951–957.
88. Meier, R.; Kruk, D.; Gmeiner, J.; and Rössler, E. A., Intermolecular relaxation in glycerol as revealed by field cycling H-1 NMR relaxometry dilution experiments. *J. Chem. Phys.* 2012, 136, 034508.
89. Kariyo, S.; Herrmann, A.; Gainaru, C.; Schick, H.; Brodin, A.; Novikov, V. N.; and Rössler, E. A., From a simple liquid to a polymer melt: NMR relaxometry study of polybutadiene. *Phys. Rev. Lett.* 2006, 97, 207803. Errata: ibid 2008, 100, 109901.

12

Applications of Relaxation-Related Measurements to Structure Determination

The most common goal of a nuclear magnetic resonance (NMR) investigation of an organic substance or of a biological macromolecule is to obtain information about its structure. Determining structural parameters often means different things to a chemist than to a structural biologist. The chemist is most likely (but not always) interested in investigating the chemical structure of a substance, whereas a structural biologist is interested in the three-dimensional (3D) structure of a molecule for which the chemical structure is already known; for instance, the amino acid sequence of a polypeptide. For a small organic molecule, questions concerning the substitution pattern or stereo-chemistry, *i.e.* the primary structure, are essential to determining the outcome of a chemical reaction. In a protein, on the other hand, typically the interesting questions concern the secondary structure and the overall fold. Despite the apparent differences between the two cases, NMR is suitable for investigating both local as well as overall structure, in the first place through the nuclear Overhauser enhancement (NOE) effect. Dipole-dipole (DD) relaxation has found widespread use in all fields of chemistry and biochemistry, since it provides a way of estimating internuclear distances in molecules. NOE, or the cross-relaxation phenomenon, has, together with *J*-coupling measurements, until recently been the most important NMR tool for investigating structure and conformation.

12.1 Applications in Organic Chemistry

Chemical structure investigations as well as conformational analyses of organic compounds are readily made by NMR methods. The two most common parameters that are easily measurable and contain conformational information are the *J*-couplings and the homonuclear ^{1}H NOE factors. For details concerning both experimental approaches, as well as examples, we refer the reader to the book by Neuhaus and Williamson (*Further reading*). The NOE contains distance information through the $1/r^6$ dependence of the square of the dipolar interaction. It would seem straightforward to deduce distances from measured cross-relaxation rates, or NOE factors, but there are several concerns that need to be addressed. The first one deals with how the separation of dynamics and structure is made. For small organic molecules, this is often not a problem, as their dynamics are well within the extreme narrowing regime. In this case, the dynamics can simply be described by an effective rotational correlation time, as discussed in Section 11.2. On the other hand, there is seldom one single conformation that describes the overall structure of a small organic molecule. NOE enhancement factors between substituents on an acyclic chain will most likely be averaged due to fast exchange between different contributing conformers. For medium-sized molecules, the dynamics start to be more complicated, as local dynamics will need to be considered, most often in terms of incorporating a motional model.

There are, however, specific structural properties that can easily be elucidated from interpreting NOEs or cross-relaxation rates. In organic chemistry, the structure determination of small molecules is mainly concerned with investigating positional substitutions, conformation around double bonds and chirality. Typically, questions regarding the position of different substituents in a molecule can be addressed by measuring NOEs, which relate to intramolecular distances. For instance, aromatic positional isomers can be detected through the analysis of NOE enhancement factors. In the same way, problems involving double bond isomerism can be solved easily using NOE-derived approaches. A simple example is given by the following reaction. The four structures shown in Figure 12.1 can be obtained, depending on the conditions of the reaction, by adding methanesulfenyl chloride to 3,3-dimethylbutyne[1]. By analysis of the NOE enhancement from one of the protons, the vinyl proton, to the two substituent groups, the first two structures can readily be distinguished from each other. The remaining two, however, will most likely display the same NOE enhancement pattern. In this case other methods, typically chemical analyses, must be employed.

Taking the analysis one step further, the interatomic distances can be measured, at least qualitatively, from cross-relaxation, or NOE between two protons. The NOE factors are usually measured by the NOE difference experiments, and the fractional enhancement of signal intensity for the different protons in a molecule, due to the irradiation of a specific proton, is reported. Given the NOE between a pair of protons, A and B, with a known distance between them, the distance between two other protons can be estimated by comparing the NOE factors from:

$$\frac{NOE_{AB}}{NOE_{CD}} = \frac{r_{CD}^6}{r_{AB}^6} \tag{12.1}$$

The symbol NOE_{AB} denotes the quantity η of Eq. (3.28), related to the steady-state enhancement of the A-spin caused by irradiation of the B-spin or vice versa. The assumptions underlying Eq. (12.1) are that the NOE arises exclusively through interaction within an isolated pair of protons, that the proton spin-lattice relaxation is at least approximately constant for all protons, and that the various proton-proton axes have similar effective correlation times.

The potential of the cross-relaxation studies increases if the cross-relaxation rates are properly evaluated and the dynamical behaviour is understood. In addition to the investigation of chemical structure, it is also feasible to obtain information about the molecular conformation of organic molecules. As an example, we shall turn to how the structure of carbohydrates can be determined. A nice compilation of NMR techniques used for structural as well as dynamic investigations of carbohydrates is provided in the book edited by Jimenez-Barbero and Peters[2].

First, the chemical structure, *i.e.* the monosaccharide composition, must be determined. This can be done by NMR methods, by which the specific sugar residues can be assigned in correlation experiments, such as two-dimensional (2D) total correlation spectroscopy (TOCSY) and ^{13}C,^{1}H-heteronuclear single quantum correlation (HSQC), but is more often complemented by chromatographic methods and a chemical analysis of the saccharide composition.

The primary structure, *i.e.* the sequence of sugar residues, can then be determined from a combination of nuclear Overhauser enhancement spectroscopy (NOESY) and ^{13}C,^{1}H-heteronuclear multiple bond connectivity (HMBC) experiments. The most important information is obtained from observing NOE connectivities between the anomeric protons of one residue and protons in the sequential residue.

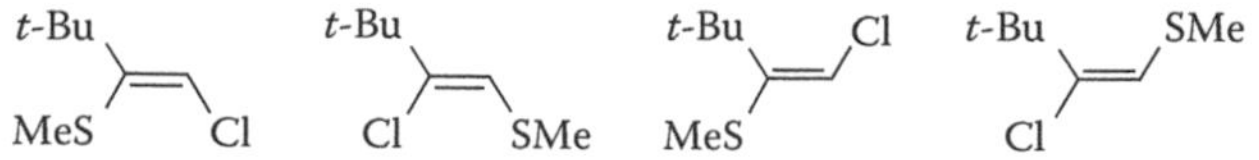

FIGURE 12.1 The four structures can be obtained, depending on the conditions of the reaction, by adding methanesulfenyl chloride to 3,3-dimethylbutyne.

By using a combination of NOE-derived information with coupling constants across the glycosidic linkage, the glycosidic linkage structure, *i.e.* α- or β-linkages, can be determined.

An example of this approach is provided by the structure determination of an extracellular polysaccharide from *Streptococcus thermophilus*[3]. The structure was determined from a combination of component and methylation analyses and NMR spectroscopy, and it was found that the polysaccharide is composed of pentasaccharide repeating units (Figure 12.2). The sequential structure was determined from ^{1}H–^{1}H NOESY data in combination with ^{13}C,^{1}H-HMBC. A part of the two-dimensional NOESY used to identify the connectivities between the sugar residues is shown in Figure 12.2.

The conformation of oligosaccharides can be studied by analyses of inter-residue cross-relaxation. The cross-relaxation rate measurement methods include the one-dimensional (1D) NOESY approach, as discussed in Section 9.4, in which one proton is selectively inverted by a shaped pulse, or by a combination of pulses, and the cross-relaxation is monitored during an arrayed mixing time. Figure 12.3 shows the build-up of magnetisation on the H3 proton in the disaccharide molecule shown in Figure 11.9b when selectively inverting the H1′ proton magnetisation[4]. The curves could be analysed to provide the cross-relaxation rate and, when combined with dynamic information from ^{13}C relaxation outside of extreme narrowing, the inter-proton distances between protons in the two residues.

Medium-sized organic molecules, such as the carbohydrates discussed in this section, and small peptides might suffer from having a close-to-critical rotational correlation time, making the cross-relaxation rate close to zero. In this case, the 1D (or 2D) rotating-frame Overhauser enhancement (ROE) measurement can be useful. The torsion angles defining the glycosidic linkage can be estimated more

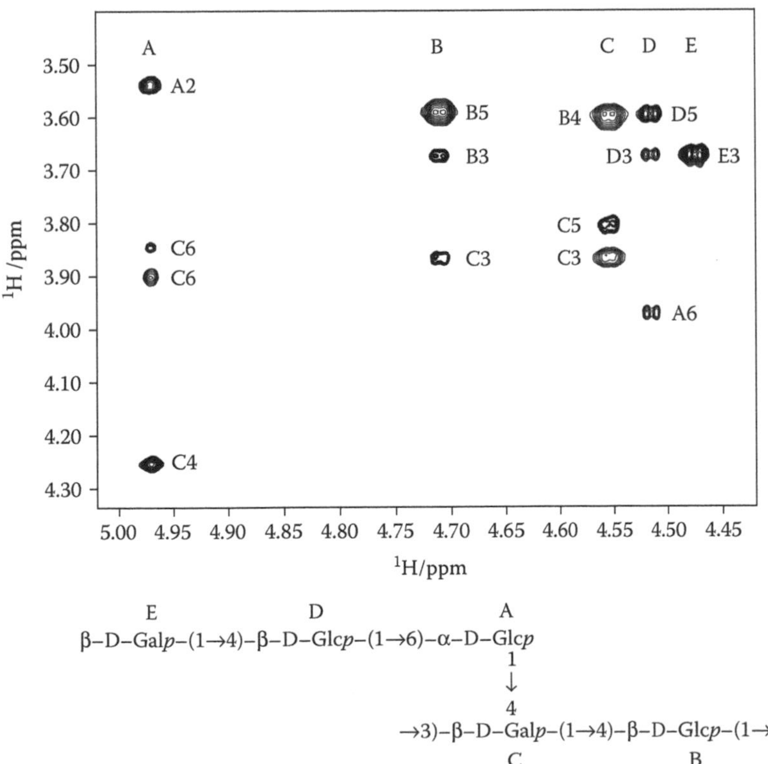

FIGURE 12.2 Two-dimensional NOESY and structure of the pentasaccharide-repeating unit in the extracellular polysaccharide (EPS) from *S. thermophilus*. (Reprinted with permission from Nordmark, E.-L. et al., *Biomacromolecules*, 6, 105–108, 2005. Copyright (2005) American Chemical Society.)

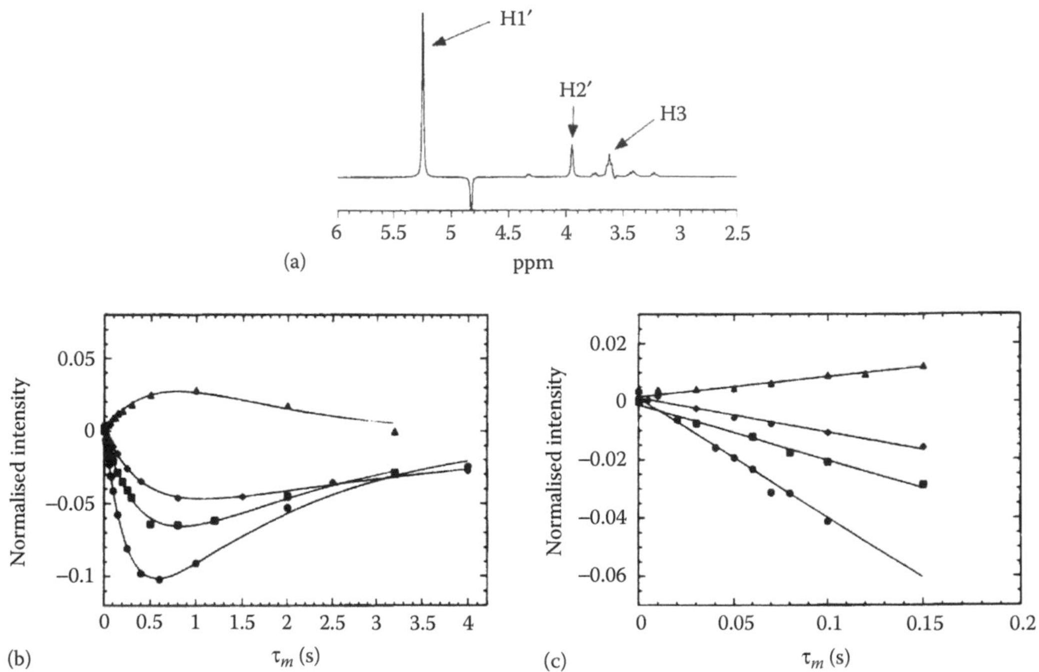

FIGURE 12.3 ^{1}H 1D NOESY of the disaccharide α-D-Man*p*-(1→3)-β-D-Glc*p*-OMe and the cross-relaxation build-up between H1′ and H3 at 9.4 T, obtained by selectively irradiating H1′ (^{1}H spectrum in (a)). Circles correspond to 268 K, squares to 282 K, diamonds to 288 K and triangles to 323 K. (b) shows the full cross-relaxation curves, while (c) shows the linear region used for determining the cross-relaxation parameters. (Reprinted with permission from Mäler, L. et al., *J. Phys. Chem.*, 100, 17103–17110, 1996. Copyright (1996) American Chemical Society.)

accurately from a combination of NOEs and molecular dynamics simulation procedures, or from combinations of *J*-couplings and cross-relaxation[2].

The drawback of this method of investigating conformation has its origin in the facts that typically only a few inter-residue NOEs can be determined, and thus the conclusions about the conformation of the molecule are to a large extent dependent on a few observations. Any errors associated with the interpretation of these NOEs will lead to large errors in the structure. Again, if the dynamics is not properly taken into account when analysing cross-relaxation build-up curves, the distances estimated using Eq. (12.1) will be erroneous. If the two proton pairs, AB and CD, in a molecule outside the extreme narrowing regime have different local dynamic properties, the simple approach offered by Eq. (12.1) will not hold. As we shall see in the next section, it is in fact easier to correctly determine the conformation, or three-dimensional structure, of a large protein from cross-relaxation, or NOEs, because of the much higher number of inter-residue NOE observations that can be made.

In Chapter 11, we quoted examples of dynamics studies in guest-host chemistry. Here, we would like to refer to structure-oriented work[5] on systems of this kind with cryptophane A (cryptophane A is similar to cryptophane E, shown in Figure 11.11, but all three linkers joining the two caps contain two rather than three methylene units), and its analogue with butoxy groups replacing the methoxy substituents, as hosts and chloromethanes (dichloromethane, chloroform) as guests. Measurements of NOE build-up curves for different proton pairs (involving spins residing in the guest and the host, as well as within the host) allowed the determination of cross-relaxation rates. From these, conclusions were drawn concerning the occurrence of different possible low-energy conformers of the host, identified based on DFT calculations.

The discussion in this section has so far concerned only the proton-proton NOEs. Heteronuclear NOEs, mainly involving ^{1}H and ^{13}C, can also be used as a tool for structural investigations and this topic has been reviewed by Canet and co-workers[6].

12.2 Applications to Three-Dimensional Structure Determination in Macromolecules

In the field of life sciences, many NMR experiments are aimed at determining the solution structure of a biological macromolecule, although it appears that the fraction of NMR-derived structures reported in the Research Collaboratory for Structural Bioinformatics (RCSB) Protein Data Bank (PDB) has decreased during the last years (www.rcsb.org). NMR is still a very valuable tool for determining the structures of proteins that cannot be crystallised, including many membrane proteins and partially unstructured or intrinsically disordered proteins. In this section, a brief summary of how a three-dimensional solution structure of a protein can be obtained will be presented. The information generally required to obtain a high-resolution solution structure is derived from many NMR sources, such as *J*-couplings, NOE and residual dipolar couplings (small dipolar couplings measured in molecules dissolved in a slightly anisotropic environment, for example, in bicelles in aqueous solution[7]).

12.2.1 NOE-Based Structure Information

Most of the NMR-derived data contain indirect information about structure, as, for instance, does the NOE or cross-relaxation rates. We shall now see how cross-relaxation is used in deriving distance constraints to be used in a structure calculation. For simplicity we will only consider the structure determination of proteins, but analogous methods are used for determining solution structures of, for example, RNA and DNA.

The basis for determining a solution structure of a protein is the same as that discussed for organic molecules in the preceding section. The goal is to measure as many inter-proton distances in the protein as possible and the fact that dipolar relaxation contains the $1/r^6$ distance dependence makes cross-relaxation measurements attractive for estimating distances. For small proteins, say less than 5–10 kDa, cross-relaxation has traditionally most often been measured from cross-peak intensities in 2D NOESY experiments. The homonuclear approach and the use of two-dimensional NOESY for generating solution structures are described in detail in the book by Wüthrich (*Further reading*).

For larger proteins, isotopic labelling, such as the incorporation of ^{15}N and/or ^{13}C, is necessary. A description of these methods can be found in a review by Clore and Gronenborn[8], and the reader interested in a general compilation of NMR methods for protein samples should consult the book by Cavanagh *et al.* (*Further reading*). Isotope-labelling enables the structural biologist to use 3D *heteronuclear-edited* NOESY for extracting distances. The general idea behind these experiments is to combine the two-dimensional NOESY experiment with a two-dimensional heteronuclear correlation experiment, such as the HSQC or transverse relaxation-optimised spectroscopy (TROSY), to obtain resolution through a third heteronuclear dimension. The information from such an experiment is, however, the same as that obtained from a conventional two-dimensional NOESY: information about proton-proton NOE or cross-relaxation rates.

It is in principle possible to monitor the build-up of magnetisation on one proton due to dipolar cross-relaxation with another proton by performing the NOESY experiment several times, incrementing the mixing time. From the initial rate of the obtained build-up curves, it would then be possible to determine the distance between two protons, much as was described for smaller organic molecules. There are two main reasons why this is seldom done. The first reason is simply that the experimental procedures would be too time-consuming and the data evaluation very complex. The second reason is connected to difficulties in determining accurate distances from cross-relaxation build-up curves, related to correctly assessing the motional contribution to the cross-relaxation, *i.e.* the spectral density function (see Chapter 11 for a description of molecular dynamics and relaxation). The dynamic behaviour of one chemical group, such as an amino acid side-chain, may be very different from another. Needless to say, a rigid backbone structure within the protein has very different dynamic properties from those of flexible regions, and even more different from those of freely rotating methyl groups. Furthermore, complex inter-proton dynamics caused by, for instance, domain flexibility and side-chain motions imposes yet another degree of complexity in interpreting cross-relaxation. Nevertheless, recent years have witnessed large progress in the methodology and applications of the so-called exact NOEs (eNOEs); see Section 12.2.4.

The $1/r^6$ distance dependence in cross-relaxation is, however, fortunate for the structural biologist. There are two main reasons for this. First, the effect of dipolar relaxation decays rapidly with increasing distance; thus, it is only possible to observe NOE effects between protons that are sufficiently close in space, around 5 Å or closer, provided that sufficiently short mixing times are used (see Chapter 9 and below). The second reason is that the errors in the measured cross-peak intensities translate to rather small errors in distance. The first reason may at first glance seem like a limitation, and to a certain degree it is. However, since the distance limit is narrow, it is possible to distinguish clearly between protons that are close, for which we observe NOE effects, and those that are not, for which we do not observe NOE effects.

This reasoning has led to a robust method for converting NOESY cross-peak intensities into distance constraints that can be used to build a model of the three-dimensional structure. NOESY data are generally acquired using a specific mixing time, which should be long enough to enable cross-peaks between the more distant protons (5 Å apart) to build up but should also be kept sufficiently short to restrict *spin-diffusion*, an effect discussed in more detail below. The cross-peak volumes are evaluated and translated into upper distance limits, which are not to be exceeded in the structural model. Due to the reasons described above, the distances may, however, be shorter and therefore the concept of an upper distance limit is useful. Usually, it is sufficient to categorise the cross-peak intensities as weak, medium and strong, which after internal calibration can be translated into specific upper distance limits, for instance, 5, 3.5 and 2.7 Å, following Williamson *et al.*[9]. The absence of NOESY cross-peaks is more difficult to interpret. In principle, a lower distance limit can be derived from an absent cross-peak, but this is very seldom done, since the internal dynamics discussed above most often leads to a reduction in cross-peak intensity and thus may cause cross-peaks to vanish.

The assigned NOE cross-peaks are usually categorised on different sub-levels depending on the sequential distance between the two protons. Intra-residue NOE cross-peaks provide distance limits for protons within one amino acid residue and contribute little to the structure determination. Sequential NOEs are between two sequential amino acid residues and are typically important in defining the local secondary structure. For helical structures, medium-range NOE cross-peaks, usually defined as NOEs between protons not more than four residues apart, are important, since there are many short inter-residue distances between residues in this range. The remaining NOEs are classified as long-range and are important in defining the overall fold of the protein.

12.2.2 Generating Structures

There are several algorithms in which NOE-derived geometry information (and other NMR-derived geometry constraints, such as those from *J*-couplings, chemical shifts and residual dipolar couplings) can be included to calculate a three-dimensional structure. The first approach that was used was *distance geometry*, based on the fact that the NOE volumes can be translated into distance restraints[10]. In theory, the Cartesian coordinates for all atoms in a molecule can be obtained if all distances between the atoms are known. In reality, only a subset of distances is determined and very crude structures are obtained. A second approach is to use a *variable target function*, in which an initial structure is fitted to the experimental restraints in a step-wise manner. The distance restraints are considered in several steps, the first being only intra-residue, the second sequential distances, and so on. This was first implemented in the program DISMAN[11] and subsequently this algorithm was developed into the more optimised program DIANA[12]. These methods are usually the basis for generating a starting structure that can be further optimised using molecular dynamics or torsion angle dynamics protocols (DYANA and later CYANA)[13,14]. The method works in torsion angle space, which is an advantage, since it keeps the covalent geometry of the system intact during the calculation.

In a molecular dynamics simulation, the classical dynamics of atoms (or perhaps molecular fragments such as CH_2 groups) is evaluated by solving Newton's equation of motion. Typically, many structures have to be calculated in order to properly evaluate the available conformational space for a given set of geometry constraints, and the need for speed requires that the simulation be kept simple and fast. A typical *simulated annealing* algorithm is performed only over a few picoseconds at very high temperatures (T = 1000 K) with a cooling phase to allow the proper fold. A large set of structures is generated in this way in order to obtain a reasonable set of acceptable conformers. A simplified potential is used in the simulated annealing protocol, containing terms to maintain covalent geometry, proper chirality and planarity, as well as simple potentials for the repulsive and electrostatic non-bonded interactions. In addition, terms for distance and torsion angle restraints are included; more recently, also residual dipolar couplings and chemical shifts are used as geometry restraints. There are several program packages available, including AMBER[15], XPLOR[16] and CHARMM[17].

The quality of an NMR structure is based on several factors. First is the amount of experimentally determined geometry information, *i.e.* the number of assigned NOE cross-peaks. The precision of the structure is generally defined by examining root mean square deviations in atomic coordinates within an ensemble of structures. The energies and constraint violation terms should be carefully inspected to identify errors in assignments and input data. No consistent violations should be observed within the ensemble of structures. The geometry, bond lengths, bond angles and chirality are checked (against the database of known structures) to search for peculiar deviations. The structures are subsequently analysed by calculating the backbone and side-chain torsion angles, and inter-atom distances indicating the possibility of hydrogen bonds. These parameters are used to assess the secondary structure of the protein. The precision of the structure thus depends on the total number of upper distance constraints that can be determined from a NOESY data set rather than the precision of the individual distances. This is what makes the NOE-based structure determination a robust and reliable method. An excellent review by Güntert on the generation of three-dimensional solution structures from NMR provides the reader with additional information concerning the structure calculation procedure[18].

12.2.3 The Calbindin D_{9k} Example

As an example, we shall examine the structure of a small protein, calbindin D_{9k}, a Ca^{2+}-binding protein for which the structure is well known from both X-ray crystallography and NMR[19–23]. The dynamics of this protein were discussed in Chapter 11. The geometry of the different structural elements in proteins, α-helices, β-sheets *etc.* is associated with very specific distances. These distances, in turn, give rise to specific cross-peak patterns in the NOESY experiment. For instance, the sequential H^N–H^N distance in an α-helical structure is very short and can be used to identify helical sequences in the protein. Figure 12.4 shows part of a ^1H–^1H NOESY for apo calbindin D_{9k}, indicating the sequential connectivities.

By analysing these cross-peak patterns, it is possible to find information about the backbone geometry, and thus the secondary structure. Figure 12.5 shows a summary of the identified short- and medium-range NOEs, *i.e.* NOE cross-peaks between protons that are not far apart in the amino acid sequence, that were found for apo calbindin D_{9k}[24]. From this information, secondary structure elements can easily be identified in the protein. Provided that a sufficient number of long-range NOEs, *i.e.* between protons that are distant in amino acid sequence, can be identified, an overall tertiary structure can be

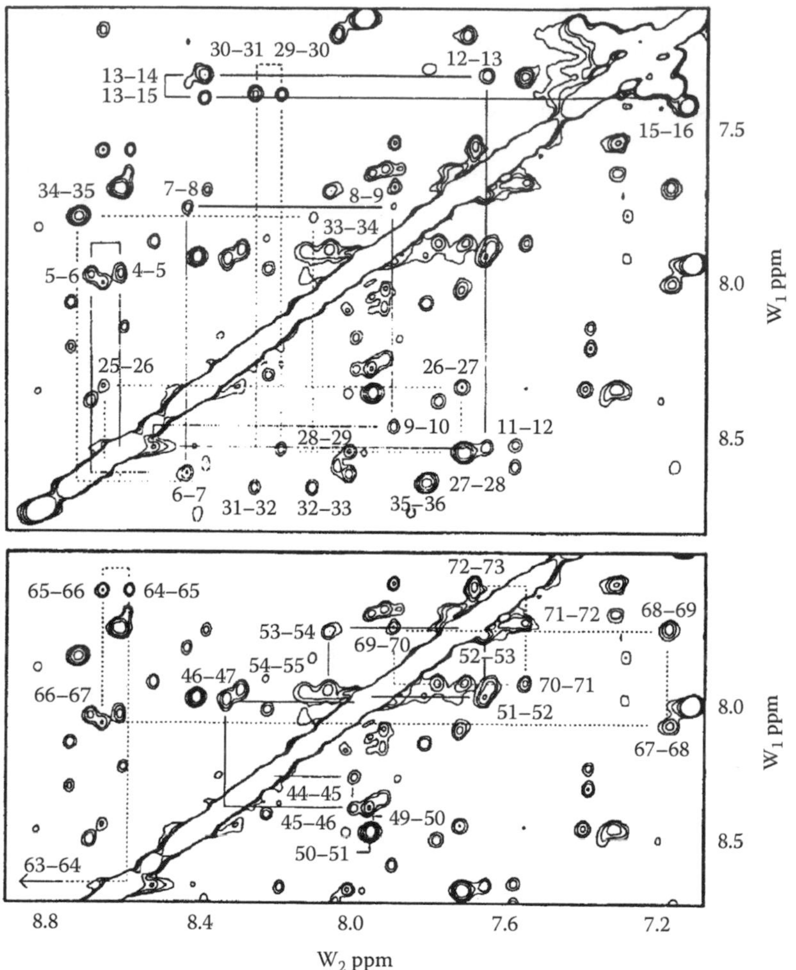

FIGURE 12.4 The H^N–H^N region of a 500 MHz NOESY recorded for calbindin D_{9k} ($\tau_m = 200$ ms). Sequential amide–amide distances are indicated. (Reprinted with permission from Skelton, N. J. et al., *Biochemistry*, 29, 5752–5761, 1990. Copyright (1990) American Chemical Society.)

determined. For calbindin D_{9k}, the overall fold was determined based on a number of inter-helical NOE cross-peaks (Figure 12.6).

A structure was subsequently calculated using a combination of distance geometry and restrained molecular dynamics (Figure 12.7). A total of 995 NOE-derived distance constraints and 122 dihedral angle constraints (based on *J*-couplings) were used in the calculation[23]. Clearly, the number of identified distance constraints defines the precision of the structure, as can be seen by comparing this number with the root mean square deviation in atomic coordinates (Figure 12.8). The less defined regions of the structure cannot arbitrarily be assigned to flexible and mobile regions of the proteins, although this is likely to be so. There may well exist other reasons for not being able to define distance constraints, such as unfortunate resonance overlap. A comparison between the definition of the structure and internal mobility, defined by the generalised order parameters discussed in Section 11.4, clearly shows that the precision of the structure is often, but not always, correlated with restriction of the internal motion.

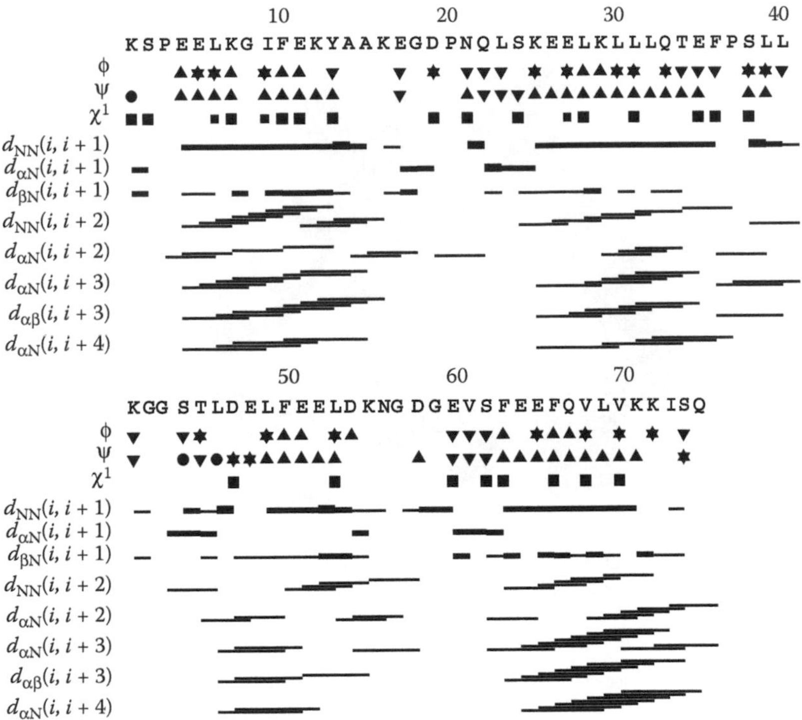

FIGURE 12.5 Overview of sequential and medium-range NOEs observed for apo calbindin D_{9k}, carrying information about the secondary structure. The thickness of the lines for the sequential distances indicates the values of upper distance bounds. In addition, torsion angle constraints are displayed. An upward triangle indicates typical helix ϕ and ψ angles, while a downward triangle indicates β-sheet values. A star indicates a constraint that is compatible with both types of structures. The different size of the squares indicates the different possible rotamers for the χ^1 angle (−60°, 60°, 180°). The constraints were obtained from the Protein Data Bank (www.rcsb.org, accession code 1CLB). The plot was generated with the program DYANA. (From Güntert, P. et al., *J. Mol. Biol.*, 273, 283–298, 1997.)

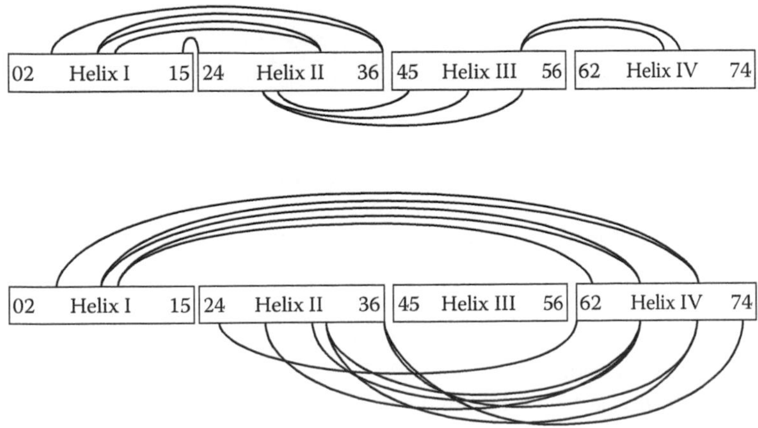

FIGURE 12.6 Summary of interhelical NOEs observed in apo calbindin D_{9k}, providing information about the overall fold. (Reprinted with permission from Skelton, N. J. et al., *Biochemistry*, 29, 5752–5761, 1990. Copyright (1990) American Chemical Society.)

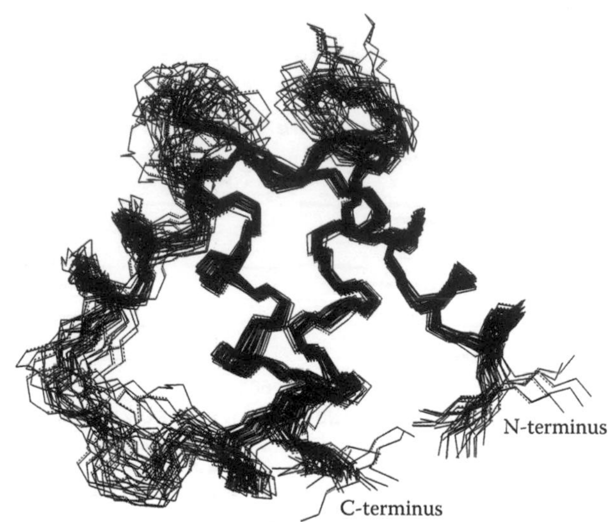

FIGURE 12.7 The solution structure of apo calbindin D_{9k} represented by an overlay of the 33 structures in the ensemble (PDB accession code 1CLB). The figure was produced using Insight (version 2000).

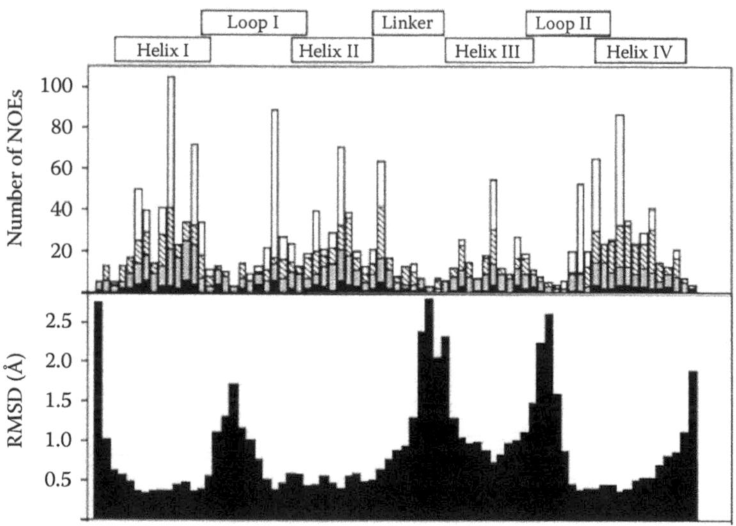

FIGURE 12.8 Plots of NOE-derived distance constraints (top panel) and average root-mean-square deviation (rmsd) of the backbone atoms of the 33 best structures (bottom) calculated for apo calbindin D_{9k}. Constraints are categorised as intra-residue (filled), sequential (shaded), medium-range (hatched) and long-range ($|j - i| \geq 5$) (open). (Reprinted with permission from Skelton, N. J. et al., *J.Mol.Biol.* 249, 441–462, 1985. Copyright (1985) Elsevier.)

The structure determination of apo calbindin D_{9k}, as well as the structure of the Ca^{2+}-bound state of the protein[22], was based on homonuclear ^1H–^1H NMR experiments only. A subsequent investigation of the N56A mutant, to which a Ca^{2+} ion was bound in the first binding site only, was made with ^{15}N-labeled protein[25]. This not only facilitated the assignment of NOESY cross-peaks but also, more importantly, enabled the backbone dynamics to be investigated (see Section 11.4). The structures of calbindin D_{9k} at all levels of Ca^{2+}-binding, together with the dynamics for all states, made it possible to investigate the cooperativity in Ca^{2+}-binding for this protein.

12.2.4 "Exact NOEs": Back to Ubiquitin

As indicated above, the last decade has witnessed a development of a new trend in NOE measurements, in which the emphasis is on precise and accurate determination of the individual homonuclear proton cross-relaxation rates and interatomic distances. The methodology makes use of the concept of exact NOEs (eNOEs). The topic was the subject of a review by Vögeli[26], in which the theoretical and experimental aspects were thoroughly covered, along with applications.

Here, we choose human ubiquitin as an example, as it has already been discussed in terms of dynamics and structure in Subsection 11.4.5. Vögeli and co-workers[27] reported an important quantitative investigation of proton cross-relaxation in this protein using a triple-labelled sample (^{2}H, ^{13}C, ^{15}N). They proposed the following approach for determining distances involving the amide protons, aiming at quantitative measurements:

1. The acquisition of data was performed using 3D ^{15}N edited ^{1}H–^{1}H NOESY, which provided sufficient resolution and sensitivity when implemented on modern spectrometers and low-noise probeheads.
2. The cross-relaxation rates were determined from fitting the build-up data obtained for short mixing times in the near-linear regime.
3. Whenever possible, the cross-relaxation rates corresponding to magnetisation transfers $i \rightarrow j$ and $j \rightarrow i$, expected to be equal, were measured. This allowed estimation of uncertainties in the cross-relaxation rates. If only one σ_{ij} was available, the expected error was assumed to be larger.
4. The spin-diffusion was corrected for, making use of approximate structural data.
5. The measured cross-relaxation rates were converted into distances between spins assuming hypothetical rigid structures and isotropic motions. The possible effects of motional anisotropy and internal motions were, however, carefully investigated and found to be small in most situations.

The experimental random error in the derived NOE distances was estimated to be 0.07 Å, much less than obtained with earlier NMR and X-ray methods. The distances were compared with those in a single structure from X-ray crystallography as well as with a conformational distribution based on a set of structures derived from residual dipolar couplings (RDC). The latter did not satisfy the eNOE-based distances better than the single structure. The authors' conclusion was that the eNOE measurements might enable detailed studies of both protein structure and dynamics.

In another paper from the same group[28], the possible error sources were investigated further, using also the data for double-labelled (^{13}C, ^{15}N) human ubiquitin. In yet another study[29], measurements of eNOEs and inter-proton distances in human ubiquitin were reported as a function of sample temperature over the range 284–326 K. By doing this, the authors were able to determine that the distances increased with increasing temperature by a few percent, *i.e.* they were able to observe increased protein breathing at elevated temperature.

Another protein, already mentioned in Chapter 11, studied extensively by Vögeli and co-workers[26,30], was GB3. As opposed to ubiquitin, the eNOE data for this system could not be satisfactorily accounted for using a single structure. Instead, the authors found that the data could be successfully interpreted using a three-state ensemble.

12.2.5 Larger Systems

The previous example was of a well-folded, small and soluble protein. The size limit imposed largely by the relaxation properties of the protein, related to the molecular size, has made it virtually impossible to investigate large proteins by NMR. The TROSY method, from the late 1990s, discussed in Chapter 10, has changed the situation dramatically[31,32]. With the aid of higher magnetic field strengths, investigations of large membrane-bound proteins have become possible. Among the first membrane proteins to be studied were the β-barrel proteins, and the two bacterial outer membrane proteins OmpA and OmpX were studied by NMR at the beginning of this century. The structure of the approximately 60 kDa complex consisting of detergents (to mimic the membrane) and the protein has been solved for these two channel-forming proteins[33–38]. Figure 12.9 shows the structure determined for OmpX in the presence of detergent micelles (DHPC), based on 526 NOE-derived distance constraints. The protein was labelled with ^{2}H, ^{13}C and ^{15}N, and a scheme utilising selective methyl group protonation was used to obtain also NOEs involving methyl groups[38]. The authors determined heteronuclear NOE factors for several amino acid residues, which carry information about the local dynamics, and they also measured amide proton exchange rates. An interesting study was made on the lipid-protein interaction, in which intermolecular NOEs were observed between the lipids and the protein[39]. The data were consistent with the hydrophobic barrel being covered with a monolayer of DHPC molecules. The hydrophobic tails of the DHPC molecule form a substitute for the hydrophobic core of a membrane bilayer; a schematic picture is shown in Figure 12.10. The example clearly shows that the molecular-size limits in NMR are

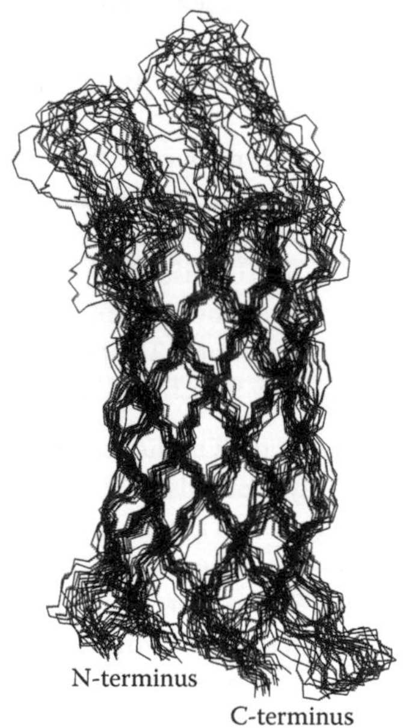

FIGURE 12.9 Solution structure of OmpX in DHPC micelles represented by an overlay of backbone atom coordinates for the 20 structures in the calculated ensemble. The coordinates were obtained from the Protein Data Bank under accession code 1Q9F. The figure was produced using Insight (version 2000).

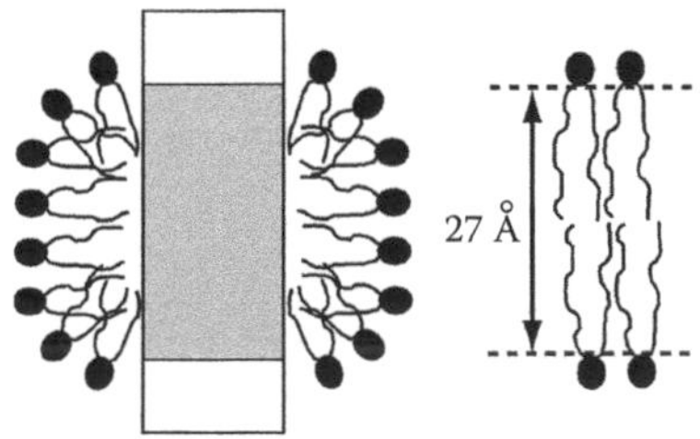

FIGURE 12.10 Hydrophobic core around OmpX as evidenced by intermolecular NOEs. The left panel shows a model of DHPC molecules surrounding the hydrophobic core of the protein. The right panel shows a cartoon of an outer membrane *E.coli* bilayer. (Reprinted with permission from Fernandez, C. et al., *Proc. Natl. Acad. Sci. USA*, 99, 13533–13537, 2002. Copyright (2002) National Academy of Sciences, U.S.A.)

being pushed forward. The TROSY, cross-correlated relaxation-induced polarisation transfer (CRIPT) and cross-correlated relaxation-enhanced polarisation transfer (CRINEPT) pulse sequences are today important tools in determining structure of larger macromolecules.

Since these structures appeared, several novel structures of helical transmembrane proteins have emerged. In 2009 Van Horn *et al.* determined the structure of the *E. coli* diacylglycerol kinase (DAGK), a homotrimer of subunits consisting of three transmembrane helices and the authors also used NMR to map out the substrate-binding sites (both the lipid- and nucleotide-binding sites) in the enzyme[40]. Shortly thereafter, structures of two seven-transmembrane (7-TM) helical proteins, sensory rhodopsin[41] and proteorhodopsin[42], were presented. Much of this work is the result of novel NMR techniques based on TROSY together with sophisticated protein labelling schemes.

12.3 Determination of Structure from Cross-Correlated Relaxation

Cross-correlated relaxation has become popular in investigating slower dynamic behaviour in biological macromolecules, such as chemical exchange. We defer the discussion of these methods until Chapter 13 and present here studies of cross-correlated relaxation rates (CCRRs) for determining structural parameters. There are two main approaches behind using cross-correlated relaxation to study structure. The first approach relies on measuring other interactions, besides the dipole-dipole interaction for an *IS* spin pair, usually consisting of one proton and a heteronucleus *S*, through cross-correlated relaxation. The second approach relates to measuring the angle between two interactions, typically two dipole-dipole interactions, through the cross-correlated relaxation. The magnitude of the interference term is related to the angle between the two interactions and thus to a dihedral angle in the molecule.

12.3.1 DD/CSA Cross-Correlation

The most common cross-correlation, or interference, effect is between the dipole-dipole and the chemical shift anisotropy (CSA) relaxation mechanism for the heteronucleus. The longitudinal cross-correlated relaxation rate is given in Eq. (10.2); the CCRR depends on the length of the internuclear vector, r_{IS}, the anisotropy of the CSA tensor for the heteronucleus, $\Delta\sigma = \sigma_{\parallel} - \sigma_{\perp}$ (assumed to be axially symmetric), the spectral density evaluated at the Larmor frequency of the *S*-spin, and the angle θ between the principal axes of the two interactions. An analogous expression for the transverse DD/CSA cross-correlation is found in Eq. (10.19). Provided that the angle between the two interactions is known, and that the dynamics of the system can be properly described, the cross-correlation rate can be used to deduce chemical shift anisotropies. Most often, a combination of different relaxation parameters are measured in order to simultaneously fit the $\Delta\sigma$ and the internal dynamics according to a suitable motional model.

Very often it is possible to characterise secondary structure, given by specific dihedral torsion angles, through the chemical shift. Much in the same way, attempts have been made to correlate structure and the values for the chemical shift anisotropies. However, uncertainties in the measured CSA values and modest site variations have made predictions about structure from CSA difficult. In addition, the commonly used approximation of axially symmetric shielding tensors may not be valid. For ^{15}N–^{1}H spin pairs in peptides and proteins, the site variation in CSA parameters seems too small to safely allow drawing any conclusions about structure (compare with the final part of Section 11.4). For C$^{\alpha}$–^{1}H spin pairs in polypeptides, on the other hand, the C$^{\alpha}$ CSAs obtained from cross-correlated relaxation have revealed a correlation between secondary structure and the CSA[43,44]. Amide proton CSA values have been shown to correlate with hydrogen bond lengths in protein structure[43,45]. It is, however, difficult to judge whether measuring CSA through cross-correlation has any advantage over measuring conventional parameters, such as isotropic chemical shifts and *J*-couplings, in deducing secondary structure.

12.3.2 DD/DD Cross-Correlation

The second way of using cross-correlated relaxation to investigate structure is to make direct use of the fact that the angle between two interactions appears in the expressions for interference terms. This is particularly useful for determining the angle between two dipolar interactions involving spins i, j and k, l in a protein backbone, such as the $i={}^{15}N$–$j={}^{1}H$ and $k={}^{13}C^{\alpha}$–$l={}^{1}H$ bond vectors[46,47]. For large, slowly tumbling macromolecules, the only important contribution to relaxation rates will be from interactions that are dependent on the spectral density function at zero frequency. The expression for transverse, dipole-dipole CCRR in such a case can be written as:

$$R_{ij,kl} = \frac{1}{2\pi}\left(\frac{\mu_0}{4\pi}\right)^2\left(\frac{\gamma_i\gamma_j\hbar}{r_{ij}^3}\right)\left(\frac{\gamma_k\gamma_l\hbar}{r_{kl}^3}\right)\frac{1}{2}(3\cos^2\theta-1)\frac{2}{5}J(0) \tag{12.2}$$

where the term $\frac{1}{2}(3\cos^2\theta-1)$ describes the dependence of the cross-correlated relaxation rate on the angle, θ, between the two dipole vectors. The spectral density function can here most often safely be approximated as:

$$J(0) = \frac{1}{2\pi}S^2\tau_M \tag{12.3}$$

where τ_M is the overall correlation time for the protein, assuming isotropic motion, and S^2 is the generalised order parameter discussed elsewhere (compare Chapters 6 and 11).

For a protein, the angle between the two dipole-dipole vectors in these equations can be expressed in terms of a dihedral torsion angle, which is useful for determining structure. The angle θ in Eq. (12.2) for the cross-correlated relaxation between the $i = {}^{15}N$–$j = {}^{1}H$ and $k={}^{13}C^{\alpha}$–$l = {}^{1}H$ dipole-dipole vectors can be related to the ψ torsion angle in the polypeptide backbone (see Figure 12.11 for a definition of the peptide geometry). The ψ torsion angle is difficult to measure by other solution NMR methods, and this CCRR approach has therefore received considerable attention. The first experiment for measuring the ψ torsion angle in the protein backbone was proposed by Reif, Henning and Griesinger in 1997[46]. They demonstrated the usefulness of this technique on the protein rhodniin, for which they showed that the experimentally obtained cross-correlated relaxation rate agreed well with the ψ torsion angles obtained from the structure determined by other means (Figure 12.12). The same approach is useful for measuring side-chain torsion angles in proteins, which can be obtained from ${}^{13}C^{\alpha}$–${}^{1}H$, ${}^{13}C^{\beta}$–${}^{1}H^{\beta}$ DD CCRRs. For a comprehensive text on various cross-correlation techniques for measuring torsion angles, as well as the dynamics, the reader is referred to the review by Schwalbe *et al.*[47].

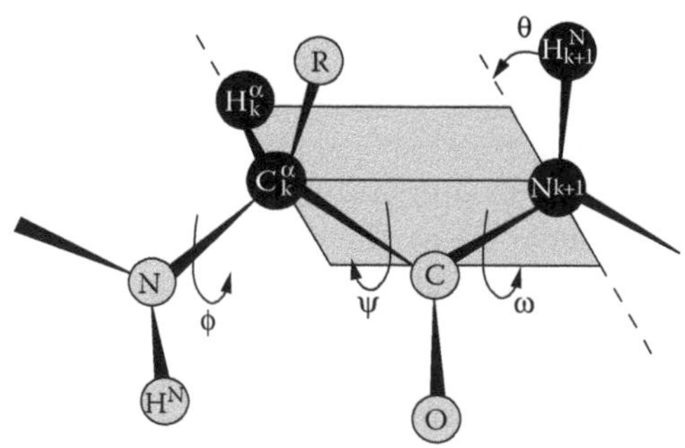

FIGURE 12.11 A schematic representation of the peptide backbone, indicating the relationship between the ψ and θ angles. (From Reif, B. et al., *Science*, 276, 1230–1233, 1997. Reprinted with permission of AAAS.)

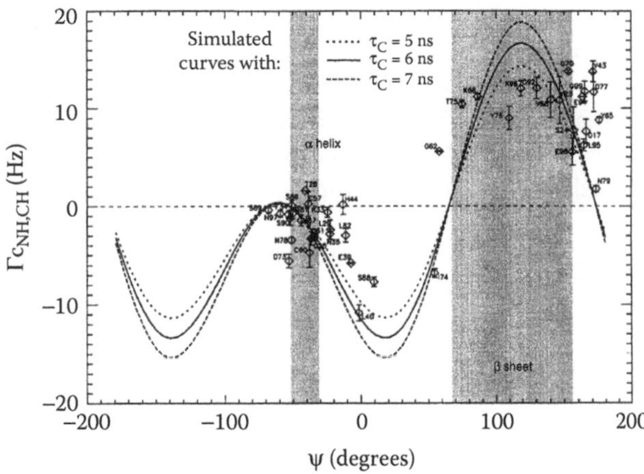

FIGURE 12.12 Simulation of cross-correlated relaxation rates as a function of the ψ angle using different values for the correlation time, τ_M, and experimentally determined values for rhodniin. (From Reif, B. et al., *Science*, 276, 1230–1233, 1997. Reprinted with permission of AAAS.)

References

1. Capozzi, G.; Caristi, C.; Lucchini, V.; and Modena, G., Control of regioselectivity in the addition of sulfenyl chlorides to 3,3-dimethylbutyne (tert-butylacetylene) as a method for differential functionalization of triple bonds. *J. Chem. Soc. Perkin Trans. I* 1982, 2197–2201.
2. Jimenez-Barbero, J.; and Peters, T., Eds., *Spectroscopy of Glycoconjugates*. Wiley-VCH: Weinheim, 2003.
3. Nordmark, E.-L.; Yang, Z. N.; Huttunen, E.; and Widmalm, G., Structural studies of an exopolysaccharide produced by Streptococcus thermophilus THS. *Biomacromolecules* 2005, 6, 105–108.
4. Mäler, L.; Widmalm, G.; and Kowalewski, J., Dynamical behavior of carbohydrates as studied by carbon-13 and proton nuclear spin relaxation. *J. Phys. Chem.* 1996, 100, 17103–17110.
5. Takacs, Z.; Steiner, E.; Kowalewski, J.; and Brotin, T., NMR investigation of chloromethane complexes of cryptophane-A and its analogue with butoxy groups. *J. Phys. Chem. B* 2014, 118, 2134–2146.
6. Canet, D.; Bouguet-Bonnet, S.; Leclerc, S.; and Yemloul, M., Carbon-13 heteronuclear longitudinal spin relaxation for geometrical (and stereochemical) determinations in small and medium size molecules. *Ann. Rep. NMR Spectr.* 2011, 74, 89–123.
7. Tjandra, N.; and Bax, A., Direct measurement of distances and angles in biomolecules by NMR in a dilute liquid crystalline medium. *Science* 1997, 278, 1111–1114.
8. Clore, G. M.; and Gronenborn, A. M., Applications of three- and four-dimensional heteronuclear NMR spectroscopy to protein structure determination. *Progr. NMR Spectr.* 1991, 23, 43–92.
9. Williamson, M. P.; Havel, T. F.; and Wüthrich, K., Solution conformation of proteinase inhibitor-IIA from bull seminal plasma by H-1 nuclear magnetic-resonance and distance geometry. *J. Mol. Biol.* 1985, 182, 295–315.
10. Braun, W.; Bosch, C.; Brown, L. R.; Go, N.; and Wüthrich, K., Combined use of proton-proton Overhauser enhancements and a distance geometry algorithm for determination of polypeptide conformations – applications to micelle-bound glycagon. *Biochim. Biophys. Acta* 1981, 667, 377–396.
11. Braun, W.; and Go, N., Calculation of protein conformations by proton proton distance constraints – A new efficient algorithm. *J. Mol. Biol.* 1985, 186, 611–626.

12. Güntert, P.; Braun, W.; and Wüthrich, K., Efficient computation of 3-dimensional protein structures in solution from nuclear-magnetic-resonance data using the program DIANA and the supporting programs CALIBA, HABAS and GLOMSA. *J. Mol. Biol.* 1991, 217, 517–530.
13. Güntert, P.; Mumenthaler, C.; and Wüthrich, K., Torsion angle dynamics for NMR structure calculation with the new program DYANA. *J. Mol. Biol.* 1997, 273, 283–298.
14. Güntert, P., Automated NMR protein structure calculation with CYANA. *Meth. Mol. Biol.* 2004, 278, 353–378.
15. Pearlman, D. A.; Case, D. A.; Caldwell, J. W.; Ross, W. S.; Cheatham, T. E.; Debolt, S.; Ferguson, D.; Seibel, G.; and Kollman, P., Amber, a package of computer-programs for applying molecular mechanics, normal-mode analysis, molecular-dynamics and free-energy calculations to simulate the structural and energetic properties of molecules. *Comp. Phys. Commun.* 1995, 91, 1–41.
16. Brünger, A. T., *X-PLOR. A System for X-ray Crystallography and NMR*. Yale University Press: New Haven, 1992.
17. Brooks, B. R.; Bruccoleri, R. E.; Olafson, B. D.; States, D. J.; Swaminathan, S.; and Karplus, M., CHARMM – A program for macromolecular energy, minimization and dynamics calculations. *J. Comput. Chem.* 1983, 4, 187–217.
18. Güntert, P., Structure calculation of biological macromolecules from NMR data. *Quart. Rev. Biophys.* 1998, 31, 145–237.
19. Szebenyi, D. M. E.; Obendorf, S. K.; and Moffat, K., Structure of vitamin D-dependent calcium-binding protein from bovine intestine. *Nature* 1981, 294, 327–332.
20. Svensson, L. A.; Thulin, E.; and Forsén, S., Proline cis-trans isomers in calbindin D9K observed by X-ray crystallography. *J. Mol. Biol.* 1992, 223, 601–606.
21. Akke, M.; Drakenberg, T.; and Chazin, W. J., 3-dimensional solution structure of Ca2+-loaded porcine calbindin-D9k determined by nuclear magnetic resonance spectroscopy. *Biochemistry* 1992, 31, 1011–1020.
22. Kördel, J.; Skelton, N. J.; Akke, M.; and Chazin, W. J., High-resolution solution structure of calcium-loaded calbindin D9k. *J. Mol. Biol.* 1993, 231, 711–734.
23. Skelton, N. J.; Kördel, J.; and Chazin, W. J., Determination of the solution structure of apo calbindin D9k by NMR spectroscopy. *J. Mol. Biol.* 1995, 249, 441–462.
24. Skelton, N. J.; Forsén, S.; and Chazin, W.J., H-1-NMR resonance assignments, secondary structure, and global fold of apo bovine calbindin-D9K. *Biochemistry* 1990, 29, 5752–5761.
25. Mäler, L.; Blankenship, J.; Rance, M.; and Chazin, W. J., Site-site communication in the EF-hand Ca2+-binding protein calbindin D9k. *Nature Struct. Biol.* 2000, 7, 245–250.
26. Vögeli, B., The nuclear Overhauser effect from a quantitative perspective. *Progr. NMR Spectr.* 2014, 78, 1–46.
27. Vögeli, B.; Segawa, T. F.; Leitz, D.; Sobol, A.; Choutko, A.; Trzesniak, D.; van Gunsteren, W.; and Riek, R., Exact distances and internal dynamics of perdeuterated ubiquitin from NOE buildups. *J. Am. Chem. Soc.* 2009, 131, 17215–17225.
28. Vögeli, B.; Friedmann, M.; Leitz, D.; Sobol, A.; and Riek, R., Quantitative determination of NOE rates in perdeuterated and protonated proteins: Practical and theoretical aspects. *J. Magn. Reson.* 2010, 204, 290–302.
29. Leitz, D.; Vögeli, B.; Greenwald, J.; and Riek, R., Temperature dependence of (1)H(N)-(1)H(N) distances in ubiquitin as studied by exact measurements of NOEs. *J. Phys. Chem. B* 2011, 115, 7648–7660.
30. Vögeli, B.; Kazemi, S.; Güntert, P.; and Riek, R., Spatial elucidation of motion in proteins by ensemble-based structure calculation using exact NOEs. *Nat. Struct. Mol. Biol.* 2012, 19, 1053–1058.
31. Pervushin, K.; Riek, R.; Wider, G.; and Wüthrich, K., Attenuated T2 relaxation by mutual cancellation of dipole-dipole coupling and chemical shift anisotropy indicates an avenue to NMR structures of very large biological macromolecules in solution. *Proc. Natl. Acad. Sci. USA* 1997, 94, 12366–12371.

32. Wüthrich, K.; and Wider, G., *Transverse Relaxation-Optimized NMR Spectroscopy with Biomacromolecular Structures in Solution*. Wiley, eMagRes, 2007.
33. Fernandez, C.; Adeishvili, K.; and Wüthrich, K., Transverse relaxation-optimized NMR spectroscopy with the outer membrane protein OmpX in dihexanoyl phosphatidylcholine micelles. *Proc. Natl. Acad. Sci. U. S. A.* 2001, 98, 2358–2363.
34. Fernandez, C.; Hilty, C.; Bonjour, S.; Adeishvili, K.; Pervushin, K.; and Wüthrich, K., Solution NMR studies of the integral membrane proteins OmpX and OmpA from Escherichia coli. *FEBS Lett.* 2001, 504, 173–178.
35. Arora, A.; Abildgaard, F.; Bushweller, J. H.; and Tamm, L. K., Structure of outer membrane protein A transmembrane domain by NMR spectroscopy. *Nature Struct. Biol.* 2001, 8, 334–338.
36. Arora, A.; and Tamm, L. K., Biophysical approaches to membrane protein structure determination. *Curr. Opin. Struct. Biol.* 2001, 11, 540–547.
37. Tamm, L. K.; Abildgaard, F.; Arora, A.; Blad, H.; and Bushweller, J. H., Structure, dynamics and function of the outer membrane protein A (OmpA) and influenza hemagglutinin fusion domain in detergent micelles by solution NMR. *FEBS Lett.* 2003, 555, 139–143.
38. Fernandez, C.; Hilty, C.; Wider, G.; Güntert, P.; and Wüthrich, K., NMR structure of the integral membrane protein OmpX. *J. Mol. Biol.* 2004, 336, 1211–1221.
39. Fernandez, C.; Hilty, C.; Wider, G.; and Wüthrich, K., Lipid-protein interactions in DHPC micelles containing the integral membrane protein OmpX investigated by NMR spectroscopy. *Proc. Natl. Acad. Sci. U. S. A.* 2002, 99, 13533–13537.
40. Van Horn, W. D.; Kim, H. J.; Ellis, C. D.; Hadziselimovic, A.; Sulistijo, E. S.; Karra, M. D.; Tian, C.; Sönnichsen, F. D.; and Sanders, C. R., Solution nuclear magnetic resonance structure of membrane-integral diacylglycerol kinase. *Science* 2009, 324, 1726–1729.
41. Gautier, A.; Mott, H. R.; Bostock, M. J.; Kirkpatrick, J. P.; and Nietlispach, D., Structure determination of the seven-helix transmembrane receptor sensory rhodopsin II by solution NMR spectroscopy. *Nature Struct. Mol. Biol.* 2010, 17, 768–774.
42. Reckel, S.; Gottstein, D.; Stehle, J.; Löhr, F.; Verhoefen, M.-K.; Takeda, M.; Silvers, R.,Kainosho, M.; Glaubitz, C.; Wachtveitl, J.; Bernhard, F.; Schwalbe, H.; Güntert, P.; and Dötsch, V. Solution NMR structure of proteorhodopsin. *Angew. Chem. Int. Ed.* 2011, 50, 11942–11946.
43. Sitkoff, D.; and Case, D. A., Theories of chemical shift anisotropies in proteins and nucleic acids. *Prog. NMR Spectr.* 1998, 32, 165–190.
44. Tjandra, N.; and Bax, A., Large variations in C-13(alpha) chemical shift anisotropy in proteins correlate with secondary structure. *J. Am. Chem. Soc.* 1997, 119, 9576–9577.
45. Wu, G.; Freure, C. J.; and Verdurand, E., Proton chemical shift tensors and hydrogen bond geometry: A H-1-H-2 dipolar NMR study of the water molecule in crystalline hydrates. *J. Am. Chem. Soc.* 1998, 120, 13187–13193.
46. Reif, B.; Hennig, M.; and Griesinger, C., Direct measurement of angles between bond vectors in high-resolution NMR. *Science* 1997, 276, 1230–1233.
47. Schwalbe, H.; Carlomagno, T.; Hennig, M.; Junker, J.; Reif, B.; Richter, C.; and Griesinger, C., Cross-correlated relaxation for measurement of angles between tensorial interactions. *Methods Enzymol.* 2001, 338, 35–81.

13

Relaxation and Chemical Exchange

In the previous chapters, we have mainly been concerned with the spectral densities for reorientation of a particular axis fixed in a molecule (or in a fragment of a molecule). Intermolecular motions can also give rise to relaxation; the case of intermolecular dipolar relaxation was considered in Section 3.5. The case of intermolecular quadrupolar relaxation, for nuclear spins residing in highly symmetric ions, will be dealt with in Chapter 14. In this chapter, we will investigate the effect of chemical exchange on relaxation.

Broadly speaking, chemical exchange can influence nuclear magnetic resonance (NMR) spectra and NMR relaxation in two ways. First, fast exchange processes can influence the time-correlation functions and spectral densities in a manner reminiscent of reorientational processes. In other words, the *chemical exchange lifetime can act as a correlation time* in connection with certain relaxation mechanisms. We describe this type of exchange effects in Section 13.1. Second, exchange processes can affect the NMR lineshapes on a more *macroscopic level*, where the exchange rate constants have an effect rather similar to relaxation rate constants. These phenomena will be discussed in Sections 13.2–13.4.

A nice presentation of the effects of chemical exchange on relaxation can be found in Woessner's review in the eMagRes[1] and parts of this chapter are based on that work. Several reviews by Palmer[2–5], focused on biomolecular systems, are also highly recommended.

13.1 Chemical Exchange as Origin of Correlation Time

Chemical exchange processes can give rise to a stochastic modulation of NMR interactions and thus act as a relaxation mechanism. Exchange processes on the microsecond to millisecond timescale, much slower than molecular tumbling, are not an efficient source of relaxation via anisotropic, zero-average interactions such as dipole-dipole or chemical shielding anisotropy (CSA). Simply speaking, these interactions are averaged out too fast for any exchange process to cause any stochastic variation (an exception to this statement will be discussed in Section 17.4). On the other hand, the exchange phenomena can efficiently modulate *isotropic interactions*, such as indirect spin-spin couplings or chemical shifts.

A particularly simple example of a relaxation effect stemming from chemical exchange is scalar relaxation, mentioned already in Section 5.4. Consider a situation with a nuclear spin I with the spin quantum number of 1/2, for example a proton in the methyl group of methanol, coupled to another spin 1/2 nucleus, for example a proton of the OH group in methanol. In the absence of chemical exchange, the spin-spin coupling to the OH protons splits the CH_3 signal into a doublet. The hydroxyl proton in alcohols tends to move from one molecule to another, in particular in the presence of traces of acid. We denote the mean lifetime of the hydroxyl proton in one molecule τ_e. If the exchange is fast, so that the inverse mean lifetime is much larger than the coupling constant, $1/\tau_e \gg J$, then the doublet will coalesce into a single line. The exchange process will make the coupling constant change randomly between the value J when the two nuclei are in the same molecule and the value 0 otherwise. The time-correlation function for the random process can be expressed:

$$G(\tau) = \langle J(t)J(t+\tau)\rangle = J^2 \exp(-\tau/\tau_e) \tag{13.1}$$

The exponential decay stems from the fact that the probability that the two spins, residing in the same molecule at time zero, still are in the same molecule after time τ decreases exponentially with the time constant τ_e. The random modulation of the scalar coupling interaction leads to a relaxation process. Abragam (*Further reading*) denotes this relaxation mechanism the *scalar relaxation of the first kind*. The mechanism leads to single exponential longitudinal relaxation with the rate constant:

$$\frac{1}{T_{1,SC}} = \frac{(2\pi J)^2}{2} \frac{\tau_e}{1+(\omega_I - \omega_S)^2 \tau_e^2} \tag{13.2a}$$

where we can recognise the Lorentzian spectral density, related to the time-correlation function of Eq. (13.1). The spectral density is evaluated at the difference between the two Larmor frequencies, originating from the $(\hat{I}_+\hat{S}_- + \hat{I}_-\hat{S}_+)$ term in the scalar coupling Hamiltonian. The corresponding transverse relaxation rate is:

$$\frac{1}{T_{2,SC}} = \frac{(2\pi J)^2}{4} \left[\tau_e + \frac{\tau_e}{1+(\omega_I - \omega_S)^2 \tau_e^2}\right] \tag{13.2b}$$

The contributions of scalar relaxation of the first kind to both spin-lattice and spin-spin relaxation are normally negligible, except for the case of the S-spin being an electron spin. We shall return to this point in Chapter 15.

There is also a related phenomenon, *scalar relaxation of the second kind* in the terminology of Abragam (*Further reading*), which can arise when the S-spin is a rapidly relaxing species, such as a nuclear spin with the quantum number S larger than 1/2 (quadrupolar nucleus) or electron spin. We shall return to these cases in Chapters 14 and 15, respectively.

In the case of scalar relaxation of the first kind, the chemical exchange lifetime acts as a correlation time, *i.e.* it enters the calculation of the time-correlation functions and spectral densities. Similar effects of exchange can also occur for isotropic chemical shifts. Consider a system with two discrete sites, with isotropic chemical shifts $+\delta$ and $-\delta$ (we choose to express δ in the angular frequency units) and equal populations. We assume that the chemical exchange is very fast compared with the relaxation processes and the difference in chemical shifts, so that $\tau_e^{-1} \gg 2\delta$. This *chemical shift modulation* is described by a time-correlation function, $G(\tau) = \delta^2 \exp(-\tau/\tau_e)^6$. The exchange contribution to spin-lattice relaxation can be neglected, while the contribution to spin-spin relaxation (exchange broadening) is given by the simple expression:

$$\frac{1}{T_2^{exch}} = \delta^2 \tau_e \tag{13.3}$$

A more complicated situation arises if the exchange becomes extremely fast, so that the lifetimes of the two sites, τ_A and τ_B (no longer assumed equal), are on the order of the rotational correlation times, τ_{cA} and τ_{cB}, characterising the sites. As an example, we can think of a water molecule exchanging on a sub-nanosecond timescale between the site B, where it is bound to a macromolecule (characterised by a long rotational correlation time), and the site A in the bulk, where it reorients very rapidly (short correlation time). The intramolecular proton-proton dipolar interaction is the main relaxation source and is equally strong at the two sites. The more efficient dipolar interaction at the bound site B will be governed by an *effective correlation time*, which because of exchange will be shorter than τ_{cB}. The quantitative result will depend on whether we assume that the orientation of the proton-proton axis is retained or randomised by the exchange event. Effects of this kind are particularly important in gels and related systems, to which we shall return in Chapter 17.

For a comprehensive discussion, the reader is referred to the important original paper by Wennerström[7]. The ideas of Wennerström (as well as those of Deverell *et al.*[6]) have been re-examined, using the stochastic Liouville equation (SLE) formalism, by Abergel and Palmer[8]. Another interesting example of exchange lifetime contributing to the correlation time, to which we shall come back in Chapter 15, is the case of labile water molecules residing for a brief period of time in the first coordination sphere of a paramagnetic metal ion.

Before leaving this section, we wish to mention another case of exchange lifetime acting as a correlation time, subject to studies during the last decade. The exchange process at hand is the proton transfer in a hydrogen-bonded system with a double-well (or possibly single-well) potential. Bernatowicz and co-workers[9] considered a situation in a $^{15}N\text{–}^{1}H\cdots{}^{15}N$ system in small molecules where the very fast proton transfer on a sub-nanosecond timescale competes with reorientation as the modulation mechanism of the $^{15}N\text{–}^{1}H$ dipolar interaction. Limiting expressions were provided for the cases of $\tau_e \ll \tau_R$ and $\tau_R \ll \tau_e$. Masuda *et al.*[10] reported a similar study of the proton transfer and proton relaxation in the $^{17}O\text{–}^{1}H\cdots{}^{17}O$ system. By comparing the proton relaxation data in natural-abundance and in ^{17}O-enriched samples, the relaxation rate contributions from the $^{1}H\text{–}^{17}O$ dipolar interaction were identified. In one of the investigated low-molecular weight systems, the analysis allowed an estimate of the proton transfer lifetimes, which were a few tens of picoseconds, similar to the reorientational correlation time. Similar work was also reported in two other papers by the same group[11,12].

13.2 Intermediate Chemical Exchange

Let us now consider a case when we still have our two sites, *A* and *B*, but now with lifetimes that are on the timescale of nuclear spin relaxation rates rather than the rotational correlation times. In such a case, the exchange process will not influence the time-correlation functions, but it may have effects on the dynamics of nuclear magnetisation. Many biological and chemical processes occur on a timescale that is usually referred to as *intermediate chemical exchange*, which is in the range of microseconds to seconds. This regime is on the timescale of the spectral frequencies or in the limiting case, on the relaxation timescale. General discussions of the effects on chemical exchange on NMR spectra can be found in standard NMR texts, *e.g.* in books by Levitt and Ernst, Bodenhausen and Wokaun (*Further reading*).

Kinetic processes, such as conformational or chemical exchange, on this timescale make the chemical shifts time dependent in different ways. One of the ways can be illustrated by the situation of hydrogen exchange in solvents, such as H_2O, in which the spins are transferred between distinct molecules (*intermolecular* exchange). A second way is by altering the magnetic environment of the spin within a molecule by, for instance, conformational exchange. We consider here mainly *intramolecular* exchange processes of the second type.

Let us look at an exchange process between two sites, *A* and *B*, with populations p_A and p_B, written as:

$$A \underset{k_{-1}}{\overset{k_1}{\rightleftharpoons}} B \tag{13.4}$$

Assume that this process can be characterised by a rate constant $k_{ex} = 1/\tau_{ex} = k_1 + k_{-1} = k_1/p_B = k_{-1}/p_A$, where k_1 and k_{-1} are assumed to be first-order rate constants or inverse lifetimes of the two sites, $k_1 = 1/\tau_A$ and $k_{-1} = 1/\tau_B$. The two sites are further characterised by unequal chemical shifts, ω_A and ω_B, where the difference between the two resonance frequencies (in angular frequency units) is $\omega_B - \omega_A = 2\delta$, and the spin-spin relaxation times are T_{2A} and T_{2B}. The spectral lineshapes are most profoundly affected by the exchange processes characterised by rate constants similar to the differences in chemical shifts. To set the ground for further discussion, we present in Figure 13.1 simulated NMR spectra[13] in the presence of chemical exchange between two equally populated sites, $p_A = p_B = 1/2$ (symmetric exchange), assuming a constant $2\delta = 300$ rad s^{-1}, $T_{2A} = T_{2B} = 1$ s and varying values of the exchange rate constant in the range 10–10^4 s^{-1}. We shall return to the question of how simulations of this type are performed in Section 13.4. We can distinguish three distinct regions in Figure 13.1. In the range where $k_{ex} \ll 2\delta$, the slow (or slow intermediate) exchange regime, we see two lines in the spectrum. The exchange processes broaden the lines according to:

$$\frac{1}{T_2'} = \frac{1}{T_2} + \frac{2}{\tau_{ex}} \tag{13.5}$$

where T_2' is related to the full linewidth at the half-height and T_2 is the spin-spin relaxation time originating from other mechanisms than exchange.

Clearly, the exchange lifetime or the exchange rate constant can be determined from the linewidth if the "exchange-free" T_2 is known and as long as $2/\tau_{ex}$ is significantly larger than the inhomogeneous broadening. If the inverse exchange lifetime is smaller than that, methods related to the longitudinal relaxation can be applied. We shall discuss this point in Section 13.3.

The opposite limiting situation, where $2\delta \ll k_{ex}$, is referred to as the *fast* (or *fast intermediate*) *exchange*. In this limit, we have only one line, positioned at the population-weighted average chemical shift, $\omega = p_A\omega_A + p_B\omega_B$, and broadened through exchange according to Eq. (13.3) or its analogue for unequal populations of the two sites. As the exchange lifetime decreases, the line broadening becomes smaller and, for sufficiently fast exchange processes, we need to resort to special, relaxation-related measuring techniques described in Section 13.4.

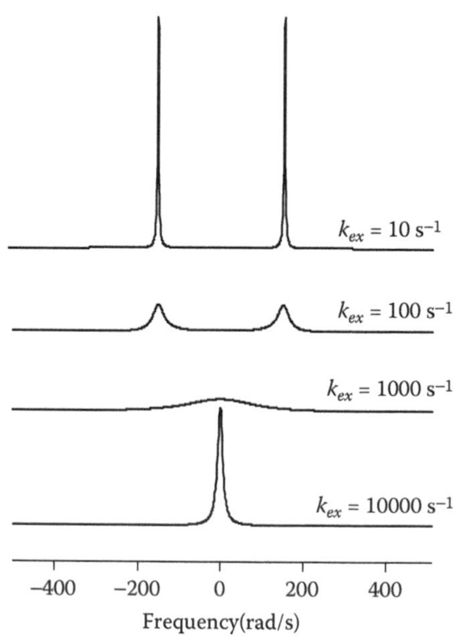

FIGURE 13.1 Simulated lineshapes in the presence of chemical exchange between two equally populated sites, $p_A = p_B = 1/2$ (symmetric exchange), assuming a constant $\delta = 150$ rad s^{-1}, $T_{2A} = T_{2B} = 1$ s and varying values of the exchange rate constant in the range 10–10^4 s^{-1}. The figure was prepared using the program QSim. (From Helgstrand, M. and Allard, P., *J. Biomol. NMR*, 30, 71–80, 2004.)

In between the two extremes, one usually speaks of the *intermediate regime*, $k_{ex}\approx 2\delta$, for which the most important effect is on the lineshape in the NMR spectrum. One commonly speaks about the *coalescence* point: if we look at a series of spectra such as those in Figure 13.1, the signal intensity in the middle of the spectrum changes at the coalescence between being a local maximum and a local minimum.

13.3 Exchange Effects on Longitudinal Magnetisation

Chemical exchange processes can lead to a flow of longitudinal magnetisation between two sites, much in the same way as the cross-relaxation phenomena. A simple modification of the Bloch equation for the longitudinal magnetisation, accommodating the effects of chemical exchange in systems without spin-spin couplings, was proposed a long time ago by McConnell[14]. This modification leads to a set of equations that can be used to evaluate the effect of exchange on the longitudinal relaxation rates.

13.3.1 The Bloch–McConnell Equations

The Bloch–McConnell equations for the longitudinal magnetisation for two species, or sites, A and B, are obtained by extending the Bloch equations (described in Section 1.3) to include exchange between A and B according to:

$$\frac{dM_{zA}}{dt}=\frac{M_{0A}-M_{zA}}{T_{1A}}-\frac{M_{zA}}{\tau_A}+\frac{M_{zB}}{\tau_B} \tag{13.6a}$$

$$\frac{dM_{zB}}{dt}=\frac{M_{0B}-M_{zB}}{T_{1B}}-\frac{M_{zB}}{\tau_B}+\frac{M_{zA}}{\tau_A} \tag{13.6b}$$

T_{1A} and T_{1B} are intrinsic spin-lattice relaxation times at the two sites. The equilibrium magnetisations M_{0A} and M_{0B} are proportional to the populations of the two sites (p_A and p_B, respectively) and follow the principle of detailed balance: $p_A/\tau_A = p_B/\tau_B$ or $M_{0A}/\tau_A = M_{0B}/\tau_B$. Eq. (13.6) may be reformulated as:

$$\frac{dM_{zA}}{dt}=\alpha_A\left(M_{0A}-M_{zA}\right)-\frac{M_{0B}-M_{zB}}{\tau_B} \tag{13.7a}$$

$$\frac{dM_{zB}}{dt}=-\frac{M_{0A}-M_{zA}}{\tau_A}+\alpha_B\left(M_{0B}-M_{zB}\right) \tag{13.7b}$$

where:

$$\alpha_A=\frac{1}{T_{1A'}}=\frac{1}{T_{1A}}+\frac{1}{\tau_A} \tag{13.8a}$$

$$\alpha_B=\frac{1}{T_{1B'}}=\frac{1}{T_{1B}}+\frac{1}{\tau_B} \tag{13.8b}$$

Eq. (13.7) can instead be represented in a compact matrix formulation:

$$\frac{d}{dt}\begin{pmatrix} M_{zA} \\ M_{zB}\end{pmatrix}=\begin{pmatrix} \alpha_A & -\tau_B^{-1} \\ -\tau_A^{-1} & \alpha_B\end{pmatrix}\begin{pmatrix} M_{0A}-M_{zA} \\ M_{0B}-M_{zB}\end{pmatrix} \tag{13.9}$$

which has the same structure as the Solomon equations, Eq. (3.21). In analogy with the Solomon equations, the solution to Eq. (13.9) for M_{zA} and M_{zB} is a sum of two exponentials with apparent relaxation rates λ_+ and λ_-. The expressions for the rate constants in the Solomon case were given in Eq. (9.2) and the corresponding eigenvalues of the present matrix in Eq. (13.9) have been given, for instance, in Woessner's review[1]:

$$\lambda_\pm = \frac{1}{2}\left\{(\alpha_A + \alpha_B) \pm \left[(\alpha_A - \alpha_B)^2 + 4\tau_A^{-1}\tau_B^{-1}\right]^{1/2}\right\} \tag{13.10}$$

For the case of equal populations (and lifetimes), Eq. (13.10) can be obtained from Eq. (9.2) by trivial substitutions. Woessner also gave the weights of the two exponentials. Here, we wish to illustrate some limiting cases. Consider first the case of very slow exchange, $\tau_A^{-1} + \tau_B^{-1} \ll T_{1A}^{-1} + T_{1B}^{-1}$. The apparent relaxation rates are then equal to α_A and α_B and very close to the intrinsic relaxation rates. The next special case is that of widely different intrinsic relaxation times and the exchange lifetimes falling between $T_{1B}^{-1} \gg \tau_A^{-1}, \tau_B^{-1} \gg T_{1A}^{-1}$. The result for this case is that the apparent relaxation time for the fast-relaxing site (B), $T_{1B'}$, is very close to T_{1B}, while $T_{1A'}$ becomes near to the exchange lifetime τ_A. Under these conditions, the measurement of spin-lattice relaxation times can be a useful tool for studying chemical kinetics. Next, when the exchange lifetimes are very short (but still long compared with the timescale of molecular motions), we obtain a single exponential solution with the relaxation rate that is a weighted mean of the relaxation rates in the two sites:

$$\frac{1}{T_{1M}} = \frac{p_A}{T_{1A}} + \frac{p_B}{T_{1B}} \tag{13.11}$$

The final limiting case is that when one of the sites, let us say B, has a very rapid relaxation and a very small population, so that $T_{1B}^{-1} \gg T_{1A}^{-1}$ and $p_A \approx 1$. The observable relaxation is approximately a single exponential with the apparent rate:

$$\frac{1}{T_{1A'}} = \frac{1}{T_{1A}} + \frac{p_B}{p_A T_{1B} + p_B \tau_A} \tag{13.12}$$

This limiting case is of great practical interest, for instance in water molecules exchanging rapidly (typically on a microsecond timescale) between a scarcely populated, rapidly relaxing bound site and the free bulk. We shall return to an example for this type of situation in the chapter on paramagnetic systems.

13.3.2 Nuclear Overhauser Enhancement Spectroscopy (NOESY) and *zz*-Exchange Methods

The limiting cases in the previous subsection are all concerned with exchange effects in experiments designed to measure spin-lattice relaxation rates, such as the inversion-recovery experiment. Another interesting experimental possibility is to monitor slow kinetic processes through the exchange of longitudinal magnetisation between the two sites, much in the same way as in cross-relaxation experiments. Following this idea, we shall next explore how the spin system described by Eq. (13.9) responds to the NOESY pulse sequence, described in Section 9.2. As discussed by Ernst, Bodenhausen and Wokaun (*Further reading*), when the sequence 90°–t_1–90°–τ_{mix}–90° is applied to a system with chemical exchange but no cross-relaxation, the experiment is denoted as *exchange spectroscopy* (EXSY). The expressions for the diagonal- and cross-peak intensities become particularly simple if we assume that the exchange is symmetric ($\tau_A = \tau_B = \tau_{ex}/2$) and that the intrinsic relaxation rates are equal ($T_{1A}^{-1} = T_{1B}^{-1} = \rho$):

$$a_{AA} = a_{BB} = \frac{1}{2}\left[1 + \exp(-\tau_{mix}/\tau_{ex})\right]\exp(-\rho\tau_{mix}) \tag{13.13a}$$

$$a_{AB} = a_{BA} = \frac{1}{2}\left[1 - \exp\left(-\tau_{mix}/\tau_{ex}\right)\right]\exp\left(-\rho\tau_{mix}\right) \tag{13.13b}$$

In reality, cross-relaxation and exchange processes often occur simultaneously and it can be difficult to separate the two in a NOESY spectrum, since they both lead to cross-peaks with the same sign. It can, however, be noted that the opposite is true in rotating-frame Overhauser effect spectroscopy (ROESY), where the cross-peaks originating from chemical exchange have a different sign, making the discrimination between the two processes easy. This has the same origin as the fact that three-spin effects, or spin-diffusion, also lead to cross-peaks of opposite sign as compared with the ROE cross-peaks.

A second problem associated with homonuclear NOESY or ROESY experiments is that exchange cross-peaks close to the diagonal tend to overlap and are therefore difficult to observe. This is particularly true for biological macromolecules undergoing conformational exchange. The chemical shift differences between the two states may not be very large, and furthermore, the linewidths become intrinsically larger. The NOESY and ROESY schemes have, however, been used successfully for investigations of exchange in small molecules.

A way to overcome problems with distinguishing between exchange and NOE cross-peaks is to instead monitor exchange through the *zz*-exchange experiment[15,16]. This method is based on separating transfer between different spin orders, *e.g.* one-spin order and two-spin order terms. The pulse sequence is shown in Figure 13.2, and, briefly, the experiment works as follows. At the beginning of the mixing time, τ_{mix}, longitudinal spin orders are created by applying a pulse with the flip angle β, different from 90°, usually 45°. The single-quantum coherences, present during the evolution period, can thus be converted into two-spin order terms:

$$\begin{aligned} 2\hat{I}_y\hat{S}_z \xrightarrow{\beta_x} & 2\hat{I}_y\hat{S}_z\cos^2\beta + 2\hat{I}_z\hat{S}_z\sin\beta\cos\beta \\ & -2\hat{I}_y\hat{S}_y\sin\beta\cos\beta - 2\hat{I}_z\hat{S}_y\sin^2\beta \end{aligned} \tag{13.14}$$

Chemical exchange will transfer the two-spin order terms to other two-spin order states during the mixing time, and these can be converted into observable antiphase coherences provided that the final pulse is also different from 90°. The cross-peak pattern in this case will be antiphase in nature and can be distinguished from NOE cross-peaks. This cross-peak pattern is, however, a problem, since broad proton signals tend to cause cancellation of antiphase cross-peaks. Several heteronuclear *zz*-exchange techniques have therefore been proposed, by which exchange in isotopically enriched compounds can be studied (see the reviews on chemical exchange by Palmer and co-workers[4] or by Palmer[5] for details). The experiment is based on the heteronuclear single quantum coherence (HSQC) experiment, and is in principle a R_1 relaxation experiment. By arraying the mixing time, it is possible to monitor a time-curve

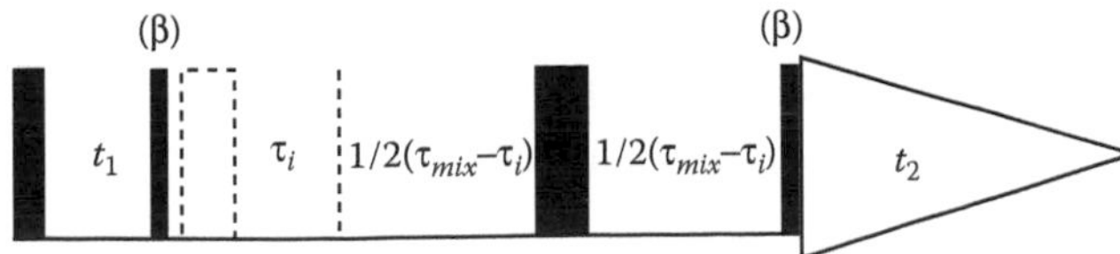

FIGURE 13.2 Homonuclear pulse sequence designed to create and detect longitudinal two-spin order (*zz*-order). The flip angle β is usually set to 45°. The two-spin order terms can be separated from the one-spin order terms, caused by the nuclear Overhauser effect, by including the hatched 180° pulse in alternating scans. The 180° pulse inserted in the middle of the mixing period partially refocuses the ZQ precession, and by varying the position of the pulse in the mixing interval (by changing τ_i) and co-adding the spectra, it is possible to reduce *J* cross-peaks.

of the build-up/decay of magnetisation, and the same experiment allows the simultaneous determination of R_1 relaxation rates for the "auto" X-^{1}H cross-peaks and exchange rates from the "exchange" cross-peaks.

We have so far discussed the relation between chemical exchange and cross-relaxation on a macroscopic level and stated that the two phenomena may manifest themselves very similarly in certain experiments, notably in the NOESY measurements. Microscopically, the chemical (conformational) exchange influences cross-relaxation rates in different ways, depending on the relation between the time-scale of the conformational exchange (which we treat as any other form of internal motion) and global reorientation. If the exchange is slow compared with the global rotational motion but fast with respect to the shift difference, then the observed cross-relaxation rate will be a population-weighted average, in analogy with Eq. (13.11):

$$\sigma_{IS} = p_A \sigma_{IS,A} + p_B \sigma_{IS,B} \tag{13.15}$$

where $\sigma_{IS,A}$ and $\sigma_{IS,B}$ are the cross-relaxation rates in the two conformations. Since each of the site-specific cross-relaxation rates is proportional to $1/r_{IS}^6$, Eq. (13.15) describes the averaging of this quantity. In the other limit, where the exchange lifetime is much shorter than the rotational correlation time, the situation becomes similar to that of the truncated Lipari–Szabo expression for the spectral density, where the measured cross-relaxation rate represents an average of the square of $1/r_{IS}^3$. The intermediate situations can be treated, as described in Section 6.3, using the concept of a radial order parameter.

13.4 Transverse Magnetisation: Measurements of Intermediate and Fast Exchange

The most profound effect of chemical reactions in NMR is on the lineshapes in the NMR spectrum at the intermediate exchange range $k_{ex} \approx 2\delta$, as seen in Figure 13.1. We formulate in this section the theory appropriate for this region as well as in the fast exchange regime.

13.4.1 Effects on Lineshape

The transverse analogues of the Bloch–McConnell equations for the evolution of longitudinal magnetisation (Eq. (13.9)), subject to free precession, relaxation and exchange, are given by:

$$\frac{d}{dt}\begin{pmatrix} M_A^+(t) \\ M_B^+(t) \end{pmatrix} = \begin{pmatrix} -i\omega_A - R_{2A}^0 - k_1 & k_{-1} \\ k_1 & -i\omega_B - R_{2B}^0 - k_{-1} \end{pmatrix} \begin{pmatrix} M_A^+(t) \\ M_B^+(t) \end{pmatrix} \tag{13.16}$$

As compared with Eq. (13.9), Eq. (13.16) is more complicated, since the matrix in the right-hand side contains both real and imaginary components. In the same way as described for the exchange of longitudinal magnetisation, a solution to this equation can be found:

$$\begin{pmatrix} M_A^+(t) \\ M_B^+(t) \end{pmatrix} = \begin{pmatrix} a_{AA}(t) & a_{AB}(t) \\ a_{BA}(t) & a_{BB}(t) \end{pmatrix} \begin{pmatrix} M_A^+(0) \\ M_B^+(0) \end{pmatrix} \tag{13.17}$$

where

$$a_{AA}(t) = \frac{1}{2}\left[(1-\Lambda)\exp(-\mu_- t) + (1+\Lambda)\exp(-\mu_+ t)\right] \tag{13.18a}$$

$$a_{BB}(t) = \frac{1}{2}\left[(1+\Lambda)\exp(-\mu_- t) + (1-\Lambda)\exp(-\mu_+ t)\right] \tag{13.18b}$$

$$a_{AB}(t) = \frac{k_{ex} p_A}{\mu_+ - \mu_-}\left[\exp(-\mu_- t) - \exp(-\mu_+ t)\right] \tag{13.18c}$$

$$a_{BA}(t) = \frac{k_{ex} p_B}{\mu_+ - \mu_-}\left[\exp(-\mu_- t) - \exp(-\mu_+ t)\right] \tag{13.18d}$$

and the parameters Λ and $\mu_\pm$ are given by:

$$\Lambda = \frac{-i2\delta + R_{2A}^0 - R_{2B}^0 + k_{ex}(p_B - p_A)}{\mu_+ - \mu_-} \tag{13.19a}$$

$$\begin{aligned} \mu_\pm &= \frac{1}{2}\left(-i\omega_A - i\omega_A + R_{2A}^0 + R_{2B}^0 + k_{ex}\right) \\ &\pm \frac{1}{2}\left[\left(-i2\delta + R_{2A}^0 - R_{2B}^0 + k_{ex}(p_A - p_B)\right)^2 + 4p_A p_B k_{ex}^2\right]^{1/2} \end{aligned} \tag{13.19b}$$

The time-domain signal is given by the sum of $M_A^+(t)$ and $M_B^+(t)$ and, consequently, the spectrum is obtained by a Fourier transform of this signal. Analyses of lineshapes using these equations can be done by simply fitting the parameters p_A, p_B, k_{ex}, R_{2A}^0 and R_{2B}^0. The lineshapes in Figure 13.1 were obtained using Eqs. (13.17)–(13.19).

In analogy with the limiting cases of solutions to Eq. (13.9), similar expressions can also be obtained for transverse relaxation. Besides the intrinsic transverse relaxation rates, T_{2A}^{-1}, T_{2B}^{-1}, and lifetimes, we need also to include the difference in chemical shifts of the two sites, $2\delta = \omega_A - \omega_B$. We refer the interested reader to Woessner's review[1] and standard texts on chemical exchange in NMR spectra (*e.g.* Levitt, *Further reading*, and the reviews by Palmer[4,5]) for detailed presentations. We wish to mention here only the case of fast exchange: $\tau_A^{-1}, \tau_B^{-1} \gg T_{2A}^{-1}, T_{2B}^{-1}, |2\delta|$ and large shift difference, $(2\delta)^2 \gg (T_{2A}^{-1} - T_{2B}^{-1})^2$. In this case, a single Lorentzian signal is obtained, with the apparent T_2 given by:

$$R_2 = R_2^0 + \frac{p_A p_B \tau_A \tau_B}{\tau_A + \tau_B}(2\delta)^2 \tag{13.20}$$

where $R_2 = 1/T_2$ is the population-weighted mean transverse relaxation rate, defined in analogy with Eq. (13.11). For equal populations of the two sites ($p_A = p_B = 1/2$, $\tau_A = \tau_B = 2\tau_{ex}$), Eq. (13.20) predicts the exchange broadening to be identical to the form of Eq. (13.3), derived in a very different way (provided that we identify τ_e with τ_{ex}).

Lineshape analysis can be generalised to a larger number of sites (multisite exchange) and is a well-established approach for investigating exchange processes. Applications range from chemical exchange in small organic or inorganic molecules to exchange due to folding, ligand binding and aggregation processes in biological macromolecules. A nice example is provided by the investigation of the *cis/trans* isomerisation of a prolyl peptide bond in a small peptide catalysed by cyclophilin, a peptidyl prolyl *cis/trans* isomerase[17]. NMR lineshape analysis allowed a complete determination of the kinetic rate constants for the four-site exchange system, involving the two free and enzyme-bound *cis* and *trans* forms, shown in Figure 13.3. It is readily apparent that by binding to the enzyme, the isomerisation rate is greatly enhanced.

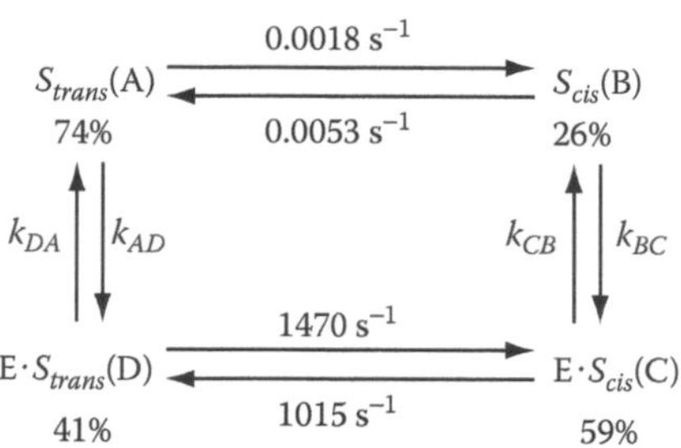

FIGURE 13.3 Schematic diagram for the four-site exchange model for the cyclophilin (E)-catalysed *cis/trans* isomerisation of a prolyl residue (S). The enhancement of the isomerisation reaction is indicated by the greatly enhanced rates, as determined from lineshape analysis of NMR spectra, when the substrate is bound to the enzyme as compared with when it is free in solution. (Adapted from Kern, D. *et al.*, *Biochemistry*, 34, 13594–13602, 1995.)

The fast intermediate exchange regime (on the chemical shift timescale) is characterised by modest line-broadening of signals, and effects may not be readily apparent directly from the NMR spectrum. Furthermore, if one of the states is sparsely populated, the corresponding components may not be visible in the NMR spectrum. Under these conditions, exchange processes can be investigated by transverse relaxation measurements. Indeed, this was mentioned already in Chapter 8, when methods for measuring R_2 relaxation rates were discussed. The transverse relaxation rate is often written as:

$$R_2 = R_2^0 + R_{ex} \tag{13.21}$$

where R_2 is the measured relaxation rate, R_2^0 is the transverse relaxation rate in the absence of exchange and R_{ex} is a phenomenological exchange term included to account for processes slower than the molecular tumbling, in the first place the relaxation by chemical shift modulation discussed in Section 13.1. In the fast intermediate exchange region, R_2^0 is the population-weighted average over various species present in solution.

13.4.2 Spin-Lock Methods

As discussed earlier (Section 11.4, see Eq. (11.13b)), the R_{ex} term in Eq. (13.21) is often needed for successful analysis of R_2 relaxation rates within the Lipari–Szabo framework and various methods exist to determine whether slow exchange processes contribute to the spectral density at zero frequency. There are, however, more direct methods to investigate specifically exchange phenomena occurring on the fast (or intermediate fast) exchange timescale. These methods are mainly based on Carr, Purcell, Meiboom and Gill (CPMG) or $R_{1\rho}$ measurements, but also cross-correlation measurements can provide insights into the contribution of exchange to R_2 relaxation rates[18].

First, we shall consider exchange effects in $R_{1\rho}$ relaxation rates. The influence of exchange on measured transverse relaxation rates was already mentioned in Section 8.2, and it was seen that the $R_{1\rho}$ depends on chemical exchange in the following way:

$$R_{1\rho} = R_1 \cos^2\theta + \left(R_2^0 + p_A p_B (2\delta)^2 \frac{k_{ex}}{k_{ex}^2 + \omega_e^2} \right) \sin^2\theta \tag{13.22}$$

where θ is the tilt angle of the effective spin-lock field and is given by $\tan\theta = \omega_1/\Delta\omega$, ω_1 is the amplitude of the radiofrequency (r.f.) field, $\Delta\omega$ is the offset from the Larmor frequency and ω_e is the effective field given by $\omega_e = ([p_A\omega_A + p_B\omega_B]^2 + \omega_1^2)^{1/2}$.

Combining Eqs. (8.11), (13.21) and (13.22), we can write:

$$R_{ex} = \frac{R_{1\rho}}{\sin^2\theta} - \frac{R_1}{\tan^2\theta} - R_2^0 = p_A p_B (2\delta)^2 \frac{k_{ex}}{k_{ex}^2 + \omega_e^2} \tag{13.23}$$

Measurements of rotating-frame relaxation rates can be performed under both on-resonance and off-resonance conditions[19–22]. On-resonance conditions typically allow rate constants on the order of 10^2–10^4 s^{-1} to be determined, whereas applying the r.f. field off-resonance allows an extension of this range to approximately 10^5 s^{-1}.[4] In practice, the term *on-resonance* is generally used for the "near on-resonance" conditions, as the r.f. frequency is positioned in the middle of the spectrum, and the ω_1 field strength is kept sufficiently strong to ensure that $\theta \gg 70°$. In the off-resonance experiment, the r.f. field is positioned far off-resonance, leading to $\theta < 70°$. In both cases, the general idea is to vary the effective field, ω_e, to monitor the field dependence of the relaxation rate given by Eq. (13.22) in a *relaxation dispersion* curve. In the on-resonance experiment, the field strength ω_1 is varied, and in the off-resonance experiment, the relaxation dispersion can be obtained by varying the field strength, ω_1, and/or the carrier offset.

If R_2^0 and R_1 can be measured independently (the value for R_2^0 can be obtained from measuring cross-correlated relaxation; see Section 10.2), we can obtain the R_{ex} through Eq. (13.23). R_{ex} can be measured at several values for ω_e^2, and fitting of the data can yield the values for k_{ex}. If $1/R_{ex}$ is plotted against ω_e^2, a straight line is obtained and k_{ex} and $p_A p_B (2\delta)^2$ can be calculated from a combination of the slope and the intercept. The populations and δ cannot be obtained separately in the fast exchange range; however, by performing a temperature-dependent experimental series and assuming that the two resonance frequencies are temperature independent, one can estimate the relative populations of the two sites. An off-resonance experiment has been proposed in which the $R_{1\rho} - R_1$ relaxation rate is measured directly, which means that the R_1 relaxation rate does not need to be measured independently[20]. A number of investigations have been performed in which exchange processes have been characterised by off-resonance relaxation. In particular, the method presented here is useful for characterising conformational exchange, fast unfolding events and rate constants for ligand-enzyme complexes. We shall return to the applications of relaxation dispersion methods for characterisation of transient, low-population states in Section 13.7.

13.4.3 Spin-Echo Methods

The $T_{1\rho}$ method has often been used to characterise chemical exchange occurring on the fast intermediate exchange timescale, but alternative, or complementary, methods exist. These methods are based on the Carr–Purcell–Meiboom–Gill R_2 experiment. The measured transverse relaxation rates depend on the spacing between consecutive echo pulses, τ_{cp} (denoted τ in Figure 8.8), and on dynamic processes that occur on the timescale given by τ_{cp}. An approximate equation for the dependence of R_2 on $1/\tau_{cp}$ in the fast exchange limit can be written[23]:

$$\begin{aligned} R_2(1/\tau_{cp}) &= R_2(1/\tau_{cp} \to \infty) + R_{ex}(1/\tau_{cp}) \\ &= R_2(1/\tau_{cp} \to \infty) + p_A p_B (2\delta)^2 \frac{1}{k_{ex}} \left[1 - \frac{2}{k_{ex}\tau_{cp}} \tanh\left(\frac{k_{ex}\tau_{cp}}{2} \right) \right] \end{aligned} \tag{13.24}$$

A plot of the measured R_2 rates as a function of $1/\tau_{cp}$ yields a curve that can be fitted to obtain the exchange rate, k_{ex}.

Carver and Richards derived a general expression relevant for the τ_{cp} dependence of the transverse relaxation rate[24]. This expression is valid for exchange on all timescales, but in the limit of fast exchange, the expression can be approximated by Eq. (13.24).

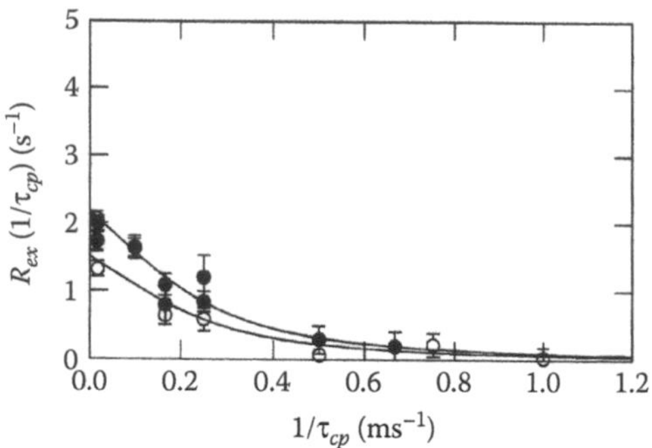

FIGURE 13.4 ^{15}N CPMG dispersion profiles for Cys 38 in bovine pancreatic trypsin inhibitor at 280 K. Open circles: 11.7 T, black circles: 14.1 T. The solid lines represent fits to Eq. (13.24). (Reprinted with permission from Grey, M.J. et al., *J. Am. Chem. Soc.*, 125, 14324–14335, 2003. Copyright (2003) American Chemical Society.)

An example of using the τ_{cp} dependence of the transverse relaxation is the measurement of water proton relaxation dispersion in protein solution[25,26]. Hills and co-workers measured the transverse relaxation of water in a solution of bovine serum albumin and found that chemical exchange between water and protein amide protons had a large influence on the relaxation rates. This was shown to have severe consequences for interpreting data concerning protein-bound water. It was suggested that only a minor effect on proton R_2 relaxation rates comes from the influence of the protein (on the bound water), while the chemical exchange effects dominate the transverse relaxation.

In order to overcome difficulties in recording CPMG experiments with sufficiently short values for τ_{cp}, required to minimise effects of antiphase coherence (compare Section 8.2), Palmer and co-workers have developed methods to measure exchange rates, and populations of the involved states, in proteins from so-called *relaxation compensated* CPMG experiments[27,28]. These methods are useful for proteins undergoing exchange in the region $k_{ex} < 10^4$ s, where the limit is set by the minimum τ_{cp} delay achievable. An illustrative example of CPMG dispersion curves for amide ^{15}N of residue cysteine 38 in bovine pancreatic trypsin inhibitor (BPTI), taken from Grey and co-workers[29], is shown in Figure 13.4. The data were measured at 280 K at the magnetic fields of 11.7 (open circles) and 14.1 T (black circles), and the solid lines were obtained by fitting the data to Eq. (13.24), valid for fast two-site exchange.

Thus, the CPMG dispersion experiments allow kinetic information (k_{ex}) to be derived along with the product $p_A p_B(2\delta)^2$. If data are available in the intermediate exchange range, it may also be possible to separate the thermodynamic parameters (populations) and the square of chemical shift differences. Skrynnikov *et al.*[30] demonstrated that it also is possible to determine the sign of δ using, for example, comparisons of chemical shifts in HSQC experiments at two different static magnetic fields.

13.5 Multiple-Quantum (MQ) Relaxation and Exchange

An interesting phenomenon concerning multiple-quantum relaxation and exchange has been discussed by Kloiber and Konrat[31] and by Früh and co-workers[32]. They have demonstrated that a difference in relaxation rates of double- and zero-quantum coherences, involving the different spins in proteins, is sensitive to the fact that relatively slow motions, such as conformational exchange, can cause cross-correlated fluctuations of the isotropic chemical shifts.

Lundström and Akke investigated conformational dynamics in the E140Q mutant of the C-terminal domain of the Ca^{2+}-binding protein calmodulin[33]. This protein is assumed to undergo conformational exchange between two states resembling the apo state and the Ca^{2+}-bound state. Conformational exchange was in this example determined by measuring differential MQ relaxation for the backbone ^{1}H-^{15}N spin pairs, which can be obtained with the experiment outlined in Section 10.4. The relevant equations are recapitulated here for the convenience of the reader. The difference in MQ relaxation can be measured from an experimental series, in which τ is arrayed, using:

$$\frac{\langle 2\hat{I}_y\hat{S}_y\rangle(\tau)}{\langle 2\hat{I}_x\hat{S}_x\rangle(\tau)} = \tanh(\Delta R_{MQ}\tau/2) \tag{13.25}$$

The difference in MQ relaxation, ΔR_{MQ}, can be used to extract information about cross-correlation between two chemical shift differences according to:

$$\Delta R_{MQ} = R_{DQ} - R_{ZQ} = R_{DQ}^0 - R_{ZQ}^0 + 4p_A p_B \Delta\omega_I \Delta\omega_S \tau_{ex} \tag{13.26}$$

where

$$R_{DQ}^0 - R_{ZQ}^0 = \eta_{cc} + \sigma_{IS} + \eta_{dd}^{remote} \tag{13.27}$$

Together with values for the ^1H and ^{15}N CSAs determined independently, separation of the contributions from dipolar-dipolar, CSA-CSA and conformational exchange cross-correlated relaxation could be made.

The procedure was used to obtain information about the exchange, in terms of differences in chemical shift for both the ^1H and ^{15}N, which in turn could be compared with actual differences in isotropic chemical shifts for the two states of the protein (Figure 13.5). The results supported the finding of conformational exchange between a presumably ion-bound and an ion-free state of the protein.

Relaxation dispersion experiments are also possible in the context of multiple-quantum coherences. Measurements of this kind can be performed in analogy with the CPMG or spin-lock methods for single-quantum coherences. The CPMG-type variety for ^1H–^{15}N spin systems was proposed by the Kay group[34,35]. Briefly, a CPMG-type block (at the proton or nitrogen-15 frequency) is inserted into an MQ

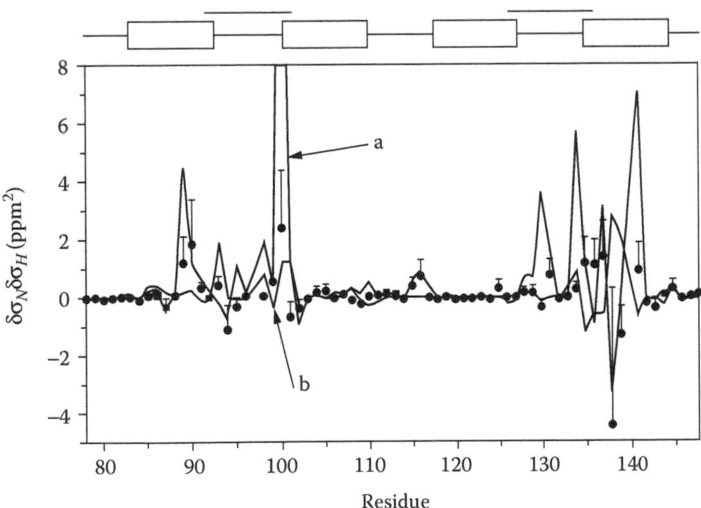

FIGURE 13.5 Differences in ^{15}N and ^1H chemical shift, $\Delta\omega_H\Delta\omega_N$, for the E140Q mutant of the C-terminal domain in calmodulin (solid circles) calculated from differential MQ relaxation. Boxes on top of the figure indicate helical regions in the protein. The lines indicate deviations in chemical shift between the wild-type apo and Ca^{2+}-bound states (a), and ring-current contribution to the $\Delta\omega_H$ (b). (Reprinted with permission from Lundström, P., and Akke, M., *J. Am. Chem. Soc.*, 126, 928–935, 2004. Copyright (2004) American Chemical Society.)

sequence and the zero- and double-quantum relaxation rates are measured as a function of the variable delay between the 180° pulses. The information from the MQ relaxation dispersion is complementary to that from the dispersion data for the two single-quantum coherences.

Ulzega, Salvi and co-workers[36,37] investigated cross-correlated cross-relaxation of MQ coherences in heteronuclear two-spin systems *IS* (for example an interconversion $X \leftrightarrow Y$, where $X = 2I_xS_x$ and $Y = 2I_yS_y$) in the presence of local dynamics under heteronuclear (^{1}H, ^{15}N) double resonance irradiation using a windowless sequence of pulses, *e.g.* Waltz 32. These experiments can be seen as multiple-quantum analogues to the single-quantum spin-lock methodology. Analytical expressions for the dependence of effective MQ cross-relaxation rates on the r.f. field strength were derived for the fast, correlated two-site exchange model with arbitrary populations. It was demonstrated that the contribution of chemical exchange to relaxation can be partly or fully quenched by the r.f. irradiation. A nice comparison of the two approaches to MQ relaxation, as applied to large proteins, was presented by Toyama and co-workers[38].

13.6 Dynamics and Exchange: An Illustrative Example

A particularly complete investigation of protein dynamics including exchange has been done on the SH3 domain of the *Drosophila* signal transduction protein drk. The protein exists in equilibrium between the folded and unfolded states in aqueous buffer solution[39]. The folding process is slow on the NMR chemical shift timescale, and separate sets of signals can be observed for the two states. The *zz*-exchange approach can thus be used to investigate the folding kinetics. Farrow *et al.*[40] developed one such method and measured exchange between the folded and unfolded states of the SH3 domain of the protein drk. This, in turn, could be used in analysing conventional R_1, R_2 and NOE data for the folded/unfolded states of the protein, and a particularly complete picture of the protein dynamics could be obtained[41,42]. The relaxation data indicated that the unfolded state of the protein was affected by slow internal dynamics in addition to the slow exchange between the folded and unfolded states.

The relaxation dispersion approach is suitable for investigating such faster conformational exchange processes, since it is sensitive to both the slow folding kinetics and fast or intermediate exchange. Consequently, relaxation dispersion was used to probe conformational dynamics present within the ensemble of unfolded states of the protein[43]. The data could be evaluated to quantify the exchange rate between the folded and unfolded states. In addition, field-dependent data revealed the presence of an exchange process on the microsecond/millisecond timescale for the unfolded state, involving one or several sub-states within the unfolded ensemble of states. As stated before, conventional NMR methods are insensitive to detecting scarcely populated states, and relaxation dispersion emerges as one of the more important methods to detect exchange between states with very different populations.

13.7 Dark State Spectroscopy

The last decade has witnessed an interesting development of new NMR methods aiming at characterising states not visible by traditional methods because of their transient nature, low equilibrium population and/or very large linewidth. These states are referred to as *dark states*[44] or *invisible states*[45]. The methods to detect and characterise the dark states can be divided into three categories: the relaxation dispersion technique, saturation transfer methods and methods using paramagnetic effects. The two former approaches will be covered here, while the last one will be discussed in Chapter 15.

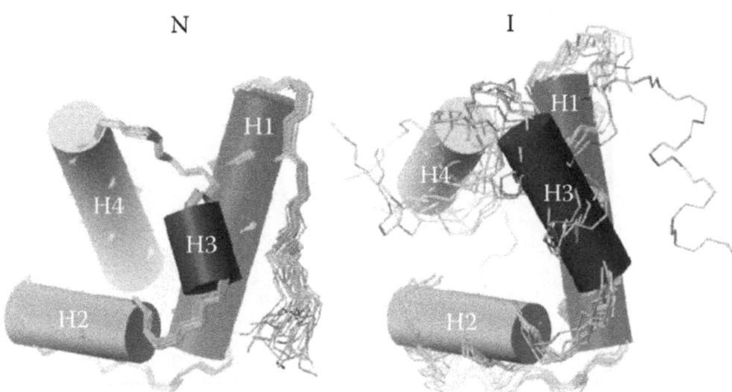

FIGURE 13.6 Comparison of the secondary structure of the native state and the invisible folding intermediate of the FF domain. H1–H4 denote the four helices. (From Korzhnev *et al.*, *Science*, 329, 1312–1316, 2010. Reprinted with permission from AAAS.)

13.7.1 Relaxation Dispersion

The relaxation dispersion methods, of both the CPMG and spin-lock varieties, covered in Section 13.4 can be used to characterise the dark states and – under favourable conditions – to determine their structure at atomic resolution. Let us illustrate the procedure with a beautiful example presented by Korzhnev and co-workers[46] dealing with a small (71 residues) protein known as the FF domain. The authors measured CPMG dispersion for a variety of nuclear species (^{15}N, $^{1}H^{N}$, $^{13}C^{\alpha}$, $^{1}H^{\alpha}$, $^{13}C^{O}$). Assuming two-site exchange between the ground state (N, native) and the invisible conformationally excited state (I, intermediate), it was possible to derive from these measurements a large number of chemical shifts for the "invisible" I-state, along with its population ($p_I \approx 3\%$) and $k_{ex} \approx 1800\ s^{-1}$. In addition, CPMG dispersion experiments were also performed on a weakly aligned sample dissolved in a dilute liquid crystalline solvent. Experiments of this kind allow the determination of the residual dipolar couplings for the dark state[47,48] and from these, the orientations of bond vectors. A combination of chemical shifts (interpreted using the Rosetta method[49,50]) and residual dipolar couplings allowed atomic resolution structure determination for the I state (excluding a few flexible residues at both termini). The secondary structures of the N and I states are compared in Figure 13.6.

13.7.2 Saturation Transfer

In order to study slow exchange processes involving dark states, one can use methods based on the phenomenon of saturation transfer. The original idea of saturation transfer can be traced to the seminal work by Forsén and Hoffman in the early 1960s[51], *i.e.* before the Fourier transform era in NMR. There are several modern varieties of the saturation transfer technique, two of which will be mentioned here: chemical exchange saturation transfer (CEST)[52–54] and dark-state exchange saturation transfer (DEST)[55,56]. We will here limit our interest to the case of two-site exchange with a large difference between the populations of the two states. In both methods, the signal intensity of a signal belonging to the major species is monitored during the application of a weak r.f. field at a variable (in separate experiments) frequency. The measured intensity is altered when the r.f. field is close to the resonance of the invisible state. The CEST experiment makes use of the difference in resonance frequencies (chemical shifts) for ^{15}N or ^{13}C between the two states and is in this sense similar to the CPMG relaxation dispersion technique. The CEST method, however, finds its application for exchange processes too slow for CPMG, typically for lifetimes of the minor state in the range 2–100 ms[44]. The DEST experiment does not require any variation of chemical shift under the exchange process between the two sites; instead,

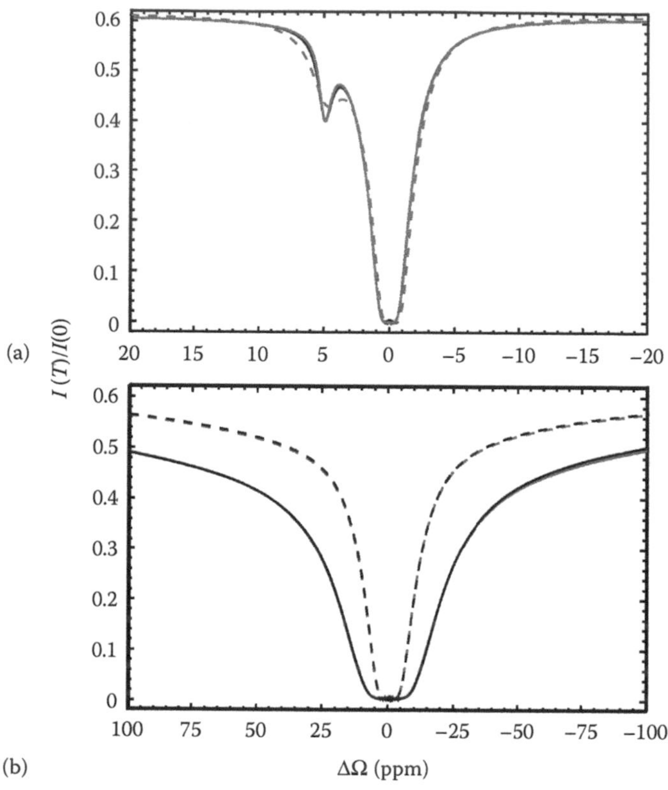

FIGURE 13.7 Theoretical (a) CEST and (b) DEST profiles. Calculations assumed k_{ex} = 50 s^{-1}, p_2 = 0.015, Ω_1 = −0.076 ppm, Ω_2 = 5 ppm and T = 0.48 s. ΔΩ is measured relative to the population-averaged resonance position. (a) $R_{11} = R_{12}$ = 1 s^{-1}, $R_{21} = R_{22}$ = 20 s^{-1}, $\omega_1/2\pi$ = 25 Hz. (b) R_{22} = 20,000 s^{-1} and (dashed) $\omega_1/2\pi$ = 150 Hz and (solid) $\omega_1/2\pi$ = 300 Hz. See the original work for details. (Reprinted with permission from Palmer, A.G., *J. Magn. Reson.*, 241, 3–17, 2014. Copyright (2014) Elsevier.)

it relies on the difference in transverse relaxation rate constant between the two states. Thus, it is useful in situations when the major state is reorienting relatively fast in solutions, while the sparsely populated state is a larger assembly, characterised by much slower tumbling. The timescale of the exchange process is in this case larger, typically between 500 μs and 1 s[44].

For details of both saturation transfer methods, the reader is referred to original papers and reviews[3,44]. Here, we choose to finish this chapter by showing theoretical CEST and DEST profiles (based on solving the Bloch–McConnell equations) for the case of k_{ex} = 50 s^{-1}, p_B = 0.015, ω_A = −0.076 ppm, ω_B = 5 ppm and the saturation pulse duration of 0.48 s. The curves, taken from Palmer[3], are displayed in Figure 13.7.

References

1. Woessner, D. E., *Relaxation effects of chemical exchange*. Wiley: eMagRes, 2007.
2. Palmer, A. G., NMR characterization of the dynamics of biomacromolecules. *Chem. Rev.* 2004, 104, 3623–3640.
3. Palmer, A. G., Chemical exchange in biomacromolecules: Past, present, and future. *J. Magn. Reson.* 2014, 241, 3–17.

4. Palmer, A. G.; Kroenke, C. D.; and Loria, J. P., Nuclear magnetic resonance methods for quantifying microsecond-to-millisecond motions in biological macromolecules. *Methods Enzymol.* 2001, 339, 204–238.
5. Palmer, A. G., *Chemical Exchange Effects in Biological Macromolecules.* Wiley: eMagRes, 2007.
6. Deverell, C.; Morgan, R. E.; and Strange, J. H., Studies of chemical exchange by nuclear magnetic relaxation in the rotating frame. *Mol. Phys.* 1970, 18, 553–559.
7. Wennerström, H., Nuclear magnetic relaxation induced by chemical exchange. *Mol. Phys.* 1972, 24, 69–80.
8. Abergel, D.; and Palmer, A. G., On the use of the stochastic Liouville equation in nuclear magnetic resonance: Application to R-1 rho relaxation in the presence of exchange. *Concepts Magn. Reson.* 2003, 19A, 134–148.
9. Bernatowicz, P.; Kowalewski, J.; and Sandström, D., NMR relaxation study of the protonated form of 1,8-bis(dimethylamino)naphthalene in isotropic solution: Anisotropic motion outside of extreme narrowing and ultrafast proton transfer. *J. Phys. Chem. A* 2005, 109, 57–63.
10. Masuda, Y.; Nakano, T.; and Sugiyama, M., First observation of ultrafast intramolecular proton transfer rate between electronic ground states in solution. *J. Phys. Chem. A* 2012, 116, 4485–4494.
11. Nakano, T.; and Masuda, Y., Application of nuclear magnetic relaxation to elucidate proton location and dynamics in N···H···O hydrogen bonds. *J. Phys. Chem. A* 2012, 116, 8409–8418.
12. Masuda, Y.; Mori, Y.; and Sakurai, K., Effects of counterion and solvent on proton location and proton transfer dynamics of N-H···N hydrogen bond of monoprotonated 1,8-bis(dimethylamino) naphthalene. *J. Phys. Chem. A* 2013, 117, 10576–10587.
13. Helgstrand, M.; and Allard, P., QSim, a program for NMR simulations. *J. Biomol. NMR* 2004, 30, 71–80.
14. McConnell, H. M., Reaction rates by nuclear magnetic resonance. *J. Chem. Phys.* 1958, 28, 430–431.
15. Bodenhausen, G.; Wagner, G.; Rance, M.; Sorensen, O. W.; Wüthrich, K.; and Ernst, R. R., Longitudinal 2-spin order in 2D exchange spectroscopy (NOESY). *J. Magn. Reson.* 1984, 59, 542–550.
16. Wagner, G.; Bodenhausen, G.; Müller, N.; Rance, M.; Sorensen, O. W.; Ernst, R. R.; and Wüthrich, K., Exchange of two-spin order in nuclear magnetic resonance: Separation of exchange and cross-relaxation processes. *J. Am. Chem. Soc.* 1985, 107, 6440–6446.
17. Kern, D.; Kern, G.; Scherer, G.; Fischer, G.; and Drakenberg, T., Kinetic-analysis of cyclophilin-catalyzed prolyl cis/trans isomerization by dynamic NMR-spectroscopy. *Biochemistry* 1995, 34, 13594–13602.
18. Kroenke, C. D.; Loria, J. P.; Lee, L. K.; Rance, M.; and Palmer, A. G., Longitudinal and transverse H-1-N-15 dipolar N-15 chemical shift anisotropy relaxation interference: Unambiguous determination of rotational diffusion tensors and chemical exchange effects in biological macromolecules. *J. Am. Chem. Soc.* 1998, 120, 7905–7915.
19. Szyperski, T.; Luginbühl, P.; Otting, G.; Güntert, P.; and Wüthrich, K., Protein dynamics studied by rotating frame N-15 spin relaxation-times. *J. Biomol. NMR* 1993, 3, 151–164.
20. Akke, M.; and Palmer, A. G., Monitoring macromolecular motions on microsecond to millisecond time scales by R(1)rho-R(1) constant relaxation time NMR spectroscopy. *J. Am. Chem. Soc.* 1996, 118, 911–912.
21. Zinn-Justin, S.; Berthault, P.; Guenneugues, M.; and Desvaux, H., Off-resonance rf fields in heteronuclear NMR: Application to the study of slow motions. *J. Biomol. NMR* 1997, 10, 363–372.
22. Palmer, A. G.; and Massi, F., Characterization of the dynamics of biomacromolecules using rotating-frame spin relaxation NMR spectroscopy. *Chem. Rev.* 2006, 106, 1700–1719.
23. Luz, Z.; and Meiboom, S., Nuclear magnetic resonance study of the protolysis of trimethylammonium ion in aqueous solutions – order of reaction with respect to solvent. *J. Chem. Phys.* 1963, 39, 366–370.
24. Carver, J. P.; and Richards, R., A general two-site solution for the chemical exchange produced dependence of T2 upon the Carr-Purcell pulse separation. *J. Magn. Reson.* 1972, 6, 89–105.

25. Hills, B. P.; Takacs, S. F.; and Belton, P. S., The effects of proteins on the proton NMR transverse relaxation times of water. I. Native bovine serum albumin. *Mol. Phys.* 1989, 67, 903–918.
26. Hills, B. P.; Takacs, S. F.; and Belton, P. S., The effect of proteins on the proton NMR transverse relaxation time of water II. Protein aggregation. *Mol. Phys.* 1989, 67, 919–937.
27. Loria, J. P.; Rance, M.; and Palmer, A. G., A relaxation-compensated Carr-Purcell-Meiboom-Gill sequence for characterizing chemical exchange by NMR spectroscopy. *J. Am. Chem. Soc.* 1999, 121, 2331–2332.
28. Wang, C. Y.; Grey, M. J.; and Palmer, A. G., CPMG sequences with enhanced sensitivity to chemical exchange. *J. Biomol. NMR* 2001, 21, 361–366.
29. Grey, M. J.; Wang, C. Y.; and Palmer, A. G., Disulfide bond isomerization in basic pancreatic trypsin inhibitor: Multisite chemical exchange quantified by CPMG relaxation dispersion and chemical shift modelling. *J. Am. Chem. Soc.* 2003, 125, 14324–14335.
30. Skrynnikov, N. R.; Dahlquist, F. W.; and Kay, L. E., Reconstructing NMR spectra of "invisible" excited protein states using HSQC and HMQC experiments. *J. Am. Chem. Soc.* 2002, 124, 12352–12360.
31. Kloiber, K.; and Konrat, R., Differential multiple-quantum relaxation arising from cross-correlated time modulation of isotropic chemical shifts. *J. Biomol. NMR* 2000, 18, 33–42.
32. Früh, D.; Tolman, J. R.; Bodenhausen, G.; and Zwahlen, C., Cross-correlated chemical shift modulation: A signature of slow internal motions in proteins. *J. Am. Chem. Soc.* 2001, 123, 4810–4816.
33. Lundström, P.; and Akke, M., Quantitative analysis of conformational exchange contributions to H-1-N-15 multiple-quantum relaxation using field-dependent measurements. Time scale and structural characterization of exchange in a calmodulin C-terminal domain mutant. *J. Am. Chem. Soc.* 2004, 126, 928–935.
34. Orekhov, V. Y.; Korzhnev, D. M.; and Kay, L. E., Double- and zero-quantum NMR relaxation dispersion experiments sampling millisecond time scale dynamics in proteins. *J. Am. Chem. Soc.* 2004, 126, 1886–1891.
35. Korzhnev, D. M.; Kloiber, K.; and Kay, L. E., Multiple-quantum relaxation dispersion NMR spectroscopy probing millisecond time-scale dynamics in proteins: Theory and application. *J. Am. Chem. Soc.* 2004, 126, 7320–7329.
36. Ulzega, S.; Salvi, N.; Segawa, T. F.; Ferrage, F.; and Bodenhausen, G., Control of cross relaxation of multiple-quantum coherences induced by fast chemical exchange under heteronuclear double-resonance irradiation. *ChemPhysChem.* 2011, 12, 333–341.
37. Salvi, N., Theoretical tools for the design of NMR relaxation dispersion pulse sequences. *Progr. NMR Spectr.* 2015, 88–89, 105–115.
38. Toyama, Y.; Osawa, M.; Yokogawa, M.; and Shimada, I., NMR method for characterizing microsecond-to-millisecond chemical exchanges utilizing differential multiple-quantum relaxation in high molecular weight proteins. *J. Am. Chem. Soc.* 2016, 138, 2302–2311.
39. Zhang, O. W.; Kay, L. E.; Olivier, J. P.; and Forman-Kay, J. D., Backbone H-1 and N-15 resonance assignments of the N-terminal Sh3 domain of drk in folded and unfolded states using enhanced-sensitivity pulsed-field gradient NMR techniques. *J. Biomol. NMR* 1994, 4, 845–858.
40. Farrow, N. A.; Zhang, O. W.; Forman-Kay, J. D.; and Kay, L. E., A heteronuclear correlation experiment for simultaneous determination of N-15 longitudinal decay and chemical-exchange rates of systems in slow equilibrium. *J. Biomol. NMR* 1994, 4, 727–734.
41. Farrow, N. A.; Zhang, O.W.; Forman-Kay, J. D.; and Kay, L. E., Comparison of backbone dynamics of a folded and an unfolded SH3 domain existing in equilibrium in aqueous buffer. *Biochemistry* 1995, 34, 868–878.
42. Farrow, N. A.; Zhang, O. W.; Forman-Kay, J. D.; and Kay, L. E., Characterization of the backbone dynamics of folded and denatured states of an SH3 domain. *Biochemistry* 1997, 36, 2390–2402.
43. Tollinger, M.; Skrynnikov, N. R.; Mulder, F. A. A.; Forman-Kay, J. D.; and Kay, L. E., Slow dynamics in folded and unfolded states of an SH3 domain. *J. Am. Chem. Soc.* 2001, 123, 11341–11352.

44. Anthis, N. J.; and Clore, G. M., Visualizing transient dark states by NMR spectroscopy. *Quart. Rev. Biophys.* 2015, 48, 35–116.
45. Sekhar, A.; and Kay, L. E., NMR paves the way for atomic level descriptions of sparsely populated, transiently formed biomolecular conformers. *Proc. Natl. Acad. Sci. USA* 2013, 110, 12867–12874.
46. Korzhnev, D. M.; Religa, T. L.; Banachewicz, W.; Fersht, A. R.; and Kay, L. E., A transient and low-populated protein-folding intermediate at atomic resolution. *Science* 2010, 329, 1312–1316.
47. Igumenova, T. I.; Brath, U.; Akke, M.; and Palmer, A. G., Characterization of chemical exchange using residual dipolar coupling. *J. Am. Chem. Soc.* 2007, 129, 13396–13397.
48. Vallurupalli, P.; Hansen, D. F.; Stollar, E.; Meirovitch, E.; and Kay, L. E., Measurement of bond vector orientations in invisible excited states of proteins. *Proc. Natl. Acad. Sci. USA* 2007, 104, 18473–18477.
49. Shen, Y.; Lange, O.; Delaglio, F.; Rossi, P.; Aramini, J. M.; Liu, G. H.; Eletsky, A. Wu, Y. B.; Singarapu, K. K.; Lemak, A.; Ignatchenko, A.; Arrowsmith, C. H.; Szyperski, T.; Montelione, G. T.; Baker, D.; and Bax, A., Consistent blind protein structure generation from NMR chemical shift data. *Proc. Natl. Acad. Sci. USA* 2008, 105, 4685–4690.
50. Das, R.; and Baker, D., Macromolecular modeling with Rosetta. *Annu. Rev. Biochem.* 2008, 77, 363–382.
51. Forsén, S.; and Hoffman, R. A., Study of moderately rapid chemical exchange reactions by means of nuclear magnetic double resonance. *J. Chem. Phys.* 1963, 39, 2892–2901.
52. Bouvignies, G.; and Kay, L. E., Measurement of proton chemical shifts in invisible states of slowly exchanging protein systems by chemical exchange saturation transfer. *J. Phys. Chem. B* 2012, 116, 14311–14317.
53. Vallurupalli, P.; Bouvignies, G.; and Kay, L. E., Studying "invisible" excited protein states in slow exchange with a major state conformation. *J. Am. Chem. Soc.* 2012, 134, 8148–8161.
54. Hansen, A. L.; Bouvignies, G.; and Kay, L. E., Probing slowly exchanging protein systems via C-13(alpha)-CEST: Monitoring folding of the Im7 protein. *J. Biomol. NMR* 2013, 55, 279–289.
55. Fawzi, N. L.; Ying, J. F.; Ghirlando, R.; Torchia, D. A.; and Clore, G. M., Atomic-resolution dynamics on the surface of amyloid-beta protofibrils probed by solution NMR. *Nature* 2011, 480, 268–272.
56. Fawzi, N. L.; Ying, J. F.; Torchia, D. A.; and Clore, G. M., Probing exchanger kinetics and atomic resolution dynamics in high-molecular-weight complexes using dark-state exchange saturation transfer NMR spectroscopy. *Nat. Protoc.* 2012, 7, 1523–1533.

14

Effects of Quadrupolar Nuclei

Most nuclear magnetic resonance (NMR) work in solution is performed on spin 1/2 nuclei, exploiting their characteristic narrow lines. However, most of the stable NMR-receptive nuclei have spin quantum numbers larger than 1/2. Besides a magnetic dipole moment, these nuclear species are characterised by an electric quadrupole moment, and are therefore called *quadrupolar*. In this chapter, we discuss spin systems containing such nuclei. We have already mentioned briefly, in Chapter 11, that the relaxation of deuterons – nuclei with $S = 1$ – and carbon-13 (spin 1/2) has several similarities. In this chapter, we shall present in more detail the relaxation of quadrupolar nuclei. In the first section, the basic properties of the interaction between the quadrupolar nuclei and the electric field gradients in a molecular system are introduced. In Section 14.2, we proceed by presenting the theory for spin relaxation of quadrupolar nuclei that are isolated, in the sense that they do not interact with other spins. In that context, we also mention special experimental techniques developed for studies of quadrupolar nuclei in solution. We then move to relaxation phenomena in more complicated spin systems, where the quadrupolar nuclei interact with $I = 1/2$ nuclei, and discuss in particular the effects of the $S \geq 1$ species on the evolution of non-equilibrium states of the $I = 1/2$ spins. We conclude the chapter with a brief presentation of the applications of relaxation studies in quadrupolar systems.

14.1 Quadrupolar Interaction

For nuclear species with the spin quantum number of one or higher, the distribution of the electric charge within the nucleus is not spherically symmetric. This means that the electric energy of a nucleus depends on its orientation with respect to the inhomogeneous electric fields existing in a molecular framework. The interaction between the electric charge distribution of the nucleus and the electrostatic potential and its derivatives can be expressed, classically as well as quantum mechanically, in terms of multipole expansion. In the language of this model, the interaction is expressed as a sum of terms involving moments of charge distribution. Thus, the first term describes how the total charge of the nucleus interacts with the electrostatic potential at the site of the nucleus – for a point charge, this is the familiar Coulomb interaction. This term does not depend on the orientation of the nucleus and has no direct influence on the spin properties. The second term in a general multipole expansion is the interaction between the electric dipole moment and the electric field (the gradient of the potential). This term vanishes in the case of atomic nuclei, because they have no electric dipoles. Third, the quadrupole moment interacts with the gradient of the electric field. This term, called the *quadrupolar interaction*, is of great importance for NMR of nuclei with $S \geq 1$. As shown in the texts by Abragam and by Slichter (*Further reading*), the interaction between the nuclear quadrupole moment and the electric field gradient (efg) can be expressed by an effective spin-Hamiltonian:

$$\hat{H}_S^Q = \hat{\mathbf{S}} \cdot \mathbf{Q} \cdot \hat{\mathbf{S}} \tag{14.1}$$

The similarity between this Hamiltonian and that for the dipole-dipole interaction (Eq. (3.4)) is worth noting. The quadrupole coupling tensor, **Q**, can further be given by:

$$\mathbf{Q} = \frac{eQ}{4S(2S-1)\hbar}\mathbf{V} \tag{14.2}$$

The symbol e is the elementary charge and Q is the nuclear electric quadrupole moment, a fundamental property of atomic nuclei.

The quadrupole moment has the dimensionality of area and some examples of this quantity were given in Table 1.1. The nuclear quadrupole moment can have either sign, which is related to the sign of the deviation from the spherical charge distribution (cigar-shaped [prolate] or disk-shaped [oblate] nuclei). We stress that the quadrupolar Hamiltonian is only present for $S \geq 1$ (this is a consequence of the Wigner–Eckart theorem; see the books by Slichter and by Brink and Satchler [*Further reading*]) and we do not need to worry about the singularity at $S = 1/2$. **V** is the electric field gradient tensor, a rank-two tensorial quantity. The Cartesian components of the field gradient are given by:

$$V_{\alpha\beta} = \left(\frac{\partial^2 V}{\partial r_\alpha \partial r_\beta}\right)_{\mathbf{r}=0} \tag{14.3}$$

i.e. they are equal to second derivatives of the electrostatic potential V with respect to the Cartesian coordinates r_α and r_β, evaluated at the site of the nucleus ($\mathbf{r} = 0$). The electric field gradient tensor is always symmetric ($V_{\alpha\beta} = V_{\beta\alpha}$) and traceless ($V_{xx} + V_{yy} + V_{zz} = 0$). It has many properties analogous to the symmetric part of the shielding tensor: it is most obviously defined in the molecular coordinate system and we can always find a molecule-fixed principal frame (principal axis system [PAS]) in which only the diagonal elements are non-zero. In the principal frame, we have thus only two parameters: the numerically largest of the three diagonal components, V_{zz}, often denoted as *field gradient*, and the asymmetry parameter, η_Q:

$$V_{zz} = eq \tag{14.4a}$$

$$\eta_Q = \left(V_{yy} - V_{xx}\right)/V_{zz} \tag{14.4b}$$

The definition of the asymmetry parameter is associated with the convention $|V_{zz}| \geq |V_{xx}| \geq |V_{yy}|$.

For the same reasons as discussed in the case of dipolar interaction or chemical shielding, it turns out to be convenient to express the quadrupolar Hamiltonian in the irreducible spherical form:

$$\hat{H}_Q = \frac{eQ}{4S(2S-1)\hbar}\sum_{q=-2}^{2}(-1)^q V_{2,q}^{PAS}\hat{T}_{2,-q} \tag{14.5}$$

The spherical tensor operators $\hat{T}_{l,-q}$ with $l = 2$ take on a form similar to that for the dipole-dipole interaction, as given in Eq. (4.43). They are listed in Table 14.1, together with the operators with other l-values, which we shall find useful later[1].

The functions $V_{2,q}^{PAS}$ are very simple:

$$\begin{aligned} V_{2,0}^{PAS} &= (6)^{1/2} eq \\ V_{2,\pm 1}^{PAS} &= 0 \\ V_{2,\pm 2}^{PAS} &= -\eta_Q eq \end{aligned} \tag{14.6}$$

TABLE 14.1 Some Irreducible Tensor Operators, $\hat{T}_{l,q}$, Useful for Discussion of Relaxation of Quadrupolar Nuclei

l	q	Operator	Interpretation
0	0	1	Identity
1	0	$\hat{S}_z$	Longitudinal magnetisation (Zeeman order)
1	±1	$\mp\frac{1}{\sqrt{2}}\hat{S}_\pm$	Rank-one single-quantum coherences
2	0	$\frac{1}{\sqrt{6}}(3\hat{S}_z^2 - S(S+1))$	Quadrupolar order
2	±1	$\mp\frac{1}{2}(\hat{S}_z\hat{S}_\pm + \hat{S}_\pm\hat{S}_z)$	Rank-two single-quantum coherences
2	±2	$\frac{1}{2}\hat{S}_\pm\hat{S}_\pm$	Rank-two double-quantum coherences
3	0	$\frac{1}{\sqrt{10}}(5\hat{S}_z^3 - (3S(S+1)-1)\hat{S}_z)$	Octopolar order
3	±1	$\mp\frac{1}{4}\sqrt{\frac{3}{10}}\left[\left(5\hat{S}_z^2 - S(S+1) - \frac{1}{2}\right)\hat{S}_\pm + \hat{S}_\pm\left(5\hat{S}_z^2 - S(S+1) - \frac{1}{2}\right)\right]$	Rank-three single-quantum coherences
3	±2	$\frac{1}{2}\sqrt{\frac{3}{4}}\left[\hat{S}_z\hat{S}_\pm^2 + \hat{S}_\pm^2\hat{S}_z\right]$	Rank-three double-quantum coherences
3	±3	$\mp\frac{1}{2\sqrt{2}}\hat{S}_\pm^3$	Rank-three triple-quantum coherences

Source: Bowden, G.J., Hutchison, W.D., *J. Magn. Reson.*, 67, 403–414, 1986.

A quadrupolar nucleus with $S \geq 1$ in a magnetic field is of course also subject to the Zeeman interaction, given by the Hamiltonian of Eq. (1.7), which leads to the splitting of the nuclear spin energy into $2S+1$ levels, characterised by the quantum number m. At high magnetic field, the Zeeman Hamiltonian usually dominates over the quadrupolar interaction, which means that the spins are quantised in the laboratory frame, m remains a "good" quantum number and the effect of a not too large quadrupolar coupling can be included by perturbation theory. As usual, the allowed transitions that can be induced by the interaction with a radiofrequency field are between the levels m and $m \pm 1$. Thus, the NMR spectrum of a nucleus with a quantum number S corresponds to $2S$ transitions. If the Zeeman interaction is dominant, then the natural coordinate frame to express spin operators is the laboratory axis system with the z-axis along the field direction. Since the operators and the functions $V_{2,q}$ in Eq. (14.5) need to be expressed in the same frame, we choose to transform the functions to the laboratory frame. The situation is completely analogous to the case of anisotropic shielding discussed in Section 5.2. The transformation becomes particularly simple if the efg tensor is axially symmetric, implying that $V_{yy} = V_{xx}$ in the PAS of the field gradient tensor, or, in other words, $\eta_Q = 0$. Using these relations, Eq. (14.5) can be re-written as:

$$\hat{H}_Q = \xi_Q \sum_{q=-2}^{2} (-1)^q V_{2,q} \hat{T}_{2,-q} \tag{14.7}$$

The operators $\hat{T}_{2,q}$ retain the form of Table 14.1, while the lattice functions $V_{2,q}$ become equal to normalised spherical harmonics (*cf.* Eq. (4.44)) and the interaction strength is given by:

$$\xi_Q = \left(\frac{3\pi}{10}\right)^{1/2} \frac{e^2qQ}{S(2S-1)\hbar} \tag{14.8}$$

The quantity $e^2qQ/\hbar$, in angular frequency units, or its analogue in Hertz, is called the *quadrupolar coupling constant* (QCC).

In the case of a non-axially symmetric efg, the analogue of Eq. (14.7) becomes much more complicated. As discussed by Werbelow[2], a good strategy is to re-write the general traceless tensor **Q** (in its PAS) as a sum of two axially symmetric tensors:

$$\begin{bmatrix} q_{xx} & 0 & 0 \\ 0 & q_{yy} & 0 \\ 0 & 0 & q_{zz} \end{bmatrix} = \Delta Q^{+} \begin{bmatrix} \frac{2}{3} & 0 & 0 \\ 0 & -\frac{1}{3} & 0 \\ 0 & 0 & -\frac{1}{3} \end{bmatrix} + \Delta Q^{-} \begin{bmatrix} -\frac{1}{3} & 0 & 0 \\ 0 & \frac{2}{3} & 0 \\ 0 & 0 & -\frac{1}{3} \end{bmatrix} \tag{14.9}$$

with

$$\Delta Q^{+} = q_{xx} - q_{zz} = -\Delta Q\left(1 + \tfrac{1}{3}\eta_Q\right) \tag{14.10a}$$

$$\Delta Q^{-} = q_{yy} - q_{zz} = -\Delta Q\left(1 - \tfrac{1}{3}\eta_Q\right) \tag{14.10b}$$

$$\Delta Q = q_{zz} - \tfrac{1}{2}\left(q_{xx} + q_{yy}\right) \tag{14.10c}$$

$$\eta_Q = \frac{3\left(q_{yy} - q_{xx}\right)}{2\Delta Q} \tag{14.10d}$$

The principal axes of the two axially symmetric tensors are mutually orthogonal. The two representations (ΔQ, η, on the one hand, and ΔQ^{+}, ΔQ^{-}, on the other) are fully equivalent.

Assuming a high magnetic field, where the Zeeman interaction dominates over the quadrupolar interaction, we can easily calculate the energy levels for a quadrupolar spin to the first order of perturbation theory:

$$E_m = -\gamma_S B_0 m + \tfrac{1}{3}\omega_Q\left[3m^2 - S(S+1)\right] \tag{14.11}$$

with

$$\omega_Q = \frac{3e^2qQ}{4S(2S-1)\hbar}\left(\frac{3\cos^2\theta - 1}{2} - \frac{\eta_Q \sin^2\theta\cos 2\phi}{2}\right) \tag{14.12}$$

where θ is the angle between the orientation of the principal z-axis of the efg and the direction of the magnetic field and ϕ is the angle between the x-axis of the PAS and the projection of the field direction on the xy-plane of the PAS. Eq. (14.11) can be used to predict NMR spectra of quadrupolar nuclei. In a single crystal, we obtain a pattern of $2S$ equidistant lines with a spacing proportional to ω_Q, dependent on the angles θ and ϕ. In other words, the spacing changes as the crystal is rotated with respect to the magnetic field. The energy level diagram and NMR spectrum for the case of $S = 3/2$ are shown in Figure 14.1. The transition between the $m = \pm 1/2$ levels (present in the spectra of all nuclei with half-integer spin quantum number) is not shifted by the quadrupolar interaction and thus occurs at the Larmor

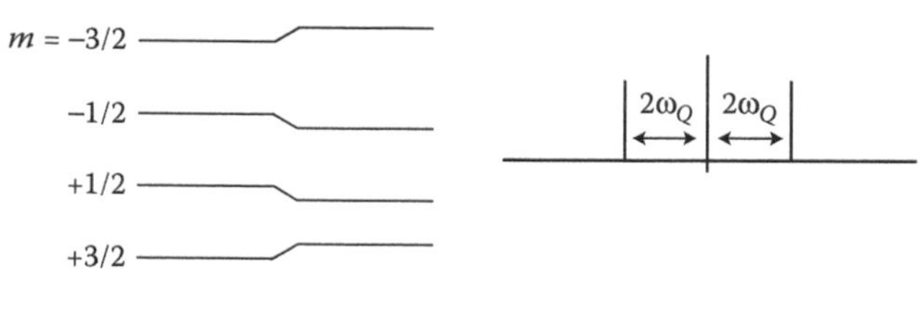

FIGURE 14.1 Energy level diagram and NMR spectrum for $S = 3/2$.

frequency (exactly as for the $S = 1/2$ case!). In the context of $S = 3/2$ and higher half-integer spins, this transition is called the *central line*, while the other transitions are called *satellite lines*. For integer spins, the $m = \pm 1/2$ levels do not exist, the central line is absent and we only have the satellite lines. In the presence of crystalline strains, the satellite transitions become broadened by a distribution of orientations of the PAS (or the q values) while the central line is not affected since, in the first-order perturbation theory, its position is independent of the quadrupolar effects. As discussed by Man in the eMagRes[3], the central line is in fact also affected by the quadrupolar interaction, but the analysis requires a treatment based on the second-order perturbation theory.

In a solid powder of a perfect crystalline material, we also have a distribution of the angles θ and ϕ. In analogy with the cases of dipolar interaction or anisotropic shielding, the shapes of the satellite transitions form characteristic powder patterns here also. In fact, assuming $\eta_Q = 0$, the powder pattern of a pair of satellite transitions is the same as the dipolar Pake doublet shown in Figure 3.4. In this case, one member of the pair might be the transition between $m = -1$ and $m = 0$ and the other member would then be the transition between $m = 0$ and $m = +1$. The width of the quadrupolar powder pattern (in frequency units) is, however, usually much larger. In a general case, the quadrupolar powder pattern also depends on the asymmetry parameter. We shall come back to the quadrupolar powder patterns for deuterons in Chapter 17.

In isotropic liquids, the quadrupolar interaction vanishes, which for the case of an axially symmetric efg can be proved by the same argument as used for the dipolar interaction in Section 3.1. In analogy with the dipolar interaction for spin 1/2 systems, the quadrupolar interaction does provide, however, a relaxation mechanism for quadrupolar spins in the liquid state.

14.2 Spin Relaxation for Nuclei with $S \geq 1$

For $S \geq 1$ spins, the quadrupolar interaction is usually much stronger than other interactions causing relaxation, discussed in Chapters 3 and 5. This means that the quadrupolar mechanism normally dominates the relaxation and we do not need to worry about other interactions if we are only concerned with the relaxation of systems containing isolated nuclei.

14.2.1 Population Kinetics and Spin-Lattice Relaxation

Spin-lattice relaxation in a system of that kind can be expressed in terms of transition probabilities, $W_{n,m}$, between the Zeeman levels[2,4]:

$$\frac{d}{dt}P_n = \sum_{m \neq n} W_{n,m}\left(\Delta P_m - \Delta P_n\right) \tag{14.13}$$

where n and m denote pairs among the $2S + 1$ levels. P_n and P_m are the corresponding populations and the symbol Δ indicates the deviation from equilibrium.

We can easily see that Eq. (14.13) is similar to Eqs. (2.1) and (3.17), which describe the population kinetics in the spin 1/2 systems. In analogy with these cases, the transition probabilities can be related to the matrix elements of the perturbation (here quadrupolar) Hamiltonian and the spectral densities. Let us assume for the time being that the quadrupolar interaction is of intramolecular origin and that its strength is constant in the molecular frame (we shall come back to the intermolecular quadrupolar relaxation at the end of this section). The transition probabilities vanish unless $m = n \pm 1$ or $m = n \pm 2$ (note that the transitions changing the total magnetic quantum number by two units were also allowed in the dipole-dipole case, discussed in Chapter 3) and are for these cases given by:

$$W_{n,n-1} = W_{-n+1,-n} = \tfrac{1}{4}(2n-1)^2(S-n+1)(S+n)\xi_Q^2 J_2(\omega_S) \tag{14.14a}$$

$$W_{n,n-2} = W_{-n+2,-n} = \tfrac{1}{4}(S-n+2)(S-n+1)(S+n-1)(S+n)\xi_Q^2 J_2(2\omega_S) \tag{14.14b}$$

where $J_2(\omega_S)$ and $J_2(2\omega_S)$ are spectral densities for the rank-two spherical harmonics (note that the functions $V_{2,q}$ in the quadrupolar Hamiltonian Eq. (14.7) are equal to spherical harmonics), evaluated at the multiples of Larmor frequency of the quadrupolar nucleus (see Table 1.1 for examples of Larmor frequencies for quadrupolar nuclei at the magnetic field of 9.4 T). The spectral densities for various motional models were discussed in Chapter 6. We remind the reader that for the simplest dynamic model, that of isotropic rotational diffusion, the spectral density is a Lorentzian (given by Eq. (3.16) or (6.20)), characterised by a rotational correlation time for rank-two spherical harmonics (τ_c, $\tau_c(l=2)$ or τ_2 in the notation of earlier chapters) or a rotational diffusion coefficient, D_R.

From the population kinetics summarised in Eq. (14.14), we can move on to the longitudinal magnetisation dynamics, recognising that each of the $2S+1$ stationary states, $|S,m\rangle$, contributes to the z-magnetisation through a product of its population and the corresponding eigenvalue of the z-component of the magnetic moment operator (compare Eq. (1.6b)), $P_m \gamma_S m$. In a more general approach, we can use the Redfield theory, in the formulation of Section 4.1 or in the operator approach described in Section 4.2, to obtain the equation of motion for $\langle \hat{S}_z \rangle(t)$. It turns out (see Werbelow[2] and Abragam [*Further reading*]) that the evolution of $\langle \hat{S}_z \rangle$ is in general coupled to expectation values of higher powers of $\hat{S}_z$, such as $\langle \hat{S}_z^2 \rangle$, $\langle \hat{S}_z^3 \rangle$, *etc.*, resulting in multiexponential longitudinal relaxation.

14.2.2 The Simple T_1 Case

We shall return to the multiexponential relaxation phenomena, but begin here by stating that a monoexponential recovery of the $\langle \hat{S}_z \rangle$ after a perturbation (*e.g.* an inversion by a non-selective π-pulse) is obtained under two sets of conditions. The first situation in which this happens is in the extreme narrowing regime, implying that $\tau_c^2 \omega_S^2 \ll 1$ for the case of isotropic small-step rotational diffusion, or that spectral densities are frequency independent in a more general case. Simple expressions are under this condition obtained in two special cases. The first of these cases applies if the molecular motion is isotropic and the spin-lattice relaxation rate, $T_{1,Q}^{-1}$, is then given by:

$$\begin{aligned} \frac{1}{T_{1,Q}} = \frac{1}{T_{2,Q}} &= \frac{3}{40}\frac{2S+3}{S^2(2S-1)}\left(\frac{e^2qQ}{\hbar}\right)^2\left(1+\frac{\eta_Q^2}{3}\right)\tau_c = \\ &= \frac{3}{40}\frac{2S+3}{S^2(2S-1)}\left(\frac{e^2qQ}{\hbar}\right)^2\left(1+\frac{\eta_Q^2}{3}\right)\frac{1}{6D_R} \end{aligned} \tag{14.15}$$

As usual in this book, the quadrupolar coupling constant $(e^2qQ/\hbar)$ is given in radians per second. Eq. (14.15) indicates that under these conditions, the spin-lattice relaxation rate is equal to the spin-spin relaxation rate, which also is well defined (see Section 14.4.2). If the motion is in the extreme narrowing regime but is not isotropic, a simple expression (again with T_1 equal to T_2) is obtained in the second special case: if the electric field gradient is axially symmetric ($\eta_Q = 0$):

$$\frac{1}{T_{1,Q}} = \frac{1}{T_{2,Q}} = \frac{3}{40}\frac{2S+3}{S^2(2S-1)}\left(\frac{e^2qQ}{\hbar}\right)^2 \tau_c \tag{14.16}$$

The effective correlation time refers to the motion of the principal z-axis of the efg. This equation is identical to Eq. (11.6). Its relation to the more fundamental dynamic properties depends on the model. For example, τ_c for a symmetric top rotational diffusion model is given by Eq. (11.4b). If the efg is not axially symmetric and the motion is not isotropic, the situation becomes a little more complicated (but still tractable in extreme narrowing). The spin-lattice relaxation rate is in such a case equal to a sum of three terms. Two of those are analogous to the right-hand side of Eq. (14.16), with q replaced by ΔQ^+ and ΔQ^- of Eq. (14.10) and the correlation time referring to the reorientation of the two orthogonal axes discussed above. The third term corresponds to the cross-correlation between the two quadrupolar interaction components.

The second situation in which longitudinal relaxation is simple exponential is if the spin-quantum number is unity. As discussed by Abragam (*Further reading*), for this value of the quantum number, one can prove the operator relation $\hat{S}_z^3 = \hat{S}_z$, which has the consequence that the set of two coupled longitudinal relaxation equations simplifies to a single one, even outside of extreme narrowing. Assuming an axially symmetric efg, the spin-lattice relaxation rate is in this case given by:

$$\frac{1}{T_{1,Q}} = \frac{3\pi}{20}\left(\frac{e^2qQ}{\hbar}\right)^2 \left(J_2(\omega_S) + 4J_2(2\omega_S)\right) \tag{14.17}$$

The spectral densities have the same meaning here as in the case of dipolar or CSA relaxation. We note that Eq. (14.17) simplifies in extreme narrowing to Eq. (14.16) with $S = 1$ or to Eq. (11.7).

When the extreme narrowing conditions do not apply, the longitudinal relaxation becomes multi-exponential for $S \geq 3/2$. In practice, the results of an inversion-recovery experiment can still often be interpreted as a simple exponential process. This situation has been investigated theoretically by Halle and Wennerström[5] using a perturbation theory approach. The result was that if the product $\tau_c^2\omega_S^2$ is on the order of unity (or even somewhat larger) and the efg is axially symmetric, a quasiexponential behaviour should be observed with an apparent rate:

$$\frac{1}{T_{1,Q}^{app}} = \frac{3\pi}{100}\frac{2S+3}{S^2(2S-1)}\left(\frac{e^2qQ}{\hbar}\right)^2 \left(J_2(\omega_S) + 4J_2(2\omega_S)\right) \tag{14.18}$$

One should notice that this equation becomes exact for $S = 1$ for any motional conditions (within the validity limit of the Redfield theory, which here can be expressed as $|(e^2qQ/\hbar)\tau_c| \ll 1$; compare Eq. (4.24)). The formulation of Halle and Wennerström has found many applications, in particular for monoatomic ions carrying quadrupolar nuclei (such as, for example, the species $^{23}Na^+$ or $^{35}Cl^-$) interacting with biological macromolecules and other polyelectrolytes. We shall come back to such an application in Section 14.4.

14.2.3 Transverse Relaxation

The transverse relaxation of quadrupolar nuclei can be described using the formulation of Section 4.1 or 4.2. As indicated in the previous subsection, the transverse relaxation is simple exponential (Lorentzian lineshape) in the extreme narrowing regime, with $T_2 = T_1$. The relaxation is usually very efficient, also in this motional regime, and results in broad lines. In such cases, the inhomogeneity of the external magnetic field can often be neglected and the relaxation rate can be obtained from measuring the full width of the NMR line at half-height, $\Delta\nu$ (in Hertz), through $T_2 = 1/\pi\Delta\nu = T_1$. The transverse relaxation is also exponential for $S = 1$, as proved by Abragam (*Further reading*). In the case of an axially symmetric efg, the rate is given by:

$$\frac{1}{T_{2,Q}} = \frac{\pi}{40}\left(\frac{e^2qQ}{\hbar}\right)^2 \left(9J_2(0) + 15J_2(\omega_S) + 6J_2(2\omega_S)\right) \tag{14.19}$$

The nearly single exponential transverse relaxation for higher values of the S quantum number has also been investigated by Halle and Wennerström[5] and the transverse analogue of Eq. (14.18) can be found in the original paper. In the condition far outside of extreme narrowing, one should be aware of one more complication, the dynamic frequency shifts. The phenomenon, mentioned already in Chapter 4, is related to the imaginary parts of the spectral densities. An NMR line of a quadrupolar nucleus in an isotropic liquid is a superposition of several degenerate transitions (between different pairs of adjacent Zeeman levels, m and $m \pm 1$) or coherences. As discussed by Werbelow[2], the dynamic frequency shifts remove this degeneracy and are another source of deviations from Lorentzian lineshapes.

14.2.4 Observation and Interpretation of Multiexponential Relaxation

We now turn to the observation and interpretation of multiexponential relaxation of quadrupolar nuclei with $S > 1$. The topic has been discussed in a general case by Werbelow[2] and the specific case of $S = 3/2$ has been described in reviews by Wimperis[6], van der Maarel[7] and Chandra Shekar *et al.*[8]. The discussion of quadrupolar relaxation is most conveniently carried out using the expansion of the density operator in terms of irreducible spherical tensor operators rather than the product operators introduced in Sections 1.4 and 7.1. The density operator is in this formalism given as:

$$\hat{\rho}(t) = \sum_{l=0}^{2S} \sum_{p=-l}^{l} b_{l,p}(t) \hat{T}_{l,p} \tag{14.20}$$

The time dependence is explicitly contained in the expansion coefficients $b_{l,p}(t)$. For the $S = 3/2$ case, the first summation proceeds from $l = 0$ to $l = 3$. The irreducible tensor operators occurring in Eq. (14.20) are listed in Table 14.1.

Consider the density operator for a system of quadrupolar nuclei immediately after a 90° y-pulse. Following Jaccard *et al.*[9] and Wimperis[6], we can write:

$$\hat{\rho}(0) = \hat{S}_x = \frac{1}{\sqrt{2}}\left(\hat{T}_{1,-1} - \hat{T}_{1,+1}\right) \tag{14.21a}$$

As a result of biexponential transverse relaxation, the density operator evolves into:

$$\hat{\rho}(t) = \frac{1}{\sqrt{2}}\left[f_{11}^{(1)}(t)\left(\hat{T}_{1,-1} - \hat{T}_{1,+1}\right) + f_{31}^{(1)}(t)\left(\hat{T}_{3,-1} - \hat{T}_{3,+1}\right)\right] \tag{14.21b}$$

where the weights $f_{11}^{(1)}(t)$ and $f_{31}^{(1)}(t)$ depend on the two relaxation rate constants and time; see Jaccard *et al.*[9] and Wimperis[6] for details. The first of these functions is a sum of the "slow" and "fast" exponentials, while $f_{31}^{(1)}(t)$ is a difference of two exponential decays, which vanishes in the extreme narrowing. We want to stress that Eq. (14.21b) illustrates an important general property of relaxation processes: relaxation can change the tensor rank of a coherence, but not its order. This can also be seen in the "Redfield kite", discussed in Chapter 4. A relation similar to Eq. (14.21) can be formulated for longitudinal relaxation after an initial 180° pulse:

$$\hat{\rho}(0) - \hat{\rho}^{eq} = -2\hat{S}_z = -2\hat{T}_{1,0} \tag{14.22a}$$

$$\hat{\rho}(t) - \hat{\rho}^{eq} = -2\left(f_{11}^{(0)}(t)\hat{T}_{1,0} + f_{31}^{(0)}(t)\hat{T}_{3,0}\right) \tag{14.22b}$$

In analogy with Eqs. (14.21), the longitudinal relaxation functions $f_{11}^{(0)}(t)$ and $f_{31}^{(0)}(t)$ depend on the two longitudinal relaxation rates. Moreover, $f_{31}^{(0)}(t)$ vanishes in extreme narrowing.

The function $f_{31}^{(1)}(t)$ can be measured by a smart experimental technique using multiple-quantum (MQ) filtration. This pulse sequence is depicted in Figure 14.2 (top panel). At the end of the interval τ_e, both the rank-one ($\hat{T}_{1,\pm1}$) and rank-three single-quantum coherences ($\hat{T}_{3,\pm1}$) are present as a result

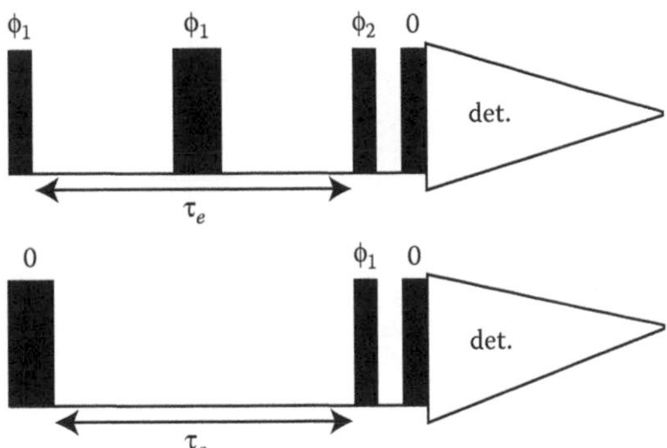

FIGURE 14.2 Pulse sequences for multiple-quantum filtered experiments for $S = 3/2$. The top panel shows the experiment for measuring transverse relaxation and the bottom panel the experiment for measuring longitudinal relaxation. The phase cycle is $\phi_1 = 30°, 90°, 150°, 210°, 270°, 330°$; $\phi_2 = 120°, 180°, 240°, 300°, 0°, 60°$; and $\phi_{rec} = 0, 180°$ for triple-quantum filtration.

of the biexponential relaxation (the 180° pulse in the middle of the interval only refocuses the possible frequency offset and does not affect relaxation). The 90° pulse converts $\hat{T}_{3,\pm1}$ into higher-order, rank-three coherences, $\hat{T}_{3,\pm2}$ and $\hat{T}_{3,\pm3}$, according to the rule that radiofrequency pulses can change the order, but not the rank, of a coherence. The triple-quantum coherence can be selected by appropriate phase cycling. The last 90° pulse reconverts the triple-quantum coherence into single-quantum coherence, $\hat{T}_{3,\pm1}$. The block containing the two final pulses, with a short interval between them, is called a *triple-quantum filter*, in analogy with what was discussed in Section 10.4. By selecting another phase cycle, a double-quantum filter can also be created. For a more general discussion of MQ filtration, the reader is referred to Ernst, Bodenhausen and Wokaun's book (*Further reading*). Note that the contributions to the final signal from the $\hat{T}_{1,\pm1}$ terms, present at the end of the τ_e period, are filtered out by the MQ filter, since this $l = 1$ term cannot be converted into $\hat{T}_{3,\pm2}$ or $\hat{T}_{3,\pm3}$. The high-rank single-quantum coherence is not directly visible, as it can be considered a single-line analogue of an antiphase multiplet (a superposition of degenerate transitions with opposite intensities). However, during the acquisition period, the transverse relaxation converts the rank-three single-quantum coherence into a mixture of rank-one and rank-three single-quantum coherences in a process analogous to that described by Eq. (14.21). The resulting FID starts at zero, builds up and eventually decays[37] (*cf.* Figure 14.3a). The corresponding spectrum is more informative (Figure 14.3b). Since it has its origin in the $f_{31}^{(1)}(t)$ function, which is a difference of two exponentials, the two components show up with opposite phases. The spectrum from a simple, single-pulse experiment is, on the other hand, dominated by $f_{11}^{(1)}(t)$ where the two Lorentzians are in phase, which makes it difficult to determine the linewidth unless the two rate constants are very different. Analogously, the sequence shown in Figure 14.2 (bottom panel) can be called a *multiple-quantum filtered inversion-recovery experiment* and allows measurements of $f_{31}^{(0)}(t)$.

Transverse relaxation of quadrupolar nuclei can also be studied by employing radiofrequency fields in the form of either multiple-echo or spin-lock experiments, related to the schemes discussed in Section 8.2 for $I = 1/2$ spins. For a detailed description of these experiments, the reader is referred to a review by van der Maarel[10].

14.2.5 Intermolecular Quadrupolar Relaxation

As was indicated at the beginning of this section, quadrupolar relaxation can have its origin in intermolecular electric field gradients, where both the efg magnitude and the orientation of the principal axis system vary with time. This occurs, for example, for monoatomic ions with quadrupolar nuclei, such as $^{23}Na^+$, $^{35}Cl^-$ or $^{79}Br^-$ (all with $S = 3/2$), in dilute aqueous solution. The electric field gradient at the site of

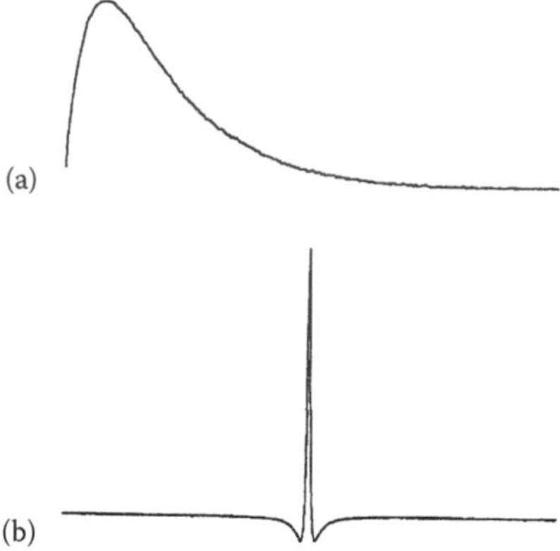

FIGURE 14.3 The FID (a) and the spectrum (b) obtained in the experiment shown in Figure 14.2 (top panel). (Reprinted with permission from Wimperis, S. and Wood, B. *J. Magn. Reson.* 95, 428–436, 1991. Copyright (1991) Elsevier.)

the nucleus in such systems fluctuates because of the motion of the surrounding water molecules and their charge distributions. A nice qualitative discussion of these phenomena can be found in a review by Woessner[11]. Since the water molecules move fast, we can safely assume that the extreme narrowing conditions pertain and formulate the relaxation rates as follows [12,13]:

$$\frac{1}{T_{1,Q}} = \frac{1}{T_{2,Q}} = \frac{3}{40} \frac{2S+3}{S^2(2S-1)} \left(\frac{eQ}{\hbar}\right)^2 \int_0^\infty \left\langle V_{zz}^{LAB}(0) V_{zz}^{LAB}(t) \right\rangle dt \tag{14.23}$$

where the time-correlation functions (tcfs) of the *zz* field gradient components, evaluated in the laboratory frame, are explicitly included. These functions can be obtained directly from combinations of quantum chemical calculations and molecular dynamics (MD) simulations, as demonstrated a long time ago by, for instance, Engström and co-workers[12]. In a somewhat simplified approach, the field gradient at the site of the nucleus can be expressed as a product of the *external field gradient*, resulting from the charges from surrounding solvent molecules, and the Sternheimer factor, $1+\gamma_\infty$, a constant characteristic for each atomic species, which describes the polarisation of the atomic charge cloud:

$$V_{\alpha\beta} = (1+\gamma_\infty) V_{\alpha\beta}^{ext} = (1+\gamma_\infty) \sum_i V_{\alpha\beta,i}^{ext} \tag{14.24}$$

In the last formulation, the external gradient (in an arbitrary frame) has been written as a sum over contributions from individual solvent molecules. The Sternheimer factor can be factored out in the correlation function in Eq. (14.23) and the field gradient correlation function can be written as a sum over individual solvent molecules and over pairs of molecules:

$$\begin{aligned} G_{zz}(t) &= \left\langle V_{zz}^{LAB}(0) V_{zz}^{LAB}(t) \right\rangle = (1+\gamma_\infty)^2 \left\langle V_{zz}^{ext}(0) V_{zz}^{ext}(t) \right\rangle \\ &= (1+\gamma_\infty)^2 \left[\sum_i \left\langle V_{zz,i}^{ext}(0) V_{zz,i}^{ext}(t) \right\rangle + \sum_{i \neq j} \left\langle V_{zz,i}^{ext}(0) V_{zz,j}^{ext}(t) \right\rangle \right] \\ &= (1+\gamma_\infty)^2 \left(G_{zz}^{auto}(t) + G_{zz}^{cross}(t) \right) \end{aligned} \tag{14.25}$$

The term $G_{zz}^{auto}(t)$ describes the correlation of the external field gradient originating from molecule i at time zero with the gradient caused by the same molecule at time t. The second term $G_{zz}^{cross}(t)$ is a cross-term in the sense that it describes the correlation between the gradients arising from charges in different molecules at different times. The integral over the full tcf in Eq. (14.23) can, in turn, be expressed as:

$$\int_0^\infty \left\langle V_{zz}^{LAB}(0)V_{zz}^{LAB}(t)\right\rangle dt = (1+\gamma_\infty)^2\left\langle \left(V_{zz}^{ext}(0)\right)^2\right\rangle \tau_Q \tag{14.26}$$

Here, $\langle (V_{zz}^{ext}(0))^2\rangle$ is the mean square fluctuation of the external gradient and τ_Q is an effective correlation time.

It is related to the integrals over $G_{zz}^{auto}(t)$ and $G_{zz}^{cross}(t)$ and can be evaluated either through a molecular dynamics simulation or in terms of molecular electrostatic models. Roberts and Schnitker[13] compared these two approaches and concluded that the common electrostatic models were not able to account for the prominent fast initial decay seen in the correlation function from the MD simulation. This decay is caused by the interplay of static field gradient cancellations and conservation of solvent-solvent cross-correlations in the hydrogen-bonded solvation sphere of the ion.

In a variety of interesting situations (*cf.* Section 14.4), monoatomic ions with quadrupolar nuclei can exchange between a bound site, where the ion is bound to a macromolecule, and a free site. The measured relaxation properties and lineshapes then reflect the inherent relaxation properties of the two sites as well as the chemical exchange dynamics (see Chapter 13).

14.3 Quadrupolar Effects on Neighbouring $I = 1/2$ Spins

The quadrupolar nuclei (S-spins) can interact with neighbouring spin 1/2 nuclei (I-spins) through both the dipole-dipole interaction and the scalar interaction (J-coupling). Usually, the dipolar interaction does not differ in any interesting way from the case described in Chapter 3. The dipole-dipole interaction with the S-spins contributes to the T_1 of the I-spins according to Eq. (3.26a). Compared with Eq. (3.31a), we only need to take into account the fact that the square of the magnetic moment of the S-spin is now proportional to $S(S+1)$, which replaces the specific constant of 3/4 corresponding to $S = 1/2$. The result is:

$$T_{1I}^{-1} = \rho_I = \tfrac{4\pi}{15} b_{IS}^2 S(S+1)\left[J(\omega_S-\omega_I)+3J(\omega_I)+6J(\omega_S+\omega_I)\right] \tag{14.27a}$$

In the same way, the dipolar contribution to the transverse relaxation rate becomes, in analogy with Eq. (3.35):

$$T_{2I}^{-1} = \tfrac{2\pi}{15} b_{IS}^2 S(S+1)\left[4J(0)+J(\omega_S-\omega_I)+3J(\omega_I)+6J(\omega_S)+6J(\omega_S+\omega_I)\right] \tag{14.27b}$$

Quadrupolar nuclei usually have an efficient relaxation of their own, which means that the cross-relaxation between the I- and S-spins is suppressed. Exceptions from this rule can occur for quadrupolar nuclei with very low quadrupole moments, such as the lithium isotopes, ^7Li and, in particular, ^6Li.

The efficient quadrupolar relaxation of the S-spins can have other consequences, not taken into consideration in the formulation of Eq. (14.27). This issue is thoroughly discussed in the book by Kruk (*Further reading*). If the S-spin relaxation can be described within the Redfield theory, it is relatively straightforward to include the S-spin relaxation effects and we shall discuss this topic for the similar case of S being the electron spin in Chapter 15. Here, we want to mention the phenomenon known as *quadrupolar relaxation enhancement* (QRE). It is observed in the magnetic field (B_0) dependence of the spin-lattice relaxation rate ("nuclear magnetic relaxation dispersion" (NMRD) experiments) for spin 1/2 nuclei (usually protons) dipole-coupled to quadrupolar spins. NMRD measurements have been mentioned earlier in this book, in Sections 3.5 and 8.1. The diagrams of relaxation rate versus B_0 (or I-spin Larmor frequency) can show sharp maxima (QRE or quadrupolar peaks) when the spin 1/2 Larmor frequency corresponds to the quadrupolar energy level differences, if the rotational motion is slow enough to bring the quadrupolar relaxation beyond the Redfield limit. The theory of QRE is an active field of research; see, for example, some recent papers from various groups[14–16]. Figure 14.4, taken from Fries and Belorizky[16], shows the predicted ^{1}H NMRD profiles for water protons interacting with amide ^{14}N ($I = 1$) nuclei in the immobilised protein bovine pancreatic trypsin inhibitor (BPTI, solid line, $\tau_c = 1.06\,\mu$s) and in a hypothetical, slowly tumbling molecule (dashed lines, $\tau_c = 500$, 100 ns). Note that the QRE disappears for the shortest correlation time.

The effects of quadrupolar spins on relaxation of spin 1/2 nuclei, mediated through the J-couplings, are more interesting in the context of liquid state NMR on small molecules. In the context of the modulation of the scalar coupling interaction through chemical exchange (scalar relaxation of the first kind), discussed in Section 13.1, we have already mentioned that a similar phenomenon can arise when the same interaction is modulated by the rapid relaxation of the S-spin. Abragam (*Further reading*) called this phenomenon *scalar relaxation of the second kind*. Let us assume first that the relaxation of the longitudinal and transverse components of the S-spin is monoexponential, with uniquely defined $T_{1S} = T_{1,Q}$ and $T_{2S} = T_{2,Q}$, much smaller than $(2\pi J_{IS})^{-1}$. The relevant time correlation functions determining the scalar relaxation of the I-spin are in this case:

$$G_z(\tau) = \left(2\pi J_{IS}\right)^2 \left\langle \hat{S}_z(0)\hat{S}_z(\tau)\right\rangle = \tfrac{1}{3}\left(2\pi J_{IS}\right)^2 S(S+1)\exp(-\tau/T_{1S}) \tag{14.28a}$$

and

$$G_\pm(\tau) = \tfrac{1}{4}\left(2\pi J_{IS}\right)^2 \left\langle \hat{S}_\pm(0)\hat{S}_\mp(\tau)\right\rangle = \tfrac{1}{6}\left(2\pi J_{IS}\right)^2 S(S+1)\exp(-\tau/T_{2S})\exp(i\omega_S\tau) \tag{14.28b}$$

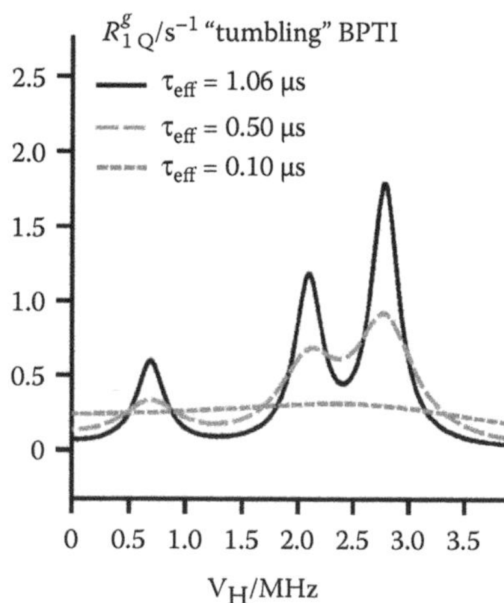

FIGURE 14.4 Predicted ^{1}H NMRD profiles for water protons interacting with amide ^{14}N ($I = 1$) nuclei in the immobilised protein BPTI (solid line, $\tau_c = 1.06\,\mu$s) and in a hypothetical, slowly tumbling molecule (dashed lines, $\tau_c = 500$ ns, 100 ns). Pay attention to the fact that QRE disappears for the shortest correlation time. (Reproduced with permission from Fries, P.H., and Belorizky, E., *J. Chem. Phys.*, 143, 044202, 2015. Copyright (2015), American Institute of Physics.)

which give rise to the relaxation rate expressions:

$$T_{1I,SC}^{-1} = \frac{2(2\pi J_{IS})^2 S(S+1)}{3} \frac{T_{2S}}{1+(\omega_I - \omega_S)^2 T_{2S}^2} \tag{14.29a}$$

$$T_{2I,SC}^{-1} = \frac{(2\pi J_{IS})^2 S(S+1)}{3} \left[T_{1S} + \frac{T_{2S}}{1+(\omega_I - \omega_S)^2 T_{2S}^2} \right] \tag{14.29b}$$

The scalar contributions to the longitudinal I-spin relaxation are rarely of importance. An interesting exception to this rule is carbon-13 relaxation in brominated compounds. ^{79}Br happens to have a magnetogyric ratio very close to that of carbon-13 (see Table 1.1). This means that the difference between the Larmor frequencies in the denominator of Eq. (14.29a) is very small and the scalar contribution to $T_{1,obs}^{-1}$ (compare Eq. (5.52)) becomes reasonably efficient. An experimental example of this phenomenon was reported by Levy, who measured carbon-13 relaxation in bromobenzene[17].

Quadrupolar effects on the transverse I-spin relaxation and lineshape are much more common. Eq. (14.29b) is valid in the limit of fast relaxation of the quadrupolar spin, but it is interesting and instructive to discuss a more general case. Consider an I-spin directly bonded and J-coupled to an $S = 1$ spin, which relaxes monoexponentially but not necessarily very fast. If the relaxation of the S-spin is slow, we expect the I-spin spectrum to be a triplet of lines at frequencies $\omega_I + 2\pi J_{IS} m_S$, with $m_S = 1, 0, -1$. Calculation of the effects of the quadrupolar relaxation of the S-spin (in the extreme narrowing regime) on the lineshape for the I-spin is a classical problem, solved a long time ago by Pople[18]. A detailed description of the solution can be found in the book by Abragam (*Further reading*). Briefly, the time-domain NMR signal is obtained by taking the trace of the product of the density operator and the $\hat{I}_-$ operator (compare Eq. (1.17)):

$$\langle \hat{I}_- \rangle (t) = \mathrm{Tr}\{\hat{\rho}(t)\hat{I}_-\} \tag{14.30}$$

The time evolution of the density operator is obtained by re-writing Eq. (4.33):

$$\frac{d}{dt}\rho_m(t) = \sum_n (i\mathbf{\Omega} + \mathbf{R})_{m,n}\rho_n(t) = \sum_n A_{m,n}\rho_n(t) \tag{14.31}$$

The symbol $\rho_m(t)$ denotes here a specific coherence (m and n replace the pairs of indices $\alpha\alpha'$ and $\beta\beta'$ in Eq. (4.33)), $\mathbf{\Omega}$ is a diagonal matrix containing the transition frequencies (compare $\alpha' - \alpha$ in Eq. (4.33)) and $\mathbf{R}$ is the relaxation supermatrix.

In the last equality, we introduce the square and symmetric matrix $\mathbf{A}$. Eq. (14.31) has a formal solution:

$$\rho_m(t) \sum_n \rho_n(0) \left[\exp(\mathbf{A}t) \right]_{m,n} \tag{14.32}$$

which results in the time-dependent signal (FID) given by:

$$\langle \hat{I}_- \rangle (t) = \mathrm{Tr}\{\hat{\rho}(t)\hat{I}_-\} = \sum_{m,n} \rho_n(0) \left[\exp(\mathbf{A}t) \right]_{m,n} \left(\hat{I}_- \right)_m \tag{14.33}$$

The symbols $(\hat{I}_-)_m = (\hat{I}_-)_{\alpha\alpha'}$ are the matrix elements of the $\hat{I}_-$ operator. In order to get the lineshape function, $F(\omega)$, we need to Fourier transform the FID:

$$F(\omega) = 2\mathrm{Re}\int_0^\infty \langle \hat{I}_- \rangle (t) \exp(-i\omega t)\, dt = 2\mathrm{Re}\sum_{m,n} \rho_n(0) \left(\tilde{\mathbf{A}}^{-1}(\omega) \right)_{m,n} \left(\hat{I}_- \right)_m \tag{14.34}$$

where $\tilde{\mathbf{A}} = \mathbf{A} + \omega\mathbf{1}$ (the frequency ω is added to the diagonal elements of the matrix $\mathbf{A}$). We deal here with the three coherences corresponding to the three I-spin transitions with $m_S = 1, 0, -1$. Since the coherences m, or the pairs of states $\beta\beta'$, differ only in the m_S quantum number, the matrix elements of the $\hat{I}_-$ operator must be the same for all three coherences. Also, the initial values, proportional to *a priori* probabilities of the three transitions, are equal. Therefore, the lineshape expression simplifies to:

$$F(\omega) \propto \mathrm{Re}\sum_{m,n}\left(\tilde{\mathbf{A}}^{-1}(\omega)\right)_{m,n} \tag{14.35}$$

Neglecting contributions to the I-spin relaxation from all other relaxation mechanisms, Abragam (*Further reading*) gives the following expression for the matrix elements $\tilde{A}_{m,n}(\omega)$:

$$\tilde{A}_{m,n}(\omega) = \left[i(\omega_I - \omega + m_S J_{IS}) - R_{m,m}\right]\delta_{m,n} + R_{m,n} \tag{14.36}$$

where the relaxation supermatrix elements $R_{m,n}$ are equal to the transition probabilities $W_{m,n}$ in Eqs. (14.13) and (14.14) and $R_{m,m} = \sum_{n \neq m} R_{m,n}$. We stress once more that the transition probabilities refer to the transitions between different states of the S-spin and are thus related to the quadrupolar spin-lattice relaxation rate. In the extreme narrowing, the 3×3 matrix takes the form:

$$\tilde{\mathbf{A}}(\omega) = \begin{bmatrix} i(\omega_I - \omega + J_{IS}) - \dfrac{3}{5T_{1,Q}} & \dfrac{1}{5T_{1,Q}} & \dfrac{2}{5T_{1,Q}} \\ \dfrac{1}{5T_{1,Q}} & i(\omega_I - \omega) - \dfrac{2}{5T_{1,Q}} & \dfrac{1}{5T_{1,Q}} \\ \dfrac{2}{5T_{1,Q}} & \dfrac{1}{5T_{1,Q}} & i(\omega_I - \omega - J_{IS}) - \dfrac{3}{5T_{1,Q}} \end{bmatrix} \tag{14.37}$$

The reader is encouraged to check that Eq. (14.36) indeed leads to Eq. (14.37) under extreme narrowing conditions. By inverting the matrix $\tilde{\mathbf{A}}(\omega)$, one obtains a lineshape function $F(x)$ described by the parameters $w = 10\pi J_{IS} T_{1,Q}$ and $x = (\omega - \omega_I)/2\pi J_{IS}$, originally derived by Pople[18] and publicised in Abragam's book:

$$F(x) = \frac{45 + w^2(5x^2 + 1)}{225x^2 + w^2(34x^4 - 2x^2 + 4) + w^4x^2(x^4 - 2x^2 + 1)} \tag{14.38}$$

The lineshapes for varying values of w are shown in Figure 14.5. Starting from a high w value (a long quadrupolar relaxation time or a fast rotational motion), a notable feature is the differential height of the triplet components. Actually, the integrated intensity of the three lines is equal but the widths differ, because the transition probabilities for $\Delta m = \pm 1$ differ from those for $\Delta m = \pm 2$. At the other end (which can perhaps be reached by lowering the temperature or increasing the viscosity of the sample), when the relaxation rate of the S-spin is much higher than $2\pi J_{IS}$, the coupling is averaged out and the I-spin signal is a single Lorentzian, with the width corresponding to $T_{2,SC}^{-1}$ of Eq. (14.29b). The approach based on Eqs . (14.30)–(14.36) is also valid outside of the extreme narrowing range for the quadrupolar spin. Werbelow and Kowalewski argued that invoking "scalar relaxation of the second kind" as a distinct relaxation mechanism is indeed unnecessary and that the same effect can be obtained by setting up the $\tilde{\mathbf{A}}(\omega)$ matrix in an appropriate basis and using time-independent perturbation theory[19].

The simple J-coupled two-spin ($I = 1/2$, $S = 1$) system contains, in fact, even more interesting and non-trivial features. When studying the carbon spectra of the $^{13}C^{\alpha}$-^{2}H system in a glycine residue in a

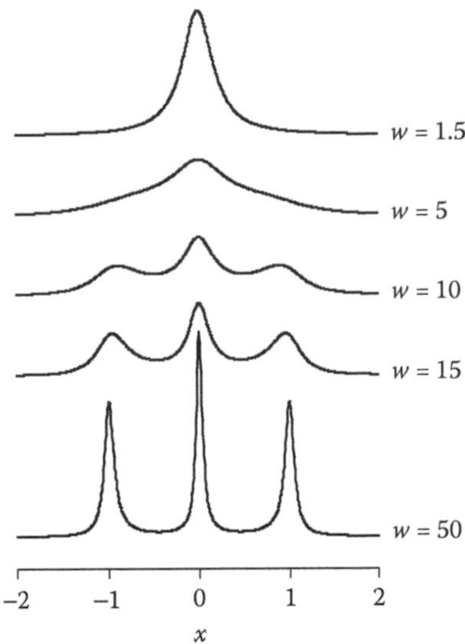

FIGURE 14.5 Lineshapes of an $I = 1/2$ spin coupled to an $S = 1$ nucleus. Simulations based on Eq. (14.38) using different values of w.

small protein, Kushlan and LeMaster noticed a correlation-time dependent asymmetry of the carbon triplet, deviating from the simple picture of Figure 14.5[20]. This observation was explained, independently, by London *et al.*[21] and by Grzesiek and Bax[22], who ascribed it to the occurrence of a dynamic frequency shift arising through cross-correlation of the quadrupolar interaction with the carbon-deuteron dipolar interaction. The phenomenon shifts the frequency of the outer lines of the triplet in one direction and the inner line in the other direction, which results in an asymmetric triplet. The effect is largest when the correlation times are long and the system is far out of extreme narrowing. Simulated lineshapes from the paper by Grzesiek and Bax[22], based on typical J_{CD} and QCC values and the magnetic field of 14.1 T, are shown in Figure 14.6. It should be noted that only the three shortest correlation times correspond to the extreme narrowing and are thus analogous to Figure 14.5. These ideas have also been extended to the case of a carbon-13 nucleus bound to two or three deuterons[23,24]. In addition to the dynamic frequency shifts, the analysis of the carbon-13 spectral lineshapes in this type of system requires consideration of one more phenomenon: the pair-wise cross-correlation of the quadrupolar interactions for different deuterons.

As a final topic in this section, we wish to mention a related case, that of two non-equivalent spin 1/2 nuclei (A- and M-spins) J-coupled to one (or more) rapidly relaxing quadrupolar spins (S). Under certain conditions, the scalar relaxation of the second kind originating from the coupling with S may cause significant broadenings in the spectra of both spin 1/2 nuclei, according to Eq. (14.29b). Following the pioneering work of Wokaun and Ernst[25] on multiple-quantum relaxation, Kowalewski and Larsson studied the scalar relaxation effects for the linewidths of the zero- and double-quantum coherences for the AM part of that AMS spin system[26]. They found that the two linewidths differed from each other by a term proportional to the cross-correlated spectral density, involving the AS and MS scalar interactions. They were able to determine the relative signs of ${}^1J_{SiCl}$ and ${}^2J_{HCl}$ in trichlorosilane, $SiHCl_3$ (with $A = {}^1H$, $M = {}^{29}Si$, $S = {}^{35}Cl$) from the linewidths of the zero-, single- and double-quantum coherences. It may be worthwhile to point out that this work has a certain similarity to the case of cross-correlated modulation of another isotropic interaction – isotropic chemical shifts – giving rise to a difference between the zero-quantum and the double-quantum relaxation rates in a system of two spin 1/2 nuclei, discussed in Section 13.1.

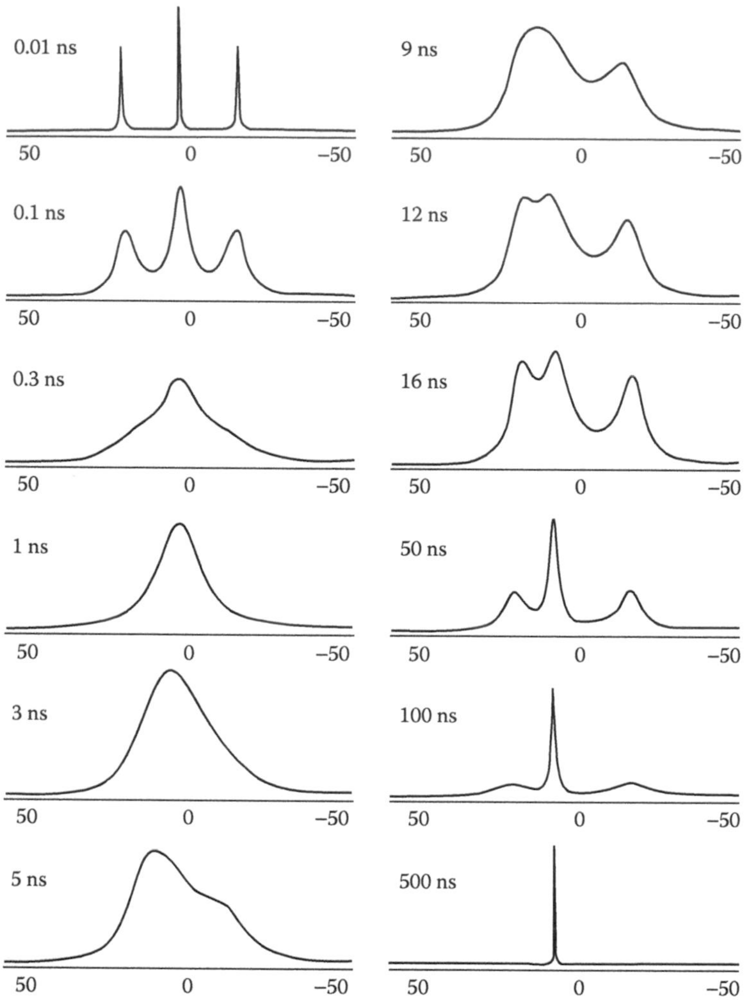

FIGURE 14.6 Simulated lineshapes for an $I = 1/2$ spin coupled to an $S = 1$ nucleus at the magnetic field of 14.1 T, based on $J_{IS} = 20$ Hz, QCC = 170 kHz and the indicated correlation times. (Reprinted with permission from Grzesiek, S, and Bax, A., *J. Am. Chem. Soc.*, 116, 10196–10201. Copyright (1994) American Chemical Society.)

14.4 Applications of Relaxation Effects of Quadrupolar Nuclei

We finish this chapter by describing some examples of applications of quadrupolar (or quadrupolar-related) relaxation. The relaxation of deuterons is, in a certain sense, similar to that of carbon-13 and some applications of deuteron measurements to molecular dynamics studies in small organic molecules, as well as in multiple isotope-labelled proteins, were mentioned in Chapter 11. We refer the reader to that chapter for such discussions. Here, we present some examples of a different kind.

14.4.1 Protein Hydration

The first group of examples is concerned with studies of protein hydration by variable-field longitudinal relaxation measurements. The proton NMRD studies of protein hydration are complicated by the interplay of chemical exchange and intermolecular dipolar relaxation processes[27]. Halle and co-workers have argued that "cleaner" information, which is more easily interpreted, can be obtained by field-dependent studies of spin-lattice relaxation of the quadrupolar nuclei of (isotope-enriched) water: oxygen-17 and deuterons. The oxygen-17 spin-lattice relaxation is unaffected by hydrogen exchange and the field-dependent measurements report exclusively on the dynamic behaviour of water molecules in the protein solution (including the exchange of water molecules between different sites), while the case of deuterons is a little more complicated. As examples, we shall discuss results for solutions of two small and very well-studied globular proteins, BPTI and ubiquitin, examples taken from work by Denisov and Halle[28,29].

The oxygen-17 field-dependent R_1 and R_2 data for BPTI show characteristic profiles, with a low-field plateau, a dispersion step (which means a significant decrease of the relaxation rate over a relatively narrow frequency range) and high-field values in excess of the relaxation rates in bulk water (Figure 14.7a). The field dependence of the spin-lattice relaxation rate can in the simplest case be described by[28]:

$$R_1(\omega_S) = R_{bulk} + \alpha + \beta\tau_c\left(\frac{0.2}{1+\omega_S^2\tau_c^2} + \frac{0.8}{1+4\omega_S^2\tau_c^2}\right) \quad (14.39)$$

A similar equation can be formulated for $R_2(\omega_S)$. The parameters in this equation are illustrated in Figure 14.7b. The rationale of this expression is based on the perturbation theory description of multiexponential relaxation (*cf.* Section 14.2). The effective correlation time, τ_c, is determined by the rotational correlation time of the macromolecule and the exchange lifetimes of various possible water sites on the surface or within the protein. A high dispersion step (large β) proves that at least some of the water molecules experience a long correlation time, on the order of 10^{-8} s or longer. Finally, a sizable α value indicates that a significant share of water molecules have their motion perturbed by the macromolecule. The most important feature in the oxygen-17 NMRD profiles for BPTI – the high β-value – is interpreted in terms of a contribution from highly ordered internal water molecules, buried within the protein and characterised by long correlation times, known to be present within the BPTI structure. This feature is practically absent in the case of ubiquitin, which has no such waters. For both proteins, the authors repeated their oxygen-17 measurements as a function of pD (the analogue of pH in deuterated water solutions) and found only a weak dependence of the profiles on that quantity. In the second paper, Denisov and Halle studied deuteron NMRD profiles in the BPTI and ubiquitin solutions in D_2O[29]. Here, the pD dependence was much stronger and the data could be interpreted in terms of known pK-values

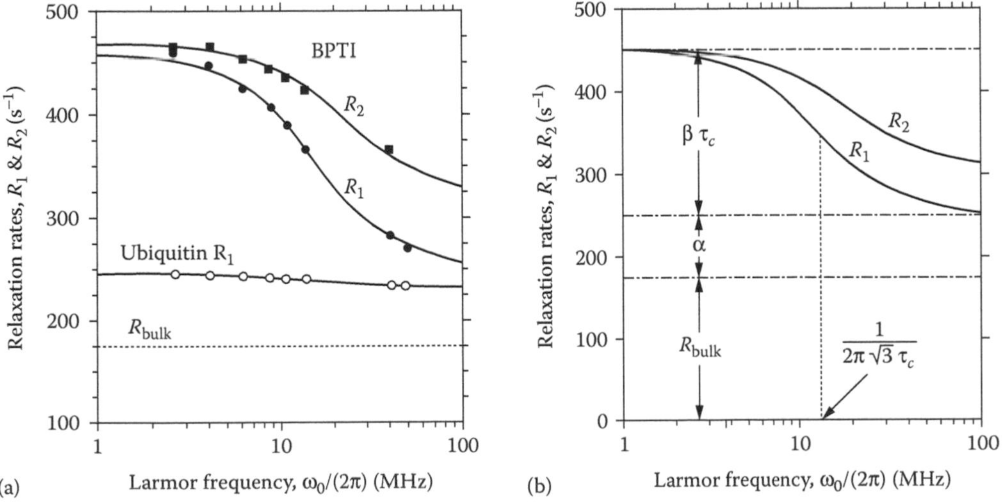

FIGURE 14.7 NMRD profiles for aqueous oxygen-17. (a) Data for BPTI and ubiquitin in aqueous solution. (b) Definition of the parameters in Eq. (14.39). (Reprinted with permission from Denisov, V. P. and Halle, B., *J. Mol. Biol.*, 245, 682–697, 1995. Copyright (1995) Elsevier.)

and hydrogen exchange rate constants. The NMRD work on BPTI and ubiquitin was followed up by Mattea *et al.*[30], who reported a variable-temperature, high-field oxygen-17 spin-lattice relaxation study of the two molecules in deeply supercooled (down to 238 K) aqueous solutions, stabilised in the form of emulsions in heptane. The main conclusions from this work were as follows: (1) the protein hydration layer is characterised by strong dynamic heterogeneity, which can be described by a power-law distribution of correlation times; (2) the long correlation time tail of the distribution is associated with a small, protein-specific population of strongly perturbed, slowly rotating water molecules; (3) the majority of hydration layer water molecules show a weak dynamic perturbation (slowed down rotation), similar for all proteins; (4) in the vicinity of room temperature, the hydration-layer-average correlation time has a stronger temperature dependence than the bulk water correlation time, while the opposite is true at low temperatures. The Halle group has reported numerous studies of hydration of other biomolecules; some of the work has been described in reviews[31,32].

14.4.2 Simple Ions in Micro-Heterogeneous Environment

The second group of examples concerns quadrupolar relaxation in simple ions in micro-heterogeneous but macroscopically isotropic solutions. The first case we wish to describe, again taken from Denisov and Halle[33], is ^{23}Na ($S = 3/2$) relaxation in B-DNA oligonucleotides. They measured the slow and fast spin-lattice and spin-spin relaxation rates for the sodium nuclei at 10 different magnetic fields (some of which differed from each other by a factor of 2, which makes the analysis a little simpler, since $J(2\omega_0)$ at one field is identical to $J(\omega_0)$ at the double field). The measurements provided spectral density functions at 14 different frequencies (including $J(0)$), which are displayed for several oligonucleotides at 4°C and pH 7.0 in Figure 14.8. The spectral densities depended strongly on the nucleotide sequence; the frequency dependence could in each case be represented by a sum of three Lorentzian components plus a frequency-independent contribution. The lowest-frequency Lorentzian, which corresponded to the sodium ions most closely associated with the DNA, was found to be most dependent on the sequence. The low-frequency Lorentzian was analysed to provide a correlation time for the sodium ions in the range of 20–40 ns. These values are in rough agreement with the rotational correlation time of the macromolecule but can also contain contributions from the exchange lifetimes out of the binding site. The other two Lorentzian components, with correlation times in the range of 1–7 ns, were related to the diffusion or jumps of the ions over the DNA surface. The conclusion of the study was that sodium ions

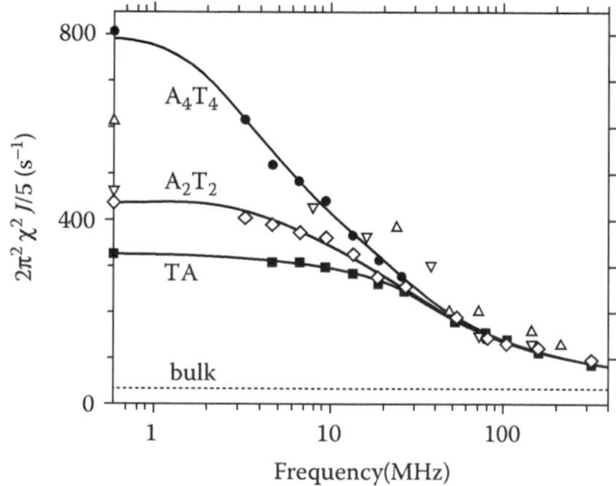

FIGURE 14.8 Maps of spectral densities obtained from ^{23}Na relaxation data in different DNA dodecamers at 277 K and pH 7. Data are shown for three different oligonucleotide sequences consisting of repeating units of A_4T_4, A_2T_2 and TA. (Reprinted with permission from Denisov, V. P. and Halle, B., *Proc. Natl. Acad. Sci. USA*, 97, 629–633, 2000. Copyright (2000) National Academy of Sciences, U.S.A.)

show specific binding to the minor groove AT tract of B-DNA. Chemically different, but NMR-wise similar, measurements were reported by Hedin and Furo for ^{81}Br in bromide ions ($S = 3/2$) in a micellar solution of hexadecyl-trimethylammonium bromide (CTAB)[34]. They measured effective spin-lattice and spin-spin relaxation rates for bromide-81 nuclei and interpreted their results in terms of the following dynamic model. The positively charged micellar surface causes a slight average deformation of the hydration shell of the bromide ions, which results in a non-zero quadrupolar coupling, persisting on a time scale of about 100 ps and with the principal direction coinciding with the local normal to the micellar surface. The residual coupling is further averaged to zero by the ion diffusion on the micellar surface and isotropic tumbling of the aggregates.

Metal ion–macromolecule interactions have also been studied by multiple-quantum filtration techniques. Chung and Wimperis studied ^{23}Na in bovine albumin solutions[35]. The ions were assumed to exchange rapidly between the free and bound sites. The fast and slow transverse relaxation rates (measured at a single magnetic field) were interpreted in terms of a correlation time for the bound ions and an effective exchange-averaged quadrupolar coupling constant, $(p_b/p_f)^{1/2}(e^2qQ/\hbar)$, where p_b and p_f are the populations of the bound and free ions. The MQ-filtered longitudinal relaxation experiments were found less informative, since they are most sensitive to shorter values of the correlation time. In another paper, the same authors investigated divalent ^{25}Mg nuclei ($S = 5/2$) in bovine albumin solutions[36]. Here, the theory of spin 5/2 relaxation was presented together with a variety of MQ-filtered and MQ-relaxation experiments. Again, the transverse relaxation data were found to be more informative than the longitudinal relaxation results. The results were interpreted in a similar way as for the ^{23}Na case. The correlation time for ^{25}Mg was found to be an order of magnitude longer than for sodium. Both the correlation time and the effective quadrupolar coupling constant determined from various experiments were reasonably consistent.

14.4.3 The $^{13}CD_3$ System

The final example concerns effects of quadrupolar nuclei – deuterons – on carbon-13 relaxation and lineshape in the $^{13}CD_3$ group in doubly isotope-labelled DL-alanine (Figure 14.9) dissolved in a cryo-mixture of D_2O and DMSO-d_6, studied by Bernatowicz and co-workers[24]. The quadrupolar relaxation

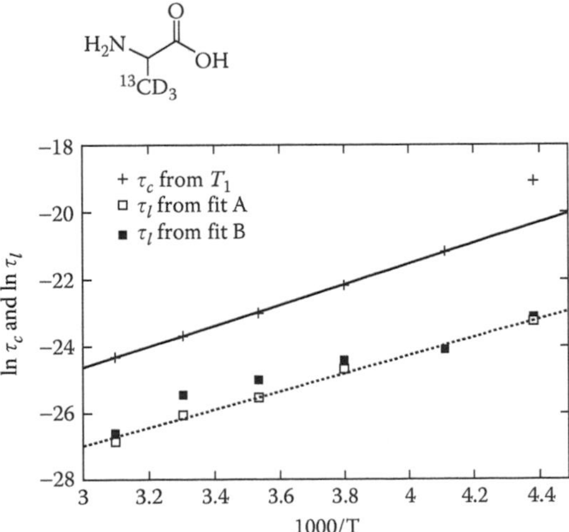

FIGURE 14.9 Arrhenius plot of logarithms of correlation times versus inverse temperature for DL-alanine labelled with ^{13}C and D in the methyl group. The open and filled squares show the internal correlation time for three-fold methyl jumps corresponding to slightly different fitting of the experimental data. Crosses show the overall rotational correlation time. (Reprinted with permission from Bernatowicz P. et al., *J. Phys. Chem. A*, 108, 9018–9025, 2004. Copyright (2004) American Chemical Society.)

of deuterons reflects a superposition of two motions: the overall reorientation of the molecule as a whole and the internal rotation of the methyl group. In this study, the authors assumed that the former motion is isotropic small-step rotational diffusion, with a correlation time that can be determined from the dipolar carbon-13 relaxation of the C^{α} carbon. Using this overall rotational correlation time, τ_c, it was possible to analyse the carbon-13 lineshapes from simple one-dimensional spectra, as well as from Carr–Purcell spin-echo experiments, and to obtain the rates of the internal motion, modelled as random 120° jumps with a characteristic time τ_I. The measurements were carried out over a broad temperature range (228–323 K), which was used to calculate the Arrhenius activation energy of 21.9 ± 1.2 kJ mol^{-1} for the jump motion. The temperature dependence of the jump rate is shown in Figure 14.9.

Exercises for Chapter 14

1. Prove that Eq. (14.36) leads to Eq. (14.37) under extreme narrowing conditions.

References

1. Bowden, G. J.; and Hutchison, W. D., Tensor operator formalism for multiple-quantum NMR. 1. Spin-1 nuclei. *J. Magn. Reson.* 1986, 67, 403–414.
2. Werbelow, L. G., *Relaxation Theory for Quadrupolar Nuclei*. Wiley, eMagRes, 2011.
3. Man, P. P., *Quadrupolar Interactions*. Wiley, eMagRes, 2011.
4. Ainbinder, N. E.; and Shaposhnikov, I. G., Transient phenomena in nuclear quadrupole resonance. *Adv. Nucl. Quadr. Reson.* 1978, 3, 67–130.
5. Halle, B.; and Wennerström, H., Nearly exponential quadrupolar relaxation. A perturbation treatment. *J. Magn. Reson.* 1981, 44, 89–100.
6. Wimperis, S., *Relaxation of Quadrupolar Nuclei Measured via Multiple Quantum Filtration*. Wiley: eMagRes, 2011.
7. Van der Maarel, J. R. C., Thermal relaxation and coherence dynamics of spin 3/2. I. Static and fluctuating quadrupolar interaction in the multipole basis. *Concepts Magn. Reson.* 2003, 19A, 97–116.
8. Chandra Shekar, S.; Tang, J. A.; and Jerschow, A., Dynamics of I=3/2 nuclei in isotropic slow motion, anisotropic and partially ordered phases. *Concepts Magn. Reson.* 2010, 36A, 362–387.
9. Jaccard, G.; Wimperis, S.; and Bodenhausen, G., Multiple-quantum NMR spectroscopy of S=3/2 spins in isotropic phase: A new probe for multiexponential relaxation. *J. Chem. Phys.* 1986, 85, 6282–6293.
10. Van der Maarel, J. R. C., Thermal relaxation and coherence dynamics of spin 3/2. II. Strong radiofrequency field. *Concepts Magn. Reson.* 2003, 19A, 117–133.
11. Woessner, D. E., NMR relaxation of spin-3/2 nuclei: Effects of structure, order, and dynamics in aqueous heterogeneous systems. *Concepts Magn. Reson.* 2001, 13, 294–325.
12. Engström, S.; Jönsson, B.; and Impey, R. W., Molecular dynamics simulation of quadrupole relaxation of atomic ions in aqueous solution. *J. Chem. Phys.* 1984, 80, 5481–5486.
13. Roberts, J. E.; and Schnitker, J., Ionic quadrupolar relaxation in aqueous solution - dynamics of the hydration sphere. *J. Phys. Chem.* 1993, 97, 5410–5417.
14. Westlund, P.-O., Quadrupole-enhanced proton spin relaxation for a slow reorienting spin pair: (I)-(S). A stochastic Liouville approach. *Mol. Phys.* 2009, 107, 2141–2148.
15. Kruk, D.; Kubica, A.; Masierak, W.; Privalov, A. F.; Wojciechowski, M.; and Medycki, W., Quadrupole relaxation enhancement - application to molecular crystals. *Solid State NMR* 2011, 40, 114–120.
16. Fries, P. H.; and Belorizky, E., Simple expressions of the nuclear relaxation rate enhancement due to quadrupole nuclei in slowly tumbling molecules. *J. Chem. Phys.* 2015, 143, 044202.
17. Levy, G. C., Carbon-13 spin-lattice relaxation: Carbon-bromine scalar and dipole-dipole interactions. *J. Chem. Soc. Chem. Commun.* 1972, 352–354.

18. Pople, J. A., Effect of quadrupolar relaxation on nuclear magnetic resonance multiplets. *Mol. Phys.* 1958, 1, 168–174.
19. Werbelow, L. G.; and Kowalewski, J., Nuclear spin relaxation of spin one-half nuclei in the presence of neighboring higher-spin nuclei. *J. Chem. Phys.* 1997, 107, 2775–2781.
20. Kushlan, D. M.; and LeMaster, D. M., H-1-detected NMR relaxation of methylene carbons via stereoselective and random fractional deuteration. *J. Am. Chem. Soc.* 1993, 115, 11026–11027.
21. London, R. E.; LeMaster, D. M.; and Werbelow, L. G., Unusual NMR multiplet structures of spin-1/2 nuclei coupled to spin-1 nuclei. *J. Am. Chem. Soc.* 1994, 116, 8400–8401.
22. Grzesiek, S.; and Bax, A., Interference between dipolar and quadrupolar interactions in the slow tumbling limit: A source of line shift and relaxation in H-2-labeled compounds. *J. Am. Chem. Soc.* 1994, 116, 10196–10201.
23. Bernatowicz, P.; Kruk, D.; Kowalewski, J.; and Werbelow, L. G., C-13 NMR lineshapes for the C-13-H-2-H-2' isotopomeric spin grouping. *ChemPhysChem.* 2002, 3, 933–938.
24. Bernatowicz, P.; Kowalewski, J.; Kruk, D.; and Werbelow, L. G., C-13 NMR line shapes in the study of dynamics of perdeuterated methyl groups. *J. Phys. Chem. A* 2004, 108, 9018–9025.
25. Wokaun, A.; and Ernst, R. R., The use of multiple quantum transitions for relaxation studies in coupled spin systems. *Mol. Phys.* 1978, 36, 317–341.
26. Kowalewski, J.; and Larsson, K., Scalar relaxation of heteronuclear multiple-quantum coherences and relative signs of nuclear spin-spin coupling constants. *Chem. Phys. Lett.* 1985, 119, 157–161.
27. Venu, K.; Denisov, V. P.; and Halle, B., Water H-1 magnetic relaxation dispersion in protein solutions. A quantitative assessment of internal hydration, proton exchange, and cross-relaxation. *J. Am. Chem. Soc.* 1997, 119, 3122–3134.
28. Denisov, V. P.; and Halle, B., Protein hydration dynamics in aqueous solution: A comparison of bovine pancreatic trypsin inhibitor and ubiquitin by oxygen-17 spin relaxation dispersion. *J. Mol. Biol.* 1995, 245, 682–697.
29. Denisov, V. P.; and Halle, B., Hydrogen exchange and protein hydration: The deuteron spin relaxation dispersion of bovine pancreatic trypsin inhibitor and ubiquitin. *J. Mol. Biol.* 1995, 245, 698–709.
30. Mattea, C.; Qvist, J.; and Halle, B., Dynamics at the protein-water interface from O-17 spin relaxation in deeply supercooled solutions. *Biophys. J.* 2008, 95, 2951–2963.
31. Halle, B.; and Denisov, V. P., Magnetic relaxation dispersion studies of biomolecular solutions. *Methods Enzymol.* 2001, 338, 178–201.
32. Halle, B., Protein hydration dynamics in solution: A critical survey. *Phil. Trans. R. Soc. Lond. B* 2004, 359, 1207–1223.
33. Denisov, V. P.; and Halle, B., Sequence-specific binding of counterions to B-DNA. *Proc. Natl. Acad. Sci. USA* 2000, 97, 629–633.
34. Hedin, N.; and Furo, I., Fast diffusion of Br- ions on a micellar surface. *J. Phys. Chem. B* 1999, 103, 9640–9644.
35. Chung, C. W.; and Wimperis, S., Optimum detection of spin-3/2 biexponential relaxation using multiple-quantum filtration technique. *J. Magn. Reson.* 1990, 88, 440–447.
36. Chung, C. W.; and Wimperis, S., Measurement of spin-5/2 relaxation in biological and macromolecular systems using multiple-quantum NMR techniques. *Mol. Phys.* 1992, 76, 47–81.
37. Wimperis, S; and Wood, B., Triple quantum sodium imaging. *J. Magn. Reson.* 1991, 95, 428–436.

15

Nuclear Spin Relaxation in Paramagnetic Systems in Solution

Paramagnetic systems are materials with positive magnetic susceptibility, associated with unpaired electrons. Paramagnetic solutions of interest to chemists usually contain free radicals or transition metal complexes, including metalloproteins. In addition, the paramagnetism of certain species with an electronic triplet ground state, such as the oxygen molecule, sometimes needs to be considered. To a certain extent, the unpaired electron spins behave in a similar way to the nuclear spins. The differences – and the profound influence of unpaired electron spins on nuclear magnetic resonance (NMR) spectra – are in the first place related to the large value of the electronic magnetic moment, about 650 times that of the proton. In the second place, we need to consider both the strong coupling of the electron spin to other degrees of freedom and the delocalised nature of the electron.

The paramagnetic species influence NMR spectra of liquids in several ways and a general account of the various phenomena can be found in the book by Bertini and co-workers[1]. The effect of greatest interest for the present book is the fact that the nuclear spin-relaxation rates are enhanced. The paramagnetic relaxation enhancement (PRE) is caused by a random variation of the electron spin–nuclear spin interactions, which open new pathways for longitudinal as well as transverse relaxation. The PRE is most often studied for $I=1/2$ nuclei, but oxygen-17 applications are not uncommon. The development of the theory of the PRE has been subject to several reviews[2–4]. Another paramagnetic effect is that the NMR signals may be shifted, provided that the relevant electron spin–nuclear spin interaction has a non-zero average. The spin-spin splittings may also be affected. In addition, paramagnetic complexes in solution often contain exchangeable ligands, and the exchange phenomena, together with the intrinsic relaxation, shift and splitting properties, contribute to the observed lineshapes and line positions in one-dimensional as well as in multidimensional NMR spectra. The paramagnetic effects have been used for several decades by chemists and biochemists as a source of structural, thermodynamic and dynamic information. Numerous applications of this type have been described in the book by Bertini and co-workers[1].

We begin this chapter by summarising some important characteristics of the electron spin and its interactions and relaxation properties. We proceed by discussing the effects of paramagnetic species in solution on the simple NMR relaxation properties T_1 and T_2. This is followed by a section describing paramagnetic influences on nuclear multispin phenomena. Next, we discuss the phenomenon of dynamic nuclear polarisation (DNP) in liquids and we conclude the chapter by describing some illustrative applications.

15.1 Electron Spin Interactions and Relaxation in Liquids

In a similar way as nuclear spins are studied by NMR, systems with unpaired electrons can be investigated by *electron spin resonance* (ESR) (also called *electron paramagnetic resonance* [EPR] and *electron magnetic resonance* [EMR]). Readers interested in ESR techniques are referred to comprehensive texts on the subject; for example, by Abragam and Bleaney[5] and Weil and co-workers[6]. In the context of the effect of electron spin on nuclear spin-relaxation, we need in the first place a particular theoretical tool of ESR spectroscopy, the *spin-Hamiltonian*. The idea behind the concept is as follows. The electronic energy level structure of molecular systems is very complicated and its description requires an approximate solution of a Schrödinger equation for a many-body system. If the electronic ground state of a molecular system with unpaired electrons is well separated from the excited states, it is often possible to describe the energy levels related to the electron spin in the magnetic field by a greatly simplified Hamiltonian, the spin-Hamiltonian, including only spin operators and parameters characterising the system under consideration. The parameters of the spin-Hamiltonian are connected to the electronic structure and can be derived from more fundamental quantities such as electronic energy levels, wave functions and their expectation values (see for example Atkins and Friedman, *Further reading*). A commonly used form of the spin-Hamiltonian can be expressed as:

$$\hat{H}_S = -(\mu_B/\hbar)\hat{\mathbf{S}}\cdot\mathbf{g}\cdot\mathbf{B}_0 + \sum_I \hat{\mathbf{S}}\cdot\mathbf{A}_I\cdot\hat{\mathbf{I}} + \hat{\mathbf{S}}\cdot\mathbf{D}\cdot\hat{\mathbf{S}} \tag{15.1}$$

Here, $\hat{\mathbf{S}}$ and $\hat{\mathbf{I}}$ are the electron and nuclear spin (vector) operators. The first term in Eq. (15.1) is the electron spin Zeeman interaction, μ_B is the Bohr magneton (it is divided by $\hbar$ to give the Hamiltonian in angular frequency units) and $\mathbf{g}$ is the electronic *g-tensor* (an ESR counterpart of the shielding tensor in NMR). Just as for the nuclear Zeeman interaction, the electron Zeeman term can be written as a sum of an isotropic part and an orientation-dependent, rank-two *g-anisotropy* part. The second term in Eq. (15.1) is a sum over relevant nuclear spins, I, with which the S-spin interacts through the *hyperfine interaction*, and $\mathbf{A}_I$ is the hyperfine coupling tensor (an ESR analogue of the J-coupling). Finally, the last term is denoted as the *zero-field splitting interaction* and $\mathbf{D}$ is called the *zero-field splitting* (*ZFS*) *tensor* (the electron spin analogue of the quadrupolar interaction). The frequencies of the ESR transitions can be derived by setting up the matrix representation of the spin-Hamiltonian in the basis set of spin functions, $|S,m_S\rangle\otimes|I,m_I\rangle$ (the symbol $\otimes$ denotes the outer product) and diagonalising it.

A few more words should be said about the physical origin of the hyperfine interaction and the ZFS. The hyperfine interaction can be derived from the relativistic Breit–Pauli Hamiltonian (see Atkins and Friedman, *Further reading*) and is in the non-relativistic limit written as a sum of a *Fermi contact* (FC) term and a *dipole-dipole* (DD) term. The electron-nuclear DD term describes the interaction between two spins at a distance r_{IS}, in full analogy with the nuclear DD interaction described in Chapter 3. The FC term is a consequence of the delocalised nature of the electron spin and takes care of the non-zero probability of having an unpaired electron spin density at the site of the nucleus. The FC term is proportional to the integral over the spin density multiplied by a delta function, $\delta(\mathbf{r}_I)$, picking out the density of unpaired electron spin at the site of the nucleus, $\mathbf{r}_I$. The hyperfine coupling tensor is thus expressed as:

$$\mathbf{A}_I = \mathbf{A}_{DD} + \mathbf{A}_{FC} \tag{15.2}$$

The two tensorial terms have different properties: $\mathbf{A}_{DD}$ is traceless and can be written as a symmetric rank-two tensor, while $\mathbf{A}_{FC}$ is a scalar (rank-zero) tensor. It is also common to denote the FC part of the hyperfine interaction as the *scalar* term.

The zero-field splitting arises only in systems with an electron spin quantum number of one or larger, corresponding to two or more unpaired electrons. The ZFS tensor is a traceless, symmetric rank-two tensor, which can conveniently be expressed in different ways. A commonly used formulation, referring to the molecule-fixed principal frame, introduces the quantities D and E:

$$D = D_{zz} - \frac{1}{2}(D_{xx} + D_{yy}) \tag{15.3a}$$

$$E = \frac{1}{2}(D_{xx} - D_{yy}) \tag{15.3b}$$

The element D_{zz} is the numerically largest principal element of the ZFS tensor. The symbol D is often called the *ZFS parameter* and is analogous to the quadrupolar coupling constant for $S \geq 1$ nuclei. It is a measure of the splitting of levels with different $|m_S|$ at zero magnetic field. It is not uncommon that this quantity (usually expressed in wavenumber units) can reach the magnitude of several cm^{-1} ($1\ cm^{-1}$ corresponds to 1.9×10^{11} radians per second). The parameter E is called the *ZFS rhombicity* and is similar to the quadrupolar asymmetry parameter. These two parameters can, in turn, be related to the components, $D_{2,0}$ and $D_{2,\pm 2}$, of the irreducible spherical tensor formulation of the ZFS tensor.

Two physically distinct mechanisms can lead to a ZFS interaction of the form of the last term in Eq. (15.1). The electron spin–electron spin dipolar interaction is one possibility. The second mechanism has its origin in the relativistic description of the electron. More specifically, it is related to the second-order effect of the spin-orbit interaction. As stated a long time ago by Griffith, the spin-orbit term is usually expected to be dominant for transition metal systems[7]. The second-order spin-orbit correction may become very large if the energy levels of the full non-relativistic Hamiltonian, coupled by the spin-orbit interaction, are not far apart. Under these conditions, the spin-Hamiltonian approach may break down. Problems with the spin-Hamiltonian also arise in the limiting case of the orbitally degenerate ground state, when the first-order spin-orbit effects do not vanish.

At sufficiently high magnetic field, the electron spin Zeeman term is always dominant and the effects of the other terms can be included by perturbation theory. To have the right perspective on the concept of "high field", it may be interesting to note that the Zeeman splitting of $1\ cm^{-1}$ is reached for a free electron ($g_e = 2.0023$) at a field of about 1 T. As already mentioned, ZFS of several cm^{-1} are not uncommon, which means that the high-field limit will not be reached in many experimental situations. The g-anisotropy, the electron spin–nuclear spin DD interaction and the ZFS are most simply defined in the molecular frame. If there is motion in the system, as is the case in liquids, and the molecular frame changes orientation with respect to the laboratory frame, all these anisotropic terms become time dependent and can lead to electron-spin relaxation processes. For $S = 1/2$ systems, the electron spin-relaxation is usually discussed in terms of g-anisotropy and anisotropic hyperfine (or electron-nuclear dipole-dipole) interaction, modulated by molecular tumbling. In addition, the spin-rotation interaction is sometimes also considered. The theory is similar to the nuclear spin case, the main difference being that the electron spin interactions are orders of magnitude stronger than their nuclear counterparts, which results in much more frequent problems with the second-order perturbation (or Redfield) theory[8]. The motional range where the perturbation approach does not apply is often referred to as the *slow-motion regime*. The slow-motion ESR spectral features of the nitroxyl radical with $S = 1/2$ were reviewed a long time ago by Freed[9].

Most paramagnetic systems studied by NMR relaxation, however, contain paramagnetic metal ions with $S \geq 1$ (an important exception is Cu(II) with $S = 1/2$). In such systems, the electron spin-relaxation is usually caused by the zero-field splitting interaction, the last term in Eq. (15.1). If we deal with octahedrally or tetrahedrally coordinated metal ions in water – a common situation for the first transition metal series ions – the ZFS interaction vanishes on average by symmetry. However, the distortions of the first coordination sphere, caused for instance by collisions with the solvent molecules outside of the first coordination sphere, give rise to a *transient* or instantaneous ZFS. Again, analogies with the case of monoatomic quadrupolar ions are quite obvious. A theory of the electron spin-relaxation in hydrated first transition metal series ions was formulated by Bloembergen and Morgan in the early 1960s[10]. For $S = 1$, there is an analogy with the discussion in Section 14.2: both the longitudinal and the transverse electron spin-relaxation are single exponential processes (at least in the perturbation regime). Bloembergen and Morgan derived the following expressions for the electron spin longitudinal and transverse relaxation rates at high magnetic field:

$$\frac{1}{T_{1e}} = \frac{\Delta_t^2}{5}\left(\frac{\tau_v}{1+\tau_v^2\omega_S^2} + \frac{4\tau_v}{1+4\tau_v^2\omega_S^2}\right) \tag{15.4a}$$

$$\frac{1}{T_{2e}} = \frac{\Delta_t^2}{10}\left(3\tau_v + \frac{5\tau_v}{1+\tau_v^2\omega_S^2} + \frac{2\tau_v}{1+4\tau_v^2\omega_S^2}\right) \tag{15.4b}$$

The similarity to Eqs. (14.17) and (14.19) should be noticed. Here, τ_v is a correlation time for the modulation of the ZFS by distortions of the symmetry of the first hydration shell, ω_S is the electron spin Larmor frequency and Δ_t^2 is the trace of the square of the transient ZFS tensor:

$$\Delta_t^2 = D_{xx}^2 + D_{yy}^2 + D_{zz}^2 = \frac{2}{3}D^2 + 2E^2 \tag{15.5}$$

where the ZFS parameter and rhombicity are employed in the last formulation. The distortional correlation time, τ_v, has a similar meaning to τ_Q in Eq. (14.26). It can often be assumed to be in the range of few picoseconds, so that the Redfield condition, here $\Delta_t^2\tau_v^2 \ll 1$, requires the quantity Δ_t (converted to the wavenumbers) to be no larger than about 0.5 cm^{-1}.

If the environment of the paramagnetic ion has, on average, lower symmetry than tetrahedral or octahedral, then we may encounter a non-zero *static* ZFS (averaged over the rapid local motions), which can be modulated by rotational tumbling. Eq. (15.4) can easily be modified for systems with only static ZFS: the only changes we need to make are to replace the Δ_t by a static counterpart, Δ_s, and the distortional correlation time by the rotational correlation time, τ_R. As the rotational correlation times are usually much longer than a few picoseconds, the Redfield condition (now $\Delta_s^2\tau_R^2 \ll 1$) becomes much more difficult to fulfil, except for ions with very highly symmetric electronic structure, such as Mn(II) (with five 3*d*-electrons distributed among the five *d*-orbitals) or Gd(III) (with seven 4*f*-electrons distributed among the seven *f*-orbitals), characterised by very small ZFS. Except for such cases, the slow-motion conditions become quite common and relaxation theory as described in Chapter 4 is not applicable.

One possible way out of the difficulties is to use the formalism of the *stochastic Liouville equation* (SLE)[2,4,8,9]. In order to get a feeling for how this formalism works, let us re-write Eq. (4.33) (augmented with the correction for finite temperature) in a superoperator form:

$$\frac{d}{dt}|\hat{\rho}(t)) = -i\hat{\hat{L}}_0|\hat{\rho}(t)) - \hat{\hat{R}}\left[|\hat{\rho}(t)) - |\hat{\rho}^{eq})\right] \tag{15.6}$$

where $\hat{\hat{R}}$ is the relaxation superoperator and $\hat{\hat{L}}_0$, the Liouvillian, is the commutator with the unperturbed Hamiltonian (compare Eq. (1.29) and Section 4.3).

The notation $|\hat{\rho}(t))$ and $|\hat{\rho}^{eq})$ refers to the density operator and its equilibrium value as kets in the Liouville space. If we cannot use the Redfield approach, because the coupling between the spins and certain classical degrees of freedom is too strong, then we can try to include them in the more carefully studied subsystem, described by the density operator, along with the spin degree of freedom. Rather than in terms of Eq. (15.6), we then formulate our "master equation" in the SLE form:

$$\frac{d}{dt}|\hat{\chi}(t)) = -i\hat{\hat{L}}|\hat{\chi}(t)) - \hat{\hat{K}}\left[|\hat{\chi}(t)) - |\hat{\chi}^{eq})\right] \tag{15.7}$$

where $\hat{\chi}$ is the density operator, here involving not only the (electron) spin but also the classical degrees of freedom to which the spin is strongly coupled. $\hat{\hat{L}}$ is the corresponding Liouvillian (depending on spin and the variables related to the important classical degrees of freedom) and $\hat{\hat{K}}$ is a stochastic operator, describing the evolution of the classical degrees of freedom under consideration.

In an important class of cases, such as systems with large static ZFS, the relevant classical degree of freedom is the orientation of the molecule with respect to the laboratory frame. The superoperator $\hat{\hat{L}}$ depends in this case on the set of angles, Ω, specifying the orientation of the principal frame of the anisotropic interaction with respect to the laboratory frame, and $\hat{\hat{K}}$ takes the form of a Markov operator (see Van Kampen, *Further reading*) describing the random variation of orientation, *e.g.* through the rotational diffusion. We are not going to pursue this discussion any further here, but we shall use the idea of the SLE description of the electron spin dynamics in Section 15.2.

The description of electron-spin-relaxation in terms of the simple Eq. (15.4) is an oversimplification also when the Redfield limit applies, if we deal with $S \geq 3/2$ systems outside of extreme narrowing for the electron spin (we remind the reader that the electronic Larmor frequency ω_S is about 650 times larger than that for protons, so that the extreme narrowing condition, $\tau_c^2\omega_S^2 \ll 1$, is violated at much shorter correlation times/lower fields). This issue was first dealt with by Rubinstein *et al.*[11], who pointed out that the electron spin-relaxation for such systems becomes multiexponential (compare Section 14.2!). In the same important paper, the authors also introduced a formal model, called the *pseudorotation* model, in which the complex distortion is described by a rotational diffusion equation with a diffusion coefficient (or a correlation time) different from that for overall reorientation. The pseudorotation model leads to Eq. (15.4) for $S=1$ and has also turned out to be very useful in a more general context[2,4]. In 2000, Rast *et al.*[12] presented an improved high-field theory of the electron spin-relaxation: two dynamic processes (distortion and rotation) were allowed to modulate the transient and static part of the ZFS, respectively. Higher-order crystal field effects and multiexponential relaxation were also included and the theory was applied for the analysis and interpretation of multiple-field, variable-temperature ESR lineshapes for Gd(III) complexes in solution.

Having made the reader aware of some of the complexities of electron spin-relaxation in solution, we can move on to the main topic of this chapter, the influence of unpaired electrons on the nuclear spin-relaxation.

15.2 Paramagnetic Enhancement of Spin-Lattice and Spin-Spin Relaxation Rates

The first issue that needs to be clarified in this section is the relation between macroscopic, observable properties of nuclear spins and their microscopic (molecular) counterparts. In solutions of transition metal ions or complexes, one can commonly encounter a situation where the ligands carrying nuclear spins can reside in two types of environment: in the coordination sphere of the paramagnetic metal ion or in the bulk. If the ligand contains only one type of magnetic nuclei (which is the case for example for the $^1H_2{}^{16}O$ isotopomer of water) or if we can disregard interactions between nuclear spins within a site, each of the two sites can be characterised by nuclear spin-lattice and spin-spin relaxation times, T_1 and T_2, respectively. By assuming that the paramagnetic species are dilute, and allowing for a change in the chemical shift for the nucleus of interest[1], Swift and Connick[13] and Luz and Meiboom[14] derived a set of equations relating the observables of NMR experiments to the relaxation, shift and lifetime properties of nuclei in the paramagnetic complexes:

$$T_{1P}^{-1} = \frac{P_M}{\tau_M + T_{1M}} \tag{15.8a}$$

$$T_{2P}^{-1} = \frac{P_M}{\tau_M}\left[\frac{T_{2M}^{-2} + (T_{2M}\tau_M)^{-1} + \Delta\omega_M^2}{\left(T_{2M}^{-1} + \tau_M^{-1}\right)^2 + \Delta\omega_M^2}\right] \tag{15.8b}$$

$$\Delta\omega_P = \frac{P_M \Delta\omega_M}{\left(\tau_M/T_{2M} + 1\right)^2 + \tau_M^2 \Delta\omega_M^2} \tag{15.8c}$$

T_{1P}^{-1}, T_{2P}^{-1} and $\Delta\omega_P$ are the excess spin-lattice relaxation rate, spin-spin relaxation rate and shift measured for the ligand in solution (a difference between the quantity of interest in a paramagnetic solution and the corresponding value in a diamagnetic reference solution), while the properties with index M refer to the ligand in the paramagnetic complex. The symbol τ_M is the lifetime of the ligand in the complex and P_M is the mole fraction of ligand nuclei in bound positions. The excess spin-lattice relaxation rate is often referred to as the PRE. The PRE given by Eq. (15.8a) is commonly denoted as the *inner-sphere PRE*. If only one type of paramagnetic species exists in solution, the PRE is proportional to the concentration of that species. The PRE normalised to a 1 mM concentration of paramagnetic complexes is called *relaxivity*.

Another limiting situation arises if the paramagnetic species interact only weakly with the molecules carrying the nuclear spins. In such a case, it is not meaningful to speak about exchange between discrete sites, but rather about free diffusion or diffusion in a potential. One speaks then about *outer-sphere PRE*, still referring to the enhancement of the spin-lattice relaxation rate. The outer-sphere PRE is also proportional to the concentration of paramagnetic material (compare Section 3.5) and the concept of relaxivity is equally useful in this case.

15.2.1 The Solomon–Bloembergen–Morgan (SBM) Theory

In the remainder of this section, we concentrate on the theory for the quantities T_{1M} and T_{2M} in Eq. (15.8), having their origin in the hyperfine interaction between the nuclear and the electron spins. For a more extensive presentation, the interested reader is referred to reviews from one of our laboratories[2,4]. The simplest and the most commonly used model employs the so-called modified Solomon–Bloembergen equations. The equation for the longitudinal relaxation rate reads:

$$\begin{aligned}
T_{1M}^{-1} &= \left(T_{1M}^{SC}\right)^{-1} + \left(T_{1M}^{DD}\right)^{-1} = \frac{2}{3}A_{SC}^2 S(S+1)\frac{\tau_{e2}}{1+(\omega_S-\omega_I)^2\tau_{e2}^2} \\
&+ \frac{2}{15}S(S+1)b_{IS}^2\left[\frac{\tau_{c2}}{1+(\omega_S-\omega_I)^2\tau_{c2}^2} + \frac{3\tau_{c1}}{1+\omega_I^2\tau_{c1}^2} + \frac{6\tau_{c2}}{1+(\omega_S+\omega_I)^2\tau_{c2}^2}\right] \\
&\approx \frac{2}{3}A_{SC}^2 S(S+1)\frac{\tau_{e2}}{1+(\omega_S-\omega_I)^2\tau_{e2}^2} + \frac{2}{15}S(S+1)b_{IS}^2\left[\frac{7\tau_{c2}}{1+\omega_S^2\tau_{c2}^2} + \frac{3\tau_{c1}}{1+\omega_I^2\tau_{c1}^2}\right]
\end{aligned} \tag{15.9}$$

In the last, approximate equality, we have neglected the nuclear Larmor frequency in comparison with its electron spin counterpart and combined the first and the third term in the second line into a combined "seven" term. The scalar part, $(T_{1M}^{SC})^{-1}$, is similar to Eq. (13.2a) for the scalar relaxation of the first kind and to Eq. (14.29a) for the scalar relaxation of the second kind. The inverse of the correlation time τ_{e2}, is in fact equal to the sum of the inverse times from the two mechanisms:

$$\tau_{e2}^{-1} = \tau_M^{-1} + T_{2e}^{-1} \tag{15.10a}$$

The symbol T_{2e} denotes the transverse electron spin-relaxation time (or a similar "effective" quantity if the relaxation is multiexponential). In analogy with the discussion in Chapters 13 and 14, the scalar relaxation is usually not too important for the longitudinal relaxation enhancement in paramagnetic systems either. A related quantity, which we shall need later, is:

$$\tau_{e1}^{-1} = \tau_M^{-1} + T_{1e}^{-1} \tag{15.10b}$$

The dipolar part of the rate in Eq. (15.9), $(T_{1M}^{DD})^{-1}$, is both important and highly informative, since the electron-nuclear dipole-dipole coupling constant, b_{IS}, carries information on the distance between the nuclear spin and the electron spin (compare Eq. (3.5)). The dipolar part contains two correlation times:

$$\tau_{cj}^{-1} = \tau_R^{-1} + \tau_M^{-1} + T_{je}^{-1}; \quad j = 1,2 \tag{15.10c}$$

The idea behind Eq. (15.10c) is similar to that leading to Eq. (15.10a): if more than one process modulates the dipole-dipole interaction and the processes are not correlated, the inverse correlation time is a sum of the three rates responsible for the (exponential) loss of correlation between the dipolar Hamiltonians at times zero and *t*. Eqs. (15.9) and (15.10) are usually combined with Eq. (15.4) for the electron spin-relaxation rates, leading to a complete (albeit approximate) theory, known in the literature as the Solomon, Bloembergen and Morgan (SBM) theory. The theory is frequently applied to interpret water proton PRE, measured as a function of magnetic field (the nuclear magnetic relaxation dispersion [NMRD] profile, discussed earlier in Chapters 3 and 14).

Determinations of the paramagnetic transverse relaxation are less common, since both the measurements and their interpretation are somewhat more complicated. Still, for the sake of completeness, we present here the transverse counterpart of Eq. (15.9):

$$\begin{aligned} T_{2M}^{-1} &= (T_{2M}^{SC})^{-1} + (T_{2M}^{DD})^{-1} = \frac{1}{3} A_{SC}^2 S(S+1) \left(\tau_{e1} + \frac{\tau_{e2}}{1 + (\omega_S - \omega_I)^2 \tau_{e2}^2} \right) \\ &+ \frac{1}{15} S(S+1) b_{IS}^2 \left[4\tau_{c1} + \frac{3\tau_{c1}}{1 + \omega_I^2 \tau_{c1}^2} \right. \\ &\left. + \frac{\tau_{c2}}{1 + (\omega_S - \omega_I)^2 \tau_{c2}^2} + \frac{6\tau_{c2}}{1 + \omega_S^2 \tau_{c2}^2} + \frac{6\tau_{c2}}{1 + (\omega_S + \omega_I)^2 \tau_{c2}^2} \right] \end{aligned} \tag{15.11}$$

where τ_{e1}^{-1} is given in Eq. (15.10b). The prediction of the SBM theory for the magnetic field dependence of the in-complex relaxation rates is shown in Figure 15.1. The scalar part is neglected and the rotation is assumed to dominate the modulation of the dipole-dipole interaction (see Eq. (15.10c)), which makes the correlation times field independent. The relative relaxation rates shown in the diagram are the dipolar rates divided with the common low-field limit. The curve for T_{1M}^{-1} contains two characteristic dispersions; the reduction of the relaxivity at a low field corresponds to the "seven" term in the last line of Eq. (15.9), decaying at $\omega_S\tau_{c2} \approx 1$, while the "three" term disperses at a higher field, when $\omega_I\tau_{c1} \approx 1$.

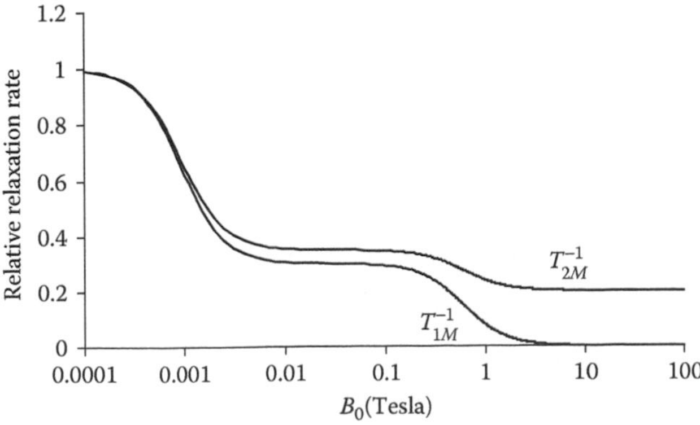

FIGURE 15.1 Plots of the field-dependence of the dipolar parts of the relaxation rates according to Eqs. (15.9) and (15.11), assuming that the reorientation is the dominant modulation mechanism. A correlation time of 1 ns was used.

15.2.2 Beyond the SBM Theory

The SBM theory includes many approximations, which impose limitations on the applicability of the approach. These limitations may be more or less serious. Two of the approximations are conceptually quite simple. The first one is the *point-dipole approximation*, usually invoked in the interpretation of the interspin distance in the dipole-dipole coupling constant in Eq. (15.9) as the distance between the nucleus I and the metal, where the electron spin is supposed to reside and to behave as a magnetic point dipole[2–4]. The approximation thus neglects the delocalisation of the electron spin density. This often works reasonably well for protons, while large deviations occur for other nuclei. To circumvent this approximation, one can formulate an expression for the effective distance between the spin I and the delocalised electron spin, which can be calculated by quantum chemistry[15–17]. The issue of deviations from point-dipole approximation is an active field of research[18–20]. The next quite obvious approximation is that the reorientation is assumed to be isotropic, characterised by a single rotational correlation time, neglecting possible rotational anisotropy and internal motions. This approximation is probably not too critical and a Lipari–Szabo-type correction has been proposed[21].

Other approximations are of a somewhat more sophisticated nature and maybe of larger importance. First, exchange, rotation and electron spin-relaxation are assumed to be uncorrelated, which allows the formulation of Eq. (15.10). This so-called *decomposition approximation* is fairly obvious for the exchange process but is problematic (and cannot be removed easily) for the rotation and electron spin-relaxation in small molecules; we shall return to this point in the context of the SLE-based "Swedish slow-motion theory". Finally, we have the most serious approximations concerning the electron spin-relaxation itself. It is assumed to be single exponential, following the Redfield theory for $S = 1$ in the high-field (Zeeman-dominated) limit, and assuming only the distortional modulation of the transient ZFS.

Among the last group of deficiencies, the simplest to repair is the assumption of single exponential relaxation. Westlund proposed replacing the SBM theory by a generalised version, in which the multi-exponential relaxation (within the Redfield limit and in the high-field limit) is taken into consideration. The resulting theory is called the *generalised SBM* (GSBM)[22].

When experimental NMRD profiles are determined, one usually includes measurements at low magnetic fields, where the Zeeman interaction may not be dominant. The theory has been gradually developed to interpret relaxation in systems with a sizable static component of the ZFS. A description of different steps and ideas in this development is beyond the scope of this book; we refer the reader to the review by Kowalewski *et al.* mentioned already[2]. Here, we describe the approach by Bertini *et al.*[23] and Kruk *et al.*[24], developed for large, slowly rotating complexes, implemented in the software package developed in Florence[25] and called the *modified Florence* approach. Briefly, in this model one assumes that the rotation is so slow that it has no effect on relaxation (which removes the problems with the decomposition approximation) and that the PRE can be obtained by powder-averaging the relaxation contributions over all possible orientations of the molecular frame with respect to the laboratory frame. The static ZFS and the Zeeman interaction are both included in the unperturbed Hamiltonian. The electron spin energy levels and eigenstates are computed for various orientations of the ZFS principal axis system with respect to the laboratory frame. For every orientation, the Redfield relaxation matrices for the electron spin molecules are constructed using the appropriate basis, obtained from the eigenstates of the combined unperturbed Hamiltonian, while assuming that the distortional modulation of the transient ZFS acts as the only source of electron spin-relaxation. Some results of such calculations will be mentioned in Section 15.5. This approach can, with some difficulties and some loss of generality, be adapted to fast-rotating systems[26] if the Redfield conditions apply to both the transient and the static ZFS ($\Delta_t^2\tau_v^2 \ll 1$, $\Delta_s^2\tau_R^2 \ll 1$). If this is not the case, the situation becomes difficult and one has a choice of either some kind of "spin dynamics" simulation, where the ZFS Hamiltonian varies pseudo-randomly when expressed in the laboratory frame (we shall return to these methods), or the SLE-based approach.

The SLE-based "Swedish slow-motion theory" was originally proposed in 1983[27] and has been developed over the last decades by groups in Sweden. The method is based on a mixed quantum mechanical and classical description of the lattice, which can be handled in the spirit of the SLE. The essential – and heavy – computational step involves setting up and inverting a very large matrix, representing the Liouville superoperator for the lattice in a basis set containing direct products of appropriate spin-space operators and Wigner rotation matrices. The most recent version of the method[28], designed for inner-shell relaxation, allows including two dynamic processes to be included: pseudorotation, which modulates the transient ZFS, and reorientation of the complex that modulates the static ZFS, with arbitrary ZFS amplitudes and correlation times. Both types of ZFS can be rhombic and the static ZFS principal axis can have an arbitrary orientation with respect to the electron-nuclear dipole-dipole axis. The approach can deal with systems with an S quantum number up to 7/2. Furthermore, the decomposition approximation is absent. The method has been used to predict and explain trends in the NMRD profiles upon variation of various parameters[28], and occasionally for fitting experimental data[29]. Probably the most important use of the "Swedish slow-motion theory" is as a benchmark for testing simpler models. To illustrate this, Figure 15.2 shows the dependence of NMRD profiles in $S=1$ systems on the angle θ between the static ZFS principal axis and the dipole-dipole axis and on the rhombicity of the static ZFS calculated using different models.

As mentioned, the situation outside of the perturbation limit can also be dealt with by using simulations in time domain. Some time ago, Odelius and co-workers[30] reported a combined quantum

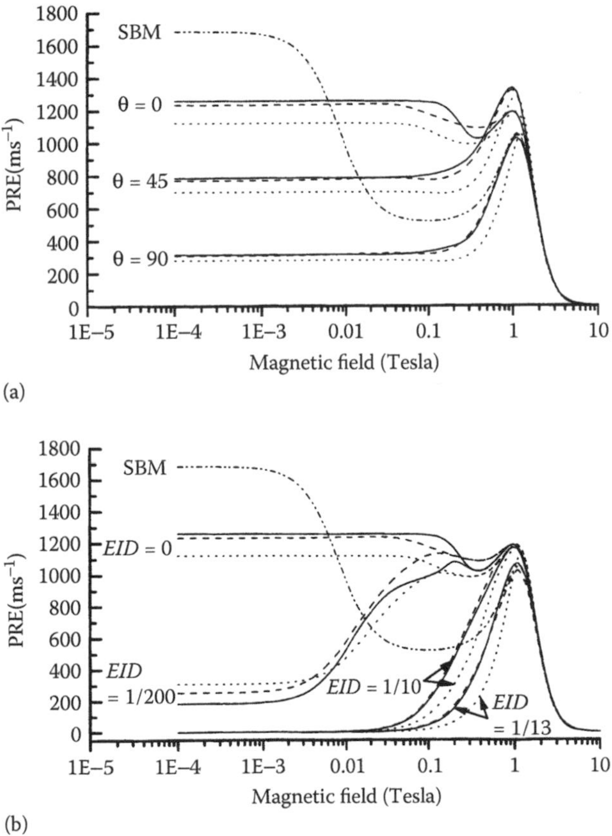

FIGURE 15.2 Calculated dependence of the NMRD profile for $S=1$ on (a) the angle θ between the principal axes of the static ZFS and the electron spin–nuclear spin dipole-dipole interaction; (b) the rhombicity of the static ZFS. Solid lines: the slow-motion theory[28], dotted lines: the original approach from the Florence group[25], dashed lines: the "modified Florence" approach[23], dashed-dotted lines: the SBM theory. (Reprinted with permission from Bertini, I. et al., *J. Chem. Phys.*, 111, 5795–5807. Copyright (1999) American Institute of Physics.)

mechanics and molecular dynamics (MD) study of fluctuations of the zero-field splitting in an aqueous solution of Ni(II). Based on simplified *ab initio* calculations for hexa-aquo Ni(II), they estimated the ZFS originating from distortions of the T_h symmetry of the complex. The time-dependent distorted complex structures were obtained through classical MD simulations, and the resulting trajectory was translated into time variation of the ZFS. Similar work on the same system, but using present-day computational technology, has also been reported[31,32]. In a subsequent study from the same group, the sequence of magnitudes and orientations of the ZFS tensor was reformulated in terms of the ZFS Hamiltonian varying randomly in small time steps[33]. This could be combined with the Zeeman interaction of arbitrary strength to provide a time-dependent spin-Hamiltonian. The time-dependent Schrödinger equation with this Hamiltonian could be solved numerically, allowing the evaluation of time-correlation functions (tcfs) for the components of the electron spin:

$$G_r(t) = \left\langle \hat{S}_r(t)\hat{S}_r(0) \right\rangle \tag{15.12}$$

where $r = x, y, z$. From the tcfs, it was possible to derive information on electronic and nuclear spin-relaxation.

It is possible to obtain an analogous expression to Eq. (15.12) using other, simpler assumptions concerning fluctuations of the ZFS. This approach was developed by Sharp and co-workers[34,35] in Ann Arbor and by Fries and Belorizky[36,37] in Grenoble. One may, for example, assume a spin-Hamiltonian with the Zeeman interaction along with the static and transient ZFS of given magnitude (and rhombicity), fluctuating in time by variation of the orientation of the ZFS principal axes, in analogy with the Swedish slow-motion theory. The three theoretical approaches (Swedish, Ann Arbor and Grenoble), presented as valid beyond the Redfield limit, were compared with each other by calculations of the ^{1}H NMRD profiles for a large number of parameter sets, mimicking the situations that may arise in complexes of Ni(II) ($S = 1$, ZFS in the range of few cm^{-1}) and Gd(III) ($S = 7/2$, small ZFS)[38]. The methods of the Swedish groups and of the Grenoble team produced results in excellent agreement with each other, while the Ann Arbor results were somewhat different. One example of the NMRD profiles calculated with the three approaches is shown in Figure 15.3. This comparative study clearly supports the credibility of the advanced theoretical methods in use today.

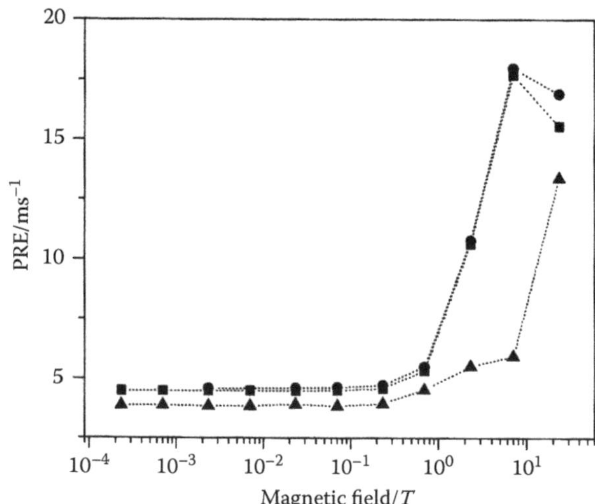

FIGURE 15.3 Proton NMRD profile for $S = 1$ in a fast rotating complex ($\tau_R = 100$ ps) with a large static ZFS ($\Delta_s = 10$ cm^{-1}), medium transient ZFS ($\Delta_t = 1$ cm^{-1}) and a short distortional correlation time ($\tau_v = 5$ ps). Triangles: the Ann Arbor method, squares: the Swedish slow motion method, circles: the Grenoble method. (Adapted with permission from Belorizky, E. et al., *J. Chem. Phys.*, 128, 052315, 2008.)

15.2.3 The Curie-Spin Relaxation

The very large magnetic moment associated with an unpaired electron spin leads to still another complication in the nuclear spin-relaxation, related to the size of the dominant Zeeman interaction. Among the three dynamic processes modulating the electron-nuclear dipole-dipole interaction, summarised in a simple form by Eqs. (15.9)–(15.11), the electron spin-relaxation is often the fastest, especially for large molecules with sluggish reorientation. This situation has been considered by Gueron[39] and by Vega and Fiat[40]. The authors write the electron spin vector as a sum of the thermal average "Curie spin", aligned along the magnetic field, and the oscillating and zero-average remainder, **s**. The magnitude of the Curie spin is:

$$S_C = \left\langle \hat{S}_z \right\rangle = g_e \mu_B S(S+1) B_0 / 3k_B T \tag{15.13}$$

where k_B is the Boltzmann constant and T is temperature, while the mean square of **s** is $S(S+1)/3 - S_C^2$. g_e is the electronic g factor and μ_B is the Bohr magneton. We note that $\hbar\gamma_S = g_e\mu_B$. The Curie spin is related to the magnetic susceptibility and Eq. (15.13) is derived under the assumption that the susceptibility is given by the Curie law[1]. The nuclear spin interacts, through the dipole-dipole interaction, with both S_C and **s**. The magnetic moment related to the Curie spin is a small fraction of the total electron spin magnetic moment, but it becomes non-negligible at sufficiently high magnetic field. At high field, we can in addition neglect the terms in Eqs. (15.9)–(15.11) that contain ω_S. Neglecting also the scalar interaction terms, the relaxation rates given by Eqs. (15.9)–(15.11) become under these conditions:

$$T_{1M}^{-1} = \frac{6}{5}\left(\frac{\mu_0}{4\pi}\right)^2 \gamma_I^2 g_e^2 \mu_B^2 r_{IS}^{-6} \left\{ S_C^2 \frac{\tau_D}{1+\omega_I^2\tau_D^2} + \left[\frac{1}{3}S(S+1) - S_C^2\right] \frac{\tau_{c1}}{1+\omega_I^2\tau_{c1}^2} \right\} \tag{15.14a}$$

and

$$T_{2M}^{-1} = \frac{1}{5}\left(\frac{\mu_0}{4\pi}\right)^2 \gamma_I^2 g_e^2 \mu_B^2 r_{IS}^{-6} \cdot \left\{ S_C^2 \left(4\tau_D + \frac{3\tau_D}{1+\omega_I^2\tau_D^2}\right) + \left[\frac{1}{3}S(S+1) - S_C^2\right] \left(4\tau_{c1} + \frac{3\tau_{c1}}{1+\omega_I^2\tau_{c1}^2}\right) \right\} \tag{15.14b}$$

The first term in both expressions corresponds to the relaxation by the Curie spin (Curie-spin relaxation [CSR]), while the second is very similar to the dipolar parts of expressions of Eqs. (15.9)–(15.11). The Curie spin is modulated by reorientation and exchange but not by electron spin-relaxation, and the corresponding correlation time is given by $\tau_D^{-1} = \tau_R^{-1} + \tau_M^{-1}$. If $\tau_D \gg \tau_{c1}$, which is a common situation for large paramagnetic molecules with rapid electron spin-relaxation, then the Curie spin terms become important and can in certain situations dominate over the modified Solomon-like terms. The Curie-spin relaxation is also called *magnetic susceptibility relaxation*. The magnetic susceptibility relaxation is usually more important for T_2 than for T_1. In fact, the proton linewidth in paramagnetic proteins at high magnetic field[1] is often determined by this relaxation mechanism.

15.2.4 The Outer-Sphere Relaxation

As mentioned in Section 3.5, the electron-nuclear dipole-dipole interaction can act as a relaxation mechanism for nuclear spins, even if they do not enter a well-defined complex with the paramagnetic species.

When compared with the inner-sphere case, the theory of this outer-sphere paramagnetic relaxation involves similar complications to those noticed in Section 3.5 when the diamagnetic intermolecular dipolar relaxation was contrasted with the (intramolecular) Solomon treatment. In principle, we have to deal with both the increased complexity caused by the stochastic variation of the spin-spin distance and the problems of electron spin-relaxation.

The intermolecular counterpart of the modified Solomon–Bloembergen equations has been presented by Freed[41], who combined the force-free diffusion model with a simple, single-parameter description of electron spin-relaxation. An improved formulation of the theory, retaining the force-free diffusion but providing a more sophisticated description of electron spin-relaxation, was reported by Kruk *et al.*[24] within the limit of slowly rotating systems. Developments allowing for electron spin-relaxation outside of the Redfield limit have been presented by Abernathy and Sharp[34], Rast, Fries and co-workers[42,43] and Kruk and Kowalewski[44]. Within the last formulation, the outer-sphere relaxation rate is obtained as an integral of a product of a tcf for translational diffusion and a function describing electronic relaxation, similar to the spectral density occurring in the inner-sphere theory. Calculations using this model involve numerical integration and are computer time consuming. In summary, the current understanding is that the outer-sphere contributions to the PRE may be important and should be included in the analysis of NMRD data.

15.3 Paramagnetic Effects on Nuclear Multispin Phenomena

The paramagnetic species in solution affect also nuclear multispin phenomena. The nuclear cross-relaxation rates are not changed by the paramagnetic effects, but the NOE, η, is reduced as the PRE increases the self-relaxation rate of the spins (*cf.* Eq. (3.28)). As a consequence of this phenomenon, the proton-proton NOE can be difficult to observe in, for instance, paramagnetic proteins in the vicinity of the paramagnetic centre. The difficulties are, in addition, aggravated by the increased linewidth of all signals.

Paramagnetic cross-correlated relaxation phenomena are more interesting. Besides causing an increased linewidth (a single-spin property), the magnetic susceptibility effects mentioned in the previous section (and reviewed in 2002 by Bertini[45,46]) can also give rise to more complicated multispin relaxation phenomena, such as differential line-broadening of multiplets.

If the magnetic susceptibility is allowed to be anisotropic, the thermally averaged electronic magnetic moment per molecule, induced by the external magnetic field, can be expressed as:

$$\langle \boldsymbol{\mu} \rangle = \frac{\boldsymbol{\chi} \cdot \mathbf{B}_0}{\mu_0} \tag{15.15}$$

where $\boldsymbol{\chi}$ is the magnetic susceptibility tensor. The Hamiltonian describing the dipolar interaction between the nuclear spin and this average magnetic moment can be written:

$$\begin{aligned} H_\chi &= -\frac{\mu_0}{4\pi}\left[\frac{3\left(\hbar\gamma_I \hat{\mathbf{I}}\cdot\mathbf{r}\right)\left(\langle\boldsymbol{\mu}\rangle\cdot\mathbf{r}\right)}{r^5} - \frac{\left(\hbar\gamma_I \hat{\mathbf{I}}\cdot\langle\boldsymbol{\mu}\rangle\right)}{r^3}\right] \\ &= -\frac{1}{4\pi}\left[\frac{3\left(\hbar\gamma_I \hat{\mathbf{I}}\cdot\mathbf{r}\right)\left(\mathbf{r}\cdot\boldsymbol{\chi}\cdot\mathbf{B}_0\right)}{r^5} - \frac{\left(\hbar\gamma_I \hat{\mathbf{I}}\cdot\boldsymbol{\chi}\cdot\mathbf{B}_0\right)}{r^3}\right] \end{aligned} \tag{15.16}$$

where the radius vector $\mathbf{r}$ is defined with respect to the paramagnetic metal ion at the origin (if more than one paramagnetic ion is present, the interaction can be expressed as a sum over terms similar to the right-hand side of Eq. (15.16)). We can note that both terms in the second line of Eq. (15.16) contain the nuclear spin operator $\hat{\mathbf{I}}$ on the left and $\mathbf{B}_0$ on the right. We can condense the terms in between the two vectors into a new tensor quantity, $\boldsymbol{\sigma}$, and obtain:

$$H_\chi = -\gamma_I \hat{\mathbf{I}} \cdot \boldsymbol{\sigma} \cdot \mathbf{B}_0 \tag{15.17}$$

This formulation is identical (except for the sign) to the anisotropic magnetic shielding term in the nuclear spin-Hamiltonian (*cf.* Eq. (5.22)). Since the origin of this shielding is the dipolar interaction with the induced magnetic moment, we call it *dipolar shielding*. In the principal frame of the susceptibility tensor, the dipolar shielding tensor is given by:

$$\sigma = \frac{1}{4\pi}\begin{pmatrix} (3x^2 - r^2)\chi_x r^{-5} & 3xy\chi_y r^{-5} & 3xz\chi_z r^{-5} \\ 3xy\chi_x r^{-5} & (3y^2 - r^2)\chi_y r^{-5} & 3yz\chi_z r^{-5} \\ 3xz\chi_x r^{-5} & 3yz\chi_y r^{-5} & (3z^2 - r^2)\chi_z r^{-5} \end{pmatrix} \tag{15.18}$$

Consider the case where the I-spin has a neighbouring nuclear spin N, with which it interacts by the J-coupling as well as the dipolar coupling. Having stated the interaction with the thermally averaged electronic magnetic moment in the form of Eq. (15.17), we can immediately use the Redfield theory for multispin relaxation phenomena as developed for diamagnetic systems (*cf.* Chapter 5). Bertini *et al.*[47] used this way of looking at the differential line-broadening (DLB) in a system of two nuclear spins (*e.g.* the amide ^{1}H and ^{15}N in a paramagnetic protein), influenced by anisotropic paramagnetic susceptibility, following the DD-CSA cross-correlation theory as described in Section 5.2. The effects of electron spin-nuclear spin cross-correlations on the differential broadening of multiplet components arising through the J-coupling between the amide ^{1}H (I-spin) and ^{15}N (N-spin) in paramagnetic proteins were studied even earlier by Ghose and Prestegard[48]. Assuming an isotropic paramagnetic susceptibility given by the Curie law, the expression for the difference in the linewidth of the two components of the I nucleus doublet was derived and can be written as[47,48]:

$$\Delta\nu = \frac{2}{15\pi}\left(\frac{\mu_0}{4\pi}\right)^2 \frac{B_0\gamma_I^2\gamma_N\mu_B^2 g_e^2 \hbar S(S+1)}{r^3 r_{IN}^3 k_B T}\left(4\tau_c + \frac{3\tau_c}{1+\omega_I^2\tau_c^2}\right)\frac{3\cos^2\theta_{SIN} - 1}{2} \tag{15.19}$$

The angle θ_{SIN} is that between the IN axis and the IS axis. The expression $(3\cos^2\theta-1)/2$ in Eq. (15.19) is characteristic of cross-correlated relaxation effects (compare Eqs. (5.44), (6.21c) *etc.*). In the context of two-dimensional heteronuclear correlations spectra, the interference terms between the dipole interaction and the dipolar shift anisotropy (DD-DSA) can give rise to a paramagnetic effect analogous to the transverse relaxation-optimised spectroscopy (TROSY) effect discussed in Section 10.3[49].

The Ghose and Prestegard paper also raises the issue of the dynamic frequency shifts in paramagnetic systems[48]. They show that the dynamic shifts change the positions of the I-spin multiplet components in opposite directions and thus influence the apparent IN spin-spin coupling. An expression for the frequency shift of the I lines was given, but it is probably incorrect, as it is not in agreement with more general expressions for the DD-CSA case as given by Werbelow[50]. The possible dynamic frequency shifts are difficult to distinguish from the residual dipolar couplings, which result from the self-orientation of paramagnetic proteins with anisotropic susceptibility in solution. Therefore, the understanding of dynamic frequency shifts is important from the point of view of paramagnetic constraints for protein structure determination[1,45,46].

The dipolar shift anisotropy can also give rise to cross-correlated phenomena with other interactions. Pintacuda and co-workers[51] discussed the interference phenomenon between DSA and CSA. This effect is related to the case, mentioned in Section 5.2, of cross-correlation between two CSA terms (originating from the decomposition of an arbitrary shielding tensor into two axially symmetric tensors).

This type of interference can influence single-spin relaxation rates and, under appropriate geometric conditions between the principal axes of the two tensors, can reduce the spin-lattice or spin-spin relaxation rates for a nuclear spin. This amounts to a negative PRE or *paramagnetically effected narrowing* (PEN). Experimental observations of this effect have been reported for ^{15}N spins in a double mutant of calbindin D_{9k} loaded with one calcium ion and one lanthanide[52]. Finally, cross-correlation effects between DSA and quadrupolar interaction have also been discussed[53].

15.4 Dynamic Nuclear Polarisation

The concept of nuclear Overhauser enhancement (NOE) has been introduced in Chapter 3 and discussed mainly in Chapters 9 and 12. The concept derives from the seminal work by Albert Overhauser from 1953[54], dealing with polarisation of nuclear spins in metals caused by irradiation of electronic spins. The phenomenon of transferring polarisation from unpaired electron spin to nuclear spins is known as *dynamic nuclear polarisation* (DNP). The DNP in liquids containing paramagnetic radicals or metal ions usually arises through a mechanism very similar to the one discussed in Chapter 3; one speaks often about the Overhauser DNP (ODNP). In the solid state, the rotational and translational motion is suppressed and the mechanism of DNP is usually different. Depending on the system and conditions, one speaks in this case about the *solid effect*, the *cross effect* and the *thermal mixing*. The solid-state DNP phenomena are beyond the scope of this book; for an introduction, we refer the reader to the books by Kruk (*Further reading*) and by Bertini and co-workers[1].

A typical sample for an ODNP experiment is a solution containing a nitroxide (or related) radical as a source of polarisation and nuclear spins (*e.g.* protons) carried by the substance of interest, which might be water. The nitroxide can be a small molecule, such as, for example, 4-hydroxy-2,2,6,6-tetramethylpiperidin-1-oxyl (TEMPOL) or 4-oxo-2,2,6,6-tetramethylpiperidine-N-oxyl (TEMPONE) (see Figure 15.4), or a related fragment attached to a macromolecule. The important requirement is that the electron spin should be characterised by slow relaxation (narrow ESR lines), so that it is possible to saturate the ESR transitions. The simplest possible expression for the ODNP enhancement factor can be obtained in full analogy with the steady-state solution given in Eq. (3.28) together with Eq. (3.22), where we assume that the S-spin is saturated:

$$\left\langle \hat{I}_z \right\rangle_{steady-state} = I_z^{eq}\left(1+\frac{\gamma_S}{\gamma_I}\frac{\sigma_{IS}}{\rho_I}\right) = I_z^{eq}\left(1+\frac{\gamma_S}{\gamma_I}\frac{W_2 - W_0}{W_0 + 2W_{1I} + W_2}\right) = I_z^{eq}(1+\varepsilon) \qquad (15.20)$$

We should notice that the ratio γ_S/γ_I is very large, about 650 if I are protons, which allows very large (and negative, but this is not important) signal enhancements $(1 + \varepsilon)$. The transition probabilities refer to Figure 3.5, modified to account for the huge difference between the I- and S-level splittings. As discussed in the classical review by Hausser and Stehlik[55], or in its modern-day counterpart[56], Eq. (15.20) needs to be corrected in two ways to account for the experimental setup. First, we need to include the possibility

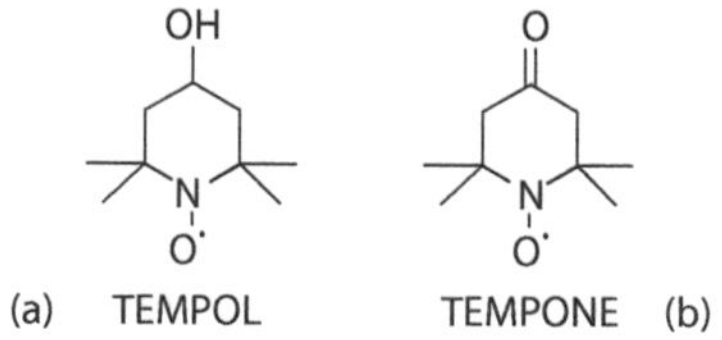

OH
O
N
O·
(a) TEMPOL
TEMPONE (b)

FIGURE 15.4 Structural formulas of two common nitroxide radicals.

of other, diamagnetic relaxation mechanisms ($R_{1,dia}$) for the I-spins. This leads to a quantity called the *leakage factor*, f, in the form:

$$f = \rho_I / (\rho_I + R_{1,dia}) \tag{15.21}$$

which usually is close to unity. The other correction, the saturation factor s, is related to the extent of saturation of the S-spin. If we deal with a nitroxide radical, its ESR line is split by rather strong hyperfine interaction with the nitrogen spin (spin 1 for ^{14}N or 1/2 for ^{15}N), which makes it difficult to fully saturate it. The expressions for the factor s can be found in Hausser and Stehlik[55] and Griesinger *et al.*,[56] and s can be reasonably close to unity. Expression 15.20 then becomes:

$$\left\langle \hat{I}_z \right\rangle_{steady-state} = I_z^{eq}\left(1 + \frac{\gamma_S}{\gamma_I}\frac{\sigma_{IS}}{\rho_I} fs\right) = I_z^{eq}\left(1 + \frac{\gamma_S}{\gamma_I}\frac{W_2 - W_0}{W_0 + 2W_{1I} + W_2} fs\right) = I_z^{eq}\left(1 + \frac{\gamma_S}{\gamma_I}\xi fs\right) \tag{15.22a}$$

with

$$\xi = \frac{W_2 - W_0}{W_0 + 2W_{1I} + W_2} \tag{15.22b}$$

The quantity ξ is called the *coupling factor* and contains the important magnetic field dependence of the DNP enhancement factor. The maximum value it can attain for pure dipole-dipole interaction is 1/2 (exactly the same as in Eq. (3.32b)), which results in maximal signal enhancement of about 330.

The transition probabilities entering the coupling factors are related to the spectral densities, as discussed in Chapter 3. It should be noted that the spectral densities corresponding to some transitions will be taken at very high frequencies, dictated by the high magnetogyric ratio for the electron spin. The transition probabilities can have their origin in the dipole-dipole nuclear spin–electron spin part of the hyperfine interaction, as well as in the scalar part. The scalar part seems to be unimportant in the case of nitroxide solutions and we neglect it in the following. Moreover, the nitroxide radicals used as the source of polarisation may not interact strongly with water (or other molecules whose nuclear spins we are interested in), as a consequence of which the spectral densities of interest can contain an important component caused by translational diffusion. The coupling factor depends in any case on the same quantities/parameters as the inner- and outer-sphere parts of the PRE, as discussed in Section 15.2. Following this line of thought, Bennati and co-workers[57] reported a variable-temperature ^1H NMRD study of ^{15}N-labelled and deuterated TEMPONE (see Figure 15.4) in water solution, analysed the field-dependent PRE in terms of superposition of a rotational and translational dynamics, and used the resulting parameters to predict the coupling factor as a function of the magnetic field. At 298 K and 0.35 T (15 MHz ^1H frequency, X-band ESR) they obtained $\xi = 0.35$, corresponding together with an estimated $f = 0.95$ to the maximum possible enhancement of over 200, to be compared with 170 obtained experimentally[58]. At the higher field of 3.3 T (140 MHz ^1H frequency, W-band ESR), the predicted coupling factor fell to 0.05, again in qualitative agreement with experimental data. The DNP experiments at both fields were carried out by placing the sample inside the electron-nuclear double resonance (ENDOR) probe head[58].

The fast decrease of the coupling factor with increasing magnetic field is a general phenomenon[56] depending on the occurrence of spectral densities at very high frequencies, corresponding to the range beyond the second dispersion in the T_{1M}^{-1} plot in Figure 15.1. This makes liquid-state DNP with direct irradiation of the ESR lines at high magnetic fields, which is required for biomolecular systems, a very demanding endeavour. An alternative experimental approach makes use of a mechanical shuttling of the sample between a lower field, where the polarisation can be carried out with a reasonably high coupling factor, and a high field, where the NMR detection is performed under high-resolution conditions.

For details of this approach, we refer the reader to the review by Griesinger *et al.*[56]. The low-field ODNP experiments and applications to water dynamics in the presence of various biomolecular systems have recently been reviewed[59].

Another DNP method for liquids was proposed some time ago by Ardenkjaer-Larsen and co-workers[60]. Their approach is called *dissolution DNP* and also makes use of sample shuttling. The polarisation step is here carried out in a separate high-field magnet, where the sample is frozen and cooled down to about 1 K, and the ESR irradiation is provided by a high-power microwave source (a *gyrotron*). The DNP step is thus carried out under a combined effect of low temperature (which in itself greatly enhances the polarisation) and the DNP mechanisms operational in the solid state. The net result is nuclear polarisation many orders of magnitude larger than that at thermal equilibrium at room temperature. When the steady-state polarisation is reached (which may take one or several hours), the sample is rapidly dissolved by an injection system inside the polarising device and transferred to a high-resolution spectrometer. Dissolution DNP is mainly used for measurements on low-γ nuclei, such as ^{13}C and ^{15}N, with slow spin-lattice relaxation.

15.5 Applications of NMR Relaxation in Paramagnetic Systems

We wish to finish this chapter by describing some selected applications of paramagnetic effects on NMR relaxation.

15.5.1 MRI Contrast Agents

The first important application of paramagnetic relaxation studies that we wish to exemplify is related to the use of gadolinium(III)-based contrast agents for magnetic resonance imaging (MRI), described in numerous reviews[61,62] and books[63,64]. The contrast agents are important diagnostic tools, used in about 30% of all clinical MRI scans. When the contrast agents accumulate in a certain tissue or region, the water protons in that region experience a stronger PRE, which can easily be detected in T_1-weighted imaging sequences. In addition, certain specific binding effects can occur. Gadolinium(III) ions are toxic and the contrast agents must be stable, efficient (in the sense of creating sizable PRE at low concentration) and easily excreted.

Water proton NMRD profiles are an important tool for characterising and optimising this type of system. Sometimes, the NMRD studies are combined with multi-frequency EPR. We wish to introduce this kind of work by sketching a case history of bis-amide derivatives of the Gd(DTPA)-complex (DTPA = diethylenetriaminepentaacetic acid). The structure of the bis-methyl amide (BMA) of the DTPA ligand is shown in Figure 15.5. The ligands bind Gd(III) in an eight-coordinate fashion, leaving

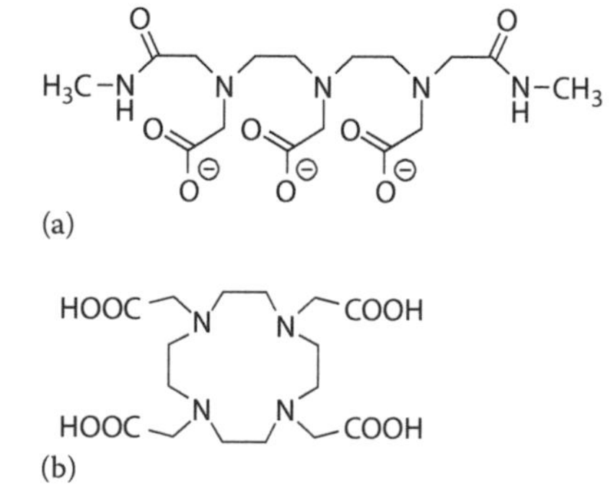

(a)

(b)

FIGURE 15.5 The structure of (a) bis-methyl amide of diethylenetriaminepentaacetic acid, DTPA-BMA; (b) 1,4,7,10-tetraazacyclododecane-1,4,7,10-tetraacetic acid, DOTA.

one coordination site for a labile water molecule, which can exchange rapidly with the bulk. The DTPA complex of Gd(III) has a charge of –2, while the bis-amide derivatives are electrically neutral. Gonzalez and co-workers studied the Gd(DTPA-BMA)(H_2O) complexes (the commercial contrast agent OMNISCAN) using oxygen-17 NMR at variable temperature, pressure and magnetic field[65]. The oxygen-17 longitudinal relaxation rate is a sum of a quadrupolar contribution and a PRE, while the transverse relaxation is dominated by the scalar relaxation and, in particular, the zero-frequency first term in Eq. (15.11). The measurements of Gonzalez *et al.* thus carry information on the reorientational dynamics, electron spin-relaxation and chemical exchange. In a later paper from the same group, the water proton NMRD and multiple-field ESR lineshapes were also reported for the same complex (and for some related systems, such as the charged Gd(DTPA)(H_2O)), and the massive experimental data sets were analysed using the SBM theory[66]. The theory is not fully adequate (see Section 15.2) and the authors tried to compensate for this by making certain *ad hoc* assumptions. The ESR data for Gd(DTPA-BMA)(H_2O) were re-interpreted more successfully by Rast and co-workers, who used a better model and reached a better understanding of the dynamic processes leading to electron spin-relaxation[12]. According to these authors, the effects of rotational modulation of the static ZFS are important for the ESR lineshapes of Gd(DTPA-BMA)(H_2O).

A fairly obvious way to increase the relaxivity of Gd(III) complexes, apparent from Eqs. (15.9) and (15.10), is to increase their rotational correlation time—the longer τ_R yields longer correlation times, τ_{cj}, which, in turn leads to enhanced T_{1M}^{-1}. This can be accomplished by forming various kinds of adducts with macromolecules. This strategy does not work with Gd(DTPA-BMA)(H_2O), which seems fairly inert towards *e.g.* proteins. This feature was in fact used in a nice way by Otting and Pintacuda, who added this complex to a protein solution and used the protein proton PREs to determine which residues were at the protein surface[67]. However, other amides behave in a chemically different manner. For example, Anelli and co-workers investigated Gd(DTPA) functionalised with the sulfonamide ethylene sulfanilamide (Gd-DTPA-SA)[68]. NMRD profiles of aqueous protons were measured in the presence and in the absence of the protein carbonic anhydrase and the low-field PRE was found to increase by a factor of about four after forming the protein adduct. The effect of the protein can be schematically illustrated as in Figure 15.6. Anelli *et al*[68]. interpreted their data using the SBM approach for the low-molecular weight complex and the method of Bertini *et al.*[25] for the adduct. The data were re-interpreted by Kruk

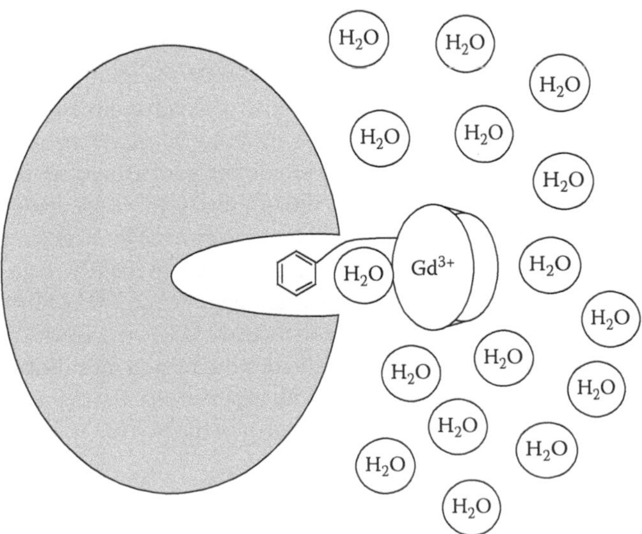

FIGURE 15.6 Schematic view of the non-covalent interactions between a Gd^{3+}-complex and a protein. (Adapted from Aime, S. et al., *Chem. Soc. Rev.*, 27, 19–29, 1998.)

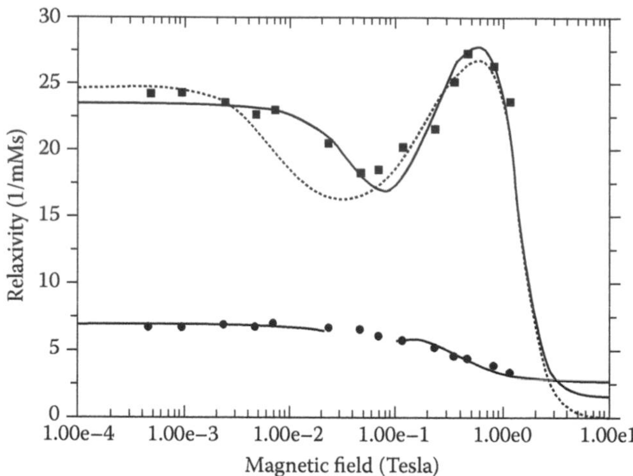

FIGURE 15.7 Experimental and calculated (solid line) NMRD profiles of aqueous solution of Gd-DTPA-SA (circles) and its adduct with carbonic anhydrase (squares). The best SBM curve for the adduct case is also shown (dashed line). (Reprinted with permission from Kruk D. and Kowalewski J., *J. Magn. Reson.* 162, 229–240, 2003. Copyright (2003) Elsevier.)

and Kowalewski using the "modified Florence" approach for the adduct and a model allowing for fast rotation for the case of Gd-DTPA-SA[26]. The results of the fits are shown in Figure 15.7, where a visibly less successful SBM fit for the adduct is also included. The authors first fitted the data for the slowly rotating adduct, for which the NMRD profile has more features and the theory is simpler. To interpret the Gd-DTPA-SA results, most of the parameters (except for τ_R) were simply assumed to be transferable from the larger complex to the smaller one. The rotation of the low–molecular weight complex is faster than the electron spin-relaxation and the results show that the formation of the protein adduct enhances the PRE by eliminating (or at least reducing) this most efficient term in the effective correlation time for the electron-nuclear dipole-dipole interaction (compare Eq. (15.10c)).

The last and most recent example we wish to provide here is concerned with Gd(III) complexed by a derivative of another frequently used ligand, 1,4,7,10-tetraazacyclododecane-1,4,7,10-tetraacetato(4-) (DOTA) (see Figure 15.5). As is the case with the DTPA analogues, the DOTA skeleton can be decorated with substituents in order to increase the molecular mass and thus, the rotational correlation time. Vander Elst and co-workers[69] investigated the Gd(DOTA) derivative with the molecular mass of 5.6 kDa, known as P760, with τ_R in water of about 2 ns at 310 K. The physico-chemical characterisation of the compound in aqueous solution was reported, including the NMRD profile analysed using a simple model. However, the relatively long rotational correlation time, combined with the expected sizable static ZFS, takes the system outside of the Redfield limit. The NMRD data were thus re-interpreted by Kruk *et al.*[70] using the Swedish slow-motion theory with both static and transient ZFS. In addition to the PRE data, the authors reported solution ESR data measured at two magnetic fields. The parameters obtained by fitting slow-motion theory to the NMRD profiles in the NMR frequency range up to about 40 MHz were used to predict high-frequency ESR spectra producing, however, too narrow lines. After allowing for a small anisotropy of the g-tensor, the simulated lineshapes reached a reasonable agreement with experiments. Further work of this kind is likely to be fruitful and to provide more definite joint interpretation of the NMRD and ESR data.

15.5.2 Paramagnetic Metalloproteins

The second type of application we wish to mention is related to the structure determination of paramagnetic metalloproteins in solution. An important difference between this case and the previous examples is the fact that in proteins one measures on the surrounding of the paramagnetic metal centre residing permanently within the molecule, *i.e. in the absence of exchange*. Thus, the paramagnetic effects are the strong in-complex properties, such as T_{1M}^{-1}, and not the quantities, such as T_{1P}^{-1}, reduced by the mole fraction, P_M, in Eq. (15.8). The use of paramagnetic constraints as a tool for protein structure determination has been subject to several reviews; for example, by the group of Bertini[45,46,71] and by Otting[72]. Briefly, four paramagnetism-based observable quantities can be useful: the pseudocontact shifts (PCSs), the residual dipolar couplings (RDCs), the PREs and the DLBs/cross-correlated relaxation rates (CCRRs). A pseudocontact shift is the change in the line position corresponding to the trace of Eq. (15.18). The PCSs have the large advantage of being very easy to measure in a simple heteronuclear single quantum coherence experiment. The RDCs make themselves visible by modifying the *J*-splittings and originate from a partial (and usually very weak) self-orientation of paramagnetic molecules with anisotropic magnetic susceptibility in the magnetic field, resulting in incomplete averaging of the dipolar interactions between nuclear spins. The use of the RDCs is, to a certain degree, complicated by their similarity to the dynamic frequency shifts, mentioned in Section 15.3. The two remaining properties, which are of main interest in the present context, were discussed in Section 15.2 (the PREs) and in Section 15.3 (the DLBs).

The first case we choose to discuss is the NMR characterisation of an example of a type II copper(II) protein, the copper-trafficking protein CopC from *Pseudomonas syringae*, reported by Arnesano and co-workers[73]. The structures of the apo protein and of the diamagnetic Cu(I) analogue were known beforehand. Copper(II) has a single unpaired electron and is thus an $S=1/2$ case. The classification in type I and type II copper(II) proteins refers to the rate of the electron spin-relaxation. Type II systems are characterised by slower electron spin-relaxation than type I; a slower electronic relaxation results in longer effective correlation times (see Eq. (15.10)), with the concomitant larger broadening of NMR signals (compare Eq. (15.11)). The electron spin-relaxation properties were investigated by NMRD measurements for the exchanging aqueous (solvent) protons, under the assumption that the correlation times in Eqs. (15.10) have an important contribution from the electronic relaxation and assuming a single, effective electron spin-relaxation time. This relaxation time was estimated to be about 3 ns at room temperature. High-resolution ^{1}H spectra are difficult to obtain for protons in residues less than about 11 Å from the metal because of extensive broadening. Since the paramagnetic relaxation rates are proportional to γ_I^2, the broadening of ^{13}C (or ^{15}N) resonances is expected to be less severe, which indeed proved to be the case for the protein under consideration. The enhancements of the longitudinal relaxation rate, dominated by the dipolar relaxation and thus proportional to r_{IS}^{-6}, were measured for 23 carbon nuclei and were used (along with 1777 proton-proton NOEs, some dihedral angle and hydrogen-bond constraints, as well as 83 proton PCSs) to determine the structure. The resulting structure is shown in Figure 15.8. It should be stressed that the location of the copper ions could only be determined by including the ^{13}C PREs in the refinement.

The second example is that of a calcium protein, Calbindin D_{9k}, discussed earlier in Chapters 11 and 12, in which one of the two calcium ions is substituted by cerium(III), a lanthanide ion characterised by a single 4*f* electron. Note that, because of the strong spin-orbit coupling, the relevant angular momentum quantum number is in this case *J* (here, equal to 5/2 in the ground state) rather than *S*[1] and that the electron spin-relaxation is very fast. Also, the electron *g*-factor (g_e) in the equations introduced in the preceding sections should be substituted by the Landé factor, g_J (see Bertini *et al.*[1] and the book by Atkins and Friedman, *Further reading*). The system was studied by Bertini and co-workers[74], who measured the DLBs associated with the cross-correlation between the ^{1}H–^{15}N dipolar interaction and the DSA/Curie-spin relaxation for more than 50 residues. Having excluded the few residues with large internal mobility, the authors used the remainder of the DLBs as structural constraints (along with a large number of "diamagnetic" constraints) in the refinement procedure. Including the relatively small number of CCRRs in the refinement provided a significant decrease of the root-mean-square deviation of the atomic coordinates in the structure. The efficiency of the DLBs as structural constraints has been compared with other paramagnetic data[74,75]. The calbindin D_{9k} example involves a certain level of protein modification by substituting the native calcium ion by a lanthanide. More sophisticated forms of protein (or nucleic acids) engineering have been developed in order to attach a nitroxide radical or a chelating

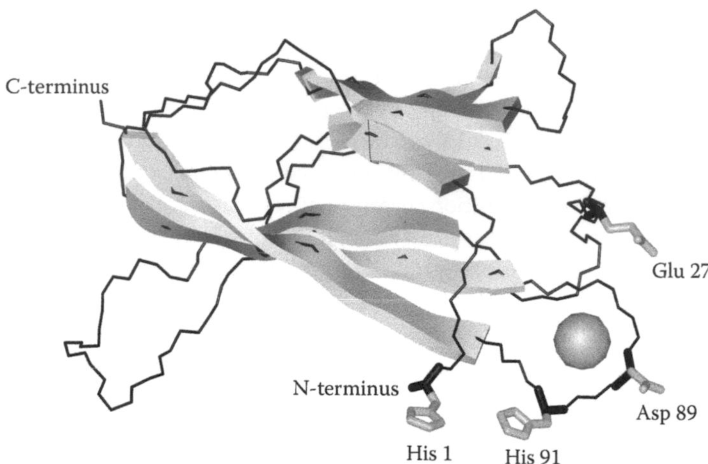

FIGURE 15.8 Ribbon diagram of the structure of the copper-trafficking protein CopC from *Pseudomonas syringae*, determined from solution NMR. The coordinates were taken from the Protein Data Bank (www.rscb.org, accession code 1OT4). The side-chains of the possible Cu(II) ligands, as determined from the structure, are shown. The figure was produced using Insight (version 2000).

group (which can be similar to the ligands shown in Figure 15.5) able to bind lanthanides to a protein with the goal of using their paramagnetism. This field was dealt with in the book by Bertini *et al.*[1] and in a review by Liu and co-workers[76].

15.5.3 Paramagnetic Dark State Spectroscopy

In Section 13.7, we have discussed the possibilities of characterising the sparsely populated transient states by transverse relaxation dispersion and saturation transfer experiments. The use of paramagnetic NMR tools opens up other possibilities to investigate this type of system, as reviewed by Clore and co-workers[77,78].

Paramagnetic dark state spectroscopy makes use of measurements of the enhancement of the transverse relaxation rate. This can give straightforward information about the transient excited states provided that two conditions are fulfilled. The first requirement is that the exchange is fast in comparison with the difference of R_2 between the minor and the major state; $k_{ex} \gg R_2^{minor} - R_2^{major}$. The second condition is that the sparsely populated state is characterised by a significantly shorter effective distance between the nuclear spin of interest and the electron spin/paramagnetic centre. Assuming that the scalar contribution to the PRE is negligible, the measured transverse PRE reports on population-weighted $\langle r_{IS}^{-6} \rangle$, which will then have an important contribution from the minor state and will allow, under advantageous conditions, its structural characteristics to be estimated. Advantageous conditions can occur, for example, if the minor state is much more compactly folded than the major state, which leads to a dark state with much shorter distances from a variety of protons to the paramagnetic centre compared with the ground state. For a more detailed description of the method and its data analysis, we wish to refer to the reviews by Clore and co-workers[77,78].

Another method for studies of dark states in paramagnetic systems is the Carr, Purcell, Meiboom and Gill (CPMG) relaxation dispersion method, discussed already in Chapter 13. Here, it can be applied to estimate the exchange rates and the pseudocontact shifts in the excited states. An early example of such a study was reported by Wang and co-workers[79], who studied a protein containing Zn(II) in the native state, substituted by paramagnetic Co(II), characterised by anisotropic magnetic susceptibility and thus giving rise to pseudocontact shifts. The authors measured the ^{15}N CPMG dispersions and found that some residues in the C-terminal domain (the metal ion is bound in the N-terminal part) of the Co(II) form displayed pronounced dispersions, which did not occur in the Zn(II) analogue. The difference in behaviour was interpreted in terms of modulation of the PCS by the domain-domain motion on the timescale of milliseconds. A more recent and more quantitatively oriented example published by Hass *et al.*[80] deals with a protein without metals in its native state, decorated with a paramagnetic lanthanide tag. The authors found protons to be more suitable than ^{15}N for CPMG dispersion studies. By measuring the dispersion for a large number of amide protons at several temperatures (covering fast and intermediate exchange ranges), they were able to present a quantitative determination of a number of PCS values for the excited dark state. The analysis led, however, to a somewhat disappointing conclusion that the observed relaxation dispersion effects could be accounted for by motion of the tag itself.

References

1. Bertini, I.; Luchinat, C.; Parigi, G.; and Ravera, E., *NMR of Paramagnetic Molecules. Applications to Metallobiomolecules and Models.* 2nd edn. Elsevier: Amsterdam, 2017.
2. Kowalewski, J.; Kruk, D.; and Parigi, G., NMR relaxation in solution of paramagnetic complexes: Recent theoretical progress for $S \geq 1$. *Adv. Inorg. Chem.* 2005, 57, 41–104.
3. Helm, L., Relaxivity in paramagnetic systems: Theory and mechanisms. *Progr. NMR Spectr.* 2006, 49, 45–64.
4. Kowalewski, J.; and Kruk, D., *Paramagnetic Relaxation in Solution.* Wiley, eMagRes, 2011.
5. Abragam, A.; and Bleaney, B., *Electron Paramagnetic Resonance of Transition Ions.* Clarendon Press: Oxford, 1970.
6. Weil, J. A.; Bolton, J. R.; and Wertz, J. E., *Electron Paramagnetic Resonance.* Wiley: New York, 1994.
7. Griffith, J. S., *The Theory of Transition-Metal Ions.* Cambridge University Press: Cambridge, 1961.
8. Muus, L. T.; and Atkins, P. W., Eds., *Electron Spin Relaxation in Liquids.* Plenum Press: New York, 1972.
9. Freed, J. H., Theory of slow tumbling ESR spectra for nitroxides. In *Spin Labeling: Theory and Applications*, Berliner, L. J., Ed. Academic Press: New York, 1976; pp. 53–132.
10. Bloembergen, N.; and Morgan, L. O., Proton relaxation times in paramagnetic solutions. Effects of electron spin relaxation. *J. Chem. Phys.* 1961, 34, 842–850.
11. Rubinstein, M.; Baram, A.; and Luz, Z., Electronic and nuclear relaxation in solutions of transition metal ions with spin S = 3/2 and 5/2. *Mol. Phys.* 1971, 20, 67–80.
12. Rast, S.; Fries, P. H.; and Belorizky, E., Static zero field splitting effects on the electronic relaxation of paramagnetic metal ion complexes in solution. *J. Chem. Phys.* 2000, 113, 8724–8735.
13. Swift, T. J.; and Connick, R. E., NMR-relaxation mechanism of O17 in aqueous solutions of paramagnetic cations and the lifetime of water molecules in the first coordination sphere. *J. Chem. Phys.* 1962, 37, 307–320.
14. Luz, Z.; and Meiboom, S., Proton relaxation in dilute solutions of cobalt(II) and nickel(II) ions in methanol and the rate of methanol exchange of the solvation sphere. *J. Chem. Phys.* 1964, 40, 2686–2692.
15. Gottlieb, H. P. W.; Barfield, M.; and Doddrell, D. M., Electron-spin density matrix description of nuclear-spin-lattice relaxation in paramagnetic molecules. *J. Chem. Phys.* 1977, 67, 3785–3794.
16. Kowalewski, J.; Laaksonen, A.; Nordenskiöld, L.; and Blomberg, M., A nonempirical SCF-MO study of the validity of the Solomon-Bloembergen equation for the hexa-aquonickel (II) ion. *J. Chem. Phys.* 1981, 74, 2927–2930.
17. Nordenskiöld, L.; Laaksonen, A.; and Kowalewski, J., Applicability of the Solomon-Bloembergen equation to the study of paramagnetic transition metal-water complexes. An ab initio SCF-MO study. *J. Am. Chem. Soc.* 1982, 104, 379–382.
18. Ma, L. X.; Jorgensen, A. M. M.; Sorensen, G. O.; Ulstrup, J.; and Led, J. J., Elucidation of the paramagnetic R-1 relaxation of heteronuclei and protons in Cu(II) plastocyanin from Anabaena variabilis. *J. Am. Chem. Soc.* 2000, 122, 9473–9485.

19. Hansen, D. F.; Gorelsky, S. I.; Sarangi, R.; Hodgson, K. O.; Hedman, B.; Christensen, H. E. M.; Solomon, E. I.; and Led, J. J., Reinvestigation of the method used to map the electronic structure of blue copper proteins by NMR relaxation. *J. Biol. Inorg. Chem.* 2006, 11, 277–285.
20. Hansen, D. F.; Westler, W. M.; Kunze, M. B. A.; Markley, J. L.; Weinhold, F.; and Led, J. J., Accurate structure and dynamics of the metal-site of paramagnetic metalloproteins from NMR parameters using natural bond orbitals. *J. Am. Chem. Soc.* 2012, 134, 4670–4682.
21. Toth, E.; Helm, L.; Kellar, K. E.; and Merbach, A. E., Gd(DTPA-bisamide)alkyl copolymers: A hint for the formation of MRI contrast agents with very high relaxivity. *Chem. Eur. J.* 1999, 5, 1202–1211.
22. Westlund, P.-O., A generalized Solomon-Bloembergen-Morgan theory for arbitrary electron spin quantum number S—the dipole-dipole coupling between a nuclear spin $I = 1/2$ and an electron spin system $S = 5/2$. *Mol. Phys.* 1995, 85, 1165–1178.
23. Bertini, I.; Kowalewski, J.; Luchinat, C.; Nilsson, T.; and Parigi, G., Nuclear spin relaxation in paramagnetic complexes of S=1: Electron spin relaxation effects. *J. Chem. Phys.* 1999, 111, 5795–5807.
24. Kruk, D.; Nilsson, T.; and Kowalewski, J., Nuclear spin relaxation in paramagnetic systems with zero-field splitting and arbitrary electron spin. *Phys. Chem. Chem. Phys.* 2001, 3, 4907–4917.
25. Bertini, I.; Galas, O.; Luchinat, C.; and Parigi, G., A computer program for the calculation of paramagnetic enhancements of nuclear-relaxation rates in slowly rotating systems. *J. Magn. Reson. Ser. A* 1995, 113, 151–158.
26. Kruk, D.; and Kowalewski, J., Nuclear spin relaxation in paramagnetic systems ($S \geq 1$) under fast rotation conditions. *J. Magn. Reson.* 2003, 162, 229–240.
27. Benetis, N.; Kowalewski, J.; Nordenskiöld, L.; Wennerström, H.; and Westlund, P.-O., Nuclear spin relaxation in paramagnetic systems. The slow motion problem for electron spin relaxation. *Mol. Phys.* 1983, 48, 329–346.
28. Nilsson, T.; and Kowalewski, J., Slow-motion theory of nuclear spin relaxation in paramagnetic low- symmetry complexes: A generalization to high electron spin. *J. Magn. Reson.* 2000, 146, 345–358.
29. Nilsson, T.; Parigi, G.; and Kowalewski, J., Experimental NMRD profiles for some low-symmetry Ni(II) complexes ($S = 1$) in solution and their interpretation using slow-motion theory. *J. Phys. Chem. A* 2002, 106, 4476–4488.
30. Odelius, M.; Ribbing, C.; and Kowalewski, J., Molecular dynamics simulation of the zero-field splitting fluctuations in aqueous Ni(II). *J. Chem. Phys.* 1995, 103, 1800–1811.
31. Rantaharju, J.; Mares, J.; and Vaara, J., Spin dynamics simulation of electron spin relaxation in Ni2+(aq). *J. Chem. Phys.* 2014, 141, 014109.
32. Rantaharju, J.; and Vaara, J., Liquid-state paramagnetic relaxation from first principles. *Phys. Rev. A* 2016, 94, 043413.
33. Odelius, M.; Ribbing, C.; and Kowalewski, J., Spin dynamics under the Hamiltonian varying with time in discrete steps: Molecular dynamics-based simulation of electron and nuclear spin relaxation in aqueous nickel(II). *J. Chem. Phys.* 1996, 104, 3181–3188.
34. Abernathy, S. M.; and Sharp, R. R., Spin dynamics calculations of electron and nuclear spin relaxation times in paramagnetic solutions. *J. Chem. Phys.* 1997, 106, 9032–9043.
35. Schaefle, N.; and Sharp, R., Four complementary theoretical approaches for the analysis of NMR paramagnetic relaxation. *J. Magn. Reson.* 2005, 176, 160–170.
36. Rast, S.; Fries, P. H.; Belorizky, E.; Borel, A.; Helm, L.; and Merbach, A. E., A general approach to the electronic spin relaxation of Gd(III) complexes in solutions. Monte Carlo simulations beyond the Redfield limit. *J. Chem. Phys.* 2001, 115, 7554–7563.
37. Fries, P. H.; and Belorizky, E., Relaxation theory of the electronic spin of a complexed paramagnetic metal ion in solution beyond the Redfield limit. *J. Chem. Phys.* 2007, 126, 204503.
38. Belorizky, E.; Fries, P. H.; Helm, L.; Kowalewski, J.; Kruk, D.; Sharp, R. R.; and Westlund, P. O., Comparison of different methods for calculating the paramagnetic relaxation enhancement of nuclear spins as a function of the magnetic field. *J. Chem. Phys.* 2008, 128, 052315.

39. Gueron, M., Nuclear relaxation in macromolecules by paramagnetic ions: A novel mechanism. *J. Magn. Reson.* 1975, 19, 58–66.
40. Vega, A. J.; and Fiat, D., Nuclear relaxation processes of paramagnetic complexes. The slow motion case. *Mol. Phys.* 1976, 31, 347–355.
41. Freed, J. H., Dynamic effects of pair correlation functions on spin relaxation by translational diffusion in liquids. II. Finite jumps and independent T1 processes. *J. Chem. Phys.* 1978, 68, 4034–4037.
42. Rast, S.; Belorizky, E.; Fries, P. H.; and Travers, J. P., Mechanisms of the intermolecular nuclear magnetic relaxation dispersion of the (CH3)(4)N+ protons in Gd3+ heavy-water solutions. Interest for the theory of magnetic resonance imaging. *J. Phys. Chem. B* 2001, 105, 1978–1983.
43. Fries, P. H.; Ferrante, G.; Belorizky, E.; and Rast, S., The rotational motion and electronic relaxation of the Gd(III) aqua complex in water revisited through a full proton relaxivity study of a probe solute. *J. Chem. Phys.* 2003, 119, 8636–8644.
44. Kruk, D.; and Kowalewski, J., General treatment of paramagnetic relaxation enhancement associated with translational diffusion. *J. Chem. Phys.* 2009, 130, 174104.
45. Bertini, I.; Luchinat, C.; and Parigi, G., Paramagnetic constraints: An aid for quick solution structure determination of paramagnetic metalloproteins. *Concepts Magn. Reson.* 2002, 14, 259–286.
46. Bertini, I.; Luchinat, C.; and Parigi, G., Magnetic susceptibility in paramagnetic NMR. *Progr. NMR Spectr.* 2002, 40, 249–273.
47. Bertini, I.; Kowalewski, J.; Luchinat, C.; and Parigi, G., Cross correlation between the dipole-dipole interaction and the Curie spin relaxation: The effect of anisotropic magnetic susceptibility. *J. Magn. Reson.* 2001, 152, 103–108.
48. Ghose, R.; and Prestegard, J. H., Electron spin-nuclear spin cross-correlation effects on multiplet splittings in paramagnetic proteins. *J. Magn. Reson.* 1997, 128, 138–143.
49. Madhu, P. K.; Grandori, R.; Hohenthanner, K.; Mandal, P. K.; and Müller, N., Geometry dependent two-dimensional heteronuclear multiplet effects in paramagnetic proteins. *J. Biomol. NMR* 2001, 20, 31–37.
50. Werbelow, L. G., *Dynamic Frequency Shift*. Wiley, eMagRes, 2011.
51. Pintacuda, G.; Kaikkonen, A.; and Otting, G., Modulation of the distance dependence of paramagnetic relaxation enhancements by CSAxDSA cross-correlation. *J. Magn. Reson.* 2004, 171, 233–243.
52. Orton, H. W.; Kuprov, I.; Loh, C. T.; and Otting, G., Using paramagnetism to slow down nuclear relaxation in protein NMR. *J. Phys. Chem. Lett.* 2016, 7, 4815–4818.
53. Ling, W.; and Jerschow, A., Relaxation-allowed nuclear magnetic resonance transitions by interference between the quadrupolar coupling and the paramagnetic interaction. *J. Chem. Phys.* 2007, 126, 064502.
54. Overhauser, A. W., Polarization of nuclei in metals. *Phys. Rev.* 1953, 92, 411–415.
55. Hausser, K. H.; and Stehlik, D., Dynamic nuclear polarization in liquids. *Adv. Magn. Reson.* 1968, 3, 79–139.
56. Griesinger, C.; Bennati, M.; Vieth, H. M.; Luchinat, C.; Parigi, G.; Höfer, P.; Engelke, F.; Glaser, S. J.; Denysenkov, V.; and Prisner, T. F., Dynamic nuclear polarization at high magnetic fields in liquids. *Progr. NMR Spectr.* 2012, 64, 4–28.
57. Bennati, M.; Luchinat, C.; Parigi, G.; and Türke, M. T., Water H-1 relaxation dispersion analysis on a nitroxide radical provides information on the maximal signal enhancement in Overhauser dynamic nuclear polarization experiments. *Phys. Chem. Chem. Phys.* 2010, 12, 5902–5910.
58. Türke, M. T.; Tkach, I.; Reese, M.; Höfer, P.; and Bennati, M., Optimization of dynamic nuclear polarization experiments in aqueous solution at 15 MHz/9.7 GHz: A comparative study with DNP at 140 MHz/94 GHz. *Phys. Chem. Chem. Phys.* 2010, 12, 5893–5901.
59. Franck, J. M.; Pavlova, A.; Scott, J. A.; and Han, S., Quantitative cw Overhauser effect dynamic nuclear polarization for the analysis of local water dynamics. *Progr. NMR Spectr.* 2013, 74, 33–56.
60. Ardenkjaer-Larsen, J. H.; Fridlund, B.; Gram, A.; Hansson, G.; Hansson, L.; Lerche, M. H.; Servin, R.; Thaning, M.; and Golman, K., Increase of signal-to-noise ratio of >10,000 times in liquid state NMR. *Proc. Natl. Acad. Sci. USA* 2003, 100, 10158–10163.

61. Caravan, P.; Ellison, J. J.; McMurry, T. J.; and Lauffer, R. B., Gadolinium(III) chelates as MRI contrast agents: Structure, dynamics, and applications. *Chem. Rev.* 1999, 99, 2293–2352.
62. Aime, S.; Botta, M.; Fasano, M.; and Terreno, E., Lanthanide(III) chelates for NMR biomedical applications. *Chem. Soc. Rev.* 1998, 27, 19–29.
63. Krause, W. E. Ed., *Contrast Agents I. Magnetic Resonance Imaging.* Springer: Berlin, 2002.
64. Toth, E.; and Merbach, A. E. Eds., *The Chemistry of Contrast Agents in Medical Magnetic Resonance Imaging.* Wiley: Chichester, 2001.
65. Gonzalez, G.; Powell, D. H.; Tissieres, V.; and Merbach, A. E., Water-exchange, electronic relaxation, and rotational dynamics of the MRI contrast agent [Gd(DTPA-BMA)(H2O)] in aqueous solution—a variable pressure, temperature, and magnetic field O-17 NMR study. *J. Phys. Chem.* 1994, 98, 53–59.
66. Powell, D. H.; Ni Dhubhghaill, O. M.; Pubanz, D.; Helm, L.; Lebedev, Y. S.; Schlaepfer, W.; and Merbach, A. E., Structural and dynamic parameters obtained from O-17 NMR, EPR, and NMRD studies of monomeric and dimeric Gd3+ complexes of interest in magnetic resonance imaging: An integrated and theoretically self consistent approach. *J. Am. Chem. Soc.* 1996, 118, 9333–9346.
67. Pintacuda, G.; and Otting, G., Identification of protein surfaces by NMR measurements with a paramagnetic Gd(III) chelate. *J. Am. Chem. Soc.* 2002, 124, 372–373.
68. Anelli, P. L.; Bertini, I.; Fragai, M.; Lattuada, L.; Luchinat, C.; and Parigi, G., Sulfonamide-functionalized gadolinium DTPA complexes as possible contrast agents for MRI: A relaxometric investigation. *Eur. J. Inorg. Chem.* 2000, 625–630.
69. Vander Elst, L.; Port, M.; Raynal, I.; Simonot, C.; and Muller, R. N., Physicochemical characterization of P760, a new macromolecular contrast agent with high relaxivity. *Eur. J. Inorg. Chem.* 2003, 2495–2501.
70. Kruk, D.; Kowalewski, J.; Tipikin, D. S.; Freed, J. H.; Moscicki, M.; Mielczarek, A.; and Port, M., Joint analysis of ESR lineshapes and ^{1}H NMRD profiles of DOTA-Gd derivatives by means of the slow motion theory. *J. Chem. Phys.* 2011, 134, 024508.
71. Bertini, I.; Luchinat, C.; Parigi, G.; and Pierattelli, R., Perspectives in paramagnetic NMR of metalloproteins. *Dalton Trans.* 2008, 3782–3790.
72. Otting, G., Protein NMR using paramagnetic ions. *Ann. Rev. Biophys.* 2010, 39, 387–405.
73. Arnesano, F.; Banci, L.; Bertini, I.; Felli, I. C.; Luchinat, C.; and Thompsett, A. R., A strategy for NMR characterization of type II copper(II) proteins: The case of copper trafficking protein CopC from Pseudomonas Syringae. *J. Am. Chem. Soc.* 2003, 125, 7200–7208.
74. Bertini, I.; Cavallaro, G.; Cosenza, M.; Kümmerle, R.; Luchinat, C.; Piccioli, M.; and Poggi, L., Cross correlation rates between Curie spin and dipole-dipole relaxation in paramagnetic proteins: The case of cerium substituted calbindin D9k. *J. Biomol. NMR* 2002, 23, 115–125.
75. Barbieri, R.; Luchinat, C.; and Parigi, G., Backbone-only protein solution structures with a combination of classical and paramagnetism-based constraints: A method that can be scaled to large molecules. *ChemPhysChem* 2004, 5, 797–806.
76. Liu, W. M.; Overhand, M.; and Ubbink, M., The application of paramagnetic lanthanoid ions in NMR spectroscopy on proteins. *Coord. Chem. Rev.* 2014, 273, 2–12.
77. Clore, G. M.; and Iwahara, J., Theory, practice, and applications of paramagnetic relaxation enhancement for the characterization of transient low-population states of biological macromolecules and their complexes. *Chem. Rev.* 2009, 109, 4108–4139.
78. Anthis, N. J.; and Clore, G. M., Visualizing transient dark states by NMR spectroscopy. *Quart. Rev. Biophys.* 2015, 48, 35–116.
79. Wang, X.; Srisailam, S.; Yee, A. A.; Lemak, A.; Arrowsmith, C.; Prestegard, J. H.; and Tian, F., Domain-domain motions in proteins from time-modulated pseudocontact shifts. *J. Biomol. NMR* 2007, 39, 53–61.
80. Hass, M. A. S.; Liu, W. M.; Agafonov, R. V.; Otten, R.; Phung, L. A.; Schilder, J. T.; Kern, D.; and Ubbink, M., A minor conformation of a lanthanide tag on adenylate kinase characterized by paramagnetic relaxation dispersion NMR spectroscopy. *J. Biomol. NMR* 2015, 61, 123–136.

16

Long-Lived Nuclear Spin States and Related Topics

One of the exciting developments in nuclear magnetic resonance (NMR) over the last decade is related to long-lived nuclear spin states. In simple cases, the long-lived states (LLS) are related to the concepts of singlet and triplet states for pairs of indistinguishable particles with spin 1/2. The ideas behind these concepts are presented in Section 16.1. The details of some experiments allowing observation of the nuclear singlet states are presented in Section 16.2, and the relaxation properties of the singlet states are described in Section 16.3. Section 16.4 deals with long-lived coherences, while Section 16.5 covers applications of the nuclear singlet states. Finally, Section 16.6 is concerned with the related topic of *para*-hydrogen induced polarisation (PHIP).

16.1 Nuclear Singlet and Triplet States

Systems of identical (indistinguishable) particles have to display distinct symmetry properties under exchange of two of the particles. The nature of the symmetry properties depends on the spin of the particles and is summarised by the Pauli principle, which can be stated as follows: when the labels of two identical particles with half-integer spin (1/2, 3/2, 5/2, …) are exchanged, the wave function of the system changes sign. Particles of this kind are called *fermions* and their wave functions are denoted as antisymmetric. For integer spin particles (*bosons*), the Pauli principle states that the wave function has to be symmetric under exchange of the labels of identical particles.

Consider a pair of electrons in the helium atom. The two identical spin 1/2 particles have to be described by an antisymmetric wave function. In the simple electron structure theory, the electrons in an atom are described by spatial functions – called *atomic orbitals* – and the spin functions α and β. The Pauli principle requires that the symmetric spatial function, for example that obtained by assigning both electrons to the 1*s* orbital (electron configuration $(1s)^2$), expressed as $\Psi^0_{sym} = \psi_{1s}(1)\psi_{1s}(2)$, has to be combined with the antisymmetric spin function:

$$\sigma_{0,0}(1,2) = (\alpha(1)\beta(2) - \beta(1)\alpha(2))/\sqrt{2} \tag{16.1}$$

We say that the wave function $\Psi^0_{sym}\sigma_{0,0}(1,2)$ describes the electronic singlet ground state for the helium atom. For excited states of the He atom, we can formulate wave functions with one electron in 1*s* and one electron in a higher orbital, *e.g.* 2*s* (electron configuration 1*s*2*s*). The spatial part of this configuration can be expressed in one of the symmetrised forms:

$$\Psi_{sym} = 1/\sqrt{2}\left\{\psi_{1s}(1)\psi_{2s}(2) + \psi_{2s}(1)\psi_{1s}(2)\right\} \tag{16.2a}$$

$$\Psi_{antisym} = 1/\sqrt{2}\left\{\psi_{1s}(1)\psi_{2s}(2) - \psi_{2s}(1)\psi_{1s}(2)\right\} \tag{16.2b}$$

In order to conform to the Pauli principle, the symmetric spatial function has to be combined with the antisymmetric spin function $\sigma_{0,0}(1,2)$. This combination describes an excited singlet state of the helium atom. In a similar way, the antisymmetric spatial function has to be combined with one of three possible symmetric spin functions:

$$\begin{aligned}\sigma_{1,1}(1,2) &= \alpha(1)\alpha(2)\\ \sigma_{1,0}(1,2) &= \frac{1}{\sqrt{2}}\{\alpha(1)\beta(2)+\beta(1)\alpha(2)\}\\ \sigma_{1,-1}(1,2) &= \beta(1)\beta(2)\end{aligned} \tag{16.3}$$

These functions correspond to the situation with the individual spins of the two electrons coupled in such a way that the total spin quantum number of the two-electron system is $S=1$, with the magnetic quantum number, M_S, indicated by the second subscript. The number of possible states (multiplicity) is $2S+1=3$, which is the origin of the notation "triplet." In the same way, the function in Eq. (16.1) corresponds to $S=0$, $Ms=0$, with the multiplicity $2S+1=1$ (singlet). The spectroscopic transitions between different electronic states of the helium atom take place as a result of interaction with the electric component of the electromagnetic radiation. The selection rules say that the transitions between the orbitals differing by one unit in the orbital angular momentum (*e.g.* between *s* and *p* or *p* and *d* orbitals, but not between *s* and *d*) are allowed and that transitions between singlets and triplets (involving a multiplicity change) are forbidden.

Let us now consider a system of two indistinguishable nuclear spins 1/2; for example, the two protons in the hydrogen molecule, H_2. In full analogy with the electron spin case in the helium atom, the two proton spins can be described by the wave function of Eq. (16.1) (a singlet state) or any of the triplet state wave functions in Eq. (16.3). The spatial part of the wave function for the nuclear motion contains the vibrational, rotational and translational degrees of freedom, with the rotational motion being characterised by wave functions with well-defined symmetry properties with respect to the exchange of the indistinguishable atomic nuclei. The rotational states corresponding to the even values ($J=0, 2, 4, \ldots$) of the spatial angular momentum quantum number for the rigid rotor are symmetric with respect to the exchange of the particle labels, while those described by odd quantum numbers ($J=1, 3, 5,\ldots$) are antisymmetric. Following the Pauli principle, the nuclear spin singlets can only be combined with even J-values. The hydrogen molecules fulfilling this condition are called *para*-hydrogen. Similarly, the nuclear spin triplet states have to combine with the odd J-values, giving rise to the *ortho*-hydrogen molecules. Transitions between the *ortho*- and *para*- forms of hydrogen are forbidden by spectroscopic selection rules and, as a consequence, the two forms can be separated and stored for long periods. We shall return to the interesting NMR applications of *para*-hydrogen in Section 16.6.

The spin functions in Eqs. (16.1) and (16.3) can be replaced by corresponding kets, $|S_0\rangle, |T_{+1}\rangle, |T_0\rangle, |T_{-1}\rangle$ where S stands for singlet and T for triplet, while the subscripts denote the magnetic quantum numbers.

16.2 Creation and Detection of Long-Lived Spin States

The experimental proof of the possibility of creating and detecting nuclear singlet states was provided in the seminal papers by Carravetta, Johannessen and Levitt[1,2]. The field has been very active ever since and has been subject to several reviews[3-6]. The key issue to being able to access and observe a singlet state is that there must exist a symmetry-breaking interaction. Yet, the presence of such an interaction causes rapid transitions between the singlet and triplets and thus hides the particular relaxation behaviour of the singlet state. The idea, realised in singlet experiments, is to switch between two different Hamiltonians for the spin system and this will be described in the following.

Consider a weakly coupled two-spin system at high field. The Hamiltonian for such a system was given by Eq. (4.66). The eigenstates of this Hamiltonian are $|\alpha\alpha\rangle, |\alpha\beta\rangle, |\beta\alpha\rangle, |\beta\beta\rangle$. We denote here this Hamiltonian as $\hat{H}^{PB}$, referring to its eigenkets, the "product basis" (PB).

The system of two equivalent (indistinguishable) spins is described by the Hamiltonian:

$$\hat{H}^{ST} = \omega_0\left(\hat{I}_z + \hat{S}_z\right) + 2\pi J \hat{\mathbf{I}} \cdot \hat{\mathbf{S}} \tag{16.4}$$

Here, we use the notation *IS* for a two-spin system, even though the two spins are equivalent. Also, we keep the full form of the scalar coupling Hamiltonian. The matrix representation of the $\hat{H}^{ST}$ operator in the singlet-triplet (ST) basis, $|S_0\rangle, |T_{+1}\rangle, |T_0\rangle, |T_{-1}\rangle$, can be written:

$$H^{ST} = \begin{pmatrix} -\frac{3}{2}\pi J & 0 & 0 & 0 \\ 0 & \omega_0 + \frac{1}{2}\pi J & 0 & 0 \\ 0 & 0 & \frac{1}{2}\pi J & 0 \\ 0 & 0 & 0 & -\omega_0 + \frac{1}{2}\pi J \end{pmatrix} \tag{16.5}$$

The reader is encouraged to derive the matrix elements of $\hat{H}^{ST}$ in the singlet-triplet basis and thus prove Eq. (16.5). The matrix is diagonal, which means that the basis kets are its eigenvectors.

The two sets of basis kets are related to each other by a simple linear transformation:

$$\begin{pmatrix} |\alpha\alpha\rangle \\ |\alpha\beta\rangle \\ |\beta\alpha\rangle \\ |\beta\beta\rangle \end{pmatrix} = \mathbf{V} \begin{pmatrix} |S_0\rangle \\ |T_{+1}\rangle \\ |T_0\rangle \\ |T_{-1}\rangle \end{pmatrix} = \begin{pmatrix} 0 & 1 & 0 & 0 \\ \frac{1}{\sqrt{2}} & 0 & \frac{1}{\sqrt{2}} & 0 \\ -\frac{1}{\sqrt{2}} & 0 & \frac{1}{\sqrt{2}} & 0 \\ 0 & 0 & 0 & 1 \end{pmatrix} \begin{pmatrix} |S_0\rangle \\ |T_{+1}\rangle \\ |T_0\rangle \\ |T_{-1}\rangle \end{pmatrix} \tag{16.6}$$

Note that the transformation matrix $\mathbf{V}$ is unitary $(\mathbf{V}^{-1} = \mathbf{V})$, *i.e.* the same matrix also transforms the product basis into the ST basis. The next step is to transfer from the Hilbert space, with its basis kets of the PB or ST type, to the corresponding Liouville space. As discussed in Chapter 1, the dimensionality of the Liouville space for a two-spin (I=1/2, S=1/2) system is 16. So far, we have found it useful to work with the product operator basis. Now, we need to specify an operator basis suitable for the singlet-triplet formulation of the eigenstates. To begin with, we can restrict our interest to the operators that correspond to the populations of the four states. Cavadini, Sarkar and co-workers[7,8] derived relations between the product operators and the set of 16 triplet-singlet type operators (corresponding to the four populations described and various types of coherences, *e.g.* $|T_0\rangle\langle S_0|$). The four ST-type populations, $|S_0\rangle\langle S_0|, |T_{+1}\rangle\langle T_{+1}|, |T_0\rangle\langle T_0|, |T_{-1}\rangle\langle T_{-1}|$, are related to the four product operator objects through the conversion matrix[8]:

$$\begin{pmatrix} |S_0\rangle\langle S_0| \\ |T_{+1}\rangle\langle T_{+1}| \\ |T_0\rangle\langle T_0| \\ |T_{-1}\rangle\langle T_{-1}| \end{pmatrix} = \begin{pmatrix} \frac{1}{2} & 0 & -\frac{1}{2} & -1 \\ \frac{1}{2} & \frac{1}{2} & \frac{1}{2} & 0 \\ \frac{1}{2} & 0 & -\frac{1}{2} & 1 \\ \frac{1}{2} & -\frac{1}{2} & \frac{1}{2} & 0 \end{pmatrix} \begin{pmatrix} \frac{1}{2}1_{op} \\ \hat{I}_z + \hat{S}_z \\ 2\hat{I}_z\hat{S}_z \\ ZQ_x \end{pmatrix} \tag{16.7}$$

Note that the right-hand side of Eq. (16.7) contains the sum $\hat{I}_z + \hat{S}_z$ since in the case of indistinguishable spins, only the total z-component is relevant. The zero-quantum operator, ZQ_x, is the same as given in Eq. (7.14c). For the interesting case of the singlet population, $|S_0\rangle\langle S_0|$, we can extract the simple relation:

$$|S_0\rangle\langle S_0| = \frac{1}{4}\hat{1}_{op} - \frac{1}{2}\left(2\hat{I}_z\hat{S}_z\right) - ZQ_x \tag{16.8}$$

The important conclusion from Eq. (16.7) or (16.8) is that if we want to manipulate the singlet population, we can do that through either the two-spin longitudinal order or the x-component of the zero-quantum coherence. As discussed earlier, the unity operator does not evolve with time and can for all practical purposes be omitted.

In the original paper by Carravetta *et al.*[1], the singlet NMR experiment was initiated by equilibration at high magnetic field (see Figure 16.1). This results in the density operator of the form $\hat{\rho}_1 = b_I(\hat{I}_z + \hat{S}_z)$ or, in the matrix representation in the PB basis:

$$\hat{\rho}_1 = \frac{1}{4}\begin{pmatrix} 1+b_I & 0 & 0 & 0 \\ 0 & 1 & 0 & 0 \\ 0 & 0 & 1 & 0 \\ 0 & 0 & 0 & 1-b_I \end{pmatrix} \tag{16.9}$$

The symbol $b_I = b_S$ (for a homonuclear IS spin system) is defined in Eq. (1.19).

Next, the spin system is exposed to two 90° pulses with phases 0 and −90° and separated by a delay $\tau_1 = |\pi/\omega_\Delta^{high}|$, where $\omega_\Delta^{high} = \omega_1 - \omega_2$ is the difference between the chemical shifts (in angular frequency units) of the two spins at high field. The carrier frequency is placed in the middle of the interval between the two chemical shifts. We can describe the evolution of the density operator as follows:

$$\hat{I}_z + \hat{S}_z \xrightarrow{(90^\circ)_x} -\hat{I}_y - \hat{S}_y \xrightarrow{\tau_1} \hat{I}_x - \hat{S}_x \xrightarrow{(90^\circ)_{-y}} \hat{I}_z - \hat{S}_z \tag{16.10}$$

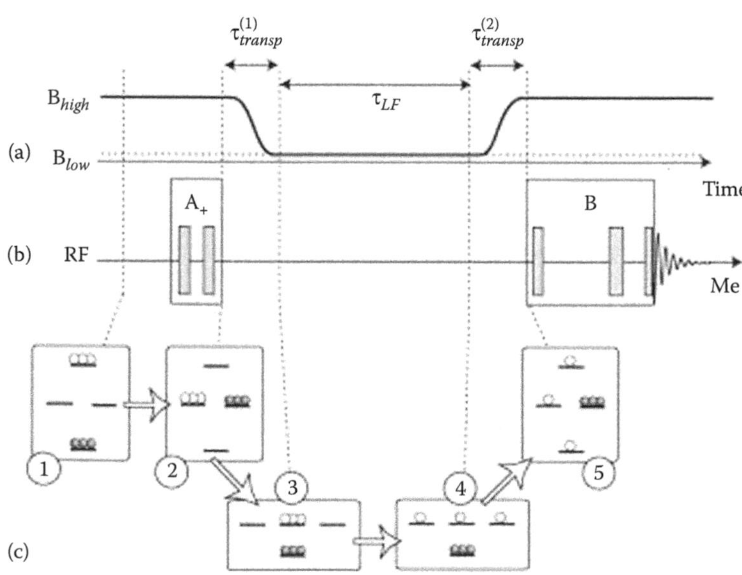

FIGURE 16.1 Field-cycling singlet experiment. (a) Sequence of magnetic fields. (b) Radiofrequency pulses at the Larmor frequency at high field. (c) Sketch of idealised spin state populations. The high-field energy levels refer to the states $|\beta\beta\rangle, |\alpha\beta\rangle, |\alpha\alpha\rangle, |\beta\alpha\rangle$, starting from the top and moving in a clockwise direction. (Reproduced with permission from Carravetta, M. and Levitt, M.H., *J. Chem. Phys.* 122, 214505, 2005. Copyright (2005) American Institute of Physics.)

We note that the same outcome could be obtained by a selective inversion of the S-spin. In terms of the PB kets and bras, the resulting density operator can be expressed as

$$\rho_2 = b_I\left(\hat{I}_z - \hat{S}_z\right) = 2b_I\left(|\alpha\beta\rangle\langle\alpha\beta| - |\beta\alpha\rangle\langle\beta\alpha|\right) \tag{16.11a}$$

With the matrix representation (still in the PB basis):

$$\hat{\rho}_2 = \frac{1}{4}\begin{pmatrix} 1 & 0 & 0 & 0 \\ 0 & 1+b_I & 0 & 0 \\ 0 & 0 & 1-b_I & 0 \\ 0 & 0 & 0 & 1 \end{pmatrix} \tag{16.11b}$$

It may be useful to think about the ρ_2 state in terms of a population diagram similar to that of Figure 3.6a, inserted also in Figure 16.1.

Next, we need to switch the physical conditions so that the Hamiltonian changes from $\hat{H}^{PB}$ (distinguishable spins) to $\hat{H}^{ST}$ (equivalent spins). In the work by Carravetta *et al.*[1], this was realised by transferring the sample mechanically to a low field (on the order of millitesla), thus suppressing the inequivalence caused by the chemical shift difference. This transfer was performed adiabatically; *i.e.* fast on the timescale of spin-lattice relaxation but slowly with respect to the J-coupling. In other words, the adiabatic transfer implies that the density operator follows changes in the Hamiltonian and commutes with it at every point in time. If the sign of the frequency difference $\omega_1 - \omega_2 = \gamma(\delta_1 - \delta_2)$ is the same as the sign of the J-coupling constant, then the high-field states correlate with the low-field states according to Figure 16.2, *i.e.* the adiabatic transfer transforms the excess population of the $|\beta\alpha\rangle$ state into the zero-field state $|S_0\rangle$, while the population-depleted $|\alpha\beta\rangle$ correlates with $|T_0\rangle$. We shall return to the correlation diagram in Figure 16.2 in Section 16.6. The population diagram after the transfer to the low field is indicated at point 3 in Figure 16.1. The corresponding density operator is given by:

$$\hat{\rho}_3 = 2b_I\left(|S_0\rangle\langle S_0| - |T_0\rangle\langle T_0|\right) \tag{16.12}$$

The matrix representation of $\hat{\rho}_3$ in the ST basis (ordered as $|S_0\rangle, |T_{+1}\rangle, |T_0\rangle, |T_{-1}\rangle$) takes the form:

$$\rho_3 = \frac{1}{4}\begin{pmatrix} 1+b_I & 0 & 0 & 0 \\ 0 & 1 & 0 & 0 \\ 0 & 0 & 1-b_I & 0 \\ 0 & 0 & 0 & 1 \end{pmatrix} \tag{16.13}$$

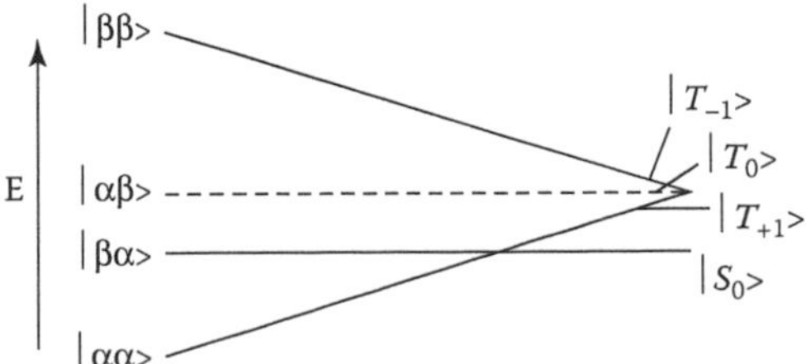

FIGURE 16.2 Correlation between high-field energy levels (left) and energy levels at zero-field (right). The correlation applies for the case where the sign of $\gamma(\delta_1 - \delta_2)$ is the same as the sign of the J-coupling. If the signs are opposite, then the $|\alpha\beta\rangle$ high-field state correlates with the $|S_0\rangle$. (Reproduced with permission from Carravetta, M. and Levitt, M.H., *J. Chem. Phys.* 122, 214505, 2005. Copyright (2005) American Institute of Physics.)

The sample is held at the low field during the period τ_{LF}, which can be used as a variable delay in a pseudo-two-dimensional (pseudo-2D) arrayed experiment series. During this period, the relaxation processes will lead the spin system towards equilibrium. Briefly, two types of processes need to be considered. First, we have the internal equilibration among the triplet states, which occurs, roughly, on the timescale of spin-lattice relaxation. Second, we have relaxation of the singlet state, typically a much slower process. We shall discuss the relaxation in larger detail in the next section. For now, it is sufficient to write the density matrix (in the ST basis) at the end of the low-field period, assumed to be much longer than the T_1 of the spins:

$$\rho_4 = \frac{1}{4}\begin{pmatrix} 1+b_I\varepsilon & 0 & 0 & 0 \\ 0 & 1-\frac{1}{3}b_I\varepsilon & 0 & 0 \\ 0 & 0 & 1-\frac{1}{3}b_I\varepsilon & 0 \\ 0 & 0 & 0 & 1-\frac{1}{3}b_I\varepsilon \end{pmatrix} \tag{16.14}$$

The factor 1/3 for the triplet state populations originates from rapid equilibration between the triplet components. The factor ε is given by $\varepsilon = \exp(-\tau_{LF}/T_S)$, where T_S is the singlet state lifetime. Before going into the theory of T_S, we want to discuss the detection of the singlet states. Again, the logic here follows the original work of Carravetta and co-workers[1]. At the end of τ_{LF}, the spin system is exposed to another adiabatic transfer (see Figure 16.1). The second adiabatic transport follows the same pathway as the first one and results in the following density matrix in the PB basis:

$$\rho_5 = \frac{1}{4}\begin{pmatrix} 1-\frac{1}{3}b_I\varepsilon & 0 & 0 & 0 \\ 0 & 1+b_I\varepsilon & 0 & 0 \\ 0 & 0 & 1-\frac{1}{3}b_I\varepsilon & 0 \\ 0 & 0 & 0 & 1-\frac{1}{3}b_I\varepsilon \end{pmatrix} \tag{16.15}$$

Note that the population of the $|\alpha\beta\rangle$ high-field state differs from the other states. Alternatively, we can express the density operator corresponding to the matrix in Eq. (16.15) as:

$$\hat{\rho}_5 = \frac{1}{4}\hat{1}_{op} - \frac{1}{6}b_I\varepsilon\left(\hat{I}_z - \hat{S}_z - 2\hat{I}_z\hat{S}_z\right) \tag{16.16}$$

A single hard 90° pulse would then convert $\hat{I}_z - \hat{S}_z$ into transverse magnetisation, with opposite signs for the two spins, while the two-spin order, $2\hat{I}_z\hat{S}_z$, would be turned into invisible multiple-quantum (MQ) coherences. This form of singlet detection is sensitive to artefacts and is rarely applied. Carravetta and co-workers[1] instead used a three-pulse sequence:

$$90^\circ_{45^\circ} - (\tau_2 + \tau_3) - 180^\circ_{-45^\circ} - \tau_3 - 90^\circ_y \tag{16.17}$$

with $\tau_2 = \tau_1/2$ and $\tau_3 = 1/4J$. As discussed by Carravetta and Levitt[2] and by Nagashima and Velan[5], this sequence transforms the two-spin order (which is indicative of the deviating population of the singlet state according to Eq. (16.8)) into antiphase transverse I-spin magnetisation. Similarly, the $\hat{I}_z$ term is transformed into antiphase transverse S-spin magnetisation, while $\hat{S}_z$ is converted into MQ coherence. The longitudinal magnetisations can occur at time point (5) through artefacts, but the sequence of Eq. (16.17) removes them from the vicinity of the I-spin resonance. As additional experimental proof of the singlet state origin of the antiphase I-spin signal, Carravetta *et al.*[1] also reported an experiment in which the first 90° pulse in Figure 16.1 had the phase of 180° rather than 0. This reverses the expected phase of the signal originating from the singlet state, which can indeed be seen experimentally in Figure 16.3, taken from Carravetta *et al.*[1].

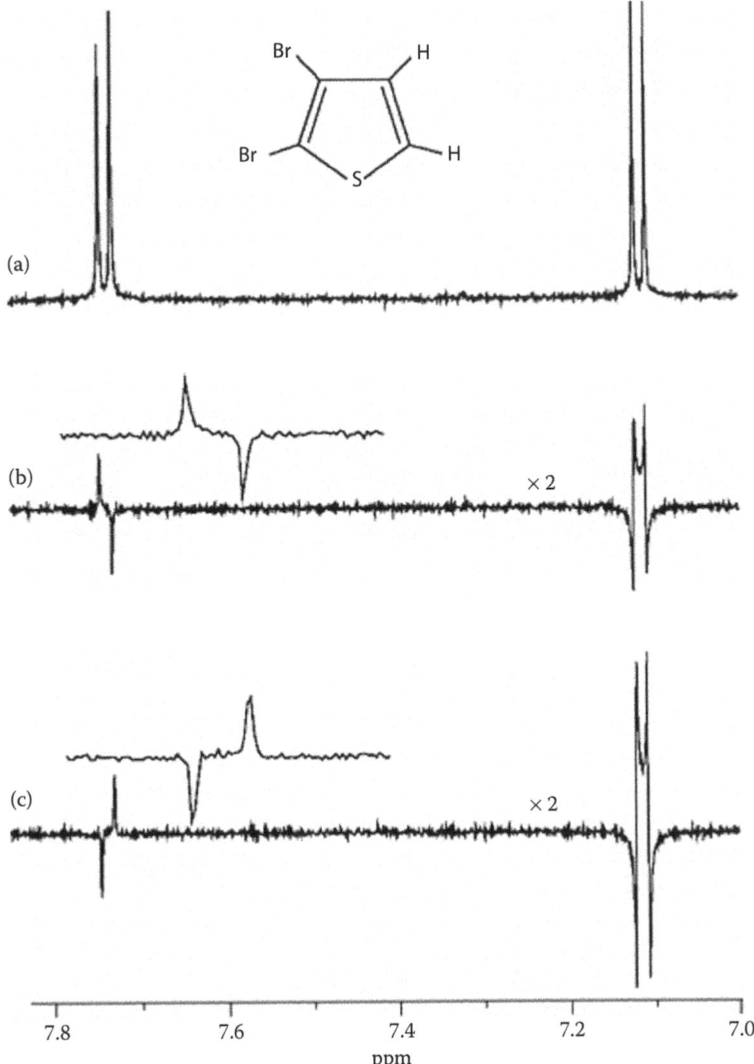

FIGURE 16.3 (a) Normal ^{1}H NMR spectrum of the solution of 2,3-dibromethiophene (inset). (b) Spectrum obtained using the sequence in Figure 16.1 with τ_{LF}=100 s. (c) Spectrum generated by changing the phase of the first 90° pulse from x to $-x$. (Reprinted with permission from Carravetta, M. et al., *Phys. Rev. Lett.*, 92, 153003, 2004. Copyright (2004) by the American Physical Society.)

The essence of the method of singlet observation discussed here is the creation of a precursor of the singlet state at high field, followed by switching the symmetry of the spin system by transporting the sample to a low field and back to the high field for detection. The crucial step of symmetry switching can also be realised without moving the sample, by the action of the resonant radiofrequency fields. The technique to be used is that of spin-locking.

The high-field, spin-locking singlet experiments do not involve adiabatic transfer and therefore require a different strategy, aiming at direct creation of the singlet population, in agreement with Eq. (16.8). We follow here again the original presentation of the high-field singlet experiment by the

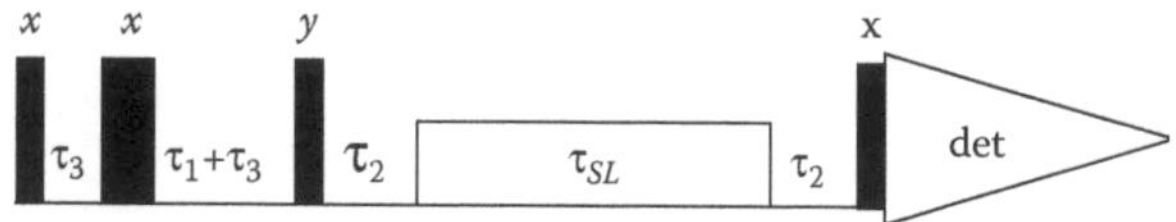

FIGURE 16.4 High-field pulse sequence for observing the decay of long-lived singlet state in the homonuclear *IS* spin system. (Adapted from Carravetta, M., and Levitt, M.H., *J. Am. Chem. Soc.*, 126, 6228–6229, 2004.)

Southampton group[9] presented in Figure 16.4. After the initial equilibration, the sample is exposed to the sequence:

$$90^\circ_x - \tau_3 - 180^\circ_x - (\tau_3 + \tau_1) - 90^\circ_y - \tau_2 \tag{16.18}$$

with the delays τ_1, τ_2, τ_3 defined below Eqs. (16.9) and (16.17). The carrier frequency is placed in the middle of the chemical shift interval. In the weak coupling limit, $|\omega_\Delta^{high}/2\pi| \gg |J|$ or $\tau_3 \gg \tau_1, \tau_2$ (which allow the *J*-coupling evolution during τ_1 and τ_2 to be neglected), the effect of this sequence is:

$$\begin{aligned} &\hat{I}_z + \hat{S}_z \xrightarrow{90^\circ_x} -\hat{I}_y - \hat{S}_y \xrightarrow{\tau_3 - 180^\circ_x - \tau_3} -2\hat{I}_x\hat{S}_z - 2\hat{I}_z\hat{S}_x \xrightarrow{\tau_1} \\ &2\hat{I}_y\hat{S}_z - 2\hat{I}_z\hat{S}_y \xrightarrow{90^\circ_y} 2\hat{I}_y\hat{S}_x - 2\hat{I}_x\hat{S}_y \equiv -(ZQ)_y \xrightarrow{\tau_2} (ZQ)_x \end{aligned} \tag{16.19}$$

In other words, the result immediately before switching on the spin-lock field is zero-quantum coherence. When the spin-lock is switched on, the chemical shift difference is suppressed, the Hamiltonian changes to $\hat{H}^{ST}$ and $(ZQ)_x$ is converted into the singlet population, $|S_0\rangle\langle S_0|$, according to Eq. (16.8).

The relaxation of the singlet state and the relaxation processes within the triplet manifold take place during the delay τ_{SL}. With a sufficiently long delay, the latter process reaches equilibrium, while the singlet population may remain non-zero. The last delay (τ_2) and the 90° pulse convert the singlet population into detectable antiphase magnetisation. In the original paper by Carravetta and Levitt[9], the spin-lock was applied using CW irradiation, but in a later paper from the same group[10] the authors investigated various spin-lock modulation schemes. Other excitation and detection methods have also been proposed and some of them are reviewed in the papers by Levitt[3], Nagashima and Velan[5] and Pileio[6].

Pileio and co-workers[11] proposed a few years ago a novel and robust technique for converting magnetisation into singlet order and back. The methods are known under the acronyms M2S (magnetisation-to-singlet) and S2M (singlet-to-magnetisation). The approach is designed to work at low magnetic field, not necessarily homogeneous. The method is outlined in Figure 16.5, reproduced from Pileio *et al.*[11]. As an initial step, the sample is magnetised at high field, transferred to the low-field region and exposed there to the M2S, a sequence of pulses of low radiofrequency, denoted as audiofrequency and corresponding to nuclear resonance frequency at low magnetic field. The sample remains then in the low field for a period τ_{LF}, after which the singlet order is transferred to magnetisation by the S2M. For the full explanation of the sequence, the reader is referred to the original work or to the review by Nagashima and Velan[5].

16.3 Relaxation Properties of Long-Lived States

The long lifetimes of the singlet states have a simple symmetry-based explanation: the singlet state is antisymmetric with respect to exchange of labels of the two spins. In order to equilibrate the singlet population with the triplet states (symmetric with respect to exchange of the two spins), we need to cause transitions between the spin states of different exchange symmetry. To induce such transitions, the interaction behind the transitions cannot be symmetric with respect to the exchange, since the transition probability in this case vanishes. The dipole-dipole (DD) interaction between spins *I* and *S* is symmetric – this can easily be seen by inspecting the form of the different terms in the dipolar alphabet (Eq. (3.11)) or the irreducible tensor operators in Eq. (4.43). Thus, the singlet–triplet transitions are not

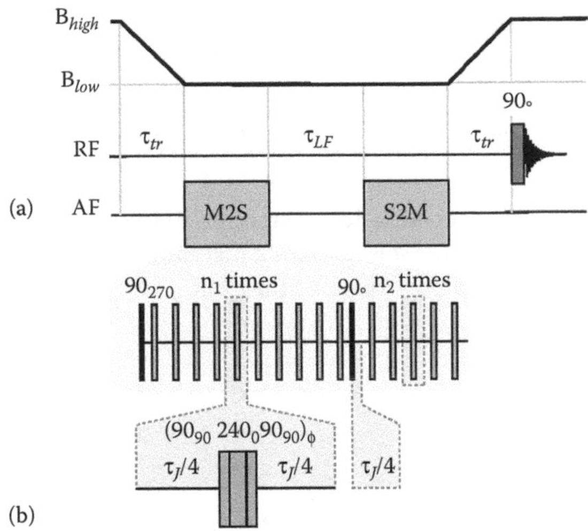

FIGURE 16.5 A robust experiment for forming and detecting singlet order. RF indicates radiofrequency and AF audio frequency, $\tau_J = J^{-1}$. (Reprinted with permission from Pileio, G. et al., *Proc. Natl. Acad. Sci. USA*, 107, 17135–17139, 2010.)

caused by the intra-pair dipole-dipole interaction, which is usually the most efficient relaxation mechanism in systems of spin 1/2 particles.

As we are going to see, the usual formulation of the Redfield theory at high field, in terms of the eigenstates of the Zeeman-plus-couplings Hamiltonian, does not provide any indications of the existence of any states with exceptionally long lifetimes. In order to obtain a quantitative description of singlet relaxation, we need to express the semiclassical, second-order perturbation theory formulation in terms of the singlet and triplet states. Following the paper by Carravetta and Levitt[2] and the review by Pileio[12], this is most conveniently done using the Liouville space formulation. The evolution of the spin system is described by the equation of motion:

$$\frac{d}{dt}\hat{\rho}(t) = \hat{\hat{L}}(t)\hat{\rho}(t) \tag{16.20}$$

with

$$\hat{\hat{L}}(t) = \hat{\hat{L}}_{coh}(t) + \hat{\hat{R}} \tag{16.21}$$

where the time-dependent Liouvillian $\hat{\hat{L}}(t)$ is a sum of the coherent part $\hat{\hat{L}}_{coh}(t)$ (expressed as a commutation superoperator with the coherent part of the Hamiltonian, $\hat{\hat{L}}_{coh} = -i[\hat{H},])$ and the relaxation superoperator $\hat{\hat{R}}$. In the usual Redfield treatment, the relaxation superoperator is represented by a relaxation supermatrix with the elements:

$$R_{st,uv} = \left(|s\rangle\langle t|\hat{\hat{R}}|u\rangle\langle v|\right) \tag{16.22}$$

The kets $|r\rangle, |s\rangle, |t\rangle, |u\rangle$ denote the eigenstates of the unperturbed Hamiltonian. As discussed in Chapter 4, the supermatrix can be blocked according to the coherence order of the operators involved. The blocking of the matrix was explained in terms of the secular approximation, but it can be shown that it is based on more general symmetry rules[2]. The block of coherence order zero corresponds to populations and zero-quantum coherences. In Chapter 5, the discussion of relaxation in a two-spin system was presented in terms of the operator basis $\{\frac{1}{2}\hat{1}_{op},\hat{I}_z,\hat{S}_z,2\hat{I}_z\hat{S}_z\}$. Following Carravetta and Levitt[2], we can expand the basis with the zero-quantum coherence. This gives:

$$B^{PB} = \left\{\frac{1}{2}\hat{1}_{op},\hat{I}_z,\hat{S}_z,2\hat{I}_z\hat{S}_z,\sqrt{2}ZQ_x,\sqrt{2}ZQ_y\right\} \tag{16.23}$$

The zero-quantum operators were defined in Eqs. (7.14c) and (7.14d). Assuming a homonuclear spin system with dipole-dipole interaction as the sole relaxation mechanism and using this basis, the coherence order zero block of the relaxation supermatrix takes the form:

$$R^{PB} = \begin{pmatrix} 0 & 0 & 0 & 0 & 0 & 0 \\ 0 & -R_{auto} & R_{cross} & 0 & 0 & 0 \\ 0 & R_{cross} & -R_{auto} & 0 & 0 & 0 \\ 0 & 0 & 0 & -R_{zz}^{auto} & R_{zz,ZQ}^{cross} & 0 \\ 0 & 0 & 0 & R_{zz,ZQ}^{cross} & -R_x^{ZQ} & 0 \\ 0 & 0 & 0 & 0 & 0 & -R_y^{ZQ} \end{pmatrix} \tag{16.24}$$

The symbol R_{auto} denotes the autorelaxation rate for z-magnetisation of the two spins (assumed equal and given in Eq. (3.31a)), while R_{cross} indicates the cross-relaxation rate (denoted in Chapter 3 as σ_{IS}) and R_{zz}^{auto} is identical to Γ_{33}, given in Eq. (5.43c). Carravetta and Levitt[2] give the following expressions for the remaining relaxation matrix elements:

$$R_{zz,ZQ}^{cross} = \frac{6\pi}{5\sqrt{2}} b_{IS}^2 J_2(\omega_I) \tag{16.25a}$$

$$R_x^{ZQ} = \frac{3\pi}{5} b_{IS}^2 J_2(\omega_I) \tag{16.25b}$$

$$R_y^{ZQ} = \frac{\pi}{5} b_{IS}^2 \left[2J_2(0) + 3J_2(\omega_I)\right] \tag{16.25c}$$

We notice that the first row and the first column contain only zeroes, which means that the unit operator (which may be identified as the sum of all level populations) is not undergoing any relaxation or is dynamically isolated. As stated earlier in this section, the remaining structure of the matrix does not indicate the existence of any long-living states. It is, however, interesting to comment on the lack of a cross-term between the two components of the zero-quantum coherence. This fact is related to the opposite parities of ZQ_x and ZQ_y under exchange of the two spins, which becomes obvious by inspection of Eqs. (7.14).

A different picture is obtained if the PB basis is replaced by the ST basis[2]. The four populations are in this case represented by the operators $|S_0\rangle\langle S_0|, |T_{+1}\rangle\langle T_{+1}|, |T_0\rangle\langle T_0|, |T_{-1}\rangle\langle T_{-1}|$ or by their linear combinations. Following Pileio and Levitt[10,12], we choose the latter option and define the coherence order zero operator basis:

$$\begin{aligned} B^{ST} = \Big\{ & \frac{1}{2}\left(|S_0\rangle\langle S_0| + |T_{+1}\rangle\langle T_{+1}| + |T_0\rangle\langle T_0| + |T_{-1}\rangle\langle T_{-1}|\right), \\ & \frac{1}{2\sqrt{3}}\left(3|S_0\rangle\langle S_0| - |T_{+1}\rangle\langle T_{+1}| - |T_0\rangle\langle T_0| - |T_{-1}\rangle\langle T_{-1}|\right), \\ & \frac{1}{\sqrt{2}}\left(|T_{+1}\rangle\langle T_{+1}| - |T_{-1}\rangle\langle T_{-1}|\right), \frac{1}{\sqrt{6}}\left(-|T_{-1}\rangle\langle T_{-1}| + 2|T_0\rangle\langle T_0| - |T_{-1}\rangle\langle T_{-1}|\right), \\ & |S_0\rangle\langle T_0|, |T_0\rangle\langle S_0| \Big\} \end{aligned} \tag{16.26}$$

The advantage of this choice of the basis is that its elements have a straightforward physical interpretation. The first element corresponds again to the sum of all populations, or to $\frac{1}{2}\hat{1}_{op}$; the second basis operator describes the population difference between the singlet state and the average triplet population, to be termed *the singlet order* in the following. The third and fourth basis operators reflect population differences within the triplet manifold, while the last two are zero-quantum coherences. Using the B^{ST} basis, the relaxation supermatrix for the intra-pair dipole-dipole interaction becomes:

$$R_{DD}^{ST} = \begin{pmatrix} 0 & 0 & 0 & 0 & 0 & 0 \\ 0 & 0 & 0 & 0 & 0 & 0 \\ 0 & 0 & -W_1^T - 2W_2^T & 0 & 0 & 0 \\ 0 & 0 & 0 & -3W_1^T & 0 & 0 \\ 0 & 0 & 0 & 0 & -R_{ZQ}^{ST} & 0 \\ 0 & 0 & 0 & 0 & 0 & -R_{ZQ}^{ST} \end{pmatrix} \tag{16.27}$$

We notice first that the chosen basis diagonalises the dipole-dipole relaxation matrix. Further, we see that now both the sum of all level populations and the deviation of singlet state population from the triplet states are dynamically isolated from the other operators. The reason for this is the observation made at the beginning of this section: the dipole-dipole interaction does not couple singlet and triplet states, because of their symmetry properties. The zero-quantum relaxation rates for the two components are here equal and given by:

$$R_{ZQ}^{ST} = \frac{\pi}{5} b_{IS}^2 \left[2J_2(0) + 3J_2(\omega_I)\right] \tag{16.28}$$

The triplet subspace is characterised by single-quantum (W_1^T) and double-quantum (W_2^T) transition probabilities. The latter is identical to W_2, introduced in the Solomon equations and given in Eq. (3.29b), which one might expect, as the outer components of the triplet are identical to the product states $|\alpha\alpha\rangle$ and $|\beta\beta\rangle$. The transition probability, W_1^T, is equal to:

$$W_1^T = \frac{3\pi}{5} b_{IS}^2 J_2(\omega_I) \tag{16.29}$$

which is similar to the quantities W_{1I} and W_{1S} introduced in Chapter 3. Under extreme narrowing conditions, the relaxation matrix takes an even simpler form[10,12]:

$$R_{DD}^{ST} = -\frac{1}{10} b_{IS}^2 \tau_c \begin{pmatrix} 0 & 0 & 0 & 0 & 0 & 0 \\ 0 & 0 & 0 & 0 & 0 & 0 \\ 0 & 0 & 15 & 0 & 0 & 0 \\ 0 & 0 & 0 & 9 & 0 & 0 \\ 0 & 0 & 0 & 0 & 5 & 0 \\ 0 & 0 & 0 & 0 & 0 & 5 \end{pmatrix} \tag{16.30}$$

It is worth noting that the relaxation rate R_{33}^{ST} is identical to the spin-lattice relaxation rate for a pair of identical spins (*cf.* Eq. (3.34)).

Under the conditions of vanishing contribution from the intra-pair DD interaction, the relaxation rate for the basis element $\frac{1}{2\sqrt{3}}(3|S_0\rangle\langle S_0| - |T_{+1}\rangle\langle T_{+1}| - |T_0\rangle\langle T_0| - |T_{-1}\rangle\langle T_{-1}|)$, corresponding to the quantity T_S introduced in the previous section, will depend on other interactions. Pileio's review[12] discusses the effects of the chemical shielding anisotropy (CSA) and spin-rotation interactions. The CSA interaction, relevant for singlet relaxation at high field, leads to the following structure of the relaxation supermatrix in the ST basis:

$$R_{CSA}^{ST} = \begin{pmatrix} 0 & 0 & 0 & 0 & 0 & 0 \\ 0 & V_S^{CSA} & 0 & V_{S/T,1}^{CSA} & 0 & 0 \\ 0 & 0 & V_{z,0}^{CSA} & 0 & V_{z,1}^{CSA} & V_{z,1}^{CSA} \\ 0 & V_{S/T,1}^{CSA} & 0 & V_T^{CSA} & 0 & 0 \\ 0 & 0 & V_{z,1}^{CSA} & 0 & V_{S/T,0}^{CSA} & V_{S/T,1}^{CSA} \\ 0 & 0 & V_{z,1}^{CSA} & 0 & V_{S/T,1}^{CSA} & V_{S/T,0}^{CSA} \end{pmatrix} \tag{16.31a}$$

We can see that the structure of the matrix in this case becomes more complicated. The ST basis operators are not eigenoperators to the R_{CSA}^{ST} supermatrix, since the matrix is not diagonal. The most interesting part of the supermatrix refers to the basis vectors number 2 and 4 and takes the form:

$$\begin{pmatrix} V_S^{CSA} & V_{S/T,1}^{CSA} \\ V_{S/T,1}^{CSA} & V_T^{CSA} \end{pmatrix} \tag{16.31b}$$

The symbol $V_{S/T,1}^{CSA}$ denotes cross-relaxation between the singlet order and one of the population imbalances among the triplet components. Diagonalisation of this matrix will provide the relaxation rate of the long-living state.

The CSA relaxation of the long-living states has been discussed by Ahuja *et al.*[13] and by Pileio and co-workers[14]. The formulations in these two papers seem to differ in the expressions for the singlet relaxation rates, but agree on the notion that the CSA mechanism is suppressed if the shielding tensors of the two spins forming the singlet coincide in magnitude and orientation of the PAS. The latter work gives compact expressions for small molecules (in the extreme narrowing limit) relating the CSA singlet relaxation to the difference between the Cartesian shielding tensors at the two nuclear sites, $\Delta\sigma = \sigma_I - \sigma_S$. The difference tensor may be decomposed into a traceless symmetric (rank-two) component, an antisymmetric part (rank-one) and a scalar shift difference (compare Eq. (5.18)). Disregarding for the moment the scalar part, we write the difference tensor as $\Delta\sigma = \Delta\sigma^{(2)} + \Delta\sigma^{(1)}$, which leads to additive relaxation rate contributions from the symmetric and antisymmetric parts:

$$R_S^{CSA,sym} = \frac{2}{9}\gamma^2 B_0^2 \tau_2 \left\|\Delta\sigma^{(2)}\right\|^2 \tag{16.32a}$$

$$R_S^{CSA,antisym} = \frac{2}{9}\gamma^2 B_0^2 \tau_1 \left\|\Delta\sigma^{(1)}\right\|^2 \tag{16.32b}$$

where $\|\mathbf{t}\|$ stands for the Frobenius norm of the tensor **t** (defined as the square root of the sum of the absolute squares of its elements). It should be noted that the two contributions contain rotational correlation times of different ranks.

Since the DD and CSA contributions can vanish or become very small, the spin-rotation relaxation mechanism may become important. Pileio[12] reported on the general form of the relaxation supermatrix in the ST basis for the spin-rotation interaction and found that the singlet order is expected to display single-exponential relaxation. Expressions for the relaxation rate are given for a symmetric top molecule with cylindrically symmetric spin-rotation coupling tensors for both spins. If the main axes of the two tensors coincide, a relatively simple expression for T_S is obtained:

$$\left(T_S^{SR}\right)^{-1} = \frac{2k_B T}{3\hbar^2}\left[\left(C_{\|,I} - C_{\|,S}\right)^2 I_\| \tau_{J,\|} + 2\left(C_{\perp,I} - C_{\perp,S}\right)^2 I_\perp \tau_{J,\perp}\right] \tag{16.33}$$

In analogy with the CSA case, the efficiency of this mechanism depends on the *difference* between the properties of the two spins. Eq. (16.33) can be compared with Eq. (5.51), which gives the spin-rotation spin-lattice relaxation rate for the even simpler case of a spherical top molecule with an isotropic spin-rotation coupling tensor and a single spin-rotation correlation time.

Pileio's review also discusses what is denoted as *coherent effects on the relaxation of singlet states* or the *singlet-triplet leakage*. This refers to the theoretical formulation involving the full Liouvillian for the system, *i.e.* a sum of the coherent interactions (such as chemical shifts and their differences, as well as *J*-couplings) and the relaxation supermatrix. Using the ST basis, one can show that the small coherent terms couple the singlet order with other basis elements within the coherence order zero block and thus influence the eigenvalues of the supermatrix, corresponding to the relaxation rates.

If the sample contains other magnetic nuclei (we call them N) than I and S, then the dipole-dipole and scalar interaction with these spins can provide a mechanism for singlet relaxation. Pileio *et al.*[15] and Grant and Vinogradov[16] investigated early on the occurrence of long-living spin states in systems of three and four spins. We do not follow this track and refer the interested reader to the original papers. We wish, however, to mention one specific aspect of the issue discussed by Tayler and co-workers[17]. They investigated the case of a pair of inequivalent protons in a CH_2 group interacting through DD interaction with other protons in the molecule while the corresponding out-of-pair *J*-couplings are negligible. In the extreme narrowing regime, the ratio T_S/T_1 is only dependent on the geometry of the molecule:

$$\frac{T_S}{T_1} = \sum_N \frac{3b_{IS}^2}{2\left(b_{IN}^2 + b_{SN}^2 - b_{IN}b_{SN}(1-3\cos^2\theta_{IS,NS})\right)} \tag{16.34}$$

For the distances, r_{IN}, r_{SN} longer than about 2 Å, the ratio decays with the inverse eighth power of the distance of the spin N from the centre of the IS pair.[16] The Southampton group has also investigated the case of intermolecular dipole-dipole relaxation of the singlet states[18] and paramagnetic effects on the singlet relaxation[19]. The effect of paramagnetic salts or dissolved oxygen on T_S (singlet relaxivity) was lower than the common T_1-related relaxivity, which was interpreted in terms of correlation between the fluctuating random fields at the two spins.

Next, we move to the scalar relaxation, another mechanism that can be operative in the presence of additional spins N. Pileio[20] reported a theoretical study of the scalar relaxation of the second kind in the nomenclature of Abragam (see *Further reading*), *i.e.* for the case when the *J*-couplings to the third spin N are modulated by efficient relaxation of the N-spin caused by another mechanism. Assuming that the T_1 and T_2 for the N-spin are well defined and that the frequency-containing terms in spectral densities can be neglected, Pileio obtained the following expression for the singlet relaxation time:

$$T_S^{-1} = \frac{2\pi^2}{3}\left(J_{IN} - J_{IS}\right)^2\left(T_1 + 2T_2\right) \tag{16.35}$$

It is noteworthy that, also here, the singlet relaxation is in a sense driven by a difference between the interactions between the two spins I and S and their environments.

Finally, we conclude this section by referring to a couple of studies of very slowly relaxing singlet states. The first example is ^{15}N-labelled nitrous oxide (N_2O), which for some time held the "world record" concerning slow spin relaxation in a liquid sample. Pileio and co-workers[21] reported measurements for this molecule dissolved in DMSO-d_6. The molecule is linear, which simplifies the analysis, and its ^{15}N spectrum is characterised by two signals with a chemical shift difference of 82.3 ppm and a ^{15}N–^{15}N scalar coupling of 8.1 Hz. These spectral parameters result, at moderately high field, in a weakly coupled ^{15}N NMR spectrum. The singlet state was excited using the field-cycling method, using the transfer time of 40 s. Making use of a variable delay at low field (compare Figure 16.1), the singlet lifetime was determined as $T_S = 1587 \pm 57$ s or 26 min. This value can be compared with the longitudinal relaxation time for the total ^{15}N magnetisation (at low field) of 197 ± 5 s. A few years later, Ghosh and co-workers[22] reported similar measurements for nitrous oxide in some other solvents (H_2O, D_2O, CS_2, n-propanol) as well as in rat blood and goose fat. The T_S values in both kinds of aqueous solvent were only slightly shorter than in DMSO-d_6, while the corresponding lifetimes in blood and fat were close to 10 min, thus providing the proof of concept for potential use of $^{15}N_2O$ as a magnetic tracer for *in vivo* NMR and magnetic resonance imaging (MRI). In a more recent paper, Ghosh *et al.*[23] extended the study of $^{15}N_2O$ in simple solvents to variable temperature measurements. The measurements were used to analyse the relaxation mechanisms and it was found that spin-rotation interaction was the dominant interaction.

The second example is a carbon-13-labelled naphthalene derivative investigated by Stevanato *et al.*[24]. Here, the singlet state is formed in the $^{13}C_2$ pair, positioned in the central part of the naphthalene skeleton, and characterised by a very small chemical shift difference of 0.06 ± 0.02 ppm. The measurements were performed as a function of the magnetic field, using the field-cycling method and transferring the sample to different positions along the magnet bore. The singlet order was generated using the M2S method before transfer to the low field and converted back to the detectable magnetisation by the inverted technique denoted S2M[11,14,25]. In the degassed sample, a T_S value of 3950 ± 220 s was obtained at 2 mT, increasing to 4250 ± 130 s at 0.4 T and decreasing to slightly less than 1000 s at 9.4 T. The signal decays at 9.4 and 0.4 T are shown in Figure 16.6. The data were carefully analysed by performing simulations involving molecular dynamics computations of a trajectory, complemented with DFT calculations of the CSA and spin-rotation tensors along the trajectory, and correlation functions for fluctuating spin interactions. The simulations were used to theoretically estimate the spectral densities and relaxation rates. The analysis identified the following relaxation mechanism for the singlet order:

1. Chemical shift anisotropy, in particular the antisymmetric component, dominates at high field.
2. Spin-rotation and spin-internal-motion interactions, associated with fluctuating magnetic fields generated by internal vibrational motions, are the most important mechanism at low magnetic fields.
3. Intramolecular dipole-dipole gives a small but significant contribution; the intermolecular dipolar interactions are clearly seen in the non-degassed sample.
4. Singlet-triplet leakage, driven by symmetry-breaking (small difference in chemical shift) coherent interaction, plays a certain role at high field.

The experimental data and estimated contributions from various mechanisms are shown in Figure 16.7.

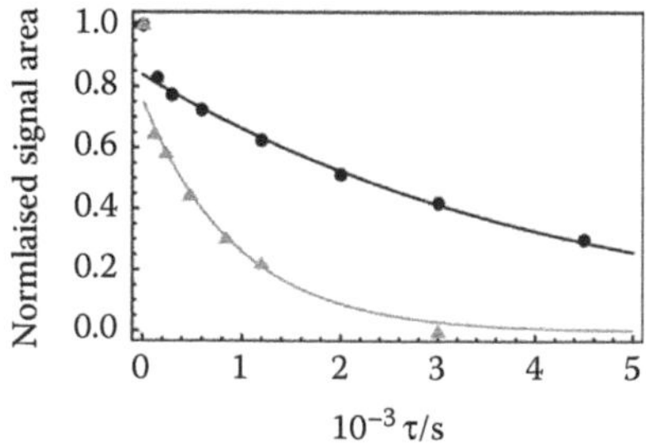

FIGURE 16.6 Experimental decay curves for a double ^{13}C-labeled naphthalene derivative (see the original paper for details) in solution, with extremely long singlet lifetimes. Grey triangles: relaxation at 9.4 T, black circles: 0.4 T. Solid lines are fits to exponential decays, ignoring the first point. (Reprinted with permission from Stevanato, G. et al., *Angew. Chem. Int. Ed.*, 54, 3740–3743, 2015. Copyright (2015) Wiley-VCG Verlag.)

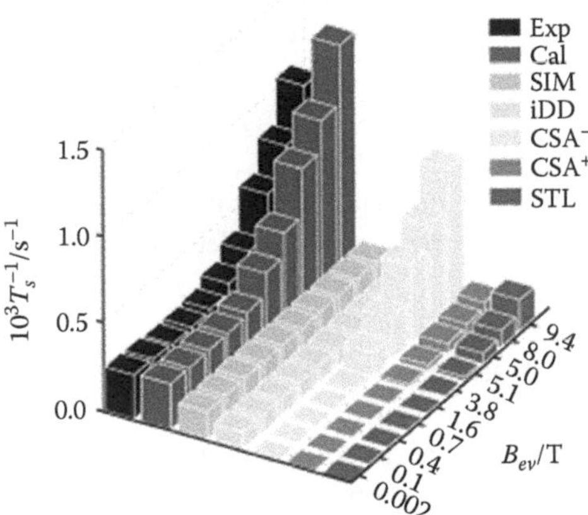

FIGURE 16.7 Experimental decay curves for a double ^{13}C-labelled naphthalene derivative (see the original paper for details) in solution, with extremely long singlet lifetimes. Black indicates experimental values, grey calculated values. The contribution of different mechanisms (spin-internal-motion [SIM], intramolecular dipolar [iDD], anti-symmetric chemical shift anisotropy [CSA$^-$], symmetric chemical shift anisotropy [CSA+] and singlet–triplet leakage [STL]) to the total decay is shown. (Reprinted with permission from Stevanato, G. et al., *Angew. Chem. Int. Ed.*, 54, 3740–3743, 2015. Copyright (2015) Wiley-VCG Verlag.)

16.4 Long-Lived Coherences

We now turn our attention to the lower right corner of the relaxation supermatrix in Eqs. (16.27), (16.30) and (16.31a), *i.e.* to the relaxation properties of the zero-quantum coherences coupling the singlet and the triplet states, $|S_0\rangle\langle T_0|, |T_0\rangle\langle S_0|$. Sarkar and co-workers[26] coined the concept of *long-lived coherences* (LLCs) for these zero-quantum coherences and the same group has provided a review of this field[27]. As we can see in Eq. (16.30), the dipolar relaxation rate of the LLC is, in the extreme narrowing limit, three times smaller than the spin-lattice relaxation rate (equal under these conditions to the spin-spin relaxation rate). In the limit of slow rigid-body rotation, when $J_2(0) \gg J_2(\omega_0), J_2(2\omega_0)$, the theoretical ratio of dipolar rates R_2/R_{ZQ}^{ST} increases to nine.

The zero-quantum coherences cannot be observed directly. Sarkar *et al.*[26] proposed a simple experimental scheme allowing the observation of LLCs using the principles of two-dimensional NMR (see Figure 16.8). The density operator at the beginning of the sequence is $\rho_1 = \hat{I}_z + \hat{S}_z$. The selective 180°_x pulse on the *S*-spin inverts the *z*-magnetisation of that spin, yielding $\rho_2 = \hat{I}_z - \hat{S}_z$, and the non-selective 90°_y pulse results in $\rho_3 = \hat{I}_x - \hat{S}_x$. After time point 3, the spin-lock field (assume for simplicity that the carrier is placed in the middle of the shift interval between *I* and *S* and that $\omega_1 = \gamma B_1$ is much larger than the offset) is switched on and two things should be considered: (1) the system should be described in the tilted rotating frame (see Section 4.4), which together with the assumption above implies that the density operator takes the form $\rho_4 = \hat{I}_Z - \hat{S}_Z$; (2) the spins are made equivalent by the pulse-sequence and

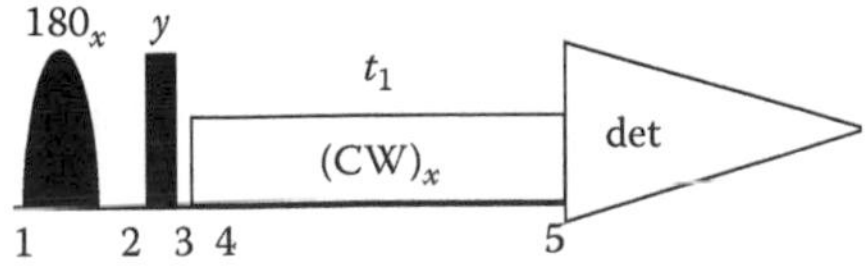

FIGURE 16.8 The pulse sequence used for detection of long-lived coherences. (Adapted from Sarkar, R. et al., *Phys. Rev. Lett.*, 104, 053001, 2010.)

the ST basis becomes the natural one to use. Applying the relationship between the operators in the PB and ST bases tabulated by Cavadini and Vasos[7], we can reformulate ρ_4 as:

$$\rho_4 = \hat{I}_Z - \hat{S}_Z = |S_0\rangle\langle T_0| + |T_0\rangle\langle S_0| \tag{16.36}$$

During the t_1 period in Figure 16.8, ρ_4 undergoes relaxation and coherent evolution under *J*-coupling (see the supplementary information of Sarkar *et al.*[26] for details). When the spin-lock is switched off, the density operator becomes[26,27]:

$$\rho_5(t_1) = \left[\left(\hat{I}_x - \hat{S}_x\right)\cos(\pi J t_1) + \left(2\hat{I}_y\hat{S}_z - 2\hat{I}_z\hat{S}_y\right)\sin(\pi J t_1)\right]\exp\left(-t_1 R_{ZQ}^{ST}\right) \tag{16.37}$$

The LLC oscillates thus at the frequency of the scalar coupling and is damped by the relaxation process with the rate given in Eq. (16.28). The density operator is sampled as a function of t_1 in the usual manner of 2D NMR.

16.5 Applications of Long-Lived States

The extended lifetime of the singlet (and related) states opens up new possibilities for a number of applications. Here, we concentrate on three examples and the first two follow a similar principle. Let us begin with measurements of slow chemical exchange. In Chapter 13, we introduced exchange spectroscopy (EXSY). The exchange-mediated cross-peaks undergo a decay as a result of exchange, superimposed on the longitudinal relaxation. If the timescale of spin-lattice relaxation is too short, then the decay caused by exchange cannot be followed, as summarised by Eqs. (13.13). Sarkar and co-workers[8] proposed a number of methods in which the chemical exchange can be measured on the much slower timescale of singlet state relaxation, as quantified by T_S. The methods to be used must fulfil certain requirements: (1) the singlet excitation efficiency must be insensitive to the chemical shifts and *J*-coupling (because the two exchanging forms must be characterised by different chemical shifts and possibly different couplings); (2) the spin-lock bringing the two spins into equivalence needs to be efficient over a range of chemical shifts and shift differences. To this end, they described a number of 2D techniques making use of transformation of either the ZQC_x or the two-spin longitudinal order into singlet states. The long-living singlet order could then undergo chemical exchange in a similar way as in the EXSY, and the singlet order at the end of the mixing time was converted into detectable magnetisation. The technique was dubbed singlet state exchange spectroscopy (SS-EXSY) and was demonstrated using a saccharide with a proton pair surrounded by a deuterated environment.

The second, related application is concerned with slow translational diffusion processes. Measurements of this kind are in general outside the scope of this book, and there exists a large body of published monographs[28] and reviews[29–31]. The principle of the pulsed-field gradient (PFG) spin-echo measurements was, however, briefly mentioned in Section 8.2. Here, we follow the discussion presented by Cavadini and co-workers in their original publication on diffusion measurement by singlet state NMR[32] and in the review a few years later[7]. The effect of a field gradient can be characterised by the quantity:

$$\kappa = \gamma p s G_{max} \delta \tag{16.38}$$

where γ is the magnetogyric ratio, p is the coherence order, s is the shape factor, G_{max} is the maximum intensity of the gradient and δ is the gradient duration. Another important quantity is the time period Δ, during which the diffusion process is taking place, after the initial encoding of the location of the spins by the gradient (or combination of gradients) and before decoding by other gradients. During the diffusion period, the spins move and relax and the NMR signal decays according to:

$$S(\kappa, \Delta) = S_0 \exp(-D\kappa^2\Delta)\exp(-R\Delta) \tag{16.39}$$

Here, D is the translational diffusion coefficient and R is the relevant relaxation rate constant.

In the simple spin-echo PFG experiment, R is equal to the transverse relaxation rate. In the stimulated echo measurement, R is given by the typically somewhat slower longitudinal relaxation rate, while in the singlet state diffusion-ordered spectroscopy (SS-DOSY) experiment R is equal to the still lower singlet state relaxation rate, T_S^{-1}. In a typical PFG experiment, the strength of the gradient G_{max} (and thus κ) is varied, while everything else is kept constant. Measuring the Stejskal–Tanner[33] attenuation factor:

$$S/S_0 = \exp\left(-D\kappa^2\Delta\right) \tag{16.40}$$

as a function of κ allows the value of the diffusion coefficient to be fitted. The constant delay Δ has to be chosen with care. It has to be long enough for the attenuation factor to change significantly as a function of the available gradient strength and yet short enough not to cause the signal to decay away due to the relaxation term in Eq. (16.39). The beauty of the SS-DOSY experiment is that it allows a big reduction of this relaxation term and thus increased Δ.

The last application of the slow singlet relaxation that we wish to mention is related to the storage of hyperpolarised magnetisation over time. This storage is a prerequisite to applications *in vivo* in the context of both NMR spectroscopy and MRI. Here, we wish to point out some important developments in the area. Vasos and co-workers[34] proposed in 2009 a method to preserve the high magnetisation obtained by means of dynamic nuclear polarisation (DNP) (see Section 15.4) by conversion into long-lived states. The experiment they designed was performed on a sample of the dipeptide Ala-Gly at natural isotope abundance and consisted of five stages: (1) DNP hyperpolarisation (using gyrotron radiation and a radical as the source of electron spin polarisation) of ^{13}C magnetisation at intermediate field (3.35 T) and low temperature (1.2 K), followed by rapid heating, dissolution and transport of the room-temperature solution to a high-resolution magnet (7.05 T); (2) reverse INEPT transfer of the enhanced ^{13}C magnetisation to the α-proton pair in glycine; (3) transformation of the enhanced proton magnetisation into a long-lived state; (4) sustaining the LLS by spin-locking and (5) detection by partial conversion into proton magnetisation. The detected magnetisation consisted of the two α-proton signals characterised by the shift difference of 0.15 ppm and a J-coupling of −17.3 Hz. For the details of the experimental procedure, the reader is referred to the original paper. The same group a few years later reported a similar study[35], but using in addition a third magnet, working at low magnetic field, and applying the field-cycling approach for preserving the LLS.

Pileio *et al.*[36] also made an important contribution to the combination of DNP for creation of enhanced magnetisation and its storage as LLS. They proposed a recycling protocol in which the dissolution DNP–generated enhanced ^{13}C polarisation is observed with full intensity and then returned to the singlet order. The method was demonstrated using ethyne derivatives characterised by nearly equivalent ^{13}C pairs with a large J-coupling. The first step of the experiment is the same as in the work of Vasos *et al.*[34]. In the second step, the sample is moved to the NMR magnet and the enhanced ^{13}C longitudinal magnetisation is subjected to the M2S sequence at high field, leading to the creation of enhanced singlet order. Next, the sample is transferred to low magnetic field where the spins are magnetically equivalent and kept there for a period t. Under these conditions, the singlet relaxation is very slow. The following step is complicated: the sample is moved to high field again, where it is subjected to a sandwich of pulses denoted S2M2S, which allows the singlet order to be converted temporarily into observable transverse magnetisation and back again into the singlet order. The storage at low field and the S2M2S observation periods can then be repeated and used for multiple observation of the long-lived singlet order with only modest signal losses. The authors offer an analogy between the near-equivalent $^{13}C_2$ spin system and a bank where the polarisation is deposited for safe-keeping. The same paper also presents how the hyperpolarised magnetisation can be used in the context of MRI.

16.6 *Para*-Hydrogen Induced Polarisation

We finish this chapter by discussing phenomena closely related to the nuclear singlet states, known under the acronym PHIP. The topic has been subject to several reviews, of which we wish to mention some[37–39]. The basic idea behind these experiments is – again – switching between the Hamiltonians for equivalent and non-equivalent spins. Here, the switching is caused by a chemical reaction involving *para*-hydrogen (equivalent spins), *e.g.* its addition to a multiple bond, and yielding two protons in non-equivalent positions in a larger molecule. If the population of *para*-hydrogen is larger than that corresponding to the thermal equilibrium (at room temperature, the equilibrium population of p-H_2 is close to 25%), the resulting signal intensities of the reaction products are greatly enhanced – the sample is hyperpolarised. The *para*-hydrogen enrichment can be carried out by cooling down hydrogen gas to cryogenic temperatures in the presence of an appropriate catalyst. At 77 K, the equilibrium percentage *para*-isomer grows to about 50%, while it reaches almost 100% at 20 K[39]. Upon removal of the catalyst, the high population of *para*-hydrogen can persist for a long time.

There are three ways of performing PHIP experiments. Two of them were proposed in the 1980s by Weitekamp and co-workers: the Parahydrogen and Synthesis Allow Dramatically Enhanced Nuclear Alignment (PASADENA)[40] and Adiabatic Longitudinal Transport After Dissociation Engenders Net Alignment (ALTADENA)[41] experiments. We shall briefly go through these methods first and return to the third one later on.

The PASADENA and ALTADENA experiments can be explained simply in terms of populations of energy levels. In the PASADENA experiment, the chemical reaction takes place at the high field, inside the NMR spectrometer. The *para*-hydrogen molecule is initially in the $|S_0\rangle$ state. The reaction is assumed to place the two protons from the hydrogen molecule in inequivalent positions (different chemical shifts). Thus, after the reaction, the appropriate product basis functions become the eigenstates and the initial *para*-population ends up in the two middle high-field states, $|\alpha\beta\rangle$ and $|\beta\alpha\rangle$ (*cf.* Figure 16.9). Figure 16.9 also shows the resulting spectral pattern. In the ALTADENA experiment, the reaction takes place outside of the magnet and the sample with the products is inserted adiabatically into the field of the NMR magnet. Under these conditions, only one of the states $|\alpha\beta\rangle$ and $|\beta\alpha\rangle$ becomes populated (we neglect here the small population differences due to the Boltzmann factors); which one it will be depends on the signs of relevant interactions. Again, the populations and spectra are shown in Figure 16.9. Concerning the experimental setup, we refer the readers to the original work and the reviews mentioned at the beginning of the section.

The PASADENA and ALTADENA experiments can, of course, also be described in terms of density operators. Again, we refer the reader to the original work and the reviews for details of the derivations. Here, we limit ourselves to quoting the final results for the very simplest case of the two-spin system. In the PASADENA experiment leading to a weakly coupled hyperpolarised *IS* spin pair, the density operator after the reaction takes the form:

$$\hat{\rho}_{PASADENA} = a2\hat{I}_z\hat{S}_z \tag{16.41}$$

where the symbol a is a proportionality constant. When exposed to a 90° pulse, $\hat{\rho}_{PASADENA}$ is transformed into invisible zero- and double-quantum coherences. If a different, smaller, flip angle is applied instead, the resulting density operator will contain a term proportional to $2\hat{I}_z\hat{S}_x + 2\hat{I}_x\hat{S}_z$. If the free induction decay is collected, these terms will yield two antiphase doublets, in agreement with Figure 16.9.

Similarly, the ALTADENA density operator after the sample is settled in the NMR magnet attains the form:

$$\hat{\rho}_{ALTADENA} = a\left(2\hat{I}_z\hat{S}_z \pm \left(\hat{I}_z - \hat{S}_z\right)\right) \tag{16.42}$$

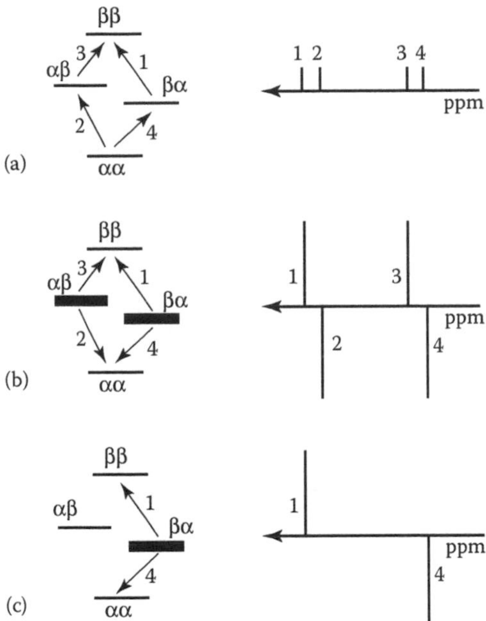

FIGURE 16.9 Schematic populations of the high-field energy levels (left, thick lines denote enhanced population) for an *IS* spin system and corresponding spectra (right). (a) normal spectrum; (b) PASADENA experiment; (c) ALTADENA experiment. (Reprinted with permission from Natterer, J. and Bargon. J., *Progr. NMR Spectrosc.* 31, 293–315, 1997. Copyright (1997) Elsevier.)

The + or − sign in Eq. (16.42) depends on the sign of the product $(\omega_I - \omega_S)J_{IS}$. If a small-angle detection pulse is applied, the magnitudes of the terms stemming from the two-spin order and from the *z*-magnetisations are similar, which will lead to partial cancellation of the inner lines in the two doublets and to a spectrum like that in Figure 16.9.

The third way of obtaining the *para*-hydrogen hyperpolarisation effects in NMR is known as Signal Amplification By Reversible Exchange (SABRE). This technique is more recent; it was proposed by Adams *et al.* in 2009[42] and has been included in the review by Green and co-workers[39]. Another review we wish to mention in this context is that by Mewis[43]. Briefly, the method makes use of the fact that the formation of a complex of the general formula $[Ir(H)_2(PR_3)\ (substrate)_3]^+$, where the hydride ligands originate from *para*-hydrogen, PR_3 is a phosphine and the substrate is a suitable nitrogen ligand (*e.g.* pyridine), can be sufficient to break the exchange symmetry of the proton pair. The PHIP effect can then be transferred reversibly to the substrate without the incorporation of the *para*-hydrogen into the molecule under investigation. The polarisation is here transferred through the *J*-coupling network and can give a signal enhancement (for NMR or MRI of, for example, protons, ^{13}C and ^{15}N) by a factor of several hundred. The mechanistic details of this process are, however, beyond the scope of this book.

Exercise for Chapter 16

1. Derive the matrix elements of $\hat{H}^{ST}$ in the singlet-triplet basis and thus prove Eq. (16.5).

References

1. Carravetta, M.; Johannessen, O. G.; and Levitt, M. H., Beyond the T-1 limit: Singlet nuclear spin states in low magnetic fields. *Phys. Rev. Lett.* 2004, 92, 153003.

2. Carravetta, M.; and Levitt, M. H., Theory of long-lived nuclear spin states in solution nuclear magnetic resonance. I. Singlet states in low magnetic field. *J. Chem. Phys.* 2005, 122, 214505.
3. Levitt, M. H., *Singlet and Other States with Extended Lifetimes*. Wiley, eMagRes, 2010.
4. Levitt, M. H., Singlet nuclear magnetic resonance. *Annu. Rev. Phys. Chem.* 2012, 63, 89–105.
5. Nagashima, K.; and Velan, S. S., Understanding the singlet and triplet states in magnetic resonance. *Concepts Magn. Reson. A* 2013, 42, 165–181.
6. Pileio, G., Singlet NMR methodology in two spin-1/2 systems. *Progr. NMR Spectr.* 2017, *98–99*, 1–19.
7. Cavadini, S.; and Vasos, P. R., Singlet states open the way to longer time-scales in the measurement of diffusion by NMR spectroscopy. *Concepts Magn. Reson. A* 2008, 32A(1), 68–78.
8. Sarkar, R.; Vasos, P. R.; and Bodenhausen, G., Singlet-state exchange NMR spectroscopy for the study of very slow dynamic processes. *J. Am. Chem. Soc.* 2007, 129, 328–334.
9. Carravetta, M.; and Levitt, M. H., Long-lived nuclear spin states in high-field solution NMR. *J. Am. Chem. Soc.* 2004, 126, 6228–6229.
10. Pileio, G.; and Levitt, M. H., Theory of long-lived nuclear spin states in solution nuclear magnetic resonance. II. Singlet spin locking. *J. Chem. Phys.* 2009, 130, 214501.
11. Pileio, G.; Carravetta, M.; and Levitt, M. H., Storage of nuclear magnetisation as long-lived singlet order in low magnetic field. *Proc. Natl. Acad. Sci. USA* 2010, 107, 17135–17139.
12. Pileio, G., Relaxation theory of nuclear singlet states in two spin-1/2 systems. *Progr. NMR Spectr.* 2010, 56, 217–231.
13. Ahuja, P.; Sarkar, R.; Vasos, P. R.; and Bodenhausen, G., Molecular properties determined from the relaxation of long-lived spin states. *J. Chem. Phys.* 2007, 127, 134112.
14. Pileio, G.; Hill-Cousins, J. T.; Mitchell, S.; Kuprov, I.; Brown, L. J.; Brown, R. C. D.; and Levitt, M. H., Long-lived nuclear singlet order in near-equivalent C-13 spin pairs. *J. Am. Chem. Soc.* 2012, 134, 17494–17497.
15. Pilelo, G.; Concistrè, M.; Carravetta, M.; and Levitt, M. H., Long-lived nuclear spin states in the solution NMR of four-spin systems. *J. Magn. Reson.* 2006, 182, 353–357.
16. Grant, A. K.; and Vinogradov, E., Long-lived states in solution NMR: Theoretical examples in three- and four-spin systems. *J. Magn. Reson.* 2008, 193, 177–190.
17. Tayler, M. C. D.; Marie, S.; Ganesan, A.; and Levitt, M. H., Determination of molecular torsion angles using nuclear singlet relaxation. *J. Am. Chem. Soc.* 2010, 132, 8225–8227.
18. Pileio, G., Singlet state relaxation via intermolecular dipolar coupling. *J. Chem. Phys.* 2011, 134, 214505.
19. Tayler, M. C. D.; and Levitt, M. H., Paramagnetic relaxation of nuclear singlet states. *Phys. Chem. Chem. Phys.* 2011, 13, 9128–9130.
20. Pileio, G., Singlet state relaxation via scalar coupling of the second kind. *J. Chem. Phys.* 2011, 135, 174502.
21. Pileio, G.; Carravetta, M.; Hughes, E.; and Levitt, M. H., The long-lived nuclear singlet state of N-15-nitrous oxide in solution. *J. Am. Chem. Soc.* 2008, 130, 12582–12583.
22. Ghosh, R. K.; Kadlecek, S. J.; Ardenkjaer-Larsen, J. H.; Pullinger, B. M.; Pileio, G.; Levitt, M. H.; Kuzma, N. N.; and Rizi, R. R., Measurements of the persistent singlet state of N_2O in blood and other solvents – potential as a magnetic tracer. *Magn. Reson. Med.* 2011, 66, 1177–1180.
23. Ghosh, R. K.; Kadlecek, S. J.; Kuzma, N. N.; and Rizi, R. R., Determination of the singlet state lifetime of dissolved nitrous oxide from high field relaxation measurements. *J. Chem. Phys.* 2012, 136, 174508.
24. Stevanato, G.; Hill-Cousins, J. T.; Håkansson, P.; Roy, S. S.; Brown, L. J.; Brown, R. C. D.; Pileio, G.; and Levitt, M. H., A nuclear singlet lifetime of more than one hour in room-temperature solution. *Angew. Chem. Int. Ed.* 2015, 54, 3740–3743.
25. Tayler, M. C. D.; and Levitt, M. H., Singlet nuclear magnetic resonance of nearly-equivalent spins. *Phys. Chem. Chem. Phys.* 2011, 13, 5556–5560.
26. Sarkar, R.; Ahuja, P.; Vasos, P. R.; and Bodenhausen, G., Long-lived coherences for homogeneous line narrowing in spectroscopy. *Phys. Rev. Lett.* 2010, 104, 053001.

27. Sarkar, R.; Ahuja, P.; Vasos, P. R.; Bornet, A.; Wagnieres, O.; and Bodenhausen, G., Long-lived coherences for line-narrowing in high-field NMR. *Progr. NMR Spectr.* 2011, 59, 83–90.
28. Callaghan, P. T., *Translational Dynamics and Magnetic Resonance: Principles of Pulsed Gradient Spin Echo NMR*. Oxford University Press: Oxford, 2014.
29. Johnson, C. S., Diffusion ordered nuclear magnetic resonance spectroscopy: Principles and applications. *Progr. NMR Spectr.* 1999, 34, 203–256.
30. Price, W. S., Pulsed-field gradient nuclear magnetic resonance as a tool for studying translational diffusion.1. Basic theory. *Concepts Magn. Reson.* 1997, 9, 299–336.
31. Price, W. S., Pulsed-field gradient nuclear magnetic resonance as a tool for studying translational diffusion: Part II. Experimental aspects. *Concepts Magn. Reson. A* 1998, *A10*, 197–237.
32. Cavadini, S.; Dittmer, J.; Antonijevic, S.; and Bodenhausen, G., Slow diffusion by singlet state NMR spectroscopy. *J. Am. Chem. Soc.* 2005, 127, 15744–15748.
33. Stejskal, E. O.; and Tanner, J. E., Spin diffusion measurements: Spin echoes in the presence of a time-dependent field gradient. *J. Chem. Phys.* 1965, 42, 288–292.
34. Vasos, P. R.; Comment, A.; Sarkar, R.; Ahuja, P.; Jannin, S.; Ansermet, J. P.; Konter, J. A.; Hautle, P.; van den Brandt, B.; and Bodenhausen, G., Long-lived states to sustain hyperpolarized magnetisation. *Proc. Natl. Acad. Sci. USA* 2009, 106, 18469-18473.
35. Bornet, A.; Jannin, S.; and Bodenhausen, G., Three-field NMR to preserve hyperpolarized proton magnetisation as long-lived states in moderate magnetic fields. *Chem. Phys. Lett.* 2011, 512, 151–154.
36. Pileio, G.; Bowen, S.; Laustsen, C.; Tayler, M. C. D.; Hill-Cousins, J. T.; Brown, L. J.; Brown, R. C. D.; Ardenkjaer-Larsen, J. H.; and Levitt, M. H., Recycling and imaging of nuclear singlet hyperpolarization. *J. Am. Chem. Soc.* 2013, 135, 5084–5088.
37. Natterer, J.; and Bargon, J., Parahydrogen induced polarization. *Progr. NMR Spectr.* 1997, 31, 293–315.
38. Bowers, C. R., Sensitivity enhancement using parahydrogen. Wiley, eMagRes, 2007.
39. Green, R. A.; Adams, R. W.; Duckett, S. B.; Mewis, R. E.; Williamson, D. C.; and Green, G. G. R., The theory and practice of hyperpolarization in magnetic resonance using parahydrogen. *Progr. NMR Spectr.* 2012, 67, 1–48.
40. Bowers, C. R.; and Weitekamp, D. P., Transformation of symmetrization order to nuclear-spin magnetisation by chemical-reaction and nuclear-magnetic-resonance. *Phys. Rev. Lett.* 1986, 57, 2645–2648.
41. Pravica, M. G.; and Weitekamp, D. P., Net NMR alignment by adiabatic transport of parahydrogen addition products to high magnetic field. *Chem. Phys. Lett.* 1988, 145, 255–258.
42. Adams, R. W.; Aguilar, J. A.; Atkinson, K. D.; Cowley, M. J.; Elliott, P. I. P.; Duckett, S. B.; Green, G. G. R.; Khazal, I. G.; Lopez-Serrano, J.; and Williamson, D. C., Reversible interactions with parahydrogen enhance NMR sensitivity by polarization transfer. *Science* 2009, 323, 1708–1711.
43. Mewis, R. E., Developments and advances concerning the hyperpolarisation technique SABRE. *Magn. Reson. Chem.* 2015, 53, 789–800.

17

Nuclear Spin-Relaxation in Other Aggregation States

The topic of this book is relaxation in isotropic liquids. For the sake of completeness, however, we wish to introduce briefly in this chapter the relaxation phenomena in aggregation states other than liquids. As both authors of this book are liquid-state nuclear magnetic resonance (NMR) spectroscopists, this will of necessity be superficial. The main objective is to point out some of the similarities and differences between NMR relaxation in liquids, on the one hand, and in gases, anisotropic fluids, solids and heterogeneous systems, on the other.

17.1 Gases

In this section dealing with relaxation in the gas phase, we shall first examine relaxation under high pressure, the conditions when the gas becomes more liquid-like. This will be followed by a brief presentation of relaxation in gases under moderate pressure and a short discussion of NMR of ^{129}Xe.

17.1.1 Supercritical Fluids

A special category of gaseous systems under high pressure is *supercritical fluids*. A supercritical fluid can be obtained when an ordinary pure liquid (in equilibrium with its vapour) is heated in a container with a constant volume larger than that of the liquid at low temperature. When the *critical temperature* is reached, the density of the liquid and the vapour phase become identical and the phase boundary disappears. The vapour pressure at the critical temperature is called the *critical pressure* and the corresponding density is denoted as the *critical density*. At and above the critical temperature, the supercritical fluid has properties intermediate between those of gases and liquids. As one might expect, NMR relaxation in gases at high density (high pressure) displays a behaviour similar to that in liquids, which gives rise to interesting applications of supercritical fluids as solvents for NMR studies, as reviewed by Taylor and Jacobson[1] in eMagRes. The supercritical conditions are not too difficult to achieve for some substances, *e.g.* for carbon dioxide, for which the critical temperature is 304.2 K and the critical pressure is 72.9 atm. We choose to exemplify NMR relaxation in gases at high pressure, including the supercritical range, by presenting data for carbon-13 spins in CO_2 and in some solutes dissolved in it.

Carbon-13 relaxation in $^{13}CO_2$ was studied by Etesse and co-workers[2]. Their measurements were performed at a low magnetic field (2.1 T), where the only efficient relaxation mechanism for the carbon-13 spins is the spin-rotation interaction (compare Section 5.3), independently of the pressure. Etesse *et al.* express the spin-lattice relaxation rate as:

$$\frac{1}{T_{1,SR}} = \frac{4C_{\perp}^{2}Ik_{B}T}{3\hbar^{2}}\frac{\tau_{K}\sigma_{Geom}}{\sigma_{J}} \qquad (17.1)$$

An important difference between the spin-rotation relaxation rate as presented in Eq. (5.51) and in Eq. (17.1) is the fact that in the latter the angular momentum correlation time, τ_J, is replaced by $\tau_K\sigma_{Geom}/\sigma_J$. The symbol τ_K is the kinetic gas theory value of the mean-free time between collisions (inverse collision frequency). The symbol σ_{Geom} is a geometric cross section of the molecule, $4\pi a^2$, where a is an appropriate molecular radius, while σ_J is a cross section for transfer of angular momentum in a collision. Both τ_K and σ_J are expected to be dependent on temperature and density. In the high-density regime, the authors use the Hubbard relation between τ_J and the rotational correlation time (*cf.* Eq. (5.50)) and the Stokes–Einstein relation between the latter quantity and viscosity (*cf.* Eq. (2.43)). According to this reasoning, $T_{1,SR}T/\eta$ should be constant, which agrees quite well with experimental data at high densities (see Figure 17.1). The data at low gas densities display a qualitatively different behaviour, to which we shall return.

The relaxation of nuclear spins in solutes dissolved in supercritical CO_2 has also been studied by Grant and co-workers, among others, and their work on methanol will be used as an example.[3,4] They measured ^{13}C T_1 (under proton decoupling) and nuclear Overhauser enhancement (NOE) for a sample with a small fraction of methanol dissolved in CO_2. For the proton-carrying methanol carbon, we need to consider both the dipole-dipole (DD) and the spin-rotation mechanisms, which under extreme narrowing conditions can be separated by quantifying the contribution from the DD mechanism from NOE measurements (compare Eq. (11.2)). Both mechanisms give comparable contributions, with the exact relation depending on the temperature-pressure conditions. The results of the T_1 measurements are summarised in Figure 17.2. We can see that the relaxation time increases slowly (the relaxation rate decreases) with increasing pressure at constant temperature. Increasing pressure reduces the mean-free time between collisions, so the trend can be explained in terms of the spin-rotation mechanism and using Eq. (17.1). The higher pressure is, at the same time, expected to lengthen the rotational correlation time and thus increase the efficiency of the dipolar relaxation[4], which explains the weakness of the variation of T_1 with pressure. We can note that the pressure effects are largest at 315.4 K, *i.e.* at a temperature close to the critical temperature. At constant pressure, we notice that the relaxation rate increases with increasing temperature, again in agreement with the expectations for the spin-rotation mechanism but not for the dipolar relaxation. The authors analysed the data in terms of two contributions to the spin-rotation relaxation rate, corresponding to the overall motion of methanol molecules and the internal motion of the methyl groups. The model allowed the detection of the presence of hydrogen-bonded methanol clusters.[3]

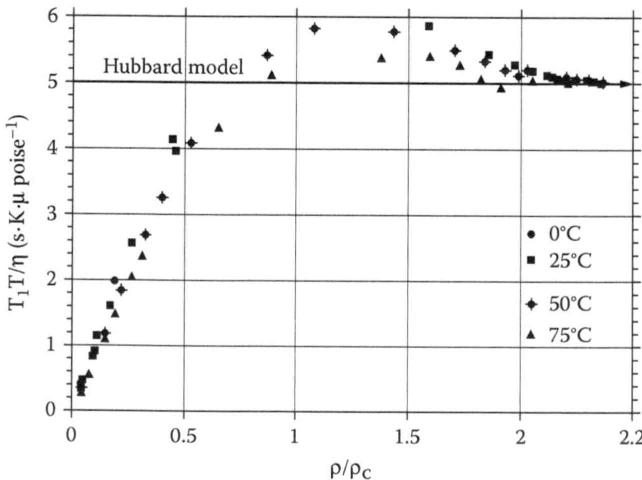

FIGURE 17.1 Carbon-13 T_1 in carbon dioxide, corrected for temperature and viscosity, versus relative density. (Reprinted with permission from Etesse, P. et al., *J. Chem. Phys.*, 97, 2022–2029, 1992. Copyright (1992) American Institute of Physics.)

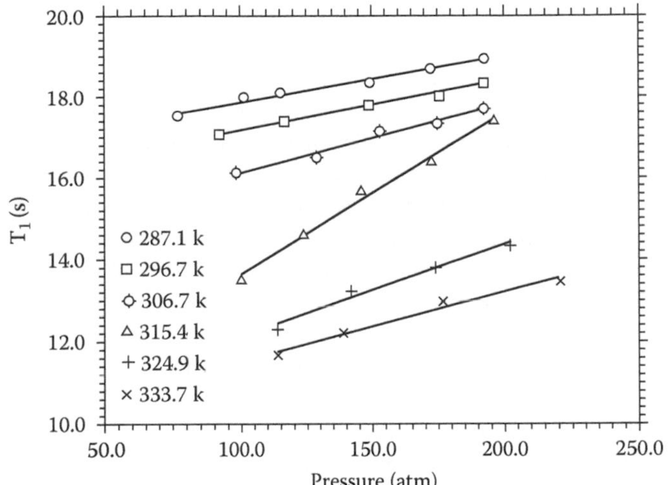

FIGURE 17.2 Spin-lattice relaxation time isotherms as a function of pressure for ^{13}C in methanol dissolved in supercritical CO_2. (Reprinted with permission from Taylor, C.M.V., et al., *J. Phys. Chem. B*, 101, 5652–5658, 1997. Copyright (1997) American Chemical Society.)

The fact that supercritical fluids are characterised by a much lower viscosity than common liquids has been used as a tool to obtain spectra with increased resolution. Robert and Evilia[5] proposed in the mid-1980s that a low viscosity (and thus a short correlation time) could lead to narrow lines for quadrupolar nuclei, such as ^{14}N and ^{17}O, and demonstrated that this might permit observing *J*-couplings to such nuclei. In a similar vein, Wand *et al.*[6] observed that a very significant narrowing of NMR signals from proteins could be obtained by encapsulating the macromolecules in reverse micelles dissolved in low-viscosity fluids. The solvents used in that work were liquid propane and larger hydrocarbons, but the authors indicated that supercritical ethane would provide even greater benefit (but also increased experimental difficulties caused by high pressure).

17.1.2 Gases under Low and Moderate Pressure

Relaxation studies under low and moderate gas pressure provide qualitatively different information than in the high-pressure (or liquid) region. The intermolecular dynamics in that case is determined by binary collisions and can be modelled more accurately, which allows investigations of pair-wise intermolecular interaction potentials and their anisotropies. Taking another look at Figure 17.1, and concentrating on data at a single temperature, we can see that the ratio $T_{1,SR}T/\eta$ increases linearly with density in the low-density range. Noting that the viscosity of an ideal gas is independent of density, we can interpret the graph as demonstrating the linear dependence of $T_{1,SR}$ on density. The subject of relaxation in gases has been reviewed by Jameson[7] and ter Horst[8], who discussed various theories explaining this observation and applications for interpreting measurements on pure gases and binary gas mixtures of very small molecules. The reader is referred to that work for further details.

17.1.3 Xenon-129

A gaseous species that has attracted a lot of attention in the context of solution and solid-state NMR is ^{129}Xe. It is a spin 1/2 nucleus and its chemical shift has been shown to be very sensitive to details of the chemical environment that xenon atoms can access. In the gas phase, the nuclear spin-relaxation of ^{129}Xe at atmospheric pressure is extremely slow, with a T_1 of hours,[9] but is accelerated dramatically by the presence of molecular oxygen, which has a paramagnetic triplet ground state (see Chapter 15)[10]. An important breakthrough in the use of xenon as an atomic resolution probe of its surrounding came in the early 1980s, when Raftery and co-workers showed that it was possible to polarise xenon-129 spins by *optical pumping*[11]. A mixture of xenon and rubidium vapour in a weak magnetic field is subjected to near-infrared laser light, which polarises the single unpaired electron of the rubidium atoms. Collisions between the rubidium and xenon atoms transfer the spin polarisation to xenon-129 nuclei and thus create a macroscopic state of high nuclear polarisation. Because of the long T_1, the nuclear polarisation is sufficiently long-lived to permit transfer to an NMR sample containing solid or liquid material that can adsorb the xenon gas. After a single radiofrequency (r.f.) pulse at the ^{129}Xe frequency, a greatly enhanced xenon-NMR signal can then be detected. In addition, when the highly polarised (hyperpolarised) ^{129}Xe nuclei are dissolved in a liquid, one can also observe a time-dependent deviation of the proton spin polarisation from its thermal equilibrium state[12]. This phenomenon is a consequence of cross-relaxation, or the NOE, between ^{129}Xe and protons and is known in the literature as *s*pin *p*olarisation-*i*nduced NOE (SPINOE). An elegant example of the SPINOE application was presented by Song and co-workers[13]. They used the pulse sequence shown in Figure 17.3 on a sample of α-cyclodextrin (α-CD, a cone-shaped molecule with a hydrophobic cavity that can accommodate a xenon atom) in dimethyl sulfoxide (DMSO)-d_6, and the result is shown in Figure 17.4. Clearly, a very strong SPINOE effect is observed for some of the protons in α-CD, notably for H3 and H5 atoms located inside the hydrophobic pocket. The SPINOE phenomenon was some time ago reviewed by Song[14].

17.2 Anisotropic Fluids

A material is called *isotropic* if its macroscopic properties do not depend on the orientation of the sample with respect to a laboratory frame, for example, that defined by the magnetic field direction of an NMR magnet. Common liquids are isotropic in that sense, even if the molecular motions are anisotropic, as was discussed in Chapters 6 and 11. Many substances, *i.e.* low-symmetry crystals, do not have this property and are called *anisotropic*. We shall come back to the case of solids in the next section. Here, we discuss anisotropic fluid phases, called *liquid crystals* (LC) or mesophases. Liquid crystals can be divided into two categories: thermotropic and lyotropic. In thermotropic samples, the liquid crystalline phases of a pure material (or a mixture) exist within a certain temperature interval, usually between the solid state and the isotropic liquid. In lyotropic systems, on the other hand, the LC phase can be obtained by varying the composition, along with the temperature, of a multicomponent system. The common feature of liquid crystalline systems is that they exhibit a certain level of orientational order, while the positional order is absent or reduced. When a molecule is dissolved in a liquid crystalline medium, it also attains a certain long-range orientational order. The simplest situation arises if we have a *uniaxial* molecule, characterised by a three-fold or higher symmetry axis (the reorientational dynamics of such symmetric rotors or symmetric diffusors was discussed in Section 6.2) dissolved in a uniaxial LC phase, which has axially symmetric properties. The principal axis of the uniaxial liquid crystal is called

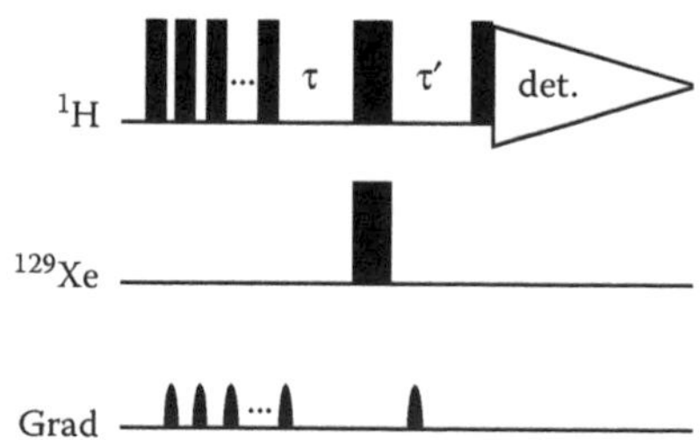

FIGURE 17.3 Heteronuclear difference NOE pulse sequence used for the SPINOE experiment. (Adapted from Song, Y.Q., et al., *Angew. Chem. Int. Ed.*, 36, 2368–2370, 1997.)

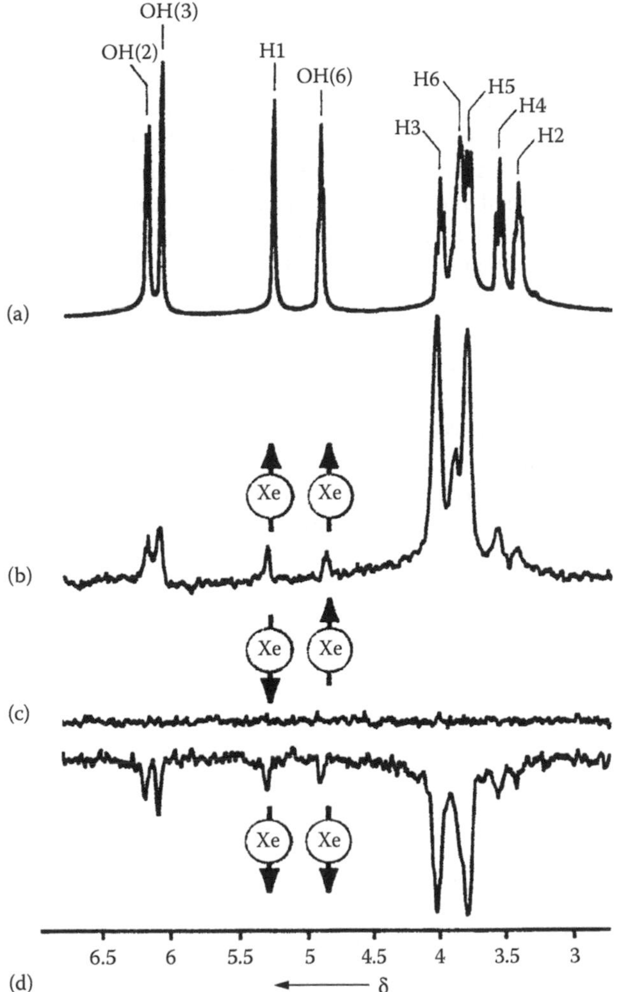

FIGURE 17.4 (a) ^{1}H spectrum of 0.05 M cyclodextrin in DMSO-d$_6$. (b) ^{1}H SPINOE spectrum obtained after introduction of negatively polarised ^{129}Xe. (c) as in (b), but omitting the 129Xe 180° pulse. (d) ^{1}H SPINOE spectrum acquired after introduction of positively polarised ^{129}Xe. (Reprinted with permission from Song, Y. Q. et al., *Angew. Chem. Int. Ed.* 1997, 36, 2368–2370. Copyright Wiley-VCH Verlag GmbH & Co. KGaA.)

the *director*. The orientational order in such a system is characterised by a single parameter, S_{zz}, called the *order parameter*:

$$S_{zz} = \left\langle D_{0,0}^{2} \right\rangle = \left\langle \frac{1}{2}\left(3\cos^2\beta - 1\right) \right\rangle \tag{17.2}$$

where $D_{0,0}^{2}$ is the Wigner rotation matrix element and β is the angle between the principal axes of the molecule and of the LC phase.

If there is no orientational order, the order parameter vanishes, while it attains the limiting value of unity when the molecular axis is perfectly ordered along the LC phase axis. In more general situations, the ordering of a probe molecule is described by the rank-two *order* or *alignment tensor*. Uniaxial mesophases often orient in the magnetic field in such a way that the director is aligned along with or perpendicular to the field. If the director is parallel to the field, the angle β in Eq. (17.2) becomes that between the principal molecular axis and the magnetic field.

In this section, we deal with mesophases oriented in the magnetic field. NMR in LC systems in which different domains have different orientations, however, instead resembles NMR of solid powders, covered in Section 17.3. As opposed to a gas phase, discussed in the previous section, NMR studies of liquid crystals are very common and have been the subject of numerous books and reviews (Dong[15], Luckhurst and Veracini[16], Burnell and de Lange[17] and Dong[18] are some examples). The most characteristic property of NMR spectra of molecules in a partially oriented medium, as opposed to isotropic solutions, is that anisotropic interactions, such as the dipole-dipole, chemical shielding anisotropy (CSA) and quadrupolar interactions, are not fully averaged out. If the director is oriented parallel to the field, the interactions are scaled by the order parameter (for uniaxial molecules in uniaxial media) or by the order tensor components (in a general case). As a consequence, the spectra of spin 1/2 nuclei contain (residual) dipolar couplings, which can be described by the dipole-dipole Hamiltonian of Eq. (3.8) (where θ is now the angle between the field direction and the director), scaled with the order parameter. Spectra of $S \geq 1$ spins display, in a similar way, quadrupolar splittings (compare Eq. (14.12)), again scaled by the order parameter. The occurrence of the splitting in the spectra of quadrupolar nuclei, for example into doublets for $S=1$ deuterons, allows new types of relaxation experiments, to which we shall return.

17.2.1 Relaxation Theory for Anisotropic Fluids

The theory of nuclear spin-relaxation in liquid crystalline solutions differs in some important points from the case of isotropic liquids. We concentrate here on the deuteron ($S=1$) relaxation, which combines the simplicity of a single (quadrupolar) relaxation mechanism with the richness of experimental possibilities, and assume that the electric field gradient (efg tensor) is axially symmetric. The topic of deuteron studies of liquid crystals (and solids) has been reviewed nicely by R.R. Vold[19]. The most important differences compared with isotropic liquids are related to the fact that additional types of dynamic processes need to be considered and that the symmetry of the medium is reduced. As shown by Jacobsen and co-workers[20], the relaxation of $S=1$ spins in an anisotropic environment is characterised by five different rate constants. The five rate constants are determined by only three *laboratory-frame spectral densities* (LFSD). For deuterons in isotropic liquids, we have already encountered three spectral densities: $J_2(0)$, $J_2(\omega_0)$ and $J_2(2\omega_0)$. In that case, the symbols refer to a single function evaluated at three different frequencies. In anisotropic liquids, the spectral densities corresponding to the autocorrelation functions for the functions $V_{2,q}$ in Eq. (14.7) and for spherical harmonics, $Y_{2,q}$, (or the corresponding Wigner matrix elements), depend on q and are inherently different. Thus, we use the symbols $J^L_{2,q}(q\omega_0)$ with $q=0, 1, 2$ for the three independent functions of frequency. The superscript L on the spectral density refers to the laboratory frame. Two of the five rates describe relaxation of longitudinal "normal magnetisation modes" (compare Section 5.1), which we call *normal* because they are not coupled to each other through relaxation. One of the longitudinal modes is $\langle \hat{S}_z \rangle$, called also the Zeeman order, with the relaxation rate given by:

$$T_{1,Z}^{-1} = \frac{3\pi}{20}\left(\frac{e^2 qQ}{\hbar}\right)^2 \left(J^L_{2,1}(\omega_S) + 4J^L_{2,2}(2\omega_S)\right) \tag{17.3}$$

The relaxation rate expression is very similar to the longitudinal quadrupolar relaxation rate in Eq. (14.17); the only difference is the occurrence of the second index on the spectral densities in Eq. (17.3). We assume implicitly that the spectral densities are normalised in the same way as in Chapter 6. The Zeeman order corresponds for $S=1$ to the difference between populations, $\rho_{1,1}$ and $\rho_{-1,-1}$, of the $|S,m\rangle$ levels with $m=1$ and $m=-1$, respectively. In terms of the NMR signal, the Zeeman order translates into the sum of the intensities in the quadrupole-split doublet. The second mode corresponds to the difference intensity of the doublet components, to the population expression $\rho_{1,1}-2\rho_{0,0}+\rho_{-1,-1}$, or to the expectation value $\langle\hat{Q}_z\rangle=\langle 3\hat{I}_z^2-I(I+1)\rangle$, and is called the *quadrupolar order*. The relaxation rate for that mode is:

$$T_{1,Q_z}^{-1}=\frac{9\pi}{20}\left(\frac{e^2qQ}{\hbar}\right)^2 J_{2,1}^L(\omega_S) \tag{17.4}$$

In Eqs. (17.3) and (17.4), the field gradient tensor is assumed to be cylindrically symmetric. Clearly, if the two rates can be measured (see Section 17.2.2), we can get the two spectral densities at non-zero frequencies, $J_{2,1}^L(\omega_S)$ and $J_{2,2}^L(2\omega_S)$. The remaining three relaxation rates refer to single- and double-quantum coherences. The double-quantum relaxation rate depends only on the high-frequency spectral densities. The relaxation rates for single-quantum coherences also depend on $J_{2,0}^L(0)$, with the functional form of this dependence varying slightly with the experimental design. It is worth noting that the number of observables in an $S=1$ system in an anisotropic medium exceeds the number of the more fundamental independent quantities, the LFSDs.

The connection between the laboratory frame spectral densities, the microstructure of the anisotropic phase and the molecular dynamics is quite complicated and has been presented in an elegant way by Halle and co-workers[21]. The spin-relaxation properties in an anisotropic medium are, in principle, dependent on the orientation of the medium with respect to the magnetic field. We can thus speak about *relaxation anisotropy*. Since the microstructure and dynamics at the molecular level are not expected to depend on the macroscopic orientation, it must be possible to relate the orientation-dependent LFSDs to more fundamental quantities called *irreducible crystal-frame spectral densities* (ICFSDs) denoted $J_\lambda^C(q\omega_0)$:

$$J_{2,q}^L(q\omega_0,\theta_{LC},\phi_{LC})=\sum_{\lambda}^{N} F_{q,\lambda}(\theta_{LC},\phi_{LC})J_\lambda^C(q\omega_0) \tag{17.5}$$

where the angular functions $F_{q,\lambda}(\theta_{LC},\phi_{LC})$ describe the orientation of the mesophase with respect to the laboratory frame. These functions depend in a known way on the mesophase symmetry. The number of independent ICFSDs, *N*, also depends on the symmetry of the LC phase. Thus, the complete information content of a spin-relaxation study at a single magnetic field consists, in principle, of *N* ICFSDs, each evaluated at three frequencies ($q=0$, 1, 2), *i.e.* 3*N* model-independent data points. To obtain this information, we need to perform a number of independent relaxation experiments at *N* (at least) different orientations of the mesophase in the magnetic field. In the simplest case, for a uniaxial, so-called *nematic*, phase, we have $N=3$, and there should be nine independent ICFSD values. However, varying the orientation of a nematic phase may often be difficult or impossible in practice, which means that the only available quantities are the LFSDs. If the director of a uniaxial nematic phase is aligned along the magnetic field, which often is the case, then the LFSDs are equal to the ICFSDs.

The spectral densities (both in the laboratory and in the crystal frame) reflect the fluctuations of the efg tensor at the site of a deuterium nucleus. In liquid crystals, these fluctuations often occur on several length- and timescales. In most cases, we need to consider at least two levels of structural organisation, and in lyotropics often three levels. The rapid motions on a molecular or local level are similar to those encountered in isotropic liquids: molecular rotations and internal motions, discussed in Chapter 6. In addition, we have to account for slow, collective motions, often referred to as *order director fluctuations* (ODFs) (see Chapter 9 of the book edited by Luckhurst and Veracini[16]). In the case of lyotropic LCs, diffusional motions over curved aggregate surfaces, occurring on an intermediate time- and length scale (*microstructure* level), are often included as a separate entity. The motions on the distinct timescales should be statistically independent, which allows the crystal-frame spectral densities to be expressed as sums of two or three contributions. Following Halle *et al.*[21], we write:

$$J_{\lambda}^{C}(q\omega_0) = J_{\lambda}^{local}(q\omega_0) + J_{\lambda}^{micro}(q\omega_0) + J_{\lambda}^{ODF}(q\omega_0) \tag{17.6}$$

Other ways of dividing the spectral densities into components from different dynamic processes have also been proposed. In some models, the $J_{\lambda}^{micro}(q\omega_0)$ term is omitted, but a cross-term between the fast motions and the ODF is included. For the laboratory-frame spectral densities, one then writes[22]:

$$J_{2,q}^{L}(q\omega_0) = J_{q}^{local}(q\omega_0) + J_{q}^{ODF}(q\omega_0) + J_{q}^{cross}(q\omega_0) \tag{17.7}$$

Assuming that the director is parallel to the magnetic field and that excursions of the director from the equilibrium are small, one can show that the ODF only contributes to the $q=1$ spectral density. At a high magnetic field, the spectral density $J_1^{ODF}(\omega)$ corresponding to these slow motions is likely to have dispersed at frequencies below the Larmor frequency, ω_0. Thus, a good way to study the order director fluctuations is the field-cycling methods, where one measures $T_{1,Z}^{-1}$ (or a corresponding quantity in spin 1/2 systems) at low magnetic fields, where $J_1^{ODF}(\omega_0)$ is expected to be proportional to $\omega_0^{-1/2}$.[18,19] This type of experiment is described, for instance, in Chapter 10 of the book edited by Luckhurst and Veracini[16] and in a review by Kimmich and Anoardo[23].

17.2.2 Experimental Methods for Anisotropic Fluids

The experimental methods for deuteron relaxation studies in LCs have been summarised by, for instance, Hoatson and Vold[24]. Here, we wish to mention two important techniques, not referred to earlier in this book: the *Jeener–Broekaert* (JB) and the *quadrupole echo* (QE) experiments. The JB experiment was originally developed for creating dipolar order in spin 1/2 systems in solids[25] but is equally applicable for creating and monitoring the quadrupolar order for deuterons in oriented mesophases[26]. The JB pulse sequence is shown in Figure 17.5a. After the initial 90_x° pulse, transverse magnetisation, $\hat{S}_y$, is created. Its evolution under the quadrupolar interaction leads to the formation of the operators $(\hat{S}_x\hat{S}_z + \hat{S}_z\hat{S}_x)$ (see Ernst *et al.*, *Further reading*). After the delay τ, the second pulse (45_y°) transforms these operators into quadrupolar order, which relaxes back to equilibrium (with the rate T_{1,Q_z}^{-1}) during the relaxation period T. As usual, T is varied in the course of an arrayed experiment to obtain the relaxation rate. The first two pulses in the sequence also produce double-quantum coherence, which, however, can be suppressed by appropriate phase cycling[26]. The final $45_{\bar{y}}^\circ$ pulse converts the quadrupolar order into detectable single-quantum coherence. The amount of quadrupolar order created by the excitation segment of the sequence depends on the product $S_{zz}\omega_Q\tau$ and can thus vary between different sites in a multiply ^{2}H-labelled molecule. A broadband version of the experiment, less sensitive to the differences in the effective interaction strength, $S_{zz}\omega_Q$, has also been described[24]. The JB experiment is often combined with inversion-recovery measurements in order to obtain both $J_{2,1}^L(\omega_S)$ and $J_{2,2}^L(2\omega_S)$. This is, however, not the only way to measure these spectral densities. Jacobsen *et al.*[20] proposed that two independent longitudinal relaxation rates, and thus the two non-zero frequency spectral densities, could be obtained by comparing the results from a selective (one component of the doublet inverted) and a non-selective inversion-recovery experiment.

The quadrupole echo (also known as the *solid echo*) experiment (see Figure 17.5b), proposed originally by Powles and Strange[27], is a basic building block of several pulse sequences used for quadrupolar nuclei. One advantage of this technique, most important in solids, is that it solves the technical problem of the receiver dead time, complicating the detection of the fast-decaying free induction decay (FID) after a single 90° pulse. The quadrupole echo block can be used in many ways. It can be inserted instead of

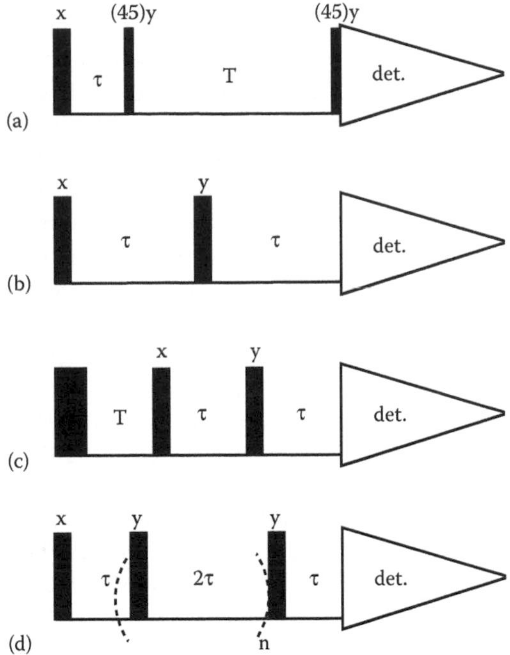

FIGURE 17.5 Pulse sequences used for ^2H in liquid crystals. (a) The Jeener–Broekaert experiment. (b) The quadrupole echo experiment. (c) The inversion-recovery experiment with quadrupole echo detection. (d) The quadrupole echo pulse train.

the 90° "read" pulse in the inversion-recovery sequence, yielding the inversion-recovery quadrupolar echo (IRQE) experiment (Figure 17.5c).[28] Using the quadrupole echo by itself, but making the delay τ the variable of an arrayed experiment, it is possible to estimate the transverse relaxation rate[24]:

$$T_{2,QE}^{-1} = \frac{3\pi}{40}\left(\frac{e^2qQ}{\hbar}\right)^2 \left(3J_{2,0}^L(0) + 3J_{2,1}^L(\omega_S) + 2J_{2,2}^L(2\omega_S)\right) \tag{17.8}$$

If the high-frequency spectral densities are known, $J_{2,0}^L(0)$ can thus be estimated. The basic QE experiment can be extended into a pulse train with a variable number of 90_y° pulses with spacing 2τ (Figure 17.5d), which allows studies of chemical exchange dynamics in liquid crystals[29]. The Carr, Purcell, Meiboom and Gill (CPMG) sequence (Section 8.2), using selective or non-selective pulses, is another useful experiment for determining $J_{2,0}^L(0)$ for deuterons in liquid crystals[24]. The knowledge of this spectral density, along with $J_{2,1}^L(\omega_S)$ and $J_{2,2}^L(2\omega_S)$, allows different models for dynamics in LCs to be tested.

17.2.3 Example: Small Molecular Probes

We wish to illustrate the use of relaxation measurements in LC systems by a couple of examples. The first of these examples concerns using a small molecular probe, chloroform, dissolved and weakly ordered in various thermotropic nematic liquid crystals (with order parameters less than 0.1 and quadrupolar splittings of a few kilohertz). Vold and Vold[30] measured the deuteron longitudinal and transverse relaxation for deuterated chloroform ($CDCl_3$) dissolved in the nematic *p*-methoxybenzylidene-*p*-butylaniline (MBBA). By combining selective and non-selective inversion-recovery measurements with selective Carr–Purcell echo experiments (all measurements at a single magnetic field), they were able to determine the laboratory frame spectral densities $J_{2,0}^{L}(0)$, $J_{2,1}^{L}(\omega_S)$ and $J_{2,2}^{L}(2\omega_S)$. They used a model for the molecular reorientation, proposed by Jacobsen *et al.*[20], in which the laboratory-frame spectral densities were dependent on the order parameter S_{zz} of Eq. (17.2) and on a corresponding average value of the $D_{0,0}^{4}$ function, and estimated the rotational correlation time for the CD-bond axis to be about 200 ps. Even earlier, Courtieu and co-workers[31] studied ^{13}C-labelled chloroform in a nematic mixture, Merck Phase 7a. From a combination of various selective and non-selective inversion-recovery experiments on the ^{13}C-^{1}H system, as well as dynamic heteronuclear NOE measurements, and by simultaneously fitting all the experimental data points, they were able to determine the spectral densities $J_{2,0}^{L}(\omega_H - \omega_C)$, $J_{2,1}^{L}(\omega_C)$, $J_{2,1}^{L}(\omega_H)$ and $J_{2,2}^{L}(\omega_H + \omega_C)$ at a single magnetic field of 2.35 T. The motion of the chloroform solute was found to be slower than in the isotropic neat liquid but still within the extreme narrowing regime. Di Bari and the authors of this book investigated ^{13}C-labelled chloroform in another nematic mixture (Merck Phase 5).[32] The two-dimensional (2D) selective nuclear Overhauser enhancement spectroscopy (NOESY) experiments described in Section 9.3 were used at three different magnetic fields to measure data that could be evaluated in terms of auto- and cross-relaxation rates, as well as DD-CSA cross-correlated relaxation rates. The dynamics was discussed in terms of Freed's model[22] allowing for a cross-term of reorientation and the ODF in the spectral densities. The spectral densities, corrected for the slightly different effects of the order parameter on the LFSDs with $q=0$, 1 and 2, were found to be dependent on frequency (*cf.* Figure 17.6). A consistent interpretation of the fairly weak, but significant, frequency dependence was obtained using the slowly relaxing local structure model, similar to the Lipari–Szabo approach. The fast motion, characterised by a correlation time of 27 ps, was associated with the reorientation of the CH-axis, while the slower motion with a correlation time of about 500 ps was assigned to the exchange of chloroform molecules between two possible residence sites.

17.2.4 Example: Model Membranes

The second example is provided by ^{2}H NMR relaxation studies of phospholipid bilayers. Brown and co-workers reported a field-dependent (between 0.38 and 9.4 T) deuteron longitudinal relaxation rate study for an isotropic suspension of small, unilamellar phospholipid vesicles of 1,2-dimyristoyl-*sn*-glycero-3-phosphocholine (DMPC)[33], while Jarrell studied the orientation dependence of the relaxation rates $T_{1,Z}^{-1}$ and T_{1,Q_z}^{-1} from aligned bilayers of a similar material at a single field of 4.7 T[34]. The two sets of data (see Figure 17.7) were modelled by Halle, who in 1991 proposed a consistent molecular interpretation[35]. The phospholipid reorientation was described as rotational diffusion of a symmetric top subject to a torque exerted by the anisotropic bilayer environment. For vesicles, the

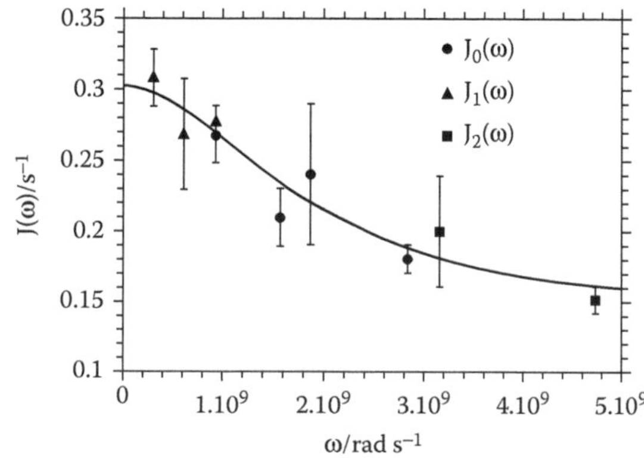

FIGURE 17.6 Corrected spectral densities for chloroform in Phase V (Merck) as functions of frequency. (Reproduced with permission from Di Bari, L. et al., *Mol. Phys.*, 84, 31–40, 1995. Copyright (1995) Taylor & Francis Ltd.)

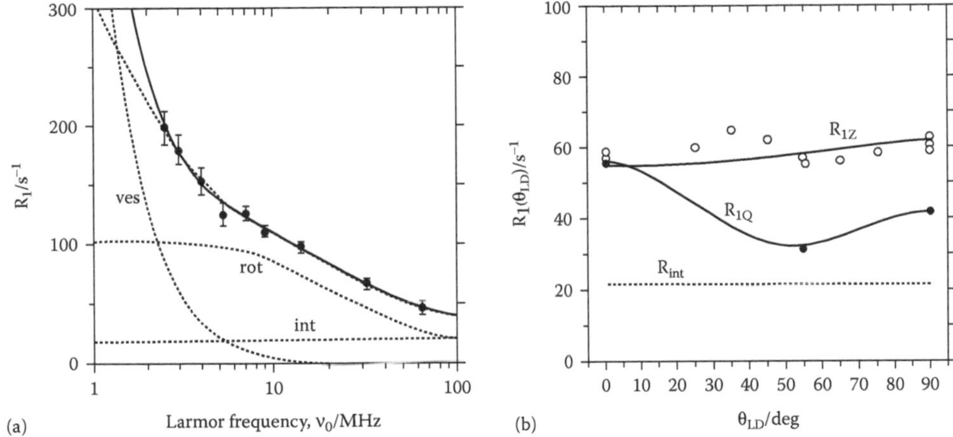

FIGURE 17.7 Fitted dependence of ^{2}H relaxation parameters for DMPC (solid lines). (a) Frequency dependence of ^{2}H spin-lattice relaxation rate in small unilamellar vesicles of 1,2-DMPC-3′,3′-d_2 at 303 K. (Experimental data from Brown, M.F. et al., *Mol. Phys.*, 69, 379–383, 1990.) (b) Orientation dependence of the relaxation rates $R_{1,Z}$ and $R_{1,Qz}$ at 30.7 MHz in aligned bilayers of 2-DMPC-4′,4′-d_2 at 303 K. (Experimental data from Jarrell, H.C. et al., *J. Chem. Phys.*, 88, 1260–1263, 1988.) (Reproduced with permission from Halle, B., *J. Phys. Chem.*, 95, 6724–6733, 1991. Copyright (1991) American Chemical Society.)

effects of the aggregate tumbling and lipid lateral diffusion were also included. The model was able to quantitatively account for the two types of data (*cf.* Figure 17.7, where effects of different motions are indicated). Similar phospholipid systems have also been subject to several more recent studies. Nevzorov and Brown[36] obtained multiple-field ^{2}H data for DMPC in small vesicles, which reorient rapidly, give liquid-like spectra and are only characterised by the $T_{1,Z}^{-1}$ rate. They also measured relaxation for DMPC in multilamellar dispersions, whose reorientation is still isotropic but much slower. In the latter case, the slower aggregate motions do not average out the quadrupolar interactions. The spectra are powder patterns (see Section 17.3) and can be interpreted in terms of an isotropic distribution of director orientations and non-zero-order parameters, different for different segments of the molecule. From these powder pattern spectra, the authors were able to determine both the $T_{1,Z}^{-1}$ and T_{1,Q_z}^{-1} rates. Moreover, they compared the deuteron data with similar carbon-13 spin-lattice relaxation rates and proposed a unified interpretation based on a model including non-collective segmental and molecular diffusion processes as well as order director fluctuations. Elastic deformations of DMPC bilayers were investigated in another paper by the same group.[37]

17.2.5 "Model-Free" Analysis of Residual Dipolar Couplings

Before leaving this section, we wish to mention a topic related to anisotropic fluids, not really involving relaxation measurements, but closely related to the Lipari–Szabo analysis of nitrogen-15 relaxation in proteins in isotropic solutions. In the mid-1990s, it was recognised that when a protein is dissolved in a weakly oriented, dilute liquid crystalline medium, the residual dipolar couplings (RDCs) for the protein can be derived from changes in the splitting patterns compared with an isotropic solution[38]. In the anisotropic medium, the splitting is given by the sum of the *J*-coupling and the RDC. The RDCs can then be used for structure refinement. A very similar phenomenon was mentioned in Chapter 15 in the context of paramagnetic proteins self-orienting in the magnetic field. This methodology has attracted a lot of attention in the biomolecular NMR community. Meiler and co-workers[39] proposed in 2001 a "model-free" dynamic interpretation of the RDCs of dipolar NH vectors. The effects of internal motions on the RDCs were analysed using a molecular dynamics simulation trajectory for the protein ubiquitin and assuming a set of alignment tensors with different orientations and rhombicities (or biaxialities, deviations from axial symmetry). The analysis makes it possible to evaluate average values of the spherical harmonics, $\langle Y_{2,m}(\theta,\phi)\rangle$, and effective orientations, $(\theta_{eff}, \phi_{eff})$, for the dipolar vectors in a molecular frame, without assuming any motional model. The $\langle Y_{2,m}(\theta,\phi)\rangle$ values can be used to calculate the RDC counterparts, S^2_{RDC}, of the generalised order parameters from the Lipari–Szabo analysis of ^{15}N relaxation data, S^2_{LS}. The Lipari–Szabo order parameter reflects motional averaging on the sub-nanosecond timescale, while the RDCs are averaged by both these fast motions and by considerably slower dynamic processes up to the millisecond timescale. By comparing these two sets of S^2 values, it becomes possible to obtain information about the slower motions and about local motional anisotropies. The ideas have, for example, been applied to the experimental data for ubiquitin in 11 different alignment media[40]. The two types of S^2 parameters were found to differ significantly from each other, in particular for the β-sheets and loop regions (the relaxation data for ubiquitin are discussed in Chapters 11 and 12 of this book).

17.3 Solid State

NMR in the solid state is a large field of science, described in numerous books, including the classical texts by Slichter and by Abragam (*Further reading*). Special techniques required to obtain high-resolution solid-state NMR spectra are described thoroughly in more recent books, *e.g.* by Schmidt-Rohr and Spiess[41] and by Stejskal and Memory[42]. In this section, we wish to touch on certain issues related to nuclear spin-relaxation in solid powders of organic materials. Good introductions to the topic can be found in a review by Duer[43].

A basic property of NMR in liquids, gases and oriented liquid crystals is relatively narrow lines, with lineshapes that are either Lorentzian or superpositions of a few Lorentzians. The definitions of transverse relaxation time (or times) are related to the widths of the Lorentzians. The situation in solid, polycrystalline or amorphous, powders is, however, very different. In the following we will exemplify solid-state relaxation by measurements on three different nuclear species commonly occurring in natural-abundance or isotopically labelled organic solids: deuterons, protons and carbon-13.

17.3.1 Deuteron NMR

Measurements on quadrupolar nuclei, such as the deuterons, can in a way be considered as the simplest case, because of the dominant role of the quadrupolar interaction related to a single spin. Several aspects of deuteron NMR in the solid state are similar to the liquid crystal case[19]. In analogy with the LC case, deuteron NMR in the solid state is an excellent tool for studies of molecular motions.[19,43,44] As opposed to the situation for oriented liquid crystals, the deuteron spectra of solid powders (and, in fact, also of non-oriented liquid crystalline materials) are broad Pake-like powder patterns. In the case of an axially symmetric efg, the powder patterns are similar to that displayed in Figure 3.4. The quadrupolar powder pattern corresponds to a distribution of orientations of microcrystallites with respect to the magnetic field, in the same way as discussed for the dipolar interaction in Section 3.1. The frequency spacing between the two "horns" for deuterons is larger than for the dipolar coupled proton pairs. In the absence of molecular motions, it amounts to 3/4 of the quadrupolar coupling constant (QCC), and with a typical QCC of 170 kHz, the spacing is thus 127.5 kHz or about 8×10^5 rad s^{-1}. Solid-state spectra of this kind are usually measured using the quadrupole echo technique (see Figure 17.5b). Molecular motions, in the case of solids commonly modelled by various jump models, have a profound effect on the deuteron powder patterns[43,44] as well as on the outcome of experiments designed to measure spin-lattice relaxation. If the jump rate is of similar magnitude to the QCC, then the shape of the pattern depends on the type of jumps considered and on their rate. This is illustrated by the quadrupole echo ^{2}H spectra in solid alanine (see Figure 14.9), perdeuterated in the methyl group, obtained at different temperatures by Batchelder *et al.*[45] and shown

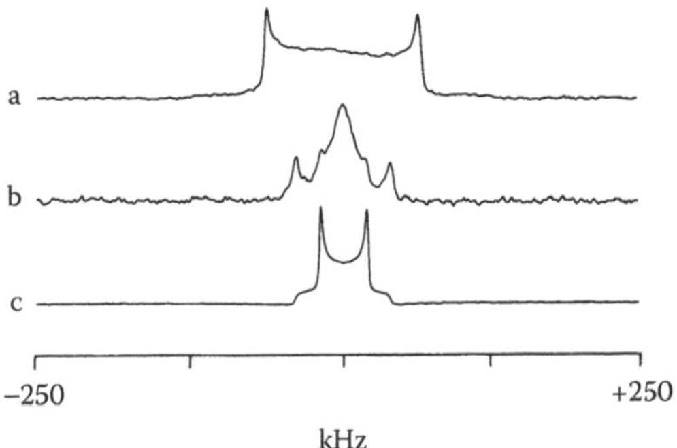

FIGURE 17.8 Quadrupole echo ^{2}H spectra of polycrystalline L-alanine, perdeuterated in the methyl group at different temperatures: (a) −150°C; (b) −96°C; (c) 20°C. (Reprinted with permission from Batchelder, L.S. et al., *J. Am. Chem. Soc.*, 105, 2228–2231, 1983. Copyright (1983) American Chemical Society.)

in Figure 17.8. At the lowest temperature of −150°C, the methyl group motion is frozen out and, essentially, a rigid-lattice powder pattern is obtained. At the intermediate temperature of −96°C, the pattern is affected by the three-site jump process with the rate on the order of inverse quadrupolar coupling constant. One can think of this phenomenon as a demonstration of the anisotropy of the spin-spin relaxation rate. In this rate regime, the spectra are also dependent on the quadrupole echo delay, τ in Figure 17.5b, an effect explained in detail by Vega and Luz[46]. At the highest temperature of 20°C, the jump rate is much larger than the QCC and we can speak about a *motionally narrowed* powder pattern, insensitive to the QE delay and to the actual jump rate. Dynamic information in this rate regime can be obtained from inversion-recovery experiments with quadrupole-echo detection (the IRQE experiment; see Figure 17.5c). The topic of spin-lattice relaxation in solids has been discussed thoroughly in the important paper by Torchia and Szabo[47]. An illustrative set of IRQE spectra, again taken from the work of Batchelder *et al.*[45], is shown in Figure 17.9. The figure also illustrates the anisotropy of the spin-lattice relaxation rate, which in the case of a solid powder

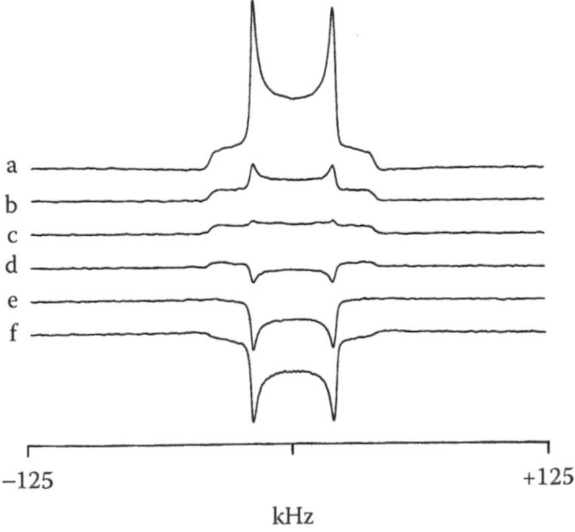

FIGURE 17.9 IRQE ^{2}H spectra of polycrystalline L-alanine perdeuterated in the methyl group obtained at 23°C. The relaxation period varies between 0.5 ms in (f) and 50 ms in (a). (Reprinted with permission from Batchelder, L.S. et al., *J. Am. Chem. Soc.*, 105, 2228–2231, 1983. Copyright (1983) American Chemical Society.)

demonstrates itself in the form of different recovery rates throughout the spectral pattern. Besides providing information on the jump rates, the IRQE data and their temperature dependence may in some situations allow differentiation between dynamic models.[45,47] By measuring the jump rate at different temperatures, Batchelder and co-workers were also able to estimate the Arrhenius activation energy for the process. The result, 22.6 kJ mol^{-1}, is actually very similar to the value obtained in solution (discussed at the end of Section 14.4). Deuteron NMR in solids can also be used for studying dynamic processes on a slower timescale than the inverse QCC. We refer the reader interested in this topic to specialised books and reviews[19,41,43,44,48].

17.3.2 Proton NMR

Proton signals of typical organic solids are more difficult to interpret. They may have a width of about 50 kHz, stemming from a large number of dipole-dipole couplings, and non-Lorentzian shapes. T_2 is thus not a useful parameter. The protons in the spin system are strongly coupled to each other, which has important consequences for their relaxation properties. In fact, in systems of this kind, one often uses a thermodynamic view on relaxation, with the spin system characterised by a *spin temperature* (see Abragam's and Slichter's books, *Further reading*), derived from the Boltzmann distribution of the spins among the energy levels of a large number of coupled spins. The spin temperature is sustained through mutual flip-flop processes, known as *spin-diffusion*. Note that the same concept, with a similar meaning, was earlier introduced in the context of cross-relaxation in macromolecules in solution (see Section 9.2). The topic of spin diffusion in solids has been reviewed by Cheung.[49]

In analogy with the case of deuterons, also proton spectra in solids are sensitive to molecular motions. The simplest tool for studying the effects of dynamics on solid-state proton spectra is the *method of moments*, described in Slichter's book. The second moment of a wide-line NMR spectrum is defined as follows:

$$\langle \omega^2 \rangle = \int_0^\infty \omega^2 F(\omega) d\omega \Big/ \int_0^\infty F(\omega) d\omega \tag{17.9}$$

where the function $F(\omega)$ describes the shape of the NMR line. Andrew and Eades presented in the early 1950s a classical study of the proton second moment of solid benzene[50]. The moment was constant at temperatures below about 100 K, dropped to about 20% of that value between about 100 and 120 K, and then remained constant when the temperature was further increased. The data were interpreted in terms of the onset of the fast rotation around the six-fold axis between 100 and 120 K. More quantitative information can also be obtained from proton relaxation studies. Spin-lattice relaxation of protons in the solid state is often measured as a function of temperature. As in the previous section, solid alanine is used as an example. Andrew and co-workers[51] reported proton inversion-recovery measurements over the temperature range 4–500 K at a single magnetic field of 1.4 T. The relaxation process was analysed as a single exponential and the plot of the obtained T_1 values versus temperature is shown in Figure 17.10. The plot shows two distinct minima, and the interpretation is that each minimum is associated with a particular type of motion, with a correlation time at the minimum matching the condition $\omega_0\tau_c \approx 1$ (compare Figure 2.6). For alanine in the zwitterionic form, the low-temperature minimum was assigned to the reorientation of the methyl group, while the minimum at a higher temperature was ascribed to the motion of the NH_3^+ moiety. Analysing the temperature dependence of the T_1 in the vicinity of the minimum, the authors estimated the methyl rotation barrier to be 22.4 kJ mol^{-1}, a value very similar to the result of the deuteron study by Batchelder *et al.*[45] mentioned earlier. Slower motion in organic solids can also be studied using spin-lock and JB experiments, as reviewed by Ailion[52].

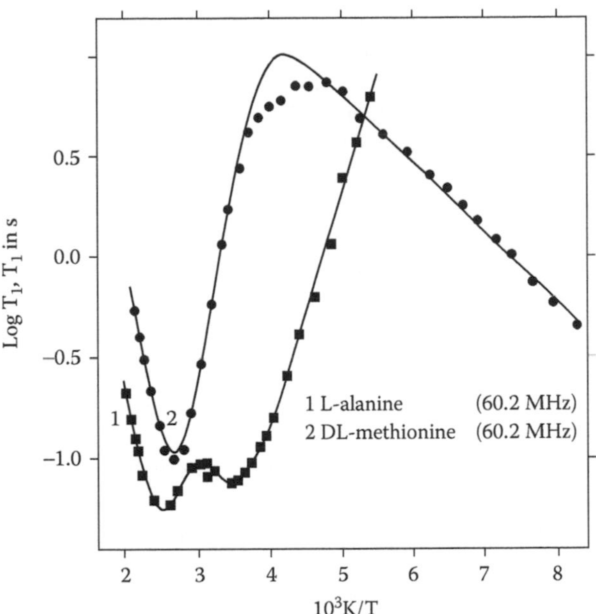

FIGURE 17.10 The variation of the proton spin-lattice relaxation time in polycrystalline amino acids with inverse temperature. (Reproduced with permission from Andrew, E.R. et al., *Mol. Phys.*, 32, 795–806, 1976. Copyright (1976) Taylor & Francis Ltd.)

17.3.3 Carbon-13 NMR

Carbon-13 solid-state spectra are also broad and featureless because of the combined effect of heteronuclear ^{13}C-^{1}H dipolar couplings and anisotropic chemical shielding (compare Section 5.2). However, for low-natural abundance species, such as carbon-13, it is in fact possible to obtain narrow, liquid-like spectra by using a combination of two techniques: high-power proton decoupling and magic angle spinning (MAS).[41,42,48] The proton decoupling suppresses heteronuclear dipolar interactions in a similar way as the broadband decoupling in liquids removes the *J*-couplings. The removal of very large heteronuclear dipolar couplings is technically difficult, but the decoupling is assisted by the high-speed spinning of the sample around an axis at the "magic angle" of $\theta = 54.7°$ with respect to the magnetic field. The angle makes the function $(3\cos^2\theta - 1)$, characteristic of rank-two tensorial interactions, vanish. Because of this, MAS also removes the effects of anisotropic chemical shift. In addition to high-power decoupling and MAS, *cross-polarisation* (CP) is the third important tool often used in solid-state ^{13}C NMR.[41,42,53] Cross-polarisation does not influence the lineshapes, but it increases the sensitivity of the experiment by transferring polarisation from protons to carbons. This effect is obtained by a simultaneous spin-lock on both carbons (the *S*-spins) and protons (the *I*-spins), with the r.f. field for the two spins fulfilling a certain condition, which for a non-spinning sample can be formulated as:

$$\gamma_S B_{1S} = \gamma_I B_{1I} \tag{17.10a}$$

so that the two frequencies match each other:

$$\omega_{1S} = \omega_{1I} \tag{17.10b}$$

The condition in Eq. (17.10) is called the *Hartmann–Hahn matching*, in honour of the inventors[54]. Under MAS, the conditions have to be modified slightly. The matching of energies in the rotating frame allows an efficient energy transfer. The polarisation of the carbon spins can, at least partly, be explained in terms of spin thermodynamics. The spin-lock on protons results in the high magnetisation, obtained in the strong B_0-field, being aligned along the much weaker B_1-field. This high polarisation in a weak field corresponds to a very low spin temperature. By the Hartmann–Hahn contacts between the large reservoir of "cold" and abundant proton spins, the rare carbon spins are also cooled, which results in a stronger signal.

If the spinning speed is much larger than the width of the powder pattern corresponding to the anisotropic chemical shielding (see Figure 5.2), CP MAS experiments yield high-resolution solid-state ^{13}C NMR spectra of liquid-like appearance, with isotropic chemical shifts usually very similar in the two aggregation states. The effect of the magic angle spinning on carbon-13 spin-lattice relaxation is non-trivial and has been analysed by Varner and co-workers[55] in a study that can be considered as an extension of the Torchia and Szabo approach to relaxation[47] from the static to the rotating solids. Briefly, the authors recognised the fact that carbon-13 spin-lattice relaxation of the narrow line in a MAS experiment is an average over a distribution of orientation-dependent relaxation processes (compare with the deuteron case in Section 17.3.1!).

The combination of CP with high-power decoupling was originally proposed by Pines *et al.*[53] and, in the same paper, a CP-based experimental method for studying proton-decoupled carbon-13 spin-lattice relaxation in solids was proposed. Carbon-13 T_1 measurements in solids are today quite common and the method of choice is usually a modification of the original scheme, less sensitive to artefacts, proposed by Torchia[56]. As an example of this type of study, we wish to mention a study of two solid monosaccharides by Wang *et al.*[57] and we focus on the results for methyl α-L-fucopyranoside (Figure 17.11). Carbon-13 T_1 values at a single magnetic field (7 T) were measured for all carbons in the sugar in the temperature range 270–320 K. It was found that both the methyl and the methoxy carbons were characterised by a rapid relaxation, which the authors described by the same expression as used in the simplest case in liquids (see Section 3.4):

$$\frac{1}{T_1}=\frac{3}{10}\left(\frac{\mu_0}{4\pi}\right)^2\frac{\gamma_H^2\gamma_C^2\hbar^2}{r_{CH}^6}\left[\frac{\tau_c}{1+(\omega_H-\omega_C)^2\tau_c^2}+\frac{3\tau_c}{1+\omega_C^2\tau_c^2}+\frac{6\tau_c}{1+(\omega_H+\omega_C)^2\tau_c^2}\right] \tag{17.11}$$

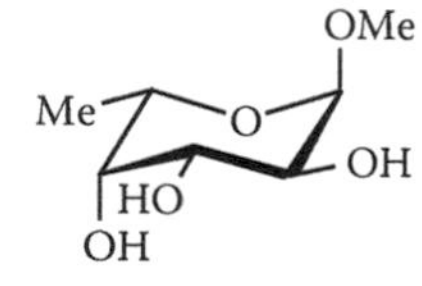

(a)

2.5
2
1.5
1
0.5
0
Ln T$_1$ (s)
● A
○ B
0.0033 0.0032 0.0034 0.0036 0.0038 0.004
1/T(K^{-1})

(b)

FIGURE 17.11 (a) The structure of methyl α-L-fucopyranoside and (b) the Arrhenius plot of logarithms of spin-lattice relaxation times for the methyl (solid circles) and methoxy (open circles) carbon-13 nuclei versus inverse temperature. (Reproduced with permission from Wang, Y.L. et al., *J. Phys. Chem. B*, 106, 12834–12840. Copyright (2002) American Chemical Society.)

The factor of three in the right-hand side accounts for three protons of the methyl group, and τ_c is an effective correlation time. The use of this equation is, strictly speaking, not fully correct in a solid, where the dynamics cannot be considered as isotropic rotational diffusion. However, the authors find the apparent correlation time for the methyl and methoxy carbon to be in the extreme narrowing regime, where the form of spectral density functions is not too important (compare Section 11.1). The other carbons relax much more slowly (on the slow-reorientation side of the T_1 minimum; compare Figure 3.8), which indicates that the motions leading to efficient relaxation for the CH_3 carbons are methyl group rotations. The logarithms of the T_1 values for both CH_3 carbons are plotted against inverse temperature in an Arrhenius plot in Figure 17.11. In the same study, the authors also measured proton T_1, $T_{1\rho}$ and the second moment over an extended temperature range. An advantage of the carbon-13 measurements is that every individual, chemically distinct site can be independently characterised, while the proton data are obtained under low-resolution conditions. Non-methyl carbon-13 nuclei in solids, such as the sugar ring carbons in the methyl α-L-fucopyranoside example, usually move relatively slowly and measurements of ^{13}C $T_{1\rho}$ can often be a better source of dynamic information. For even slower dynamics, on the millisecond or longer timescale, Spiess and his group developed multidimensional ^{13}C (and ^{2}H) exchange experiments, which can be considered as solid-state analogues of the exchange spectroscopy (EXSY) experiment discussed in Chapter 13. The reader is referred to the book by Schmidt-Rohr and Spiess[41] and the reviews by Duer[43] and by Spiess[48] for a comprehensive presentation of these techniques.

17.4 Heterogeneous Systems

Over the last decades, NMR relaxation has been used to study a large variety of problems in complicated, heterogeneous systems within the fields as diverse as geology, biophysics and food science. Here, we present two examples, water in porous media and water in immobilised gels, with importance within experimental methodology as well as in theory development.

17.4.1 Water in Porous Rocks – Inverse Laplace Transform Methods

Consider a system consisting of a rigid solid porous material with the pores filled with a simple liquid, *e.g.* water. The solid material is assumed not to give rise to any NMR signal either because of the lack of magnetic nuclei or because of excessive line-broadening. The liquid component, on the other hand, can give rise to an NMR signal. Assumc further that the protons in the liquid are characterised by a common chemical shift/Larmor frequency (which may be the result of the use of a low magnetic field) but not necessarily by a single spin-spin relaxation time. The spread of the latter property may be related to the water molecules occurring in pores of different sizes; according to current theory of NMR in porous media, the relaxation properties of aqueous protons are enhanced in the vicinity of the solid surfaces, among other reasons because of the possible presence of surface-bound paramagnetic ions[58,59], and are thus dependent on the surface-to-volume ratio. Assuming a finite number of exponential decays, the time-dependent magnetisation (the free induction decay after a 90° pulse on resonance) can be expressed as:

$$M(t) = \sum_i m_i \exp\left(-t/T_{2i}\right) \tag{17.12}$$

where m_i is the amplitude of the ith component, with spin-spin relaxation time T_{2i}. Here, the signal from a specific site (specific type of pore) is a real decaying exponential, as opposed to the complex exponential arising in the presence of characteristic Larmor frequencies (compare with the multifrequency analogue of Eq. (7.19)). Different frequencies present in that case allow different signals to be separated by Fourier transformation. The separation of different decay rates in Eq. (17.12), denoted sometimes as *inverse Laplace transform* (ILT), is mathematically much more difficult (the concept of an ill-posed problem is often used) if the signal is superimposed by noise. Similarly to the two-dimensional NMR experiments discussed in Section 7.2, it is possible to design two-dimensional experiments in which different dimensions are defined in terms of different relaxation processes[60]. For example, we consider the T_1–T_2 correlation experiment[61], illustrated in Figure 17.12. The experiment consists of two segments: the inversion-recovery (180°–τ_1–90°) sequence, followed by the CPMG segment, with the length τ_2. The experiment produces a data matrix $S(\tau_1, \tau_2)$ that depends on both relaxation processes. Song and co-workers[61,62] described an algorithm (fast two-dimensional Laplace inversion) allowing a translation of this type of data into a distribution of relaxation times. The mathematics of the ILT procedure is described in detail in the review by Mitchell and co-workers[63]. The stacked plot of T_1 versus T_2 for a water-saturated sample of Indiana limestone is shown in Figure 17.13.

Another two-dimensional experiment frequently used for porous media is the T_2-exchange experiment, proposed by Washburn and Callaghan[64]. The experiment is reminiscent of EXSY (see Chapter 13) and consists of a CPMG block followed by a 90° pulse (bringing the transverse magnetisation to the z-axis) and a mixing period. During the mixing period, the system undergoes exchange and longitudinal relaxation. After the mixing period, another CPMG sequence is applied. The result of the experiment on water-saturated Castlegate sandstone is shown in Figure 17.14. Here, the 2D ILT procedure leads to the detection of four discrete types of pores. The cross-peaks indicate water exchange between the different pore types/sites. We can thus see in the figure that water molecules in three of those sites are in exchange, while the fourth seems not to participate in the exchange network. The analysis of this type of multisite relaxation exchange data has been discussed by Van Landeghem *et al.*[65].

Instead of correlating two types of relaxation decay, it is also possible to create experiments correlating decays originating from diffusion processes under the influence of field gradients with each other, or

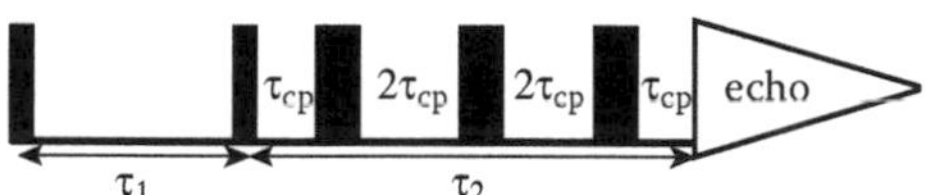

FIGURE 17.12 Pulse sequence for $T_1 - T_2$ correlation experiment. (Adapted from Song, Y.Q. et al., *J. Magn. Reson.*, 154, 261–268, 2002.) For phase cycling, please consult the original work.

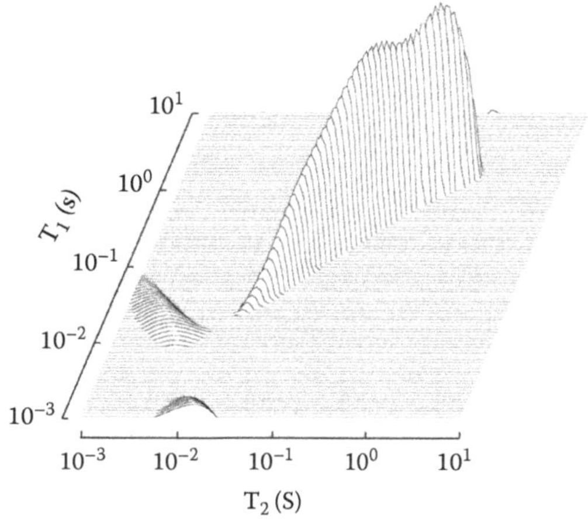

FIGURE 17.13 The T_1–T_2 correlation spectrum for water in Indiana limestone. (Reprinted with permission from Song, Y.-Q. et al., *J. Magn. Reson.*, 154, 261–268, 2002. Copyright (2002) Elsevier.)

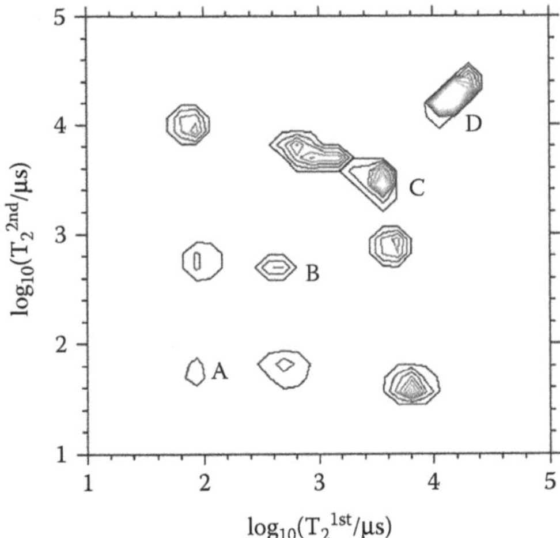

FIGURE 17.14 2D relaxation exchange distribution for water-saturated Castlegate sandstone, corresponding to the mixing time of 160 ms. A, B, C, D refer to different pore types in the sandstone in which water molecules can reside. (Reprinted with permission from Washburn, K.E. and Callaghan, P.T., *Phys. Rev. Lett.*, 97, 175502, 2006. Copyright (2006) by the American Physical Society.)

with relaxation decays[60]. Moreover, one should also stress that the usefulness of relaxation correlation and relaxation exchange experiments is not limited to porous rocks. A nice overview of work of this kind for various kinds of soft matter has been reported by Bernin and Topgaard[66].

17.4.2 Water in Immobilised Protein Gels – the EMOR Theory

The second example of heterogeneous systems we wish to discuss here is polymer gels. Gels are formed when polymers dispersed in liquid are cross-linked. Here, we wish to present examples related to water dynamics in immobilised protein gels, and effects on spin-relaxation, in such systems.

We start by presenting the theoretical approach developed over time by Halle[67] and presented in its modern form by Nilsson and Halle[68], known as the *exchange-mediated orientational randomisation* (EMOR) theory of quadrupolar relaxation. The theory applies to internal deuterated water molecules, D_2O (or labile deuterons), that exchange between long-living anisotropic A-sites within the macromolecular network and the isotropic (I) bulk, where the quadrupolar interaction is averaged to zero. The exchange model stipulates that the exchange between anisotropic sites occurs only through the I state. The theory is designed to interpret the dependence of the deuteron ($I=1$) spin-lattice relaxation on the applied magnetic field. The process of release of water molecules from the internal sites has a dual role in the theory: it mixes the two water populations (internal, bulk) and also induces the relaxation process. Since the protein is immobilised and does not tumble, the exchange (survival) time of the internal water molecules replaces the rotational counterpart as the characteristic motional correlation time. Since the internal water survival times can be in the microsecond range[67,69], the usual perturbative treatment of relaxation, as described in Chapter 4, cannot be used. Instead, Nilsson and Halle[68] describe a more general approach based on the stochastic Liouville equation (compare Section 15.1).

For a thorough presentation of the EMOR theory, we refer the reader to the original papers.[67,68] Here, we mention some of the highlights of the model and its results. The longitudinal relaxation of the deuterons is calculated numerically, or by analytical expressions in certain limiting cases, and depends on the external magnetic field (Larmor frequency, ω_I) and on a number of parameters characterising the system: (1) the quadrupole coupling constant (or the related quantity in angular frequency units, ω_Q) and the asymmetry parameter (η) of the quadrupolar interaction in the anisotropic site; (2) the mean survival time in the anisotropic site, τ_A; (3) the relative population of the anisotropic site, P_A. The model predicts, in a general case, that the longitudinal relaxation will not necessarily be exponential. To facilitate the data analysis, the authors propose using an integral longitudinal relaxation rate, which (in slightly simplified notation) can be expressed as:

$$\tilde{R}_1 = \left[\int_0^{\infty} \frac{\Delta M_z(t)}{\Delta M_z(0)} dt \right]^{-1} \tag{17.13}$$

where $\Delta M_z(t)$ denotes the deviation of the longitudinal magnetisation from equilibrium. Parameters of the model can be fitted to the integrated rate and its field dependence. An important special case, for which the model simplifies considerably, is the dilute regime, where the relative population of the anisotropic site is very small, $P_A \ll 1$. Different regions of the approach can be defined depending on the dimensionless quantities $Q = \omega_Q \tau_A$ and $L = \omega_L \tau_A$. In the motional-narrowing range, where $Q^2 \ll 1 + L^2$ and for which the perturbation theory is valid, the prediction is exponential relaxation with the rate similar to Eq. (14.17), weighted by P_A (this is valid for fast exchange and negligible relaxation in the I state). The low-field regime, $L^2 \ll 1 + Q^2$, results in another simple expression, while the intersection of the two ranges corresponds to the extreme narrowing case, where a counterpart of Eq. (14.15) is obtained. The important property of the full EMOR formulation is that it is valid also beyond the motional-narrowing regime. An example of the predicted field dependence (nuclear magnetic resonance dispersion [NMRD] profiles) of the integral relaxation rate in the dilute regime is shown in Figure 17.15. We can note that, with the chosen parameter set, the curve is highly sensitive to the asymmetry parameter.

The EMOR approach can easily be generalised to allow for partial averaging of quadrupole coupling by rapid internal motions and for kinetic heterogeneity (more than one τ_A value).[68] The theory has found numerous applications in the interpretation of variable-field deuteron spin-lattice relaxation data, for example, to cross-linked solutions of bovine pancreatic trypsin inhibitor and ubiquitin[69], where the authors were able to estimate the exchange survival lifetimes for water molecules in the microsecond range. More recently, the EMOR theory was also developed for NMRD profiles of the dipole-coupled spin 1/2 pairs[70–72].

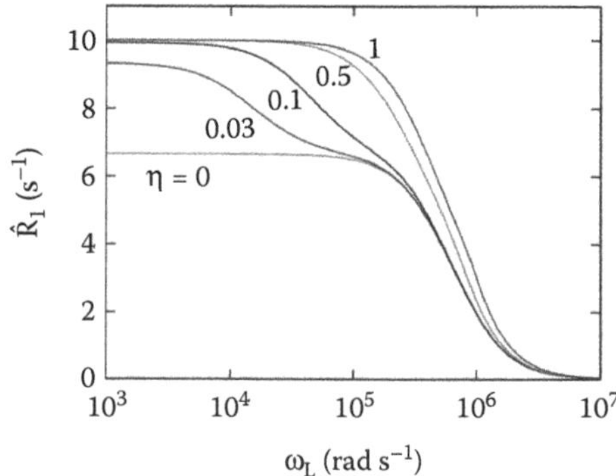

FIGURE 17.15 NMRD profile for the dilute-regime integral relaxation rate for $P_A = 10^{-3}$, $\tau_A = 100$ μs, $\omega_Q = 10^6$ rad s^{-1} and indicated values of asymmetry parameter. (Reprinted with permission from Nilsson, T. and Halle, B., *J. Chem. Phys.*, 137, 054503, 2012. Copyright (2012) American Institute of Physics.)

References

1. Taylor, C. M. V.; and Jacobson, G. B., *Supercritical Fluids*. Wiley, eMagRes, 2007.
2. Etesse, P.; Zega, J. A.; and Kobayashi, R., High pressure nuclear magnetic resonance measurement of spin lattice relaxation and self-diffusion in carbon dioxide. *J. Chem. Phys.* 1992, 97, 2022–2029.
3. Taylor, C. M. V.; Bai, S.; Mayne, C. L.; and Grant, D. M., Hydrogen bonding of methanol in supercritical CO_2 studied by C-13 nuclear spin-lattice relaxation. *J. Phys. Chem. B* 1997, 101, 5652–5658.
4. Bai, S.; Taylor, C. M. V.; Liu, F.; Mayne, C. L.; Pugmire, R. J.; and Grant, D. M., CO_2 clustering of 1-decanol and methanol in supercritical fluids by C-13 nuclear spin-lattice relaxation. *J. Phys. Chem. B* 1997, 101, 2923–2928.
5. Robert, J. M.; and Evilia, R. F., High-resolution nuclear magnetic resonance spectroscopy of quadrupolar nuclei: Nitrogen-14 and oxygen-17 examples. *J. Am. Chem. Soc.* 1985, 107, 3733–3735.
6. Wand, A. J.; Ehrhardt, M. R.; and Flynn, P. F., High-resolution NMR of encapsulated proteins dissolved in low-viscosity fluids. *Proc. Natl. Acad. Sci. USA* 1998, 95, 15299–15302.
7. Jameson, C. J., *Gas Phase Studies of Intermolecular Interactions and Relaxation*. Wiley, eMagRes, 2007.
8. Ter Horst, M. A., Quadrupolar spin relaxation in the gas phase: Experiments and calculations. *Magn. Reson. Rev.* 1998, 17, 195–261.
9. Hunt, E. R.; and Carr, H. Y., Nuclear magnetic resonance of Xe129 in natural xenon. *Phys. Rev.* 1963, 130, 2302–2305.
10. Jameson, C.; Jameson, A. K.; and Hwang, J. K., Nuclear spin relaxation by intermolecular dipolar coupling in the gas phase. ^{129}Xe in oxygen. *J. Chem. Phys.* 1988, 89, 4074–4081.
11. Raftery, D.; Long, H.; Meersmann, T.; Grandinetti, P. J.; Reven, L.; and Pines, A., High-field NMR of adsorbed xenon polarized by laser pumping. *Phys. Rev. Lett.* 1991, 66, 584–587.
12. Navon, G.; Song, Y. Q.; Room, T.; Appelt, S.; Taylor, R. E.; and Pines, A., Enhancement of solution NMR and MRI with laser-polarized xenon. *Science* 1996, 271, 1848–1851.
13. Song, Y. Q.; Goodson, B. M.; Taylor, R. E.; Laws, D. D.; Navon, G.; and Pines, A., Selective enhancement of NMR signals for α-cyclodextrin with laser-polarized xenon. *Angew. Chem. Int. Ed.* 1997, 36, 2368–2370.
14. Song, Y. Q., Spin polarization-induced nuclear Overhauser effect: An application of spin-polarized xenon and helium. *Concepts Magn. Reson.* 2000, 12, 6–20.
15. Dong, R. Y., *Nuclear Magnetic Resonance of Liquid Crystals*. 2nd edn; Springer: New York, 1994.
16. Luckhurst, G. R.; and Veracini, C. A., Eds., *The Molecular Dynamics of Liquid Crystals*. Kluwer: Dordrecht, 1994.
17. Burnell, E. E.; and de Lange, C. E., *NMR of Orientationally Ordered Liquids*. Kluwer: Dordrecht, 2002.
18. Dong, R. Y., Relaxation and the dynamics of molecules in the liquid crystalline phases. *Progr. NMR Spectr.* 2002, 41, 115–151.
19. Vold, R. R., Deuterium NMR studies of dynamics in solids and liquid crystals. In *Nuclear Magnetic Resonance Probes of Molecular Dynamics*, Tycko, R., Ed. Kluwer: Dordrecht, 1994; pp. 27–112.
20. Jacobsen, J. P.; Bildsoe, H.; and Schaumburg, K., Application of density matrix formulation in NMR spectroscopy. II. The one-spin-1 case in anisotropic phase. *J. Magn. Reson.* 1976, 23, 153–164.
21. Halle, B.; Quist, P. O.; and Furo, I., Microstructure and dynamics in lyotropic liquid crystals – principles and applications of nuclear spin relaxation. *Liq. Cryst.* 1993, 14, 227–263.
22. Freed, J. H., Stochastic-molecular theory of spin-relaxation for liquid crystals. *J. Chem. Phys.* 1977, 66, 4183–4199.
23. Kimmich, R.; and Anoardo, E., Field-cycling NMR relaxometry. *Progr. NMR Spectr.* 2004, 44, 257–320.

24. Hoatson, G. L.; and Vold, R. L., *Deuteron Relaxation Rates in Liquid Crystalline Samples: Experimental Methods*. Wiley, eMagRes, 2007.
25. Jeener, J.; and Broekaert, P., Nuclear magnetic resonance in solids: Thermodynamic effects of a pair of rf pulses. *Phys. Rev.* 1967, 157, 232–240.
26. Vold, R. L.; Dickerson, H.; and Vold, R. R., Application of the Jeener-Broekaert pulse sequence to molecular dynamics studies in liquid crystals. *J. Magn. Reson.* 1981, 43, 213–223.
27. Powles, J. G.; and Strange, J. H., Zero time resolution nuclear magnetic resonance transients in solid. *Proc. Phys. Soc.* 1963, 82, 6–15.
28. Müller, K.; Meier, P.; and Kothe, G., Multipulse dynamic NMR of liquid crystal polymers. *Progr. NMR Spectr.* 1985, 17, 211–239.
29. Ahmad, S. B.; Packer, K. J.; and Ramdsen, J. M., The dynamics of water in heterogeneous systems. II. Nuclear magnetic relaxation of the protons and deuterons of water molecules in a system with identically oriented planar interfaces. *Mol. Phys.* 1977, 33, 857–874.
30. Vold, R. R.; and Vold, R. L., Deuterium relaxation of chloroform dissolved in a nematic liquid crystal. *J. Chem. Phys.* 1977, 66, 4018–4024.
31. Courtieu, J.; Mayne, C. L.; and Grant, D. M., Nuclear relaxation in coupled spin systems dissolved in a nematic phase - the AX and A_2 spin systems. *J. Chem. Phys.* 1977, 66, 2669–2677.
32. Di Bari, L.; Mäler, L.; and Kowalewski, J., Nuclear spin relaxation in chloroform dissolved in a nematic phase - spectral densities at high frequencies. *Mol. Phys.* 1995, 84, 31–40.
33. Brown, M. F.; Salmon, A.; Henriksson, U.; and Söderman, O., Frequency dependent ^{2}H NMR relaxation rates of small unilamellar phospholipid vesicles. *Mol. Phys.* 1990, 69, 379–383.
34. Jarrell, H. C.; Smith, I. C. P.; Jovall, P. A.; Mantsch, H. H.; and Siminovitch, D. J., Angular dependence of ^{2}H NMR relaxation rates in lipid bilayers. *J. Chem. Phys.* 1988, 88, 1260–1263.
35. Halle, B., H-2 NMR relaxation in phospholipid bilayers - toward a consistent molecular interpretation. *J. Phys. Chem.* 1991, 95, 6724–6733.
36. Nevzorov, A. A.; and Brown, M. F., Dynamics of lipid bilayers from comparative analysis of H-2 and C-13 nuclear magnetic resonance relaxation data as a function of frequency and temperature. *J. Chem. Phys.* 1997, 107, 10288–10310.
37. Brown, M. F.; Thurmond, R. L.; Dodd, S. W.; Otten, D.; and Beyer, K., Elastic deformation of membrane bilayers probed by deuterium NMR relaxation. *J. Am. Chem. Soc.* 2002, 124, 8471–8484.
38. Tolman, J. R.; Flanagan, J. M.; Kennedy, M. A.; and Prestegard, J. H., Nuclear magnetic dipole interactions in field-oriented proteins: Information for structure determination in solution. *Proc. Natl. Acad. Sci. USA* 1995, 92, 9279–9283.
39. Meiler, J.; Prompers, J. J.; Peti, W.; Griesinger, C.; and Brüschweiler, R., Model-free approach to the dynamic interpretation of residual dipolar couplings in globular proteins. *J. Am. Chem. Soc.* 2001, 123, 6098–6107.
40. Peti, W.; Meiler, J.; Brüschweiler, R.; and Griesinger, C., Model-free analysis of protein backbone motion from residual dipolar couplings. *J. Am. Chem. Soc.* 2002, 124, 5822–5833.
41. Schmidt-Rohr, K.; and Spiess, H. W., *Multidimensional Solid-State NMR and Polymers*. Academic Press: London, 1994.
42. Stejskal, E. O.; and Memory, J. D., *High Resolution NMR in the Solid State*. Oxford University Press: New York, 1994.
43. Duer, M. J., Solid-state NMR studies of molecular motion. *Ann. Rep. NMR Spectr.* 2000, 43, 1–58.
44. Batchelder, L. S., *Deuterium NMR in Solids*. Wiley, eMagRes, 2007.
45. Batchelder, L. S.; Niu, C. H.; and Torchia, D. A., Methyl reorientation in polycrystalline amino acids and peptides: A ^{2}H NMR spin-lattice relaxation study. *J. Am. Chem. Soc.* 1983, 105, 2228–2231.
46. Vega, A. J.; and Luz, Z., Quadrupole echo distortion as a tool for dynamic NMR: Application to molecular reorientation in solid trimethylamine. *J. Chem. Phys.* 1987, 86, 1803–1813.
47. Torchia, D. A.; and Szabo, A., Spin-lattice relaxation in solids. *J. Magn. Reson.* 1982, 49, 107–121.

48. Spiess, H. W., *Polymer Dynamics and Order From Multidimensional Solid State NMR*. Wiley, eMagRes, 2007.
49. Cheung, T. T. P., *Spin Diffusion in Solids*. Wiley, eMagRes, 2007.
50. Andrew, E. R.; and Eades, R. G., A nuclear magnetic resonance investigation of three solid benzenes. *Proc. Roy. Soc. (London)* 1953, A218, 537–552.
51. Andrew, E. R.; Hinshaw, W.; Hutchins, M. G.; and Sjöblom, R. O. I., Proton magnetic relaxation and molecular motion in polycrystalline amino acids. II. Alanine, isoleucine, leucine, methionine, norleucine, threonine and valine. *Mol. Phys.* 1976, 32, 795–806.
52. Ailion, D. C., *Ultraslow Motions in Solids*. Wiley, eMagRes, 2007.
53. Pines, A.; Gibby, M. G.; and Waugh, J. S., Proton-enhanced NMR of dilute spins in solids. *J. Chem. Phys.* 1973, 59, 569–590.
54. Hartmann, S. R.; and Hahn, E. L., Nuclear double resonance in the rotating frame. *Phys. Rev.* 1962, 128, 2042–2053.
55. Varner, S. J.; Vold, R. L.; and Hoatson, G. L., Characterization of molecular motion in the solid state by carbon-13 spin-lattice relaxation times. *J. Magn. Reson.* 2000, 142, 229–240.
56. Torchia, D. A., The measurement of proton-enhanced carbon-13 T_1 values by a method which suppresses artifacts. *J. Magn. Reson.* 1978, 30, 613–616.
57. Wang, Y. L.; Tang, H. R.; and Belton, P. S., Solid state NMR studies of the molecular motions in the polycrystalline alpha-L-fucopyranose and methyl alpha-L- fucopyranoside. *J. Phys. Chem. B* 2002, 106, 12834–12840.
58. Song, Y. Q., Magnetic resonance of porous media (MRPM): A perspective. *J. Magn. Reson.* 2013, 229, 12–24.
59. Korb, J. P., Nuclear magnetic relaxation of liquids in porous media. *New J. Phys.* 2011, 13, 035016.
60. Galvosas, P.; and Callaghan, P. T., Multi-dimensional inverse Laplace spectroscopy in the NMR of porous media. *Compt. Rend. Physique* 2010, 11, 172–180.
61. Song, Y. Q.; Venkataramanan, L.; Hürlimann, M. D.; Flaum, M.; Frulla, P.; and Straley, C., T-1-T-2 correlation spectra obtained using a fast two-dimensional Laplace inversion. *J. Magn. Reson.* 2002, 154, 261–268.
62. Venkataramanan, L.; Song, Y. Q.; and Hürlimann, M. D., Solving Fredholm integrals of the first kind with tensor product structure in 2 and 2.5 dimensions. *IEEE Trans. Signal Proc.* 2002, 50, 1017–1026.
63. Mitchell, J.; Chandrasekera, T. C.; and Gladden, L. F., Numerical estimation of relaxation and diffusion distributions in two dimensions. *Progr. NMR Spectr.* 2012, 62, 34–50.
64. Washburn, K. E.; and Callaghan, P. T., Tracking pore to pore exchange using relaxation exchange spectroscopy. *Phys. Rev. Lett.* 2006, 97, 175502.
65. Van Landeghem, M.; Haber, A.; de Lacaillerie, J. B. D.; and Blümich, B., Analysis of multisite 2D relaxation exchange NMR. *Concepts Magn. Reson.* 2010, 36A, 153–169.
66. Bernin, D.; and Topgaard, D., NMR diffusion and relaxation correlation methods: New insights in heterogeneous materials. *Curr. Opin. Colloid Interface Sci.* 2013, 18, 166–172.
67. Halle, B., Spin dynamics of exchanging quadrupolar nuclei in locally anisotropic systems. *Progr. NMR Spectr.* 1996, 28, 137–159.
68. Nilsson, T.; and Halle, B., Nuclear magnetic relaxation induced by exchange-mediated orientational randomization: Longitudinal relaxation dispersion for spin I = 1. *J. Chem. Phys.* 2012, 137, 054503.
69. Persson, E.; and Halle, B., Nanosecond to microsecond protein dynamics probed by magnetic relaxation dispersion of buried water molecules. *J. Am. Chem. Soc.* 2008, 130, 1774–1787.
70. Chang, Z. W.; and Halle, B., Nuclear magnetic relaxation induced by exchange-mediated orientational randomization: Longitudinal relaxation dispersion for a dipole-coupled spin-1/2 pair. *J. Chem. Phys.* 2013, 139, 144203.
71. Chang, Z. W.; and Halle, B., Nuclear magnetic relaxation by the dipolar EMOR mechanism: General theory with application to two-spin systems. *J. Chem. Phys.* 2016, 144, 084202.
72. Chang, Z. W.; and Halle, B., Nuclear magnetic relaxation by the dipolar EMOR mechanism: Three-spin systems. *J. Chem. Phys.* 2016, 145, 034202.

Index

A

B

C

D

E

F

G

H

I

J

K

L

N

O

P

Q

R

S

T

U

V

W

X

Z

Milton Keynes UK
Ingram Content Group UK Ltd.
UKHW051945071024
449327UK00026B/2172